Rock Mechanics

Frontispiece Post-pillar mining in a jointed and faulted rock mass at the Dolphin Mine, King Island, Australia (photograph by permission of King Island Scheelite and CSIRO Division of Geomechanics).

Rock
Mechanics
For underground mining

Second edition

B.H.G. Brady

Manager, Applied Mechanics, Dowell Schlumberger Inc., Tulsa, USA

E.T. Brown

Deputy Vice-Chancellor, University of Queensland, Australia

CHAPMAN & HALL
London · Glasgow · New York · Tokyo · Melbourne · Madras

Published by Chapman & Hall, 2–6 Boundary Row, London SE1 8HN

Chapman & Hall, 2–6 Boundary Row, London SE1 8HN, UK

Blackie Academic & Professional, Wester Cleddens Road, Bishopbriggs, Glasgow G64 2NZ, UK

Chapman & Hall, 29 West 35th Street, New York NY10001, USA

Chapman & Hall Japan, Thomson Publishing Japan, Hirakawacho Nemoto Building, 6F, 1–7–11 Hirakawa-cho, Chiyoda-ku, Tokyo 102, Japan

Chapman & Hall Australia, Thomas Nelson Australia, 102 Dodds Street, South Melbourne, Victoria 3205, Australia

Chapman & Hall India, R. Seshadri, 32 Second Main Road, CIT East, Madras 600 035, India

First edition 1985

Second edition 1993

© 1985, 1992 B.H.G. Brady and E.T. Brown

Typeset in 10/12 pt. Times by Pure Tech Corporation, Pondicherry, India
Printed in Great Britain at the University Press, Cambridge

ISBN 0 412 47550 2

A catalogue record for this book is available from the British Library

Library of Congress Cataloging-in-Publication data available

Contents

Preface to the second edition

Since the publication of the first edition, several developments in rock mechanics have occurred which justified a comprehensive revision of the text. In the field of solid mechanics, major advances have been observed in understanding the fundamental modes of deformation, failure and stability of rock under conditions where rock stress is high in relation to rock strength. From the point of view of excavation design practice, a capacity for computational analysis of rock stress and displacement is more widely distributed at mine sites than at the time of preparing the first edition. In rock engineering practice, the development and demonstration of large-scale ground control techniques has resulted in modification of operating conditions, particularly with respect to maintenance of large stable working spans in open excavations. Each of these advances has major consequences for rock mechanics practice in mining and other underground engineering operations.

The advances in solid mechanics and geo-materials science have been dominated by two developments. First, strain localisation in a frictional, dilatant solid is now recognised as a source of excavation and mine instability. Second, variations in displacement-dependent and velocity-dependent frictional resistance to slip are accepted as controlling mechanisms in stability of sliding of discontinuities. Rockbursts may involve both strain localisation and joint slip, suggesting mitigation of this pervasive mining problem can now be based on principles derived from the governing mechanics. The revision has resulted in increased attention to rockburst mechanics and to mine design and operating measures which exploit the state of contemporary knowledge.

The development and deployment of computational methods for design in rock is illustrated by the increased consideration in the text of topics such as numerical methods for support and reinforcement design, and by discussion of several case studies of numerical simulation of rock response to mining. Other applications of numerical methods of stress and displacement analysis for mine layout and design are well established. Nevertheless, simple analytical solutions will continue to be used in preliminary assessment of design problems and to provide a basis for engineering judgement of mine rock performance. Several important solutions for zone of influence of excavations have been revised to provide a wider scope for confident application.

Significant improvements in ground control practice in underground mines are represented by the engineered use of backfill in deep-level mining and in application of long, grouted steel tendons or cable bolts in open stoping. In both cases, the engineering practices are based on analysis of the interaction between the host rock and the support or reinforcement system. Field demonstration exercises which validate these ground control methods and the related design procedures provide an assurance of their technical soundness and practical utility.

In the course of the revision, the authors have deleted some material they considered to be less rigorous than desirable in a book of this type. They have also corrected several errors brought to their attention by a perceptive and informed readership, for which they record their gratitude. Their hope is that the current version will be subject to the same rigorous and acute attention as the first edition.

B. H. G. B.
E. T. B.

Preface to the first edition

Rock mechanics is a field of applied science which has become recognised as a coherent engineering discipline within the last two decades. It consists of a body of knowledge of the mechanical properties of rock, various techniques for the analysis of rock stress under some imposed perturbation, a set of established principles expressing rock mass response to load, and a logical scheme for applying these notions and techniques to real physical problems. Some of the areas where application of rock mechanics concepts have been demonstrated to be of industrial value include surface and subsurface construction, mining and other methods of mineral recovery, geothermal energy recovery and subsurface hazardous waste isolation. In many cases, the pressures of industrial demand for rigour and precision in project or process design have led to rapid evolution of the engineering discipline, and general improvement in its basis in both the geosciences and engineering mechanics. An intellectual commitment in some outstanding research centres to the proper development of rock mechanics has now resulted in a capacity for engineering design in rock not conceivable two decades ago.

Mining engineering is an obvious candidate for application of rock mechanics principles in the design of excavations generated by mineral extraction. A primary concern in mining operations, either on surface or underground, is loosely termed 'ground control', i.e. control of the displacement of rock surrounding the various excavations generated by, and required to service, mining activity. The particular concern of this text is with the rock mechanics aspects of underground mining engineering, since it is in underground mining that many of the more interesting modes of rock mass behaviour are expressed. Realisation of the maximum economic potential of a mineral deposit frequently involves loading rock beyond the state where intact behaviour can be sustained. Therefore, underground mines frequently represent ideal sites at which to observe the limiting behaviour of the various elements of a rock mass. It should then be clear why the earliest practitioners and researchers in rock mechanics were actively pursuing its mining engineering applications.

Underground mining continues to provide strong motivation for the advancement of rock mechanics. Mining activity is now conducted at depths greater than 4000 m, although not without some difficulty. At shallower depths, single mine excavations greater than 350 m in height, and exceeding 500 000 m^3 in volume, are not uncommon. In any engineering terms, these are significant accomplishments, and the natural pressure is to build on them. Such advances are undoubtedly possible. Both the knowledge of the mechanical properties of rock, and the analytical capacity to predict rock mass performance under load, improve as observations are made of *in-situ* rock behaviour, and as analytical techniques evolve and are verified by practical application.

This text is intended to address many of the rock mechanics issues arising in underground mining engineering, although it is not exclusively a text on mining applications. It consists of four general sections, viz. general engineering mechanics relevant to rock mechanics; mechanical properties of rock and rock masses;

underground design and design of various types and associated components of a mine structure; and several topics related to rock mechanics practice. The material presented is an elaboration of a course of lectures originally prepared for undergraduate rock mechanics instruction for mining students at the Royal School of Mines, Imperial College, London. Some subsequent additions to this material, made by one of the authors while at the University of Minnesota, are also included. The authors believe that the material is suitable for presentation to senior undergraduate students in both mining and geological engineering, and for the initial stages of post-graduate instruction in these fields. It should also be of interest to students of other aspects of geomechanics, notably civil engineers involved in subsurface construction, and engineering geologists interested in mining and underground excavation design. Practising mining engineers and rock mechanics engineers involved in mine design may use the book profitably for review purposes, or perhaps to obtain an appreciation of the current state of engineering knowledge in their area of specialisation.

Throughout the text, and particularly in those sections concerned with excavation design and design of a mine structure, reference is made to computational methods for the analysis of stress and displacement in a rock mass. The use of various computation schemes, such as the boundary element, finite element and distinct element methods, is now firmly and properly embedded in rock mechanics practice. The authors have not listed computer codes in this book. They are now available in most program libraries, and are transported more appropriately on magnetic storage media than as listings in text.

The preparation of this book was assisted considerably by the authors' colleagues and friends. Part of the contribution of Dr John Bray of Imperial College is evident in the text, and the authors record their gratitude for his many other informal contributions made over a period of several years. Dr John Hudson of Imperial College and Gavin Ferguson of Seltrust Engineering Ltd read the text painstakingly and made many valuable suggestions for improvement. Professor Charles Fairhurst supported preparation activities at the University of Minnesota, for which one of the authors is personally grateful. The authors are also indebted to Moira Knox, Carol Makkyla and Colleen Brady for their work on the typescript, to Rosie and Steve Priest who prepared the index, and to Laurie Wilson for undertaking a range of tedious, but important, chores. The authors are also pleased to be able to record their appreciation of the encouragement and understanding accorded them by the publisher's representatives, Roger Jones, who persuaded them to write the book, and Geoffrey Palmer, who expertly supervised its production. Finally, they also thank the many individuals and organisations who freely gave permission to reproduce published material.

B. H. G. B.
E. T. B.

Acknowledgements

We would like to thank the following people and organisations for permission to re-produce previously published material:

Mount Isa Mines Limited (Cover photograph); King Island Scheelite and CSIRO Division of Geomechanics (Frontispiece); Soc. Min. Met. & Expl. (Figures 1.4 & 5, 13.16, 19, 20 & 21, 15.13, 15, 16, 27, 28, 29 & 32, 16.10, Tables 12.1 & 15.2); Canadian Inst. Min. Metall. (Figures 13.17 & 18); G.V. Borquez (Figure 1.4); J. C. Folinsbee (Figure 1.5); M. H. de Freitas (Figure 3.2); Elsevier (Figures 3.3, 4.8); Goldfields of S. Afr. (Figure 3.5); Pergamon Press (Figures 3.7, 8, 9, 10, 11, 12, 16, 17 & 21, 4.11, 12, 13, 19, 21, 43, 46 and 50, 11.1, 15.19, 21, 22, 17.3); Z. T. Bieniawski (Figure 3.30, Tables 3.5 & 6); Instn Min. Metall. (Figures 3.31, 4.17, 8.8 & 9, 11.13 & 30, 16.13 & 14, 18.5, 6, 7, 8 & 19, A3.5, Tables 3.8 & 9, 11.2); ELE Int. (Figure 4.14). Figure 4.20 reprinted from *Q. Colo. School Mines*, **54**(3), 177–99 (1959), L. H. Robinson, by permission of the Colorado School of Mines. Figure 4.31b–d reproduced from *J. Engng Industry*, **89**, 62–73 (1967) by permission of R. McLamore, K. E. Gray and Am. Soc. Mech. Engrs. Australasian Inst. Min. Metall. (Figures 4.34 & 36); Thomas Telford (Figures 4.35 & 37); R. E. Goodman (Figures 4.42, 43 & 45); N. R. Barton (Figure 4.46); E. Hoek (Figure 4.48); J.R. Enever (Figure 5.8); Association of Engineering Geologists (Figure 8.6); G. E. Blight and Am. Soc. Civ. Engrs (Figures 10.5c and d, 14.3a); N. G. W. Cook (Figure 10.24); J. R. Rice and Birkhauser Verlag (Figure 10.25); A. McGarr and South Afr. Inst. Min. & Metall. (Figure 10.26); Elsevier (Figures 11.17, 18, 19, 20 and 22, 14.11); W. D. Ortlepp (Figure 11.32); Chamber of Mines of South Africa (Figure 11.33); H. O. Hamrin and Soc. Min. Metall. & Expl. (Figures 12.1, 2, 5, 6, 7, 8, 9, 10, 11 & 12); Dravo Corp (Figure 12.3); H. Wagner and South Afr. Inst. Min. & Metall. (Figure 13.9); D. G. F. Hedley (Figures 13.15, 17, 18, 19 & 20); I.A. Goddard (Figure 13.21); M. F. Lee (Figures 13.22 & 23, 18.19); G. Swan (Figures 14.3b & 14.4); V. A. Koskela (Figure 14.9); P. Lappalainen (Figure 14.10); J. A. Ryder (Figure 15.1 & 2); M.D.G. Salamon (Figures 15.4, 18.13 & 14, Table 18.2); Instn Min. Engrs (Figures 15.5 & 9, 16.24); B.N. Whittaker (Figures 15.5, 9 & 12); National Coal Board and A. H. Wilson (Figures 15.6 & 7, Table 15.1); National Coal Board (Figures 15.8, 16.17, 18, 19, 20 & 23); L. J. Thomas (Figures 15.10 & 11); Figure 15.25 reproduced from *Storage in Excavated Rock Caverns*, (ed. M. Bergman) by permission of Pergamon Press. Figure 15.30 is reproduced from *Proc. 4th Canadian Rock Mech. symp.* (1968) by permission of the Minister of Supply and Services Canada. Mining Journal and G. A. Ferguson (Figure 15.31); University of Toronto Press (Figures 4.48, 16.8); D. S. Berry (Figure 16.21); C. K. McKenzie and Julius Kruttschnitt Mineral Research Centre (Figures 17.14, 15 & 16); C. K. McKenzie (Figure 17.17); K. Kovari and A. A. Balkema, Rotterdam (Figure 18.4); P. Londe and Am. Soc. Civ. Engrs (Figure 18.5); Glotzl Gesellschaft fur Baumesstechnik mbH (Figure 18.6); Am. Min. Congr. (Figure 18.9); H. F. Bock (Figure 18.18); D. H. Laubscher (Tables 3.8 and 3.9, 3.31, 11.3, 12.1, 15.2); Mining Journal (Table 11.4); E. G. Thomas and the Australian Mineral Foundation (Tables 14.1 & 2); A. A. Balkema (Tables 3.5 and 3.6, 5.8, 13.22 & 23, 14.3b, 4, 9 & 10, 15.1 & 2, 17.17)

1 Rock mechanics and mining engineering

1.1 General concepts

The engineering mechanics problem posed in all structural design is the prediction of the performance of the structure under the loads imposed on it during its prescribed functional operation. The subject of **engineering rock mechanics**, as applied in mining engineering practice, is concerned with the application of the principles of engineering mechanics to the design of the rock structures generated by mining activity. The discipline is closely related to the main streams of classical mechanics and continuum mechanics, but several specific factors identify it as a distinct and coherent field of engineering.

A widely accepted definition of **rock mechanics** is that first offered by the US National Committee on Rock Mechanics in 1964, and subsequently modified in 1974:

> Rock mechanics is the theoretical and applied science of the mechanical behaviour of rock and rock masses; it is that branch of mechanics concerned with the response of rock and rock masses to the force fields of their physical environment.

Clearly, the subject as defined is of fundamental relevance to mining engineering because the act of creating mining excavations changes the force fields of the rock's physical environment. As will be demonstrated throughout this text, the study of the response of the rock to these changes requires the application of analytical techniques developed specifically for the purpose, and which now form part of the corpus of the subject. Rock mechanics itself forms part of the broader subject of **geomechanics** which is concerned with the mechanical responses of all geological materials, including soils.

Application of rock mechanics principles in underground mine engineering is based on simple and, perhaps, self-evident premises. First, it is postulated that a rock mass can be ascribed a set of mechanical properties which can be measured in standard tests. Second, it is asserted that the process of underground mining generates a rock structure consisting of voids, support elements and abutments, and that the mechanical performance of the structure is amenable to analysis using the principles of classical mechanics. The third proposition is that the capacity to predict and control the mechanical performance of the host rock mass in which mining proceeds can assure or enhance the economic performance of the mine. This may be expressed in practice by the efficiency of resource recovery, measured in terms of volume extraction ratio, mine productivity or direct economic profitability. These ideas may seem rather elementary. However, even limited application of the concepts of mechanics in mine excavation or mine structural design is a comparatively recent innovation.

1

It is instructive to consider briefly some of the mechanical processes which occur as rock is excavated during underground mining. Figure 1.1a represents a cross section through a flat-lying, uniform orebody. ABCD and EFGH represent blocks of ore that are to be mined. Prior to mining, the material within the surfaces ABCD and EFGH exerts a set of support forces on the surrounding rock. Excavation of the orebody rock to produce the rock configuration of Figure 1.1b eliminates the support forces; i.e. the process of mining is statically equivalent to introducing a set of forces on the surfaces ABCD and EFGH equal in magnitude but opposite in sense to those acting originally. Under the action of these mining-induced forces, the following mechanical perturbations are imposed in the rock medium. Displacements of the adjacent country rock occur into the mined void. Stresses and displacements are induced in the central pillar and abutments. Total, final stresses in the pillar and abutments are derived from both the induced stresses and the initial state of stress in the rock mass. Finally, the induced surface forces acting through the induced surface displacements result in an increase of strain energy in the rock mass. The strain energy is stored locally, in the zones of increased stress concentration.

The ultimate objective in the design of a mine structure, such as the simple one being considered here, is to control rock displacements into and around mine excavations. Elastic displacements around mine excavations are typically small. Rock displacements of engineering consequence may involve such processes as fracture of intact rock, slip on a geological feature such as a fault, excessive deflections of roof and floor rocks (due, for example, to their detachment from adjacent rock), or unstable failure in the system. The latter process is expressed physically as a sudden release of stored potential energy, and significant change in the equilibrium configuration of the structure. These potential modes of rock response immediately define some of the components of a methodology intended to provide a basis for geomechanically sound excavation design. The methodology includes the following elements. The strength and deformation properties of the orebody and adjacent country rock must be determined in some accurate and reproducible way. The geological structure of the rock mass, i.e. the location, persistence and mechanical properties of all faults and other fractures of geologic

Figure 1.1 (a) Pre-mining conditions around an orebody, and (b) mechanical consequences of mining excavations in the orebody.

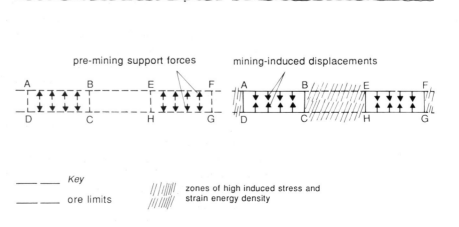

age, which occur in the zone of influence of mining activity, is to be defined, by suitable exploration and test procedures. Since the potential for slip on planes of weakness in the rock mass is related to fissure water pressure, the groundwater pressure distribution in the mine domain must be established. Finally, analytical techniques are required to evaluate each of the possible modes of response of the rock mass, for the given mine site conditions and proposed mining geometry.

The preceding brief discussion indicates that mining rock mechanics practice invokes quite conventional engineering concepts and logic. It is perhaps surprising, therefore, that implementation of recognisable and effective geomechanics programmes in mining operations is limited to the past 30 or so years. Prior to this period, there were, of course, isolated centres of research activity, and some attempts at translation of the results of applied research into mining practice. However, design by precedent appears to have had a predominant rôle in the design of mine structures. The recent appearance and recognition of the specialist rock mechanics engineer have resulted from industrial perception of the importance of the discipline in mining engineering practice.

A number of factors have contributed to the relatively recent emergence of rock mechanics as a mining technology. A major cause is the increased dimensions and production rates required of underground mining operations. These in turn are associated with pursuit of the economic goal of improved profitability with increased scale of production. Since increased capitalisation of a project requires greater assurance of its satisfactory performance in the long term, more formal and rigorous techniques are required in mine design, planning and scheduling practices. The increasing physical scale of underground mining operations has also had a direct effect on the need for effective mine structural design, since the possibility of extensive failure can be reckoned as being in some way related to the size of the active mine domain: The need to exploit mineral resources in unfavourable mining environments has also provided a significant impetus to geomechanics research. In particular, the increased depth of mining in such areas as the Witwatersrand basin of South Africa, the Kolar Gold Fields in India and the Coeur d'Alene area of the United States, has stimulated research into various aspects of rock performance under high stress. Finally, more recent social concerns with resource conservation and industrial safety have been reflected in mining as attempts to maximise the recovery from any mineral reserve, and by closer study of practices and techniques required to maintain safe and secure work places underground. Both of these concerns have resulted in greater demands being placed on the engineering skills and capacities of mining corporations and their service organisations.

In the evolution of rock mechanics as a field of engineering science, there has been a tendency to regard the field as a derivative of, if not a subordinate discipline to, soil mechanics. In spite of the commonality of some basic principles, there are key issues which arise in rock mechanics distinguishing it from soil mechanics. The principal distinction between the two fields is that failure processes in intact rock involve fracture mechanisms such as crack generation and growth in a pseudo-continuum. In soils, failure of an element of the medium typically does not affect the mechanical integrity of the individual grains. In both diffuse and locally intense deformation modes, soil failure is associated with processes such as dilatation, particle rotation and alignment. This distinction between the different media has other consequences. For example, soils in their operating engineering environments are always subject to relatively low states of stress. The opposite is frequently true for rock. Further differences arise from the relatively high elastic moduli, and the relatively low material permeabilities of rocks compared with soils. The latter distinction is important. In most rock formations, fluid flow occurs via fissures and channels, while in soils fluid migration involves movement through the pore space of the particulate assembly. It appears, therefore, that rock and soil mechanics should be regarded as complementary rather than mutually inclusive disciplines.

Having suggested that rock mechanics is a distinct engineering discipline, it is clear that its effective practical application demands an appreciation of its philosophic integration with other areas of geomechanics. Rock mechanics, soil mechanics, groundwater hydrology and structural geology are, in the authors' opinions, the kernels of the scientific basis of mining engineering. Together, they constitute the conceptual and factual base from which procedures can be developed for the control and prediction of rock behaviour during mining activity.

1.2 Inherent complexities in rock mechanics

It has been observed that rock mechanics represents a set of principles, a body of knowledge and various analytical procedures related to the general field of applied mechanics. The question that arises is – what constituent problems arise in the mechanics of geologic media, sufficient to justify the formulation or recognition of a coherent, dedicated engineering discipline? The following five issues are postulated to be paradigmatic components of the subject, that present the need for a singular research effort and a specific methodology in industrial applications.

1.2.1 Rock fracture
Fracture of conventional engineering material occurs in a tensile stress field, and sophisticated theories have been postulated to explain the pre-failure and post-failure performance of these media. The stress fields operating in rock structures are pervasively compressive, so that the established theories are not immediately applicable to the fracture of rock. A particular complication in rock subject to compression is associated with friction mobilised between the surfaces of the microcracks which are the sites for fracture initiation. This causes the strength of rock to be highly sensitive to confining stress, and introduces doubts concerning the relevance of such notions as the normality principle, associated flow and plasticity theories generally, in analysing the strength and post-failure deformation properties of rock. A related problem is the phenomenon of localisation, in which rupture in a rock medium is expressed as the generation of bands of intensive shear deformation, separating domains of apparently unaltered rock material.

1.2.2 Size effects
The response of rock to imposed load shows a pronounced effect of the size of the loaded volume. This effect is related in part to the discontinuous nature of a

(a)

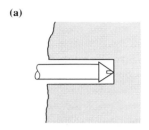

(b)

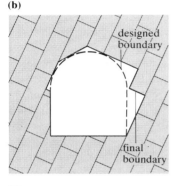

designed boundary

final boundary

(c)

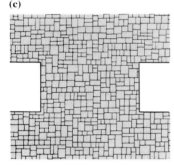

Figure 1.2 The effect of scale on rock response to imposed loads: (a) rock material failure in drilling; (b) discontinuities controlling the final shape of the excavation; (c) a mine pillar operating as a pseudo-continuum.

rock mass. Joints and other fractures of geological origin are ubiquitous features in a body of rock, and thus the strength and deformation properties of the mass are influenced by both the properties of the rock material (i.e. the continuous units of rock) and those of the various structural geological features. These effects may be appreciated by considering various scales of loading to which a rock mass is subjected in mining practice. The process of rock drilling will generally reflect the strength properties of the intact rock, since the process operates by inducing rock material fracture under the drilling tool. Mining a drive in jointed rock may reflect the properties of the joint system. In this case, the final cross section of the opening will be defined by the joint attitudes. The behaviour of the rock around the periphery of the drive may reflect the presence of discrete blocks of rock, whose stability is determined by frictional and other forces acting on their surfaces. On a larger scale, e.g. that of a mine pillar, the jointed mass may demonstrate the properties of a pseudo-continuum. Scale effects as described here are illustrated schematically in Figure 1.2.

These considerations suggest that the specification of the mechanical properties of a rock mass is not a simple matter. In particular, the unlikely possibility of testing jointed rock specimens, at scales sufficient to represent the equivalent continuum satisfactorily, indicates that it is necessary to postulate and verify methods of synthesising rock mass properties from those of the constituent elements.

1.2.3 Tensile strength

Rock is distinguished from all other common engineering materials, except concrete, by its low tensile strength. Rock material specimens tested in uniaxial tension fail at stresses an order of magnitude lower than when tested in uniaxial compression. Since joints and other fractures in rock can offer little or no resistance to tensile stresses, the tensile strength of a rock mass can be assumed to be non-existent. Rock is therefore conventionally described as a 'no-tension' material, meaning that tensile stresses cannot be generated or sustained in a rock mass. The implication of this property for excavation design in rock is that any zone identified by analysis as being subject to tensile stress will, in practice, be de-stressed, and cause local stress redistribution. De-stressing may result in local instability in the rock, expressed as either episodic or progressive detachment of rock units from the host mass.

1.2.4 Effect of groundwater

Groundwater may affect the mechanical performance of a rock mass in two ways. The most obvious is through the operation of the effective stress law (section 4.2). Water under pressure in the joints defining rock blocks reduces the normal effective stress between the rock surfaces, and therefore reduces the potential shear resistance which can be mobilised by friction. In porous rocks, such as sandstones, the effective stress law is obeyed as in granular soils. In both cases, the effect of fissure or pore water under pressure is to reduce the ultimate strength of the mass, when compared with the drained condition.

A more subtle effect of groundwater on rock mechanical properties may arise from the deleterious action of water on particular rocks and minerals. For example, clay seams may soften in the presence of groundwater, reducing the strength and increasing the deformability of the rock mass. Argillaceous rocks, such as shales and argillitic sandstones, also demonstrate marked reductions in material strength following infusion with water.

The implications of the effect of groundwater on rock mass strength are considerable for mining practice. Since rock behaviour may be determined by its geohydrological environment, it may be essential in some cases to maintain close control of groundwater conditions in the mine area. Further, since backfill is an important element in many mining operations, the lithologies considered for stope filling operations must be considered carefully from the point of view of strength properties under variable groundwater conditions.

1.2.5 Weathering

Weathering may be defined as the chemical or physical alteration of rock at its surface by its reaction with atmospheric gas and aqueous solutions. The process is analogous to corrosion effects on conventional materials. The engineering interest in weathering arises because of its influence on the mechanical properties of the intact material, as well as the potential for significant effect on the coefficient of friction of the rock surface. It appears that whereas weathering causes a steady reduction in rock properties, the coefficient of friction of a surface may suffer a step reduction (Boyd, 1975).

Although physical processes such as thermal cycling and insolation may be important in surface mining, underground weathering processes are chiefly chemical in origin. These include dissolution and ion exchange phenomena, oxidation and hydration. Some weathering actions are readily appreciated, such as the dissolution of limestone in an altered groundwater environment, or softening of marl due to sulphate removal. In others, such as the oxidation of pyrrhotite, the susceptibility of some forms of the mineral to rapid chemical attack is not fully understood. A weathering problem of particular concern is presented by basic rocks containing minerals such as olivine and pyroxenes. A hydrolysis product is montmorillonite, which is a swelling clay with especially intractable mechanical behaviour.

This discussion does not identify all of the unique issues to be considered in rock mechanics. However, it is clear that the subject transcends the domain of traditional applied mechanics, and must include a number of topics that are not of concern in any other engineering discipline.

1.3 Underground mining

Ore extraction by an underground mining method involves the generation of different types of openings, with a considerable range of functions. The schematic cross section and longitudinal section through an operating mine, shown in Figure 1.3, illustrate the different rôles of various excavations. The main shaft, level drives and cross cuts, ore haulages, ventilation shafts and airways constitute the mine access and service openings. Their duty life is comparable with, or exceeds, the mining life of the orebody and they are usually developed in barren ground. Service and operating openings directly associated with ore recovery

Longitudinal projection

Cross section

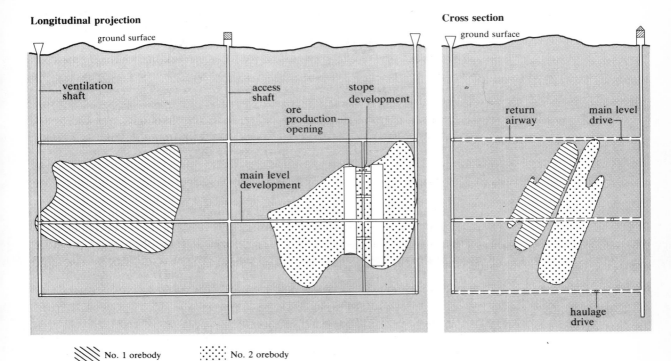

Figure 1.3 The principal types of excavation involved in underground mining by some stoping method.

consist of the access cross cuts, drill headings, access raises, extraction headings and ore passes, from, or in which, various ore production operations are undertaken. These openings are developed in the orebody, or in country rock close to the orebody boundary, and their duty life is limited to the duration of mining activity in their immediate vicinity. Many openings are eliminated by the mining operation. The third type of excavation is the ore source. It may be a stope, with well-defined, free-standing rock walls forming the geometric limits for the mined void, which increases in size with the progress of ore extraction. Alternatively, the ore source may be a rubble-filled space with fairly well-defined lower and lateral limits, usually coincident with the orebody boundaries. The rubble is generated by inducing disintegration of the rock above the crown of the orebody, which fills the mined space as extraction proceeds. The lifetime of these different types of ore source openings is defined by the duration of active ore extraction.

It is clear that there are two geomechanically distinct techniques for underground ore extraction. Each technique is represented in practice by a number of different mining methods. The particular method chosen for the exploitation of an orebody is determined by such factors as its size, shape and geometric disposition, the distribution of values within the orebody, and the geotechnical environment. The last factor takes account of such issues as the *in situ* mechanical properties of the orebody and country rocks, the geological structure of the rock mass, the ambient state of stress and the geohydrological conditions in the zone of potential mining influence.

Later chapters will be concerned with general details of mining methods, and the selection of a mining method to match the dominant orebody geometric, geological and geomechanical properties. It is sufficient to note here that mining methods may be classified on the basis of the type and degree of support provided in the mine structure created by ore extraction (Thomas, 1978). Supported mine structures are generated by methods such as open stoping and room-and-pillar mining, or cut-and-fill stoping and shrinkage stoping. In the former methods, natural support is provided in the structures by ore remnants located throughout the stoped region. In the latter methods, support for the walls of the mined void is provided by either introduced fill or by fractured ore temporarily retained in contact with mined stope walls. The second type of mine conformation recognised by Thomas is a caving structure, generated by mining methods such as block caving and sublevel caving. In these cases, no support is provided in the mined space, which fills spontaneously with fragmented and displaced orebody and cover rock.

From a rock mechanics point of view, discrimination between the two generic mining techniques, and the structures they generate, may be made on the basis of the displacements induced in the country rock and the energy redistribution which accompanies mining. In the technique of mining with support, the objective is to restrict displacements of the country rock to elastic orders of magnitude, and to maintain, as far as possible, the integrity of both the country rock and the unmined remnants within the orebody. This typically results in the accumulation of strain energy in the structure, and the mining problem is to

Figure 1.4 Principal features of a caving operation (after Borquez, 1981)

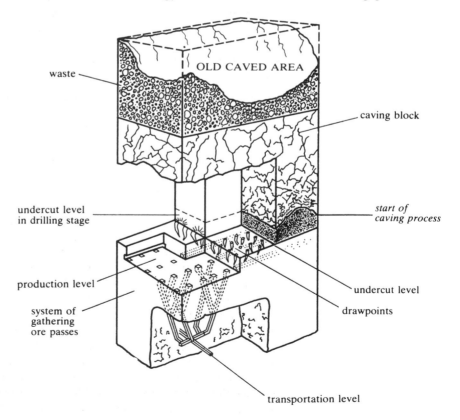

OLD CAVED AREA

waste

caving block

undercut level
in drilling stage

*start of
caving process*

production level

undercut level

system of
gathering
ore passes

drawpoints

transportation level

ensure that unstable release of energy cannot occur. The caving technique is intended to induce large-scale, rigid-body displacements of rock above the crown of the orebody, with the displacement field propagating through the cover rock as mining progresses. The principle is illustrated schematically in Figure 1.4. The process results in energy dissipation in the caving rock mass, by slip, crushing and grinding. The mining requirement is to ensure that steady displacement of the caving mass occurs, so that the mined void is continuously self-filling, and unstable voids are not generated in the interior of the caving material. The aim is therefore to achieve a temporal rate of energy dissipation in the caving medium which matches the rate of ore extraction from the orebody.

This distinction between different mining techniques does not preclude a transition from one technique to the other in the life of an orebody. In fact, the distinction is useful in that it conveys the major mechanical ramifications in any change of mining strategy.

Irrespective of the mining technique adopted for ore extraction, it is possible to specify four common rock mechanics objectives for the performance of a mine structure, and the three different types of mine openings described previously. These are:

(a) to ensure the overall stability of the complete mine structure, defined by the main ore sources and mined voids, ore remnants and adjacent country rock;
(b) to protect the major service openings throughout their designed duty life;
(c) to provide secure access to safe working places in and around the centres of ore production;
(d) to preserve the mineable condition of unmined ore reserves.

These objectives are not mutually independent. Also, the typical mine planning and design problem is to find a stope or ore block excavation sequence that satisfies these objectives simultaneously, as well as fulfilling other operational and economic requirements. The realisation of the rock mechanics objectives requires a knowledge of the geotechnical conditions in the mine area, and a capacity for analysis of the mechanical consequences of the various mining options. An appreciation is also required of the broad management policies, and general mining concepts, which have been adopted for the exploitation of the particular mineral resource.

It is instructive to define the significant difference in operational constraints between underground excavations designed for civil engineering purposes, and those types of excavations involved in mining engineering practice subject to entry by mine personnel. In the latter case, the use of any opening is entirely in the control of the mine operator, and during its active utilisation the surfaces of an excavation are subject to virtually continuous inspection by mine personnel. Work to maintain or reinstate secure conditions around an opening, ranging from surface scaling (barring down) to support emplacement, can be undertaken at any stage, at the direction of the mine management. These conditions rarely apply to excavations subject to civil engineering operating practice. Another major difference is that most mine excavations have duty lives that are significantly less than those of excavations used for civil purposes. It is not surprising, therefore, that mine excavation design reflects the degree of immediate control over

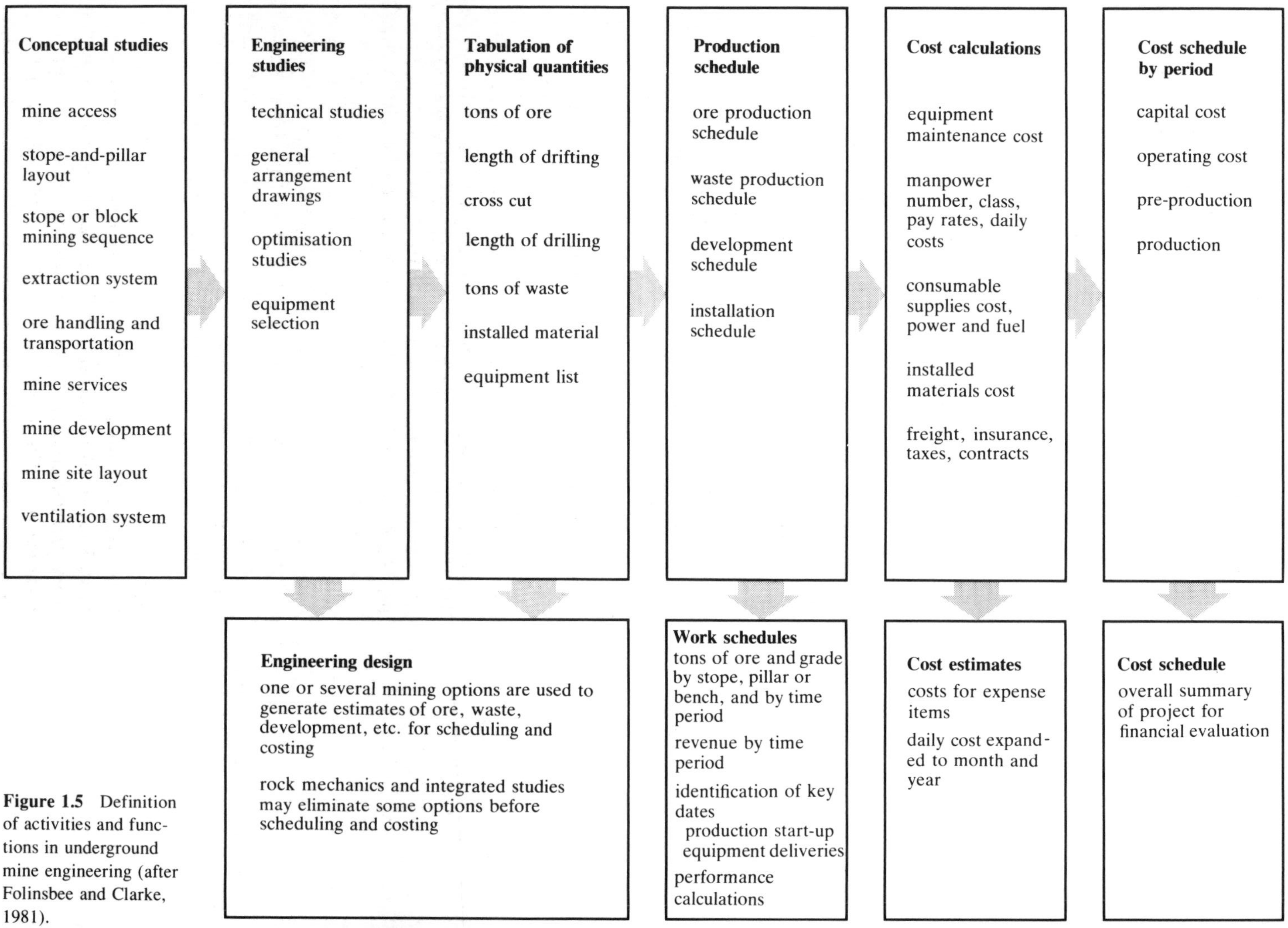

Conceptual studies

mine access

stope-and-pillar layout

stope or block mining sequence

extraction system

ore handling and transportation

mine services

mine development

mine site layout

ventilation system

Engineering studies

technical studies

general arrangement drawings

optimisation studies

equipment selection

Tabulation of physical quantities

tons of ore

length of drifting

cross cut

length of drilling

tons of waste

installed material

equipment list

Production schedule

ore production schedule

waste production schedule

development schedule

installation schedule

Cost calculations

equipment maintenance cost

manpower number, class, pay rates, daily costs

consumable supplies cost, power and fuel

installed materials cost

freight, insurance, taxes, contracts

Cost schedule by period

capital cost

operating cost

pre-production

production

Engineering design

one or several mining options are used to generate estimates of ore, waste, development, etc. for scheduling and costing

rock mechanics and integrated studies may eliminate some options before scheduling and costing

Work schedules

tons of ore and grade by stope, pillar or bench, and by time period

revenue by time period

identification of key dates

production start-up equipment deliveries

performance calculations

Cost estimates

costs for expense items

daily cost expanded to month and year

Cost schedule

overall summary of project for financial evaluation

Figure 1.5 Definition of activities and functions in underground mine engineering (after Folinsbee and Clarke, 1981).

opening utilisation, inspection, maintenance and support emplacement afforded the mine operator.

In addition to the different operating constraints for mining and civil excavations, there are marked differences in the nature of the structures generated and these directly affect the design philosophy. The principal difference is that a civil engineering rock structure is essentially fixed, whereas a mine structure continues to develop throughout the life of the mine. In the latter case, stope or ore block extraction sequences assume great importance. Decisions made early in the mine life can limit the options, and the success of mining, when seeking to establish an orderly and effective extraction strategy, or to recover remnant ore.

1.4 Functional interactions in mine engineering

The purpose of this section is to explore the roles of various engineering disciplines in the planning and design of an underground mine. The particular concern is to define the interaction of planning engineers, geologists and rock mechanics engineers in the pre-production and operating phases of mining activity.

The scope of engineering activity to be undertaken preceding and during the productive life of a mine is illustrated in the design task definition chart shown in Figure 1.5. The overall aim of the various components of engineering activity (e.g. mine access design, ventilation system) is the development of sustainable production and cost schedules for the operation. The specific rock mechanics contributions to the mine engineering programme, and its interface with other planning functions, occur primarily in tasks related to mine access, mining method development and mine layout, mining sequence and ore extraction design. Method development, mine layout and sequencing usually constitute the majority of initial and continuing geomechanics activity.

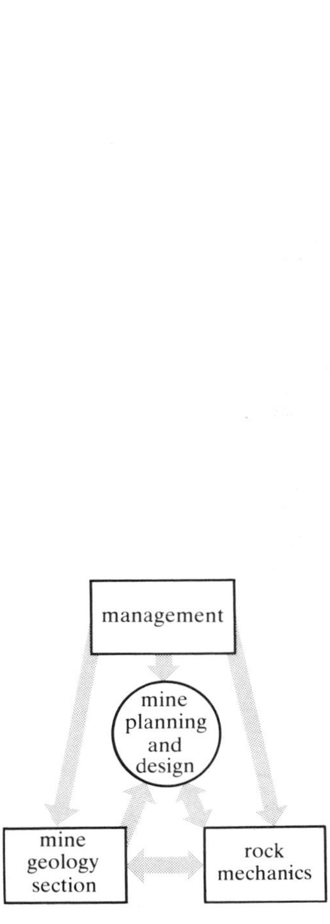

Rock mechanics design activities need to be conducted in an organisational environment which permits the integration of concepts, information and analytical activity required from management, planning engineers, geologists, and rock mechanics engineers. The logic of an integrated mine engineering philosophy is illustrated in Figure 1.6. The principles implicit in this logical scheme are, first, the mutual dependence of each functional group on the information provided by the others, and, second, that it is ultimately mine planning engineers who transform the individual technical contributions into the working drawings, production schedules and cost estimates for subsequent implementation.

Considering Figure 1.6 from a rock mechanics perspective, it is useful to summarise the engineering information that can be reasonably expected from the other functional groups, and the information that should be delivered by a rock mechanics group to planning engineers.

Figure 1.6 Interaction between technical groups involved in mine engineering.

1.4.1 Management
Information from management is a key element which is frequently not available to rock mechanics specialists. The general requirement is that the broad framework of management policy and objectives for the exploitation of a particular resource be defined explicitly. This should include such details as the volume extraction ratio sought for the orebody and how this might change in response

11

to changing product prices. The company investment strategy should be made known, if only to indicate the thinking underlying the decision to mine an orebody. Particular corporate constraints on mining technique, such as policy on disturbance of the local physical environment above the mine area, and restrictions on geohydrological disturbance, should be defined. Further, restrictions on operating practices, such as men working in vertical openings or under unsupported, temporary roof spans, need to be specified.

1.4.2 Geology

In defining the geomechanics role of exploration and engineering geologists in mine engineering, it is assumed that, at all stages of the geological exploration of an orebody, structural and geohydrological data will be logged and processed on a routine basis. A Geology Section can then provide information ranging from a general description of the regional geology, particularly the structural geology, to details of the dominant and pervasive structural features in the mine area. A comprehensive geological description would also include the distribution of lithologies in and around the orebody, the distribution of values throughout the orebody, and the groundwater hydrology of the mine area. In the last case, the primary need is to identify aquifers in the zone of possible influence of mining which might deliver groundwater into any part of the mining domain. Finally, specific geological investigations would identify sources of potential mining problems in the mine area. These include zones of shattered ground, leached zones, cavernous ground (vughs), rocks and minerals with adverse weathering properties, and major structural features such as faults and clay seams which transgress the orebody and are expressed on the ground surface.

It is clear, from this specification of duties, that mine geological activity should produce a major component of engineering geological data. It implies that successful execution of the engineering exploration of an orebody and environs requires the active co-operation of geologists and rock mechanics personnel. It may be necessary, for example, for the latter to propose drilling programmes and targets to clarify site conditions of particular mining consequence.

1.4.3 Planning personnel

Mine planning and design engineers are responsible for the eventual definition of all components of an engineering study of a prospective mining operation. Their role is initiative as well as integrative. In their interaction with rock mechanics engineers, their function is to contribute information which can usefully delineate the scope of any geomechanical analysis. Thus they may be expected to define the general mining strategy, such as one-pass stoping (no pillar recovery), or stoping and subsequent pillar extraction, and other limitations on mining technique. Details of anticipated production rates, economic sizes of stopes, and the number of required sources of ore production, can be used to define the extent of the active mine structure at any time. The possibility of using backfills of various types in the production operation should be established. Finally, the constraints imposed on future mining by the current location of mine accesses, stoping activity, permanent openings and exploration drives should be specified.

1.4.4 Rock mechanics

It has been noted that the mine engineering contribution of a Rock Mechanics Section relates to design tasks concerned principally with permanent mine openings, mining layout and sequencing, and extraction design. Specific activities associated with each of these tasks are now detailed. Design issues related to permanent mine openings include siting of service and ventilation shafts, siting, dimensioning and support specification of level main development, and detailed design of major excavations such as crusher excavations, interior shaft hoist chambers, shaft bottom facilities and workshop installations. The demand for these services is, of course, episodic, being mainly concentrated in the pre-production phase of mine operations.

The majority of rock mechanics activity in mining operations is devoted to resolution of questions concerned with the evolutionary design of the mine structure. These questions include: dimensions of stopes and pillars; layout of stopes and pillars within the orebody, taking due account of their location and orientation relative to the geological structure and the principal stress directions; the overall direction of mining advance through an orebody; the sequence of extraction of stope blocks and pillar remnants, simultaneously noting the need to protect service installations, maintain access and preserve mine structural stability; and the need for and specification of the strength parameters of any backfill in the various mined voids. In all of these design activities, effective interaction must be maintained with planning personnel, since geomechanics issues represent only part of the complete set of engineering information required to develop an operationally acceptable mining programme.

Extraction system design is concerned with the details of stope configuration and ore recovery from the stope. This involves, initially, consideration of the stability of stope boundaries throughout the stope working life, and requires close examination of the possibility of structurally controlled failures from stope and pillar surfaces. The preferred direction of stope retreat may be established from such studies. The design of the extraction horizon requires consideration of the probable performance of stope drawpoints, tramming drives, slusher drives and ore-flow control raises, during the stope life. Particular problems can occur on the extraction horizon due to the density of openings, resulting in stressed remnants, and the potential for damage by secondary breakage of oversize rock during ore recovery. A final issue in this segment of stope design is primary blast design. The issue here is blasting effects on remnant rock around the stope periphery, as well as the possibility of damage to access and adjacent service openings, under the transient loads associated with closely sequenced detonations of relatively large explosive charges.

1.5 Implementation of a rock mechanics programme

It has been stated that an effective rock mechanics programme should be thoroughly integrated with other mine technical functions in the development of a coherent mining plan for an orebody. However, the successful accomplishment of the goals of the programme requires the commitment of sufficient resources, on a continuous basis, to allow rational analysis of the range of problems posed by the various phases of mining activity.

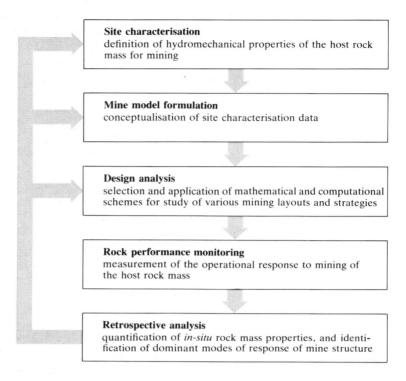

Site characterisation
definition of hydromechanical properties of the host rock mass for mining

Mine model formulation
conceptualisation of site characterisation data

Design analysis
selection and application of mathematical and computational schemes for study of various mining layouts and strategies

Rock performance monitoring
measurement of the operational response to mining of the host rock mass

Retrospective analysis
quantification of *in-situ* rock mass properties, and identification of dominant modes of response of mine structure

Figure 1.7 Components and logic of a rock mechanics programme.

A methodology for the implementation of a rock mechanics programme is illustrated schematically in Figure 1.7. Five distinct components of the programme are identified, and they are postulated to be logically integrated, i.e. deletion of any component negates the overall operating philosophy. Another point to be observed from Figure 1.7 is that the methodology implies that the programme proceeds via a multi-pass loop. There are two main reasons for this. First, the site characterisation phase never generates a sufficiently comprehensive data base from which to develop a unique plan for the complete life of the mine. Second, mine design is itself an evolutionary process in which engineering responses are formulated to reflect the observed performance of the mine structure under actual operating conditions. These issues are clarified in the following discussion of the component phases of the programme.

1.5.1 Site characterisation

The objective of this phase, in the first pass through the loop, is to define the mechanical properties and state of the medium in which mining is to occur. It involves determination of the strength and deformation properties of the various lithological units represented in and around the orebody, definition of the geometric and mechanical properties of pervasive jointing, and location and description of the properties of discrete structural features. An estimate of the *in situ* strength of the medium may then be made from the properties of the constituent elements of the mass. This phase also includes determination of the *in situ* state of stress in the mine area, and investigation of the hydrogeology of the orebody and environs.

14

The difficulty in site characterisation lies in achieving representative data defining geomechanical conditions throughout the rock medium. Under conditions of limited physical access, yielding small numbers of small rock specimens, with no unifying theory to relate the specimen properties with those of the host rock medium, a first-pass site characterisation is intrinsically deficient.

1.5.2 Mine model formulation

Formulation of a mine model represents the simplification and rationalisation of the data generated by the site characterisation. The aim is to account for the principal geomechanical features which will be expressed in the deformational behaviour of the prototype. For example, lithological units are ascribed average 'representative' strength and deformation properties, major structural features are assigned a regular geometry and average shear strength properties, and a representative specification is accepted for the pre-mining state of stress. The need for this phase arises from the limited details that can be accommodated in most of the analytical or computational methods used in design.

It is clear that significant discrepancies may be introduced at this stage, by failure to recognise the engineering significance of particular features of the mine geomechanical setting.

1.5.3 Design analysis

Having defined the prevailing conditions in the rock mass in an analytically tractable way, the mechanical performance of selected mining configurations and excavation geometries can be predicted using appropriate mathematical or numerical techniques. The analytical tools may be relatively primitive (e.g. the tributary area theory for pillar design) or advanced, employing, for example, computational schemes which may model quite complex constitutive behaviour for both the rock mass and various fabric elements. In any event, the design analyses represent the core of rock mechanics practice. Recent rapid development in the power of available computational schemes has been responsible for significant advances, and improved confidence, in the quality of rock structural design.

1.5.4 Rock performance monitoring

The objective of this phase of rock mechanics practice is to characterise the operational response of the rock mass to mining activity. The intention is to establish a comprehension of the rôles of the various elements of the rock mass in the load-deformational behaviour of the rock medium. The data required to generate this understanding are obtained by displacement and stress measurements made at key locations in the mine structure. These measurements include closures across pillars, slip on faults, and levelling and horizontal displacement measurements in and around the active mining zone. States of stress may be measured in pillars, abutments and in the interior of any rock units showing signs of excessive stress. Visual inspections must be undertaken regularly to locate any structurally controlled failures and areas of anomalous response, and these should be mapped routinely. Finally, data should be collected on the production performance of each stope, and the final configuration of each stope should be surveyed and mapped. The aim in this case is to seek any correlation between rock mass local performance and stope productivity.

1.5.5 Retrospective analysis

The process of quantitative analysis of data generated by monitoring activity is intended to reassess and improve knowledge of the *in situ* mechanical properties of the rock mass, as well as to review the adequacy of the postulated mine model. Review of the conceptualisation of the host rock mass involves analysis of the role of major structural features on the performance of the structures, and identification of the key geomechanical parameters determining the deformational response of the medium. Particularly valuable data are generated by the analysis of local failures in the system. These provide information about the orientations, and possibly relative magnitudes of the *in situ* field stresses, as well as high quality information on the *in situ* rock mass strength parameters. Subsequently, stope mechanical and production performance data can be assessed with a view to formulating detailed stope design and operating criteria. This might involve establishment of rules specifying, for example, stope shape relative to geological structure, stope blasting practice, and drawpoint layouts and designs for various types of structural and lithological conditions.

Figure 1.7 indicates that data generated by retrospective analysis are used to update the site characterisation data, mine model and design process, via the iterative loop. This procedure represents no more than a logical formalisation of the observational principle long used in soil mechanics practice (Peck, 1969). It is a natural engineering response to the problems posed by basic limitations in site characterisation and conceptualisation associated with excavation design in geologic media.

2 Stress and infinitesimal strain

2.1 Problem definition

The engineering mechanics problem posed by underground mining is the prediction of the displacement field generated in the orebody and surrounding rock by any excavation and ore extraction processes. The rock in which excavation occurs is stressed by gravitational, tectonic and other forces, and methods exist for determining the ambient stresses at a mine site. Since the areal extent of any underground mine openings is always small relative to the Earth's surface area, it is possible to disregard the sphericity of the Earth. Mining can then be considered to take place in an infinite or semi-infinite space, which is subject to a definable initial state of stress.

An understanding of the notions of force, stress and strain is fundamental to a proper and coherent appreciation of the response of a rock mass to mining activity. It was demonstrated in Chapter 1 that excavating (or enlarging) any underground opening is mechanically equivalent to the application, or induction, of a set of forces distributed over the surfaces generated by excavation. Formation of the opening also induces a set of displacements at the excavation surface. From a knowledge of the induced surface forces and displacements, it is possible to determine the stresses and displacements generated at any interior point in the rock medium by the mining operation.

Illustration of the process of underground excavation in terms of a set of applied surface forces is not intended to suggest that body forces are not significant in the performance of rock in a mine structure. No body forces are induced in a rock mass by excavation activity, but the behaviour of an element of rock in the periphery of a mine excavation is determined by its ability to withstand the combined effect of body forces and internal, post-excavation surface forces. However, in many mining problems, body force components are relatively small compared with the internal surface forces, i.e. the stress components.

Some mine excavation design problems, such as those involving a jointed rock mass and low-stress environments, can be analysed in terms of block models and simple statics. In most deep mining environments, however, the rock mass behaves as a continuum, at least initially. Prediction of rock mass response to mining therefore requires a working understanding of the concepts of force, traction and stress, and displacement and strain. The following discussion of these issues follows the treatments by Love (1944) and Jaeger (1978).

In the discussion, the usual engineering mechanics convention is adopted, with tensile normal stresses considered positive, and the sense of positive shear stress on any surface determined by the sense of positive normal stress. The geomechanics convention for the sense of positive stresses will be introduced subsequently.

17

2.2 Force and stress

The concept of stress is used to describe the intensity of internal forces set up in a body under the influence of a set of applied surface forces. The idea is quantified by defining the state of stress at a point in a body in terms of the area intensity of forces acting on the orthogonally oriented surfaces of an elementary free body centred on the point. If a Cartesian set of reference axes is used, the elementary free body is a cube whose surfaces are oriented with their outward normals parallel with the co-ordinate axes.

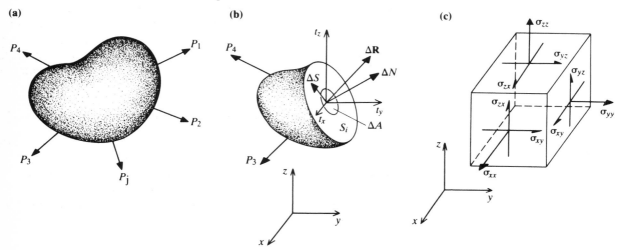

(a) **(b)** **(c)**

Figure 2.1 (a) A finite body subject to surface loading; (b) determination of the forces, and related quantities, operating on an internal surface; (c) specification of the state of stress at a point in terms of the traction components on the face of a cubic free body.

Figure 2.1a illustrates a finite body in equilibrium under a set of applied surface forces, P_j. To assess the state of loading over any interior surface, S_i, one could proceed by determining the load distribution over S_i required to maintain equilibrium of part of the body. Suppose, over an element of surface ΔA surrounding a point O, the required resultant force to maintain equilibrium is $\Delta \mathbf{R}$, as shown in Figure 2.1b. The magnitude of the **resultant stress** $\boldsymbol{\sigma}_r$ at O, or the **stress vector**, is then defined by

$$\boldsymbol{\sigma}_r = \lim_{\Delta A \to 0} \frac{\Delta \mathbf{R}}{\Delta A}$$

If the vector components of $\Delta \mathbf{R}$ acting normally and tangentially to ΔA are ΔN, ΔS, the **normal stress** component, σ_n, and the resultant **shear stress** component, τ, at O are defined by

$$\sigma_n = \lim_{\Delta A \to 0} \frac{\Delta N}{\Delta A}, \quad \tau = \lim_{\Delta A \to 0} \frac{\Delta S}{\Delta A}$$

The stress vector, $\boldsymbol{\sigma}_r$, may be resolved into components t_x, t_y, t_z directed parallel to a set of reference axes x, y, z. The quantities t_x, t_y, t_z, shown in Figure 2.1b are called **traction components** acting on the surface at the point O. As with the stress vector, the normal stress, σ_n, and the resultant shear stress, τ, the traction components are expressed in units of force per unit area. A case of particular interest occurs when the outward normal to the elementary surface ΔA is oriented parallel to a co-ordinate axis, e.g. the x axis. The traction components acting on the surface whose normal is the x axis are then used to define three components of the state of stress at the point of interest,

$$\sigma_{xx} = t_x, \quad \sigma_{xy} = t_y, \quad \sigma_{xz} = t_z \tag{2.1}$$

In the doubly-subscripted notation for stress components, the first subscript indicates the direction of the outward normal to the surface, the second the sense of action of the stress component. Thus σ_{xz} denotes a stress component acting on a surface whose outward normal is the x axis, and which is directed parallel to the z axis. Similarly, for the other cases where the normals to elements of surfaces are oriented parallel to the y and z axes respectively, stress components on these surfaces are defined in terms of the respective traction components on the surfaces, i.e.

$$\sigma_{yx} = t_x, \quad \sigma_{yy} = t_y, \quad \sigma_{yz} = t_z \tag{2.2}$$

$$\sigma_{zx} = t_x, \quad \sigma_{zy} = t_y, \quad \sigma_{zz} = t_z \tag{2.3}$$

The senses of action of the stress components defined by these expressions are shown in Figure 2.1c, acting on the visible faces of the cubic free body.

It is convenient to write the nine stress components, defined by equations 2.1, 2.2, 2.3, in the form of a stress matrix $[\boldsymbol{\sigma}]$, defined by

$$[\boldsymbol{\sigma}] = \begin{bmatrix} \sigma_{xx} & \sigma_{xy} & \sigma_{xz} \\ \sigma_{yx} & \sigma_{yy} & \sigma_{yz} \\ \sigma_{zx} & \sigma_{zy} & \sigma_{zz} \end{bmatrix} \tag{2.4}$$

The form of the stress matrix defined in equation 2.4 suggests that the state of stress at a point is defined by nine independent stress components. However, by consideration of moment equilibrium of the free body illustrated in Figure 2.1c, it is readily demonstrated that

$$\sigma_{xy} = \sigma_{yx}, \quad \sigma_{yz} = \sigma_{zy}, \quad \sigma_{zx} = \sigma_{xz}$$

Thus only six independent stress components are required to define completely the state of stress at a point. The stress matrix may then be written

$$[\boldsymbol{\sigma}] = \begin{bmatrix} \sigma_{xx} & \sigma_{xy} & \sigma_{zx} \\ \sigma_{xy} & \sigma_{yy} & \sigma_{yz} \\ \sigma_{zx} & \sigma_{yz} & \sigma_{zz} \end{bmatrix} \tag{2.5}$$

2.3 Stress transformation

The choice of orientation of the reference axes in specifying a state of stress is arbitrary, and situations will arise in which a differently oriented set of reference axes may prove more convenient for the problem at hand. Figure 2.2 illustrates a set of old (x, y, z) axes and new (l, m, n) axes. The orientation of a particular axis, e.g. the l axis, relative to the original x, y, z axes may be defined by a row vector (l_x, l_y, l_z) of direction cosines. In this vector, l_x represents the projection on the x axis of a unit vector oriented parallel to the l axis, with similar definitions for l_y and l_z. Similarly, the orientations of the m and n axes relative to the original axes are defined by row vectors of direction cosines, (m_x, m_y, m_z) and (n_x, n_y, n_z) respectively. Also, the state of stress at a point may be expressed, relative to the l, m, n axes, by the stress matrix $[\boldsymbol{\sigma}^*]$, defined by

$$[\boldsymbol{\sigma}^*] = \begin{bmatrix} \sigma_{ll} & \sigma_{lm} & \sigma_{nl} \\ \sigma_{lm} & \sigma_{mm} & \sigma_{mn} \\ \sigma_{nl} & \sigma_{mn} & \sigma_{nn} \end{bmatrix}$$

The analytical requirement is to express the components of $[\boldsymbol{\sigma}^*]$ in terms of the components of $[\boldsymbol{\sigma}]$ and the direction cosines of the l, m, n axes relative to the x, y, z axes.

Figure 2.2 shows a tetrahedral free body, $Oabc$, generated from the elementary cubic free body used to define the components of the stress matrix. The material removed by the cut abc has been replaced by the equilibrating force, of magnitude $\mathbf{t}$ per unit area, acting over abc. Suppose the outward normal OP to the surface abc is defined by a row vector of direction cosines $(\lambda_x, \lambda_y, \lambda_z)$. If the area of abc is A, the projections of abc on the planes whose normals are the x, y, z axes are given, respectively, by

Figure 2.2 Free-body diagram for establishing the stress transformation equations, principal stresses and their orientations.

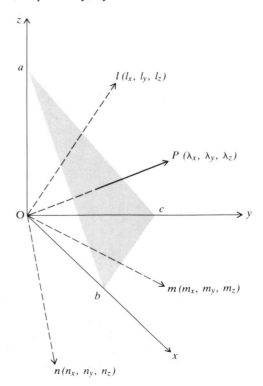

$$\text{Area } Oac = A_x = A\lambda_x$$
$$\text{Area } Oab = A_y = A\lambda_y$$
$$\text{Area } Obc = A_z = A\lambda_z$$

Suppose the traction vector **t** has components t_x, t_y, t_z. Application of the equilibrium requirement for the x direction, for example, yields

$$t_x A - \sigma_{xx} A\lambda_x - \sigma_{xy} A\lambda_y - \sigma_{zx} A\lambda_z = 0$$

or (2.6)

$$t_x = \sigma_{xx} \lambda_x + \sigma_{xy} \lambda_y + \sigma_{zx} \lambda_z$$

Equation 2.6 represents an important relation between the traction component, the state of stress, and the orientation of a surface through the point. Developing the equilibrium equations, similar to equation 2.6, for the y and z directions, produces analogous expressions for t_y and t_z. The three equilibrium equations may then be written

$$\begin{bmatrix} t_x \\ t_y \\ t_z \end{bmatrix} = \begin{bmatrix} \sigma_{xx} & \sigma_{xy} & \sigma_{zx} \\ \sigma_{xy} & \sigma_{yy} & \sigma_{yz} \\ \sigma_{zx} & \sigma_{yz} & \sigma_{zz} \end{bmatrix} \begin{bmatrix} \lambda_x \\ \lambda_y \\ \lambda_z \end{bmatrix} \qquad (2.7)$$

or

$$[\mathbf{t}] = [\boldsymbol{\sigma}] [\boldsymbol{\lambda}] \qquad (2.8)$$

Proceeding in the same way for another set of co-ordinate axes l, m, n, maintaining the same global orientation of the cutting surface to generate the tetrahedral free body, but expressing all traction and stress components relative to the l, m, n axes, yields the relations

$$\begin{bmatrix} t_l \\ t_m \\ t_n \end{bmatrix} = \begin{bmatrix} \sigma_{ll} & \sigma_{lm} & \sigma_{nl} \\ \sigma_{lm} & \sigma_{mm} & \sigma_{mn} \\ \sigma_{nl} & \sigma_{mn} & \sigma_{nn} \end{bmatrix} \begin{bmatrix} \lambda_l \\ \lambda_m \\ \lambda_n \end{bmatrix} \qquad (2.9)$$

or

$$[\mathbf{t}^*] = [\boldsymbol{\sigma}^*] [\boldsymbol{\lambda}^*] \qquad (2.10)$$

In equations 2.8 and 2.10, $[\mathbf{t}]$, $[\mathbf{t}^*]$, $[\boldsymbol{\lambda}]$, $[\boldsymbol{\lambda}^*]$ are vectors, expressed relative to the x, y, z and l, m, n co-ordinate systems. They represent traction components acting on, and direction cosines of the outward normal to, a surface with fixed spatial orientation. From elementary vector analysis, a vector $[\mathbf{v}]$ is transformed from one set of orthogonal reference axes x, y, z to another set, l, m, n, by the transformation equation

$$\begin{bmatrix} v_l \\ v_m \\ v_n \end{bmatrix} = \begin{bmatrix} l_x & l_y & l_z \\ m_x & m_y & m_z \\ n_x & n_y & n_z \end{bmatrix} \begin{bmatrix} v_x \\ v_y \\ v_z \end{bmatrix}$$

or

$$[\mathbf{v}^*] = [\mathbf{R}] [\mathbf{v}] \tag{2.11}$$

In this expression, $[\mathbf{R}]$ is the rotation matrix, whose rows are seen to be formed from the row vectors of direction cosines of the new axes relative to the old axes. As discussed by Jennings (1977), a unique property of the rotation matrix is that its inverse is equal to its transpose, i.e.

$$[\mathbf{R}]^{-1} = [\mathbf{R}]^{\mathrm{T}} \tag{2.12}$$

Returning now to the relations between $[\mathbf{t}]$ and $[\mathbf{t}^*]$, and $[\boldsymbol{\lambda}]$ and $[\boldsymbol{\lambda}^*]$, the results expressed in equations 2.11 and 2.12 indicate that

$$[\mathbf{t}^*] = [\mathbf{R}] [\mathbf{t}]$$

or

$$[\mathbf{t}] = [\mathbf{R}]^{\mathrm{T}} [\mathbf{t}^*]$$

and

$$[\boldsymbol{\lambda}^*] = [\mathbf{R}] [\boldsymbol{\lambda}]$$

or

$$[\boldsymbol{\lambda}] = [\mathbf{R}]^{\mathrm{T}} [\boldsymbol{\lambda}^*]$$

Then

$$[\mathbf{t}^*] = [\mathbf{R}] [\mathbf{t}]$$

$$= [\mathbf{R}] [\boldsymbol{\sigma}] [\boldsymbol{\lambda}]$$

$$= [\mathbf{R}] [\boldsymbol{\sigma}] [\mathbf{R}]^{\mathrm{T}} [\boldsymbol{\lambda}^*]$$

but since

$$[\mathbf{t}^*] = [\boldsymbol{\sigma}^*] [\boldsymbol{\lambda}^*]$$

then

$$[\boldsymbol{\sigma}^*] = [\mathbf{R}] [\boldsymbol{\sigma}] [\mathbf{R}]^{\mathrm{T}} \tag{2.13}$$

Equation 2.13 is the required **stress transformation equation**. It indicates that the state of stress at a point is transformed, under a rotation of axes, as a second-order tensor.

Equation 2.13 when written in expanded notation becomes

$$\begin{bmatrix} \sigma_{ll} & \sigma_{lm} & \sigma_{nl} \\ \sigma_{lm} & \sigma_{mm} & \sigma_{mn} \\ \sigma_{nl} & \sigma_{mn} & \sigma_{nn} \end{bmatrix} = \begin{bmatrix} l_x & l_y & l_z \\ m_x & m_y & m_z \\ n_x & n_y & n_z \end{bmatrix} \begin{bmatrix} \sigma_{xx} & \sigma_{xy} & \sigma_{zx} \\ \sigma_{xy} & \sigma_{yy} & \sigma_{yz} \\ \sigma_{zx} & \sigma_{yz} & \sigma_{zz} \end{bmatrix} \begin{bmatrix} l_x & m_x & n_x \\ l_y & m_y & n_y \\ l_z & m_z & n_z \end{bmatrix}$$

Proceeding with the matrix multiplication on the right-hand side of this expression, in the usual way, produces explicit formulae for determining the stress components under a rotation of axes, given by

$$\sigma_{ll} = l_x^2\,\sigma_{xx} + l_y^2\,\sigma_{yy} + l_z^2\,\sigma_{zz} + 2(l_x l_y \sigma_{xy} + l_y l_z \sigma_{yz} + l_z l_x \sigma_z x) \qquad (2.14)$$

$$\sigma_{lm} = l_x m_x \sigma_{xx} + l_y m_y \sigma_{yy} + l_z m_z \sigma_{zz} + (l_x m_y + l_y m_x)\sigma_{xy}$$

$$+ (l_y m_z + l_z m_y)\sigma_{yz} + (l_z m_x + l_x m_z)\sigma_{zx} \qquad (2.15)$$

Expressions for the other four components of the stress matrix are readily obtained from these equations by cyclic permutation of the subscripts.

2.4 Principal stresses and stress invariants

The discussion above has shown that the state of stress at a point in a medium may be specified in terms of six components, whose magnitudes are related to arbitrarily selected orientations of the reference axes. In some rock masses, the existence of a particular fabric element, such as a foliation or a schistosity, might define a suitable direction for a reference axis. Such a feature might also determine a mode of deformation of the rock mass under load. However, in an isotropic rock mass, any choice of a set of reference axes is obviously arbitrary, and a non-arbitrary way is required for defining the state of stress at any point in the medium. This is achieved by determining **principal stresses** and related quantities which are invariant under any rotations of reference axes.

In section 2.2 it was shown that the resultant stress on any plane in a body could be expressed in terms of a normal component of stress, and two mutually orthogonal shear stress components. A **principal plane** is defined as one on which the shear stress components vanish, i.e. it is possible to select a particular orientation for a plane such that it is subject only to normal stress. The magnitude of the principal stress is that of the normal stress, while the normal to the principal plane defines the direction of the principal stress axis. Since there are, in any specification of a state of stress, three reference directions to be considered, there are three principal stress axes. There are thus three principal stresses and their orientations to be determined to define the state of stress at a point.

Suppose that in Figure 2.2, the cutting plane *abc* is oriented such that the resultant stress on the plane acts normal to it, and has a magnitude σ_p. If the vector $(\lambda_x, \lambda_y, \lambda_z)$ defines the outward normal to the plane, the traction components on *abc* are defined by

$$\begin{bmatrix} t_x \\ t_y \\ t_z \end{bmatrix} = \sigma_p \begin{bmatrix} \lambda_x \\ \lambda_y \\ \lambda_z \end{bmatrix} \qquad (2.16)$$

The traction components on the plane *abc* are also related, through equation 2.7, to the state of stress and the orientation of the plane. Subtracting equation 2.16 from equation 2.7 yields the equation

$$\begin{bmatrix} \sigma_{xx} - \sigma_p & \sigma_{xy} & \sigma_{zx} \\ \sigma_{xy} & \sigma_{yy} - \sigma_p & \sigma_{yz} \\ \sigma_{zx} & \sigma_{yz} & \sigma_{zz} - \sigma_p \end{bmatrix} \begin{bmatrix} \lambda_x \\ \lambda_y \\ \lambda_z \end{bmatrix} = [\mathbf{0}] \qquad (2.17)$$

The matrix equation 2.17 represents a set of three simultaneous, homogeneous, linear equations in λ_x, λ_y, λ_z. The requirement for a non-trivial solution is that the determinant of the coefficient matrix in equation 2.17 must vanish. Expansion of the determinant yields a cubic equation in σ_p, given by

$$\sigma_p^3 - I_1\,\sigma_p^2 + I_2\,\sigma_p - I_3 = 0 \tag{2.18}$$

In this equation, the quantities I_1, I_2 and I_3 are called the **first, second and third stress invariants**. They are defined by the expressions

$$I_1 = \sigma_{xx} + \sigma_{yy} + \sigma_{zz}$$

$$I_2 = \sigma_{xx}\,\sigma_{yy} + \sigma_{yy}\,\sigma_{zz} + \sigma_{zz}\,\sigma_{xx} - (\sigma_{xy}^2 + \sigma_{yz}^2 + \sigma_{zx}^2)$$

$$I_3 = \sigma_{xx}\,\sigma_{yy}\,\sigma_{zz} + 2\sigma_{xy}\,\sigma_{yz}\,\sigma_{zx} - (\sigma_{xx}\,\sigma_{yx}^2 + \sigma_{yy}\,\sigma_{zx}^2 + \sigma_{zz}\,\sigma_{xy}^2)$$

It is to be noted that since the quantities I_1, I_2, I_3 are invariant under a change of axes, any quantities derived from them are also invariants.

Solution of the characteristic equation 2.18 by some general method, such as a complex variable method, produces three real solutions for the principal stresses. These are denoted σ_1, σ_2, σ_3, in order of decreasing magnitude, and are identified respectively as the **major, intermediate** and **minor principal stresses**.

Each principal stress value is related to a principal stress axis, whose direction cosines can be obtained directly from equation 2.17 and a basic property of direction cosines. The dot product theorem of vector analysis yields, for any unit vector of direction cosines (λ_x, λ_y, λ_z), the relation

$$\lambda_x^2 + \lambda_y^2 + \lambda_z^2 = 1 \tag{2.19}$$

Introduction of a particular principal stress value, e.g. σ_1, into equation 2.17, yields a set of simultaneous, homogeneous equations in λ_{x1}, λ_{y1} λ_{x1}. These are the required direction cosines for the major principal stress axis. Solution of the set of equations for these quantities is possible only in terms of some arbitrary constant K, defined by

$$\frac{\lambda_{x1}}{A} = \frac{\lambda_{y1}}{B} = \frac{\lambda_{z1}}{C} = K$$

where

$$A = \begin{vmatrix} \sigma_{yy} - \sigma_1 & \sigma_{yz} \\ \sigma_{yz} & \sigma_{zz} - \sigma_1 \end{vmatrix}$$

$$B = - \begin{vmatrix} \sigma_{xy} & \sigma_{yz} \\ \sigma_{zx} & \sigma_{zz} - \sigma_1 \end{vmatrix} \tag{2.20}$$

$$C = \begin{vmatrix} \sigma_{xy} & \sigma_{yy} - \sigma_1 \\ \sigma_{zx} & \sigma_{yz} \end{vmatrix}$$

Substituting for λ_{x1}, λ_{y1}, λ_{z1} in equation 2.19, gives

$$\lambda_{x1} = A/(A^2 + B^2 + C^2)^{1/2}$$

$$\lambda_{y1} = B/(A^2 + B^2 + C^2)^{1/2}$$

$$\lambda_{z1} = C/(A^2 + B^2 + C^2)^{1/2}$$

Proceeding in a similar way, the vectors of direction cosines for the intermediate and minor principal stress axes, i.e. $(\lambda_{x2}, \lambda_{y2}, \lambda_{z2})$ and $(\lambda_{x3}, \lambda_{y3}, \lambda_{z3})$, are obtained from equations 2.20 by introducing the respective values of σ_2 and σ_3.

The procedure for calculating the principal stresses and the orientations of the principal stress axes is simply the determination of the eigenvalues of the stress matrix, and the eigenvector for each eigenvalue. Some simple checks can be performed to assess the correctness of solutions for principal stresses and their respective vectors of direction cosines. The condition of orthogonality of the principal stress axes requires that each of the three dot products of the vectors of direction cosines must vanish, i.e.

$$\lambda_{x1} \lambda_{x2} + \lambda_{y1} \lambda_{y2} + \lambda_{z1} \lambda_{z2} = 0$$

with a similar result for the (2,3) and (3,1) dot products. Invariance of the sum of the normal stresses requires that

$$\sigma_1 + \sigma_2 + \sigma_3 = \sigma_{xx} + \sigma_{yy} + \sigma_{zz}$$

In the analysis of some types of behaviour in rock, it is usual to split the stress matrix into two components – a **spherical** or **hydrostatic component** $[\sigma_m]$, and a **deviatoric component** $[\sigma_d]$. The spherical stress matrix is defined by

$$[\sigma_m] = \sigma_m [\mathbf{I}] = \begin{bmatrix} \sigma_m & 0 & 0 \\ 0 & \sigma_m & 0 \\ 0 & 0 & \sigma_m \end{bmatrix}$$

where

$$\sigma_m = I_1/3.$$

The deviator stress matrix is obtained from the stress matrix $[\sigma]$ and the spherical stress matrix, and is given by

$$[\sigma_d] = \begin{bmatrix} \sigma_{xx} - \sigma_m & \sigma_{xy} & \sigma_{zx} \\ \sigma_{xy} & \sigma_{yy} - \sigma_m & \sigma_{yz} \\ \sigma_{zx} & \sigma_{yz} & \sigma_{zz} - \sigma_m \end{bmatrix}$$

25

Principal deviator stresses S_1, S_2, S_3 can be established either from the deviator stress matrix, in the way described previously, or from the principal stresses and the hydrostatic stress, i.e.

$$S_1 = \sigma_1 - \sigma_m, \text{ etc.}$$

where S_1 is the major principal deviator stress.

The principal directions of the deviator stress matrix $[\boldsymbol{\sigma}_d]$ are the same as those of the stress matrix $[\boldsymbol{\sigma}]$.

2.5 Differential equations of static equilibrium

Problems in solid mechanics frequently involve description of the stress distribution in a body in static equilibrium under the combined action of surface and body forces. Determination of the stress distribution must take account of the requirement that the stress field maintains static equilibrium throughout the body. This condition requires satisfaction of the equations of static equilibrium for all differential elements of the body.

Figure 2.3 shows a small element of a body, in which operate body force components with magnitudes X, Y, Z per unit volume, directed in the positive x, y, z co-ordinate directions. The stress distribution in the body is described in terms of a set of stress gradients, defined by $\partial\sigma_{xx}/\partial x$, $\partial\sigma_{xy}/\partial y$, etc. Considering the condition for force equilibrium of the element in the x direction yields the equation

$$\frac{\partial\sigma_{xx}}{\partial x} \cdot dx \cdot dy\, dz + \frac{\partial\sigma_{xy}}{\partial y} \cdot dy \cdot dx\, dz + \frac{\partial\sigma_{zx}}{\partial z} \cdot dz \cdot dx\, dy + X\, dx\, dy\, dz = 0$$

Applying the same static equilibrium requirement to the y and z directions, and eliminating the term $dx\, dy\, dz$, yields the differential equations of equilibrium:

$$\frac{\partial\sigma_{xx}}{\partial x} + \frac{\partial\sigma_{xy}}{\partial y} + \frac{\partial\sigma_{zx}}{\partial z} + X = 0$$

$$\frac{\partial\sigma_{xy}}{\partial x} + \frac{\partial\sigma_{yy}}{\partial y} + \frac{\partial\sigma_{yz}}{\partial z} + Y = 0 \tag{2.21}$$

$$\frac{\partial\sigma_{zx}}{\partial x} + \frac{\partial\sigma_{yz}}{\partial y} + \frac{\partial\sigma_{zz}}{\partial z} + Z = 0$$

These expressions indicate that the variations of stress components in a body under load are not mutually independent. They are always involved, in one form or another, in determining the state of stress in a body. A purely practical application of these equations is in checking the admissibility of any closed-form solution for the stress distribution in a body subject to particular applied loads. It is a straightforward matter to determine if the derivatives of expressions describing a particular stress distribution satisfy the equalities of equation 2.21.

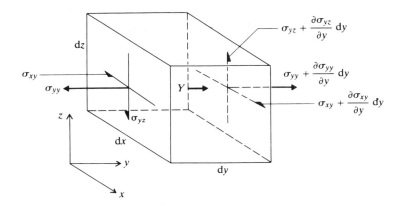

Figure 2.3 Free-body diagram for development of the differential equations of equilibrium.

2.6 Plane problems and biaxial stress

Many underground excavation design analyses involving openings where the length to cross section dimension ratio is high, are facilitated considerably by the relative simplicity of the excavation geometry. For example, an excavation such as a tunnel of uniform cross section along its length might be analysed by assuming that the stress distribution is identical in all planes perpendicular to the long axis of the excavation. Suppose a set of reference axes, x, y, z, is established for such a problem, with the long axis of the excavation parallel to the z axis, as shown in Figure 2.4. As shown above, the state of stress at any point in the medium is described by six stress components. For plane problems in the x, y plane, the six stress components are functions of (x, y) only. In some cases, it may be more convenient to express the state of stress relative to a different set of reference axes, such as the l, m, z axes shown in Figure 2.4. If the angle lOx is α, the direction cosines of the new reference axes relative to the old set are given by

Figure 2.4 A long excavation, of uniform cross section, for which a contracted form of the stress transformation equations is appropriate.

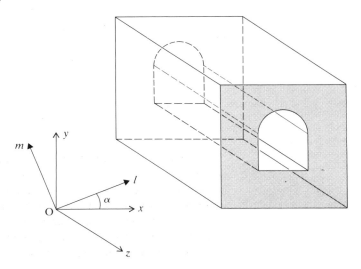

$$l_x = \cos\alpha, \qquad l_y = \sin\alpha, \qquad l_z = 0$$

$$m_x = -\sin\alpha, \qquad m_y = \cos\alpha, \qquad m_z = 0$$

Introducing these values into the general transformation equations, i.e. equations 2.14 and 2.15, yields

$$\sigma_{ll} = \sigma_{xx}\cos^2\alpha + \sigma_{yy}\sin^2\alpha + 2\sigma_{xy}\sin\alpha\cos\alpha$$

$$\sigma_{mm} = \sigma_{xx}\sin^2\alpha + \sigma_{yy}\cos^2\alpha - 2\sigma_{xy}\sin\alpha\cos\alpha$$

$$\sigma_{lm} = \sigma_{xy}(\cos^2\alpha - \sin^2\alpha) - (\sigma_{xx} - \sigma_{yy})\sin\alpha\cos\alpha \qquad (2.22)$$

$$\sigma_{mz} = \sigma_{yz}\cos\alpha - \sigma_{zx}\sin\alpha$$

$$\sigma_{zl} = \sigma_{yz}\sin\alpha + \sigma_{zx}\cos\alpha$$

and the σ_{zz} component is clearly invariant under the transformation of axes. The set of equations 2.22 is observed to contain two distinct types of transformation: those defining σ_{ll}, σ_{mm}, σ_{lm}, which conform to second-order tensor transformation behaviour, and σ_{mz} and σ_{zl}, which are obtained by an apparent vector transformation. The latter behaviour in the transformation is due to the constancy of the orientation of the element of surface whose normal is the z axis. The rotation of the axes merely involves a transformation of the traction components on this surface.

For problems which can be analysed in terms of plane geometry, equations 2.22 indicate that the state of stress at any point can be defined in terms of the plane components of stress $(\sigma_{xx}, \sigma_{yy}, \sigma_{xy})$ and the antiplane components $(\sigma_{zz}, \sigma_{yz}, \sigma_{zx})$. In the particular case where the z direction is a principal axis, the antiplane shear stress components vanish. The plane geometric problem can then be analysed in terms of the plane components of stress, since the σ_{zz} component is frequently neglected. A state of biaxial (or two-dimensional) stress at any point in the medium is defined by three components, in this case σ_{xx}, σ_{yy}, σ_{xy}.

The stress transformation equations related to σ_{ll}, σ_{mm}, σ_{lm} in equation 2.22, for the biaxial state of stress, may be recast in the form

$$\sigma_{ll} = \tfrac{1}{2}(\sigma_{xx} + \sigma_{yy}) + \tfrac{1}{2}(\sigma_{xx} - \sigma_{yy})\cos 2\alpha + \sigma_{xy}\sin 2\alpha$$

$$\sigma_{mm} = \tfrac{1}{2}(\sigma_{xx} + \sigma_{yy}) - \tfrac{1}{2}(\sigma_{xx} - \sigma_{yy})\cos 2\alpha - \sigma_{xy}\sin 2\alpha \qquad (2.23)$$

$$\sigma_{lm} = \sigma_{xy}\cos 2\alpha - \tfrac{1}{2}(\sigma_{xx} - \sigma_{yy})\sin 2\alpha$$

In establishing these equations, the x, y and l, m axes are taken to have the same sense of 'handedness', and the angle α is measured from the x to the l axis, in a sense that corresponds to the 'handedness' of the transformation. There is no inference of clockwise or anticlockwise rotation of axes in establishing these transformation equations. However, the way in which the order of the terms is specified in the equations, and related to the sense of measurement of the rotation angle α, should be examined closely.

Consider now the determination of the magnitudes and orientations of the plane principal stresses for a plane problem in the x, y plane. In this case, the σ_{zz}, σ_{yz}, σ_{zx} stress components vanish, the third stress invariant vanishes, and the characteristic equation, 2.18, becomes

$$\sigma_p^2 - (\sigma_{xx} + \sigma_{yy})\,\sigma_p + \sigma_{xx}\sigma_{yy} - \sigma_{xy}^2 = 0$$

Solution of this quadratic equation yields the magnitudes of the plane principal stresses as

$$\sigma_{1,2} = \tfrac{1}{2}(\sigma_{xx} + \sigma_{yy}) \pm \left[\tfrac{1}{4}(\sigma_{xx} - \sigma_{yy})^2 + \sigma_{xy}^2\right]^{1/2} \qquad (2.24a)$$

The orientations of the respective principal stress axes are obtained by establishing the direction of the outward normal to a plane which is free of shear stress. Suppose ab, shown in Figure 2.5, represents such a plane. The outward normal to ab is Ol, and therefore defines the direction of a principal stress, σ_p. Considering static equilibrium of the element aOb under forces operating in the x direction:

$$\sigma_p\,ab\cos\alpha - \sigma_{xx}\,ab\cos\alpha - \sigma_{xy}\,ab\sin\alpha = 0$$

or

$$\tan\alpha = \frac{\sigma_p - \sigma_{xx}}{\sigma_{xy}}$$

i.e.

$$\alpha = \tan^{-1}\frac{\sigma_p - \sigma_{xx}}{\sigma_{xy}} \qquad (2.24b)$$

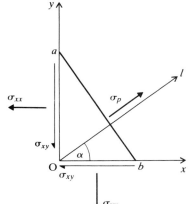

Figure 2.5 Problem geometry for determination of plane principal stresses and their orientations.

Substitution of the magnitudes σ_1, σ_2, determined from equation 2.24a, in equation 2.24b yields the orientations α_1, α_2 of the principal stress axes relative to the positive direction of the x axis. Calculation of the orientations of the major and minor plane principal stresses in this way associates a principal stress axis unambiguously with a principal stress magnitude. This is not the case with other methods, which employ the last of equations 2.23 to determine the orientation of a principal stress axis.

It is to be noted that in specifying the state of stress in a body, there has been no reference to any mechanical properties of the body which is subject to applied load. The only concept invoked has been that of static equilibrium of all elements of the body.

2.7 Displacement and strain

Application of a set of forces to a body, or change in its temperature, changes the relative positions of points within it. The change in loading conditions from the initial state to the final state causes a displacement of each point relative to all other points. If the applied loads constitute a self-equilibrating set, the problem is to determine the equilibrium displacement field induced in the body by the loading. A particular difficulty is presented in the analysis of displacements for a loaded body where boundary conditions are specified completely in terms of surface tractions. In this case, unique determination of the absolute displacement field is impossible, since any set of rigid-body displacements can be superimposed on a particular solution, and still satisfy the equilibrium condition. Difficulties of this type are avoided in analysis by employing displacement gradients as the field variables. The related concept of strain is therefore introduced to make basically indeterminate problems tractable.

Figure 2.6 shows the original positions of two adjacent particles $P(x, y, z)$ and $Q(x + dx, y + dy, z + dz)$ in a body. Under the action of a set of applied loads, P moves to the point $P^*(x + u_x, y + u_y, z + u_z)$, and Q moves to the point $Q^*(x + dx + u_x^*, y + dy + u_y^*, z + dz + u_z^*)$. If $u_x = u_x^*$, etc., the relative displacement between P and Q under the applied load is zero, i.e. the body has been subject to a rigid-body displacement. The problem of interest involves the case where $u_x \neq u_x^*$, etc. The line element joining P and Q then changes length in the process of load application, and the body is said to be in a state of strain.

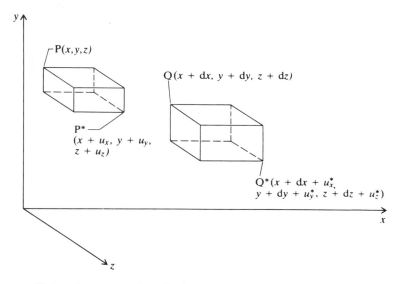

Figure 2.6 Initial and final positions of points P, Q, in a body subjected to strain.

In specifying the state of strain in a body, the objective is to describe the changes in the sizes and shapes of infinitesimal elements in the loaded medium. This is done by considering the displacement components (u_x, u_y, u_z) of a particle P, and (u_x^*, u_y^*, u_z^*) of the adjacent particle Q. Since

$$u_x^* = u_x + du_x, \quad \text{where } du_x = \frac{\partial u_x}{\partial x} dx + \frac{\partial u_x}{\partial y} dy + \frac{\partial u_x}{\partial z} dz$$

and

$$u_y^* = u_y + du_y, \quad \text{where } du_y = \frac{\partial u_y}{\partial x} dx + \frac{\partial u_y}{\partial y} dy + \frac{\partial u_y}{\partial z} dz$$

$$u_z^* = u_z + du_z, \quad \text{where } du_z = \frac{\partial u_z}{\partial x} dx + \frac{\partial u_z}{\partial y} dy + \frac{\partial u_z}{\partial z} dz$$

30

the incremental displacements may be expressed by

$$
\begin{bmatrix} du_x \\ du_y \\ du_z \end{bmatrix} = \begin{bmatrix} \dfrac{\partial u_x}{\partial x} & \dfrac{\partial u_x}{\partial y} & \dfrac{\partial u_x}{\partial z} \\ \dfrac{\partial u_y}{\partial x} & \dfrac{\partial u_y}{\partial y} & \dfrac{\partial u_y}{\partial z} \\ \dfrac{\partial u_z}{\partial x} & \dfrac{\partial u_z}{\partial y} & \dfrac{\partial u_z}{\partial z} \end{bmatrix} \begin{bmatrix} dx \\ dy \\ dz \end{bmatrix} \tag{2.25a}
$$

or

$$[d\boldsymbol{\delta}] = [\mathbf{D}]\,[d\mathbf{r}] \tag{2.25b}$$

In this expression, $[d\mathbf{r}]$ represents the original length of the line element PQ, while $[d\boldsymbol{\delta}]$ represents the relative displacement of the ends of the line element in deforming from the unstrained to the strained state.

The infinitesimal relative displacement defined by equation 2.25 can arise from both deformation of the element of which PQ is the diagonal, and a rigid-body rotation of the element. The need is to define explicitly the quantities related to deformation of the body. Figure 2.7 shows the projection of the element, with diagonal PQ, on to the yz plane, and subject to a rigid body rotation Ω_x about the x axis. Since the side dimensions of the element are dy and dz, the relative displacement components of Q relative to P are

$$du_y = -\,\Omega_x\,dz$$
$$du_z = \;\;\Omega_x\,dy \tag{2.26}$$

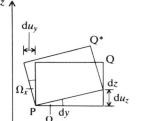

Figure 2.7 Rigid-body rotation of an element producing component displacements of adjacent points.

Considering rigid-body rotations Ω_y and Ω_z about the y and z axes, the respective displacements are

$$du_z = -\,\Omega_y\,dx$$
$$du_x = \;\;\Omega_y\,dz \tag{2.27}$$

and

$$du_x = -\,\Omega_z\,dy$$
$$du_y = \;\;\Omega_z\,dx \tag{2.28}$$

The total displacement due to the various rigid-body rotations is obtained by addition of equations 2.26, 2.27 and 2.28, i.e.

$$du_x = -\,\Omega_z\,dy + \Omega_y\,dz$$

$$du_x = \;\;\Omega_z\,dx - \Omega_x\,dz$$

$$du_z = -\,\Omega_y\,dx + \Omega_x\,dy$$

31

These equations may be written in the form

$$\begin{bmatrix} du_x \\ du_y \\ du_z \end{bmatrix} = \begin{bmatrix} 0 & -\Omega_z & \Omega_y \\ \Omega_z & 0 & -\Omega_x \\ -\Omega_y & \Omega_x & 0 \end{bmatrix} \begin{bmatrix} dx \\ dy \\ dz \end{bmatrix} \qquad (2.29a)$$

or ·

$$[d\delta'] = [\mathbf{\Omega}]\,[d\mathbf{r}] \qquad (2.29b)$$

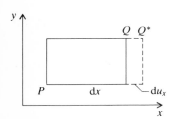

Figure 2.8 Displacement components produced by pure longitudinal strain.

The contribution of deformation to the relative displacement $[d\delta]$ is determined by considering elongation and distortion of the element. Figure 2.8 represents the elongation of the block in the x direction. The element of length dx is assumed to be homogeneously strained, in extension, and the normal strain component is therefore defined by

$$\varepsilon_{xx} = \frac{du_x}{dx}$$

Considering the y and z components of elongation of the element in a similar way, gives·the components of relative displacement due to normal strain as

$$du_x = \varepsilon_{xx}\,dx$$

$$du_y = \varepsilon_{yy}\,dy \qquad (2.30)$$

$$du_z = \varepsilon_{zz}\,dz$$

The components of relative displacement arising from distortion of the element are derived by considering an element subject to various modes of pure shear strain. Figure 2.9 shows such an element strained in the x, y plane. Since the angle α is small, pure shear of the element results in the displacement components

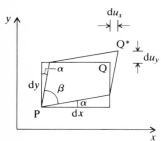

Figure 2.9 Displacement produced by pure shear strain.

$$du_x = \alpha\,dy$$

$$du_y = \alpha\,dx$$

Since shear strain magnitude is defined by

$$\gamma_{xy} = \frac{\pi}{2} - \beta = 2\alpha$$

then

$$du_x = \tfrac{1}{2}\,\gamma_{xy}\,dy$$

$$du_y = \tfrac{1}{2}\,\gamma_{xy}\,dx \qquad (2.31)$$

Similarly, displacements due to pure shear of the element in the y, z and z, x planes are given by

32

$$du_y = \frac{1}{2}\gamma_{yz}\,dz$$

$$du_z = \frac{1}{2}\gamma_{yz}\,dy$$

(2.32)

and

$$du_z = \frac{1}{2}\gamma_{zx}\,dx$$

$$du_x = \frac{1}{2}\gamma_{zx}\,dz$$

(2.33)

The total displacement components due to all modes of infinitesimal strain are obtained by addition of equations 2.30, 2.31, 2.32 and 2.33, i.e.

$$du_x = \varepsilon_{xx}\,dx + \frac{1}{2}\gamma_{xy}\,dy + \frac{1}{2}\gamma_{zx}\,dz$$

$$du_y = \frac{1}{2}\gamma_{xy}\,dx + \varepsilon_{yy}\,dy + \frac{1}{2}\gamma_{yz}\,dz$$

$$du_z = \frac{1}{2}\gamma_{zx}\,dx + \frac{1}{2}\gamma_{yz}\,dy + \varepsilon_{zz}\,dz$$

These equations may be written in the form

$$
\begin{bmatrix} du_x \\ du_y \\ du_z \end{bmatrix} =
\begin{bmatrix}
\varepsilon_{xx} & \frac{1}{2}\gamma_{xy} & \frac{1}{2}\gamma_{zx} \\
\frac{1}{2}\gamma_{xy} & \varepsilon_{yy} & \frac{1}{2}\gamma_{yz} \\
\frac{1}{2}\gamma_{zx} & \frac{1}{2}\gamma_{yz} & \varepsilon_{zz}
\end{bmatrix}
\begin{bmatrix} dx \\ dy \\ dz \end{bmatrix}
$$

(2.34a)

or

$$[d\boldsymbol{\delta}''] = [\boldsymbol{\varepsilon}]\ [d\mathbf{r}]$$

(2.34b)

where $[\boldsymbol{\varepsilon}]$ is the strain matrix.

Since

$$[d\boldsymbol{\delta}] = [d\boldsymbol{\delta}'] + [d\boldsymbol{\delta}'']$$

equations 2.25a, 2.29a and 2.34a yield

$$
\begin{bmatrix}
\dfrac{\partial u_x}{\partial x} & \dfrac{\partial u_x}{\partial y} & \dfrac{\partial u_x}{\partial z} \\[2mm]
\dfrac{\partial u_y}{\partial x} & \dfrac{\partial u_y}{\partial y} & \dfrac{\partial u_y}{\partial z} \\[2mm]
\dfrac{\partial u_z}{\partial x} & \dfrac{\partial u_z}{\partial y} & \dfrac{\partial u_z}{\partial z}
\end{bmatrix} =
\begin{bmatrix}
\varepsilon_{xx} & \frac{1}{2}\gamma_{xy} & \frac{1}{2}\gamma_{zx} \\
\frac{1}{2}\gamma_{xy} & \varepsilon_{yy} & \frac{1}{2}\gamma_{yz} \\
\frac{1}{2}\gamma_{zx} & \frac{1}{2}\gamma_{yz} & \varepsilon_{zz}
\end{bmatrix} +
\begin{bmatrix}
0 & -\Omega_z & \Omega_y \\
\Omega_z & 0 & -\Omega_x \\
-\Omega_y & \Omega_x & 0
\end{bmatrix}
$$

Equating corresponding terms on the left-hand and right-hand sides of this equation, gives for the normal strain components

$$\varepsilon_{xx} = \frac{\partial u_x}{\partial x}, \quad \varepsilon_{yy} = \frac{\partial u_y}{\partial y}, \quad \varepsilon_{zz} = \frac{\partial u_z}{\partial z}$$

(2.35)

and

$$\frac{\partial u_x}{\partial y} = \frac{1}{2}\gamma_{xy} - \Omega_z$$

$$\frac{\partial u_y}{\partial x} = \frac{1}{2}\gamma_{xy} - \Omega_z$$

Thus expressions for shear strain and rotation are given by

$$\gamma_{xy} = \frac{\partial u_x}{\partial y} + \frac{\partial u_y}{\partial x}, \quad \Omega_z = \frac{1}{2}\left(\frac{\partial u_y}{\partial x} - \frac{\partial u_x}{\partial y}\right)$$

and, similarly,

$$\gamma_{yz} = \frac{\partial u_y}{\partial z} + \frac{\partial u_z}{\partial y}, \quad \Omega_x = \frac{1}{2}\left(\frac{\partial u_z}{\partial y} - \frac{\partial u_y}{\partial z}\right)$$

$$(2.36)$$

$$\gamma_{zx} = \frac{\partial u_z}{\partial x} + \frac{\partial u_x}{\partial z}, \quad \Omega_y = \frac{1}{2}\left(\frac{\partial u_x}{\partial z} - \frac{\partial u_z}{\partial x}\right)$$

Equations 2.35 and 2.36 indicate that the state of strain at a point in a body is completely defined by six independent components, and that these are related simply to the displacement gradients at the point. The form of equation 2.34a indicates that a state of strain is specified by a second-order tensor.

2.8 Principal strains, strain transformation, volumetric strain and deviator strain

Since a state of strain is defined by a strain matrix or second-order tensor, determination of principal strains, and other manipulations of strain quantities, are completely analogous to the processes employed in relation to stress. Thus principal strains and principal strain directions are determined as the eigenvalues and associated eigenvectors of the strain matrix. Strain transformation under a rotation of axes is defined, analogously to equation 2.13, by

$$[\varepsilon^*] = [\mathbf{R}]\,[\varepsilon]\,[\mathbf{R}]^{\mathrm{T}}$$

where $[\varepsilon]$ and $[\varepsilon^*]$ are the strain matrices expressed relative to the old and new sets of co-ordinate axes.

The volumetric strain, Δ, is defined by

$$\Delta = \varepsilon_{xx} + \varepsilon_{yy} + \varepsilon_{zz}$$

The deviator strain matrix is defined in terms of the strain matrix and the volumetric strain by

$$[\varepsilon] = \begin{bmatrix} \varepsilon_{xx} - \Delta/3 & \gamma_{xy} & \gamma_{zx} \\ \gamma_{xy} & \varepsilon_{yy} - \Delta/3 & \gamma_{yz} \\ \gamma_{zx} & \gamma_{yz} & \varepsilon_{zz} - \Delta/3 \end{bmatrix}$$

Plane geometric problems, subject to biaxial strain in the xy plane, for example, are described in terms of three strain components, $\varepsilon_{xx}, \varepsilon_{yy}, \gamma_{xy}$.

2.9 Strain compatibility equations

Equations 2.35 and 2.36, which define the components of strain at a point, suggest that the strains are mutually independent. The requirement of physical continuity of the displacement field throughout a continuous body leads automatically to analytical relations between the displacement gradients, restricting the degree of independence of strains. A set of six identities can be established readily from equations 2.35 and 2.36. Three of these identities are of the form

$$\frac{\partial^2 \varepsilon_{xx}}{\partial y^2} + \frac{\partial^2 \varepsilon_{yy}}{\partial x^2} = \frac{\partial^2 \gamma_{xy}}{\partial x\,\partial y}$$

and three are of the form

$$2\frac{\partial^2 \varepsilon_{xx}}{\partial y\,\partial z} = \frac{\partial}{\partial x}\left(-\frac{\partial \gamma_{yz}}{\partial x} + \frac{\partial \gamma_{zx}}{\partial y} + \frac{\partial \gamma_{xy}}{\partial z}\right)$$

These expressions play a basic role in the development of analytical solutions to problems in deformable body mechanics.

2.10 Stress-strain relations

It was noted previously that an admissible solution to any problem in solid mechanics must satisfy both the differential equations of static equilibrium and the equations of strain compatibility. It will be recalled that in the development of analytical descriptions for the states of stress and strain at a point in a body, there was no reference to, nor exploitation of, any mechanical property of the solid. The way in which stress and strain are related in a material under load is described qualitatively by its **constitutive behaviour**. A variety of idealised constitutive models has been formulated for various engineering materials, which describe both the time-independent and time-dependent responses of the material to applied load. These models describe responses in terms of elasticity, plasticity, viscosity and creep, and combinations of these modes. For any constitutive model, stress and strain, or some derived quantities, such as stress and strain rates, are related through a set of constitutive equations. **Elasticity** represents the most common constitutive behaviour of engineering materials, including many rocks, and it forms a useful basis for the description of more complex behaviour.

In formulating constitutive equations, it is useful to construct column vectors from the elements of the stress and strain matrices, i.e. stress and strain vectors are defined by

$$[\boldsymbol{\sigma}] = \begin{bmatrix} \sigma_{xx} \\ \sigma_{yy} \\ \sigma_{zz} \\ \sigma_{xy} \\ \sigma_{yz} \\ \sigma_{zx} \end{bmatrix} \qquad \text{and} \qquad [\boldsymbol{\varepsilon}] = \begin{bmatrix} \varepsilon_{xx} \\ \varepsilon_{yy} \\ \varepsilon_{zz} \\ \gamma_{xy} \\ \gamma_{yz} \\ \gamma_{zx} \end{bmatrix}$$

The most general statement of linear elastic constitutive behaviour is a generalised form of Hooke's Law, in which any strain component is a linear function of all the stress components, i.e.

$$
\begin{bmatrix} \varepsilon_{xx} \\ \varepsilon_{yy} \\ \varepsilon_{zz} \\ \gamma_{xy} \\ \gamma_{yz} \\ \gamma_{zx} \end{bmatrix} = \begin{bmatrix} S_{11} & S_{12} & S_{13} & S_{14} & S_{15} & S_{16} \\ S_{21} & S_{22} & S_{23} & S_{24} & S_{25} & S_{26} \\ S_{31} & S_{32} & S_{33} & S_{34} & S_{35} & S_{36} \\ S_{41} & S_{42} & S_{43} & S_{44} & S_{45} & S_{46} \\ S_{51} & S_{52} & S_{53} & S_{54} & S_{55} & S_{56} \\ S_{61} & S_{62} & S_{63} & S_{64} & S_{65} & S_{66} \end{bmatrix} \begin{bmatrix} \sigma_{xx} \\ \sigma_{yy} \\ \sigma_{zz} \\ \sigma_{xy} \\ \sigma_{yz} \\ \sigma_{zx} \end{bmatrix} \tag{2.37a}
$$

or

$$
[\varepsilon] = [S][\sigma] \tag{2.37b}
$$

Each of the elements S_{ij} of the matrix $[S]$ is called a **compliance** or an **elastic modulus**. Although equation 2.37a suggests that there are 36 independent compliances, a reciprocal theorem, such as that due to Maxwell (1864), may be used to demonstrate that the compliance matrix is symmetric. The matrix therefore contains only 21 independent constants.

In some cases it is more convenient to apply equation 2.37 in inverse form, i.e.

$$
[\sigma] = [D][\varepsilon] \tag{2.38}
$$

The matrix $[D]$ is called the **elasticity matrix** or the matrix of **elastic stiffnesses**. For general anisotropic elasticity there are 21 independent stiffnesses.

Equation 2.37a indicates complete coupling between all stress and strain components. The existence of axes of elastic symmetry in a body de-couples some of the stress–strain relations, and reduces the number of independent constants required to define the material elasticity. In the case of **isotropic elasticity**, any arbitrarily oriented axis in the medium is an axis of elastic symmetry. Equation 2.37a, for isotropic elastic materials, reduces to

$$
\begin{bmatrix} \varepsilon_{xx} \\ \varepsilon_{yy} \\ \varepsilon_{zz} \\ \gamma_{xy} \\ \gamma_{yz} \\ \gamma_{zx} \end{bmatrix} = \frac{1}{E} \begin{bmatrix} 1 & -\nu & -\nu & 0 & 0 & 0 \\ -\nu & 1 & -\nu & 0 & 0 & 0 \\ -\nu & -\nu & 1 & 0 & 0 & 0 \\ 0 & 0 & 0 & 2(1+\nu) & 0 & 0 \\ 0 & 0 & 0 & 0 & 2(1+\nu) & 0 \\ 0 & 0 & 0 & 0 & 0 & 2(1+\nu) \end{bmatrix} \begin{bmatrix} \sigma_{xx} \\ \sigma_{yy} \\ \sigma_{zz} \\ \sigma_{xy} \\ \sigma_{yz} \\ \sigma_{zx} \end{bmatrix} \tag{2.39}
$$

The more common statements of Hooke's Law for isotropic elasticity are readily recovered from equation 2.39, i.e.

$$
\varepsilon_{xx} = \frac{1}{E} \left[\sigma_{xx} - \nu(\sigma_{yy} + \sigma_{zz}) \right], \text{ etc.}
$$

$$
\gamma_{xy} = \frac{1}{G} \sigma_{xy}, \text{ etc.} \tag{2.40}
$$

where

$$
G = \frac{E}{2(1+\nu)}
$$

The quantities E, G, and v are **Young's modulus**, the **modulus of rigidity** (or **shear modulus**) and **Poisson's ratio**. Isotropic elasticity is a two-constant theory, so that determination of any two of the elastic constants characterises completely the elasticity of an isotropic medium.

The inverse form of the stress–strain equation 2.39, for isotropic elasticity, is given by

$$
\begin{bmatrix} \sigma_{xx} \\ \sigma_{yy} \\ \sigma_{zz} \\ \sigma_{xy} \\ \sigma_{yz} \\ \sigma_{zx} \end{bmatrix} = \frac{E(1-v)}{(1+v)(1-2v)} \begin{bmatrix} 1 & v/(1-v) & v/(1-v) & 0 & 0 & 0 \\ v/(1-v) & 1 & v/(1-v) & 0 & 0 & 0 \\ v/(1-v) & v/(1-v) & 1 & 0 & 0 & 0 \\ 0 & 0 & 0 & \frac{(1-2v)}{2(1-v)} & 0 & 0 \\ 0 & 0 & 0 & 0 & \frac{(1-2v)}{2(1-v)} & 0 \\ 0 & 0 & 0 & 0 & 0 & \frac{(1-2v)}{2(1-v)} \end{bmatrix} \begin{bmatrix} \varepsilon_{xx} \\ \varepsilon_{yy} \\ \varepsilon_{zz} \\ \gamma_{xy} \\ \gamma_{yz} \\ \gamma_{zx} \end{bmatrix} \quad (2.41)
$$

The inverse forms of equations 2.40, usually called Lamé's equations, are obtained from equation 2.41, i.e.

$$\sigma_{xx} = \lambda\Delta + 2G\varepsilon_{xx}, \text{ etc.}$$

$$\sigma_{xy} = G\gamma_{xy}, \text{ etc.}$$

where λ is Lamé's constant, defined by

$$\lambda = \frac{2vG}{(1-2v)} = \frac{vE}{(1+v)(1-2v)}$$

and Δ is the volumetric strain.

Transverse isotropic elasticity ranks second to isotropic elasticity in the degree of expression of elastic symmetry in the material behaviour. Media exhibiting transverse isotropy include artificially laminated materials and stratified rocks, such as shales. In the latter case, all lines lying in the plane of bedding are axes of elastic symmetry. The only other axis of elastic symmetry is the normal to the plane of isotropy. In Figure 2.10, illustrating a stratified rock mass, the plane of isotropy of the material coincides with the x, y plane. The elastic constitutive equations for this material are given by

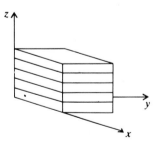

Figure 2.10 A transversely isotropic body for which the x, y plane is the plane of isotropy.

$$
\begin{bmatrix} \varepsilon_{xx} \\ \varepsilon_{yy} \\ \varepsilon_{zz} \\ \gamma_{xy} \\ \gamma_{yz} \\ \gamma_{zx} \end{bmatrix} = \frac{1}{E_1} \begin{bmatrix} 1 & -v_1 & -v_2 & 0 & 0 & 0 \\ -v_1 & 1 & -v_2 & 0 & 0 & 0 \\ -v_2 & -v_2 & E_1/E_2 & 0 & 0 & 0 \\ 0 & 0 & 0 & 2(1+v_1) & 0 & 0 \\ 0 & 0 & 0 & 0 & E_1/G_2 & 0 \\ 0 & 0 & 0 & 0 & 0 & E_1/G_2 \end{bmatrix} \begin{bmatrix} \sigma_{xx} \\ \sigma_{yy} \\ \sigma_{zz} \\ \sigma_{xy} \\ \sigma_{yz} \\ \sigma_{zx} \end{bmatrix} \quad (2.42)
$$

It appears from equation 2.42 that five independent elastic constants are required to characterise the elasticity of a transversely isotropic medium: E_1 and v_1 define properties in the plane of isotropy, and E_2, v_2, G_2 properties in a plane containing the normal to, and any line in, the plane of isotropy. Inversion of the compliance matrix in equation 2.42, and putting $E_1/E_2 = n, G_2/E_2 = m$, produces the elasticity matrix given by

$$[\mathbf{D}] = \frac{E_2}{(1+v_1)(1-v_1-2n\,v_2^2)}
\begin{bmatrix}
n(1-n\,v_2^2) & n(v_1+n\,v_2^2) & n\,v_2(1+v_1) & 0 & 0 & 0 \\
 & n(1-n\,v_2^2) & n\,v_2(1+v_1) & 0 & 0 & 0 \\
 & & (1-v_1^2) & 0 & 0 & 0 \\
\text{symmetric} & & & \begin{array}{c}0.5*n* \\ *(1-v_1-2n\,v_2^2)\end{array} & 0 & 0 \\
 & & & & \begin{array}{c}m(1+v_1)* \\ *(1-v_1-2nv_2^2)\end{array} & 0 \\
 & & & & & \begin{array}{c}m(1+v_1)* \\ *(1-v_1-2nv_2^2)\end{array}
\end{bmatrix}$$

Although it might be expected that the modulus ratios, n and m, and Poisson's ratios, v_1 and v_2, may be virtually independent, such is not the case. The requirement for positive definiteness of the elasticity matrix, needed to assure a stable continuum, restricts the range of possible elastic ratios. Gerrard (1977) has summarised the published experimental data on elastic constants for transversely isotropic rock materials and rock materials displaying other forms of elastic anisotropy, including orthotropy for which nine independent constants are required.

2.11 Cylindrical polar co-ordinates

A Cartesian co-ordinate system does not always constitute the most convenient system for specifying the state of stress and strain in a body, and problem geometry may suggest a more appropriate system. A cylindrical polar co-ordinate system is used frequently in the analysis of axisymmetric problems. Cartesian (x, y, z) and cylindrical polar (r, θ, z) co-ordinate systems are shown in Figure 2.11, together with an elementary free body in the polar system. To operate in the polar system, it is necessary to establish equations defining the co-ordinate transformation between Cartesian and polar co-ordinates, and a complete set of differential equations of equilibrium, strain displacement relations and strain compatibility equations.

The co-ordinate transformation is defined by the equations

$$r = (x^2 + y^2)^{1/2}$$

$$\theta = \arctan\left(\frac{y}{x}\right)$$

and

$$x = r \cos \theta$$
$$y = r \sin \theta$$

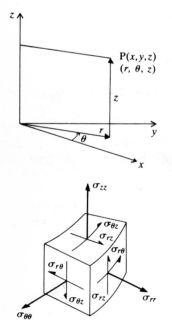

Figure 2.11 Cylindrical polar co-ordinate axes, and associated free-body diagram.

If R, θ, Z are the polar components of body force, the differential equations of equilibrium, obtained by considering the condition for static equilibrium of the element shown in Figure 2.11, are

$$\frac{\partial \sigma_{rr}}{\partial r} + \frac{1}{r}\frac{\partial \sigma_{r\theta}}{\partial \theta} + \frac{\partial \sigma_{rz}}{\partial z} + \frac{\sigma_{rr} - \sigma_{\theta\theta}}{r} + R = 0$$

$$\frac{\partial \sigma_{r\theta}}{\partial r} + \frac{1}{r}\frac{\partial \sigma_{\theta\theta}}{\partial \theta} + \frac{\partial \sigma_{\theta z}}{\partial z} + \frac{2\sigma_{r\theta}}{r} + \theta = 0$$

$$\frac{\partial \sigma_{rz}}{\partial r} + \frac{1}{r}\frac{\partial \sigma_{\theta z}}{\partial \theta} + \frac{\partial \sigma_{zz}}{\partial z} + \frac{\sigma_{zz}}{r} + Z = 0$$

For axisymmetric problems, the tangential shear stress components, $\sigma_{r\theta}$ and $\sigma_{\theta z}$, and the tangential component of body force, θ, vanish. The equilibrium equations reduce to

$$\frac{\partial \sigma_{rr}}{\partial r} + \frac{\partial \sigma_{rz}}{\partial z} + \frac{\sigma_{rr} - \sigma_{\theta\theta}}{r} + R = 0$$

$$\frac{\partial \sigma_{rz}}{\partial r} + \frac{\partial \sigma_{zz}}{\partial z} + \frac{\sigma_{rz}}{r} + Z = 0$$

For the particular case where r, θ, z are principal stress directions, i.e. the shear stress component σ_{rz} vanishes, the equations become

$$\frac{\partial \sigma_{rr}}{\partial r} + \frac{\sigma_{rr} - \sigma_{\theta\theta}}{r} + R = 0$$

$$\frac{\partial \sigma_{zz}}{\partial z} + Z = 0$$

Displacement components in the polar system are described by u_r, u_θ, u_z. The elements of the strain matrix are defined by

$$\varepsilon_{rr} = \frac{\partial u_r}{\partial r}$$

$$\varepsilon_{\theta\theta} = \frac{1}{r}\frac{\partial u_\theta}{\partial \theta} + \frac{u_r}{r}$$

$$\varepsilon_{zz} = \frac{\partial u_z}{\partial z}$$

$$\gamma_{\theta z} = \frac{1}{r}\frac{\partial u_z}{\partial \theta} + \frac{\partial u_\theta}{\partial z}$$

$$\gamma_{rz} = \frac{\partial u_r}{\partial z} + \frac{\partial u_z}{\partial r}$$

$$\gamma_{r\theta} = \frac{1}{r}\left(-u_\theta + \frac{r\partial u_\theta}{\partial r} + \frac{\partial u_r}{\partial \theta}\right)$$

The volumetric strain is the sum of the normal strain components, i.e.

$$\Delta = \varepsilon_{rr} + \varepsilon_{\theta\theta} + \varepsilon_{zz}$$

When the principal axes of strain coincide with the directions of the co-ordinate axes, i.e. the shear strain components vanish, the normal strains are defined by

$$\varepsilon_{rr} = \frac{du_r}{dr}$$

$$\varepsilon_{\theta\theta} = \frac{u_r}{r}$$

$$\varepsilon_{zz} = \frac{du_z}{dz}$$

The compatibility equations for strains are

$$\frac{\partial^2 (r\gamma_{r\theta})}{\partial r\,\partial\theta} = r\frac{\partial^2 (r\varepsilon_{\theta\theta})}{\partial r^2} - r\frac{\partial\varepsilon_{rr}}{\partial r} + \frac{\partial^2\varepsilon_{rr}}{\partial\theta^2}$$

$$\frac{\partial^2\gamma_{rz}}{\partial r\,\partial z} = \frac{\partial^2\varepsilon_{rr}}{\partial z^2} + \frac{\partial^2\varepsilon_{zz}}{\partial r^2}$$

$$\frac{\partial^2\gamma_{\theta z}}{\partial\theta\,\partial z} = \frac{\partial^2 (r\varepsilon_{\theta\theta})}{\partial z^2} + \frac{1}{r}\frac{\partial^2\varepsilon_{zz}}{\partial\theta^2} + \frac{\partial\varepsilon_{zz}}{\partial z} - \frac{\partial\gamma_{zr}}{\partial z}$$

The case where $\gamma_{r\theta} = \gamma_{\theta z} = \gamma_{rz} = 0$ yields only one compatibility equation, i.e.

$$\frac{d}{dr}(r\varepsilon_{\theta\theta}) = \varepsilon_{rr}$$

Stress components expressed relative to the Cartesian axes are transformed to the polar system using equations 2.22, with r and θ replacing l and m and θ replacing α. An analogous set of equations can be established for transformation of Cartesian strain components to the polar system.

2.12 Geomechanics convention for displacement, strain and stress

The convention used until now in the discussion of displacement, strain and stress has been the usual engineering mechanics one. Under this convention, force and displacement components are considered positive if directed in the positive directions of the co-ordinate axes. Extensile normal strains and tensile normal stresses are treated as positive. Finally, the sense of positive shear stress on a surface of the elementary free body is outward, if the outward normal to the surface is directed outward relative to the co-ordinate origin, and conversely. The sense of positive stress components, defined in this way, is illustrated in Figures 2.1c and 2.11, for Cartesian and polar co-ordinate systems. This convention has been followed in this introductory material since important notions such as traction retain their conceptual basis, and since practically significant numerical methods of stress analysis are usually developed employing it.

States of stress occurring naturally, and generated and sustained in a rock mass by excavation activity, are pervasively compressive. If the usual engineering mechanics convention for stresses were followed, all numerical manipulations related to stress and strain in rock would involve negative quantities. Although this presents no conceptual difficulties, convenience and accuracy in calculations are served by adopting the following convention for stress and strain analysis in rock mechanics:

(a)   positive force and displacement components act in the positive directions of the co-ordinate axes;
(b)   contractile normal strains are taken as positive;
(c)   compressive normal stresses are taken as positive;
(d)   the sense of positive shear stress on a surface is inward relative to the co-ordinate origin, if the inward normal to the surface acts inwards relative to the co-ordinate origin, and conversely.

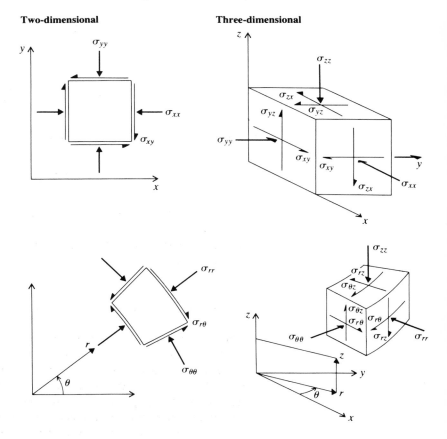

Two-dimensional

Three-dimensional

Figure 2.12 Two- and three-dimensional free bodies, for specification of the state of stress relative to Cartesian and polar co-ordinate axes, using the geomechanics convention for the sense of positive stresses.

41

The senses of positive stress components defined by this convention, for Cartesian and polar co-ordinate systems, and biaxial and triaxial states of stress, are shown in Figure 2.12. Some minor changes are required in some of the other general relations developed earlier, and these are now defined.

2.12.1 Stress-traction relations
If the outward normal to a surface has direction cosines $(\lambda_z, \lambda_y, \lambda_z)$, traction components are determined by

$$t_x = -(\sigma_{xx}\lambda_x + \sigma_{xy}\lambda_y + \sigma_{zx}\lambda_z), \text{ etc.}$$

2.12.2 Strain-displacement relations
Strain components are determined from displacement components using the expressions

$$\varepsilon_{xx} = -\frac{\partial u_x}{\partial x}$$

$$\gamma_{xy} = -\left(\frac{\partial u_y}{\partial x} + \frac{\partial u_x}{\partial y}\right), \text{ etc.}$$

2.12.3 Differential equations of equilibrium
The change in the sense of positive stress components yields equations of the form

$$\frac{\partial \sigma_{xx}}{\partial x} + \frac{\partial \sigma_{xy}}{\partial y} + \frac{\partial \sigma_{zx}}{\partial z} - X = 0, \text{ etc.}$$

All other relations, such as strain compatibility equations, transformation equations and stress invariants, are unaffected by the change in convention.

2.13 Graphical representation of biaxial stress

Analytical procedures for plane problems subject to biaxial stress have been discussed above. Where equations or relations appropriate to the two-dimensional case have not been proposed explicitly, they can be established from the three-dimensional equations by deleting any terms or expressions related to the third co-ordinate direction. For example, for biaxial stress in the x, y plane, the differential equations of static equilibrium, in the geomechanics convention, reduce to

$$\frac{\partial \sigma_{xx}}{\partial x} + \frac{\partial \sigma_{xy}}{\partial y} - X = 0$$

$$\frac{\partial \sigma_{xy}}{\partial x} + \frac{\partial \sigma_{yy}}{\partial y} - Y = 0$$

One aspect of biaxial stress that requires careful treatment is graphical representation of the state of stress at a point, using the **Mohr circle diagram**. In particular, the geomechanics convention for the sense of positive stresses introduces some subtle difficulties which must be overcome if the diagram is to provide correct determination of the sense of shear stress acting on a surface.

Correct construction of the Mohr circle diagram is illustrated in Figure 2.13. The state of stress in a small element abcd is specified, relative to the x, y co-ordinate axes, by known values of $\sigma_{xx}, \sigma_{yy}, \sigma_{xy}$. A set of reference axes for the circle diagram construction is defined by directions σ_n and τ, with the sense of the positive τ axis directed downwards. If O is the origin of the $\sigma_n - \tau$ co-ordinate system, a set of quantities related to the stress components is calculated from

$$OC = \frac{1}{2}(\sigma_{xx} + \sigma_{yy})$$

$$CD = \frac{1}{2}(\sigma_{xx} - \sigma_{yy})$$

$$DF = -\sigma_{xy}$$

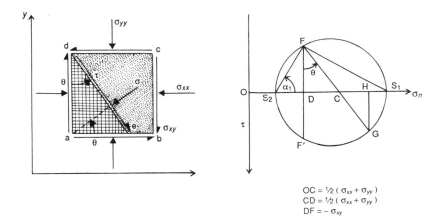

Figure 2.13 Construction of a Mohr circle diagram, appropriate to the geomechanics convention of stresses.

$OC = \frac{1}{2}(\sigma_{xx} + \sigma_{yy})$
$CD = \frac{1}{2}(\sigma_{xx} + \sigma_{yy})$
$DF = -\sigma_{xy}$

Points corresponding to C, D, F are plotted in the σ, τ plane as shown in Figure 2.13, using some convenient scale. In the circle diagram construction, if σ_{xy} is positive, the point F plots above the σ_n axis. Construction of the line FDF′ returns values of $\tau = \sigma_{xy}$ and $\sigma_n = \sigma_{xx}$ which are the shear and normal stress components acting on the surface cb of the element. Suppose the surface ed in Figure 2.13 is inclined at an angle θ to the negative direction of the y axis, or, alternatively, its outward normal is inclined at an angle θ to the x axis. In the circle diagram, the ray FG is constructed at an angle θ to FDF′, and the normal GH constructed. The scaled distances OH and HG then represent the normal and shear stress components on the plane ed.

A number of useful results can be obtained or verified using the circle diagram. For example, OS_1 and OS_2 represent the magnitudes of the major and minor principal stresses σ_1, σ_2. From the geometry of the circle diagram, they are given by

43

$$\sigma_{1,2} = OC \pm OF$$

$$= \tfrac{1}{2}(\sigma_{xx} + \sigma_{yy}) \pm [\sigma_{xy}^2 + \tfrac{1}{4}(\sigma_{xx} - \sigma_{yy})^2]^{1/2}$$

confirming the solution given in equation 2.24a. The ray FS_1 defines the orientation of the major principal plane, so FS_2, normal to FS_1, represents the orientation of the major principal axis. If this axis is inclined at an angle α_1, to the x axis, the geometry of the circle diagram yields

$$\tan \alpha_1 = \frac{(OS_1 - OD)}{DF'}$$

$$= \frac{(\sigma_1 - \sigma_{xx})}{\sigma_{xy}}$$

This expression is completely consistent with that for orientations of principal axes established analytically (equation 2.24b).

Problems

(The geomechanics convention for stress and strain is to be assumed in the following exercises.)

1   The rectangular plate shown in the figure below has the given loads uniformly distributed over the edges. The plate is 50 mm thick, AB is 500 mm and BC is 400 mm.

(a)   Determine the shear forces which must operate on the edges BC, DA, to maintain the equilibrium of the plate.

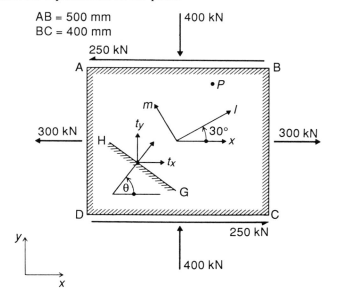

(b) Relative to the x, y reference axes, determine the state of stress at any point P in the interior of the plate.

(c) For the l, m axes oriented as shown, determine the stress components σ_{ll}, σ_{mm}, σ_{lm}.

(d) Determine the magnitudes of the principal stresses, and the orientation of the major principal stress axis to the x axis.

(e) For the surface GH, whose outward normal is inclined at $\theta°$ to the x axis, determine expressions for the component tractions, t_x, t_y, operating on it as a function of σ_{xx}, σ_{yy}, σ_{xy} and θ. Determine values of t_x, t_y for $\theta = 0°$, $60°$, $90°$, respectively. Determine the resultant stress on the plane for which $\theta = 60°$.

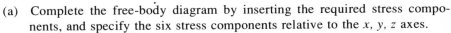

2   The unit free body shown in the figure (left) is subject to the stress components shown acting parallel to the given reference axes, on the visible faces of the cube.

(a) Complete the free-body diagram by inserting the required stress components, and specify the six stress components relative to the x, y, z axes.

(b) The l, m, n reference axes have direction cosines relative to the x, y, z axes defined by

$$(l_x, l_y, l_z) = \quad (0.281, \ 0.597, \quad 0.751)$$

$$(m_x, m_y, m_z) = \quad (0.844, \ 0.219, \ -0.490)$$

$$(n_x, n_y, n_z) = (-0.457, \ 0.771, \ -0.442)$$

Write down the expressions relating σ_{mm}, σ_{nl} to the x, y, z components of stress and the direction cosines, and calculate their respective values.

(c) From the stress components established in (a) above, calculate the stress invariants, I_1, I_2, I_3, write down the characteristic equation for the stress matrix, and determine the principal stresses and their respective direction angles relative to the x, y, z axes.

Demonstrate that the principal stress directions define a mutually orthogonal set of axes.

3   A medium is subject to biaxial loading in plane strain. Relative to a set of x, y, co-ordinate axes, a load imposed at the co-ordinate origin induces stress components defined by

$$\sigma_{xx} = \frac{1}{r^2} - \frac{8y^2}{r^4} + \frac{8y^4}{r^6}$$

$$\sigma_{yy} = \frac{1}{r^2} + \frac{4y^2}{r^4} - \frac{8y^4}{r^6}$$

$$\sigma_{xy} = \frac{2xy}{r^4} - \frac{8xy^3}{r^6}$$

where $r^2 = x^2 + y^2$

45

Verify that the stress distribution described by these expressions satisfies the differential equations of equilibrium. Note that

$$\frac{\partial}{\partial x}\left(\frac{1}{r}\right) = -\frac{x}{r^3} \quad \text{etc.}$$

4 A medium is subject to plane strain loading by a perturbation at the origin of the x, y co-ordinate axes. The displacements induced by the loading are given by

$$u_x = \frac{1}{2G}\left[\frac{xy}{r^2} + C_1\right]$$

$$u_y = \frac{1}{2G}\left[\frac{y^2}{r^2} - (3 - 4\dot{v}) \ln r + C_2\right]$$

where C_1, C_2 are indefinite constants.

(a) Establish expressions for the normal and shear strain components, ε_{xx}, ε_{yy}, γ_{xy}.

(b) Verify that the expressions for the strains satisfy the strain compatibility equations.

(c) Using the stress–strain relations for isotropic elasticity, establish expressions for the stress components induced by the loading system.

5 The body shown in the figure below is subject to biaxial loading, with stress components given by $\sigma_{xx} = 12$, $\sigma_{yy} = 20$, $\sigma_{xy} = 8$.

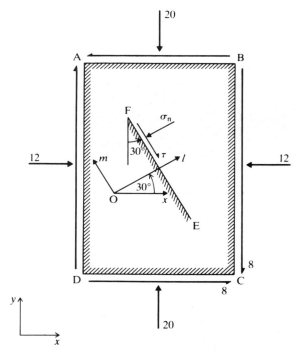

(a) Construct the circle diagram representing this state of stress. Determine, from the diagram, the magnitudes of the principal stresses, and the inclination of the major principal stress axis relative to the x reference direction. Determine, from the diagram, the normal and shear stress components σ_n and τ on the plane EF oriented as shown.

(b) Noting that the outward normal, OL, to the surface EF is inclined at an angle of $30°$ to the x axis, use the stress transformation equations to determine the stress components σ_{ll} and σ_{lm}. Compare them with σ_n and τ determined in (a) above.

(c) Determine analytically the magnitudes and orientations of the plane principal stresses, and compare them with the values determined graphically in (a) above.

3 Rock mass structure

3.1 Introduction

Rock differs from most other engineering materials in that it contains fractures of one type or another which render its structure discontinuous. Thus a clear distinction must be made between the rock element or rock material on the one hand and the rock mass on the other. **Rock material** is the term used to describe the intact rock between discontinuities; it might be represented by a hand specimen or piece of drill core examined in the laboratory. The **rock mass** is the total *in situ* medium containing bedding planes, faults, joints, folds and other structural features. Rock masses are discontinuous and often have heterogeneous and anisotropic engineering properties.

The nature and distribution of structural features within the rock mass is known as the **rock structure**. Obviously, rock structure can have a dominant effect on the response of a rock mass to mining operations. It can influence the choice of a mining method and the design of mining layouts because it can control stable excavation spans, support requirements, subsidence, cavability and fragmentation characteristics. At shallow depths and in de-stressed areas, structurally controlled failures may be the prime concern in excavation design (Figure 3.1). At depth

Figure 3.1 Sidewall failure in a mine haulage aligned parallel to the line of intersection of two major discontinuities (photograph by E. Hoek).

and in areas of high stress concentration, the influence of structure may be less marked, and limiting the induced boundary stresses or energy release rates may be more important considerations (Chapters 7 and 10).

This chapter describes the types and important properties of structural features found in rock masses, methods of collecting, processing and presenting data on rock structure, and the incorporation of such data into rock mass classification schemes. The uses of these data and rock mass classifications in selecting mining methods and designing excavations will be described in subsequent chapters.

3.2 Major types of structural features

Structural features and their origins are well described in several textbooks on general, structural and engineering geology. From an engineer's point of view, the accounts given by Price (1966), Hills (1972), Blyth and de Freitas (1974) and Hobbs (1976) are particularly helpful. The reader who is not familiar with the elements of structural geology should study one of these texts. All that will be given here is a catalogue of the major types of structural feature and brief descriptions of their key engineering properties.

Bedding planes divide sedimentary rocks into beds or strata. They represent interruptions in the course of deposition of the rock mass. Bedding planes are generally highly persistent features, although sediments laid down rapidly from heavily laden wind or water currents may contain cross or discordant bedding. Bedding planes may contain parting material of different grain size from the sediments forming the rock mass, or may have been partially healed by low-order metamorphism. In either of these two cases, there would be some 'cohesion' between the beds; otherwise, shear resistance on bedding planes would be purely frictional. Arising from the depositional process, there may be a preferred orientation of particles in the rock, giving rise to planes of weakness parallel to the bedding.

Folds are structures in which the attitudes of the beds are changed by flexure resulting from the application of post-depositional tectonic forces. They may be

Figure 3.2 Jointing in a folded stratum (after Blyth and de Freitas, 1974).

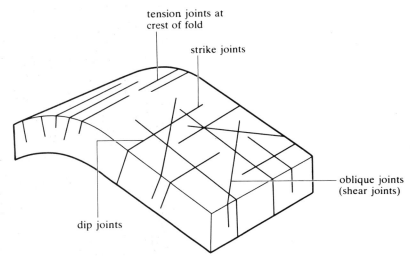

49

major structures on the scale of a mine or mining district or they may be on a smaller local scale. Folds are classified according to their geometry and method of formation (Hills, 1972, for example).

The major effects of folds are that they alter the orientations of beds locally, and that certain other structural features are associated with them. In particular, well-defined sets of joints may be formed in the crest or trough and in the limbs of a fold. Figure 3.2 shows the typical development of jointing in one stratum in an anticline. During the folding of sedimentary rocks, shear stresses are set up between the beds where slip may occur. Consequently, the bedding plane shear strength may approach, or be reduced to, the residual (section 4.7.2). Axial-plane or fracture cleavage may also develop as a series of closely spaced parallel fractures resulting from the shear stresses associated with folding.

Faults are fractures on which identifiable shear displacement has taken place. They may be recognised by the relative displacement of the rock on opposite sides of the fault plane. The sense of this displacement is often used to classify faults (Hills, 1972, for example). Faults may be pervasive features which traverse a mining area or they may be of relatively limited local extent on the scale of metres; they often occur in echelon or in groups. Fault thickness may vary from metres in the case of major, regional structures to millimetres in the case of local faults. This fault thickness may contain weak materials such as fault gouge (clay), fault breccia (recemented), rock flour or angular fragments. The wall rock is frequently slickensided and may be coated with minerals such as graphite and chlorite which have low frictional strengths. The ground adjacent to the fault may be disturbed and weakened by associated structures such as drag folds or secondary faulting (Figure 3.3). These factors result in faults being zones of low shear strength on which slip may readily occur.

Shear zones are bands of material, up to several metres thick, in which local shear failure of the rock has previously taken place. They represent zones of stress relief in an otherwise unaltered rock mass throughout which they may occur irregularly. Fractured surfaces in the shear zone may be slickensided or coated with low-friction materials, produced by the stress relief process or weathering. Like faults, shear zones have low shear strengths but they may be

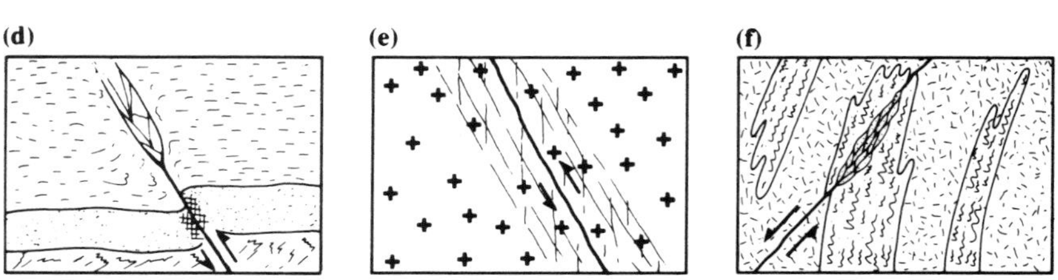

Figure 3.3 Secondary structures associated with faulting: (a) bedding plane fault in brittle rock develops associated shear and tension (gash) fractures; (b) bedding plane fault in closely bedded shale develops closely spaced, intersecting shears; (c) bedding plane fault in poorly stratified, partially ductile rock produces a wide zone of drag folds; (d) fault in competent, brittle rock dies out in weak shale; (e) fault in crystalline igneous rock develops subsidiary inclined shears and parallel sheeting; (f) a fault in an igneous rock changes character in passing through a mica-rich metamorphic rock (after Wahlstrom, 1973).

much more difficult to identify visually. Deere (1979) has described the nature of shear zones and discussed the engineering problems associated with them. Salehy *et al.* (1977) have described their occurrence in coal measures rocks as **intraformational shears**.

Dykes are long, narrow intrusions of generally fine-grained igneous rock with steep or vertical and approximately parallel sides. They may vary in width from a few centimetres to several metres and may appear as dyke swarms. Dykes may also be of considerable length. The Great Dyke of Rhodesia, for example, is some 500 km long. It is a flat, trough-like structure which is extensively mineralised, particularly on the margins. Some dyke rocks are more resistant to weathering than the country rock, but the basic igneous dyke rocks such as dolerite can weather to montmorillonite clays which are noted for their swelling characteristics. The dyke margins are frequently fractured and altered during the intrusion. They form potential seepage paths and zones of low stiffness and shear strength in which movements will tend to be concentrated. Because of their high stiffnesses, unweathered dyke rocks can develop high stresses and so be susceptible to stress-induced failure or, as in the deep-level gold mines of South Africa, be associated with rockburst conditions. Figure 3.4 (after Cook, N.G.W. *et al.*,

Figure 3.4 The effect of dyke proximity on rockburst incidence, East Rand Proprietary Mines. (a) Large dykes in continent abutments; (b) large dykes in island or remnant abutments (after Cook, N.G.W. *et al.*, 1966).

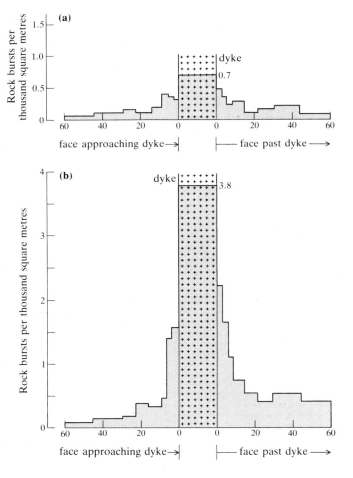

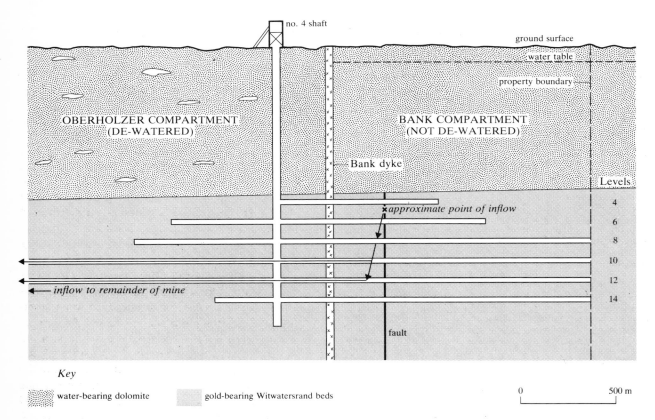

no. 4 shaft

ground surface

water table

property boundary

OBERHOLZER COMPARTMENT
(DE-WATERED)

BANK COMPARTMENT
(NOT DE-WATERED)

Bank dyke

Levels

approximate point of inflow

4

6

8

10

inflow to remainder of mine

12

14

fault

Key

water-bearing dolomite   gold-bearing Witwatersrand beds

0   500 m

Figure 3.5 Diagrammatic longitudinal section illustrating inrush of water from Bank compartment, West Dreifontein Mine, 26 October 1968. Total inflow was approximately 100 000 gal/day ($\simeq 45.5 \times 10^4$ 1/day) (after Cartwright, 1969).

1966) shows the effect of dyke proximity on rockburst incidence at East Rand Proprietary Mines. Another major mining problem caused by dykes in South African gold mines is the compartmentalisation of water-bearing dolomites causing severe differences in head between adjacent compartments after the water level in one has been drawn down during mining operations. At the West Driefontein Mine in 1968, a stope hangingwall failure adjacent to a fault in the compartment on the non-dewatered side of a major vertical dyke triggered the flooding of a portion of the mine (Figure 3.5).

Joints are the most common and generally the most geotechnically significant structural features in rocks. Joints are breaks of geological origin along which there has been no visible displacement. A group of parallel joints is called a **joint set**, and joint sets intersect to form a **joint system**. Joints may be open, filled or healed. They frequently form parallel to bedding planes, foliations or slaty cleavage, when they may be termed bedding joints, foliation joints or cleavage joints. Sedimentary rocks often contain two sets of joints approximately orthogonal to each other and to the bedding planes (Figure 3.2). These joints sometimes end at bedding planes, but others, called master joints, may cross several bedding planes.

It is common in rock mechanics to use the term **discontinuity** as a collective term for all fractures or features in a rock mass such as joints, faults, shears, weak bedding planes and contacts that have zero or relatively low tensile strengths (e.g. Priest and Hudson, 1976, 1981, International Society for Rock Mechanics,

1978a). This terminology will be used here and will be departed from only when it is necessary to identify the geological origin of the structural feature being discussed.

3.3 Important geomechanical properties of discontinuities

This section lists and discusses briefly the most important of those properties of discontinuities that influence the engineering behaviour of rock masses. For a fuller discussion of these properties, the reader should consult the document 'Suggested methods for the quantitative description of discontinuities in rock masses' prepared by the Commission on Standardization of Laboratory and Field Tests, International Society for Rock Mechanics (1978a), subsequently referred to as the ISRM Commission (1978a).

Orientation, or the attitude of a discontinuity in space, is described by the **dip** of the line of maximum declination on the discontinuity surface measured from the horizontal, and the **dip direction** or azimuth of this line, measured clockwise from true north (Figure 3.6). Some geologists record the **strike** of the discontinuity rather than the dip direction, but this approach can introduce some ambiguity and requires that the sense of the dip must also be stated for unique definition of discontinuity orientation. For rock mechanics purposes, it is usual to quote orientation data in the form of dip direction (three digits)/dip (two digits) thus, 035/70, 290/15. Obviously, the orientations of discontinuities relative to the faces of excavations have a dominant effect on the potential for instability due to falls of blocks of rock or slip on the discontinuities (Chapter 9). The mutual orientations of discontinuities will determine the shapes of the blocks into which the rock mass is divided.

Spacing is the perpendicular distance between adjacent discontinuities, and is usually expressed as the mean spacing of a particular set of joints. The spacing of discontinuities determines the sizes of the blocks making up the rock mass. The mechanism of deformation and failure can vary with the ratio of discontinuity spacing to excavation size. Engineering properties such as cavability, fragmentation characteristics and rock mass permeability also vary with discontinuity spacing.

It is to be expected that, like all other characteristics of a given rock mass, discontinuity spacings will not have uniquely defined values but, rather, will take a range of values, possibly according to some form of statistical distribution. Priest and Hudson (1976) made measurements on a number of sedimentary rock masses in the United Kingdom and found that, in each case, the discontinuity spacing histogram gave a probability density distribution that could be approximated by the negative exponential distribution. Thus the frequency, $f(x)$, of a given discontinuity spacing value, x, is given by the function

$$f(x) = \lambda e^{-\lambda x} \tag{3.1}$$

where $\lambda \simeq 1/\bar{x}$ is the mean discontinuity frequency of a large discontinuity population and $\bar{x}$ is the mean spacing.

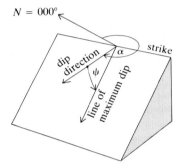

Figure 3.6 Definition of dip direction (α) and dip (ψ).

53

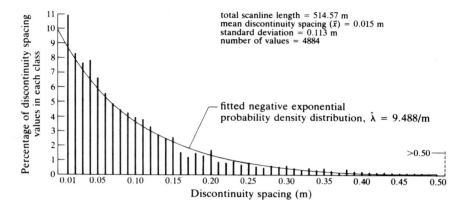

Figure 3.7 Discontinuity spacing histogram, Lower Chalk, Chinnor, Oxfordshire (after Priest and Hudson, 1976).

A discontinuity spacing histogram and the corresponding negative exponential distribution calculated from equation 3.1 are shown for the Lower Chalk, Chinnor, Oxfordshire, UK, in Figure 3.7. Priest and Hudson's findings have since been verified for a wider range of rock types. The use of frequency distributions such as that given by equation 3.1 permits statistical calculations to be made of such factors as probable block sizes and the likelihood that certain types of intersection will occur.

In classifying rock masses for engineering purposes, it is common practice to quote values of **Rock Quality Designation** (*RQD*), a concept introduced by Deere (1968) in an attempt to quantify discontinuity spacing. *RQD* is determined from drill core and is given by

$$RQD = \frac{100 \, \Sigma \, x_i}{L} \tag{3.2}$$

where x_i are the lengths of individual pieces of core in a drill run having lengths of 0.1 m or greater and L is the total length of the drill run.

Priest and Hudson (1976) found that an estimate of *RQD* could be obtained from discontinuity spacing measurements made on core or an exposure using the equation

$$RQD = 100e^{-0.1\lambda} (0.1 \, \lambda + 1) \tag{3.3}$$

For values of λ in the range 6 to 16/m, a good approximation to measured *RQD* values was found to be given by the linear relation

$$RQD = -3.68 \, \lambda + 110.4 \tag{3.4}$$

Figure 3.8 shows the relations obtained by Priest and Hudson (1976) between measured values of *RQD* and λ, and the values calculated using equations 3.3 and 3.4.

Discontinuity spacing is a factor used in many rock mass classification schemes. Table 3.1 gives the terminology used by the ISRM Commission (1978a).

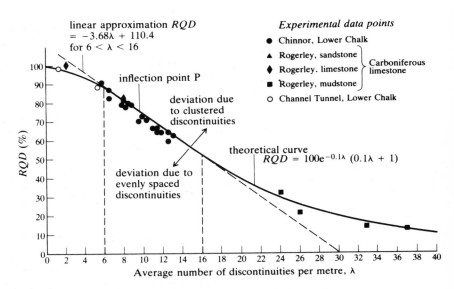

Figure 3.8 Relation between *RQD* and mean discontinuity frequency (after Priest and Hudson, 1976).

linear approximation *RQD*
= −3.68λ + 110.4
for 6 < λ < 16

Experimental data points

● Chinnor, Lower Chalk
▲ Rogerley, sandstone
◆ Rogerley, limestone } Carboniferous limestone
■ Rogerley, mudstone
○ Channel Tunnel, Lower Chalk

inflection point P

deviation due to clustered discontinuities

theoretical curve
$RQD = 100e^{-0.1\lambda}(0.1\lambda + 1)$

deviation due to evenly spaced discontinuities

RQD (%)

Average number of discontinuities per metre, λ

Table 3.1 Classification of discontinuity spacing.

Description	Spacing (mm)
extremely close spacing	<20
very close spacing	20–60
close spacing	60–200
moderate spacing	200–600
wide spacing	600–2000
very wide spacing	2000–6000
extremely wide spacing	>6000

Figure 3.9 Illustration of persistence of various sets of discontinuities (after ISRM Commission, 1978a).

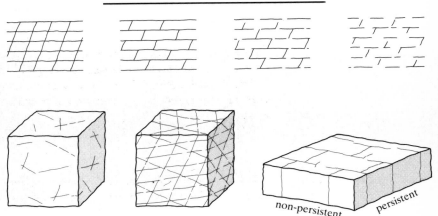

non-persistent   persistent

Persistence is the term used to describe the areal extent or size of a discontinuity within a plane. It can be crudely quantified by observing the trace lengths of discontinuities on exposed surfaces. It is one of the most important rock mass parameters but one of the most difficult to determine. Figure 3.9 shows a set of simple plane sketches and block diagrams used to help indicate the persistence of various sets of discontinuities in a rock mass. Clearly, the persistence of discontinuities will have a major influence on the shear strength developed in the plane of the discontinuity and on the fragmentation characteristics, cavability and permeability of the rock mass.

The ISRM Commission (1978a) uses the most common or modal trace lengths of each set of discontinuities measured on exposures (section 3.4.2) to classify persistence according to Table 3.2.

Roughness is a measure of the inherent surface unevenness and waviness of the discontinuity relative to its mean plane. The wall roughness of a discontinuity has a potentially important influence on its shear strength, especially in the case of undisplaced and interlocked features (e.g. unfilled joints). The importance of roughness declines with increasing aperture, filling thickness or previous shear displacement. The important influence of roughness on discontinuity shear strength is discussed in section 4.7.2.

When the properties of discontinuities are being recorded from observations made on either drill core or exposures, it is usual to distinguish between small-scale surface irregularity or **unevenness** and larger-scale undulations or **waviness** of the surface (Figure 3.10). Each of these types of roughness may be quantified

Table 3.2 Classification of discontinuity persistence.

Description	Modal trace length (m)
very low persistence	< 1
low persistence	1–3
medium persistence	3–10
high persistence	10–20
very high persistence	20

Table 3.3 Classification of discontinuity roughness.

Class	Description
I	rough or irregular, stepped
II	smooth, stepped
III	slickensided, stepped
IV	rough or irregular, undulating
V	smooth, undulating
VI	slickensided, undulating
VII	rough or irregular, planar
VIII	smooth, planar
IX	slickensided, planar

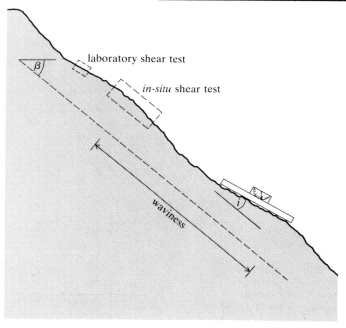

Figure 3.10 Different scales of discontinuity roughness sampled by different scales of shear test. Waviness can be characterised by the angle i (after ISRM Commission, 1978a).

Figure 3.11 Typical roughness profiles and suggested nomenclature. Profile lengths are in the range 1 to 10 m; vertical and horizontal scales are equal (after ISRM Commission, 1978a).

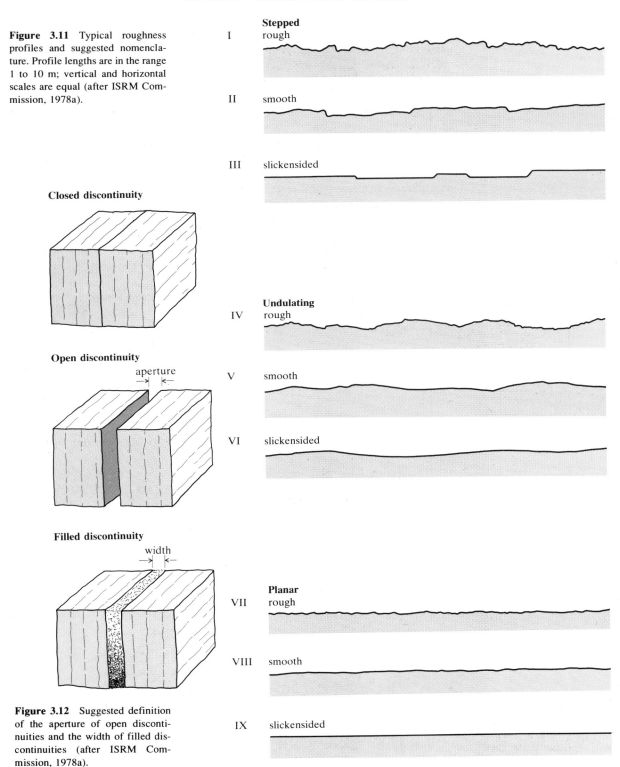

Closed discontinuity

Open discontinuity

Filled discontinuity

Figure 3.12 Suggested definition of the aperture of open discontinuities and the width of filled discontinuities (after ISRM Commission, 1978a).

Stepped

I   rough

II   smooth

III   slickensided

Undulating

IV   rough

V   smooth

VI   slickensided

Planar

VII   rough

VIII   smooth

IX   slickensided

on an arbitrary scale of, say, one to five. Descriptive terms may also be used particularly in the preliminary stages of mapping (e.g. during feasibility studies). For example, the ISRM Commission (1978a) suggests that the terms listed in Table 3.3 and illustrated in Figure 3.11 may be used to describe roughness on two scales – the small scale (several centimetres) and the intermediate scale (several metres). Large-scale waviness may be superimposed on such small- and intermediate-scale roughness (Figure 3.10).

Aperture is the perpendicular distance separating the adjacent rock walls of an open discontinuity in which the intervening space is filled with air or water. Aperture is thereby distinguished from the width of a filled discontinuity (Figure 3.12). Large apertures can result from shear displacement of discontinuities having appreciable roughness, from outwash of filling materials (e.g. clay), from solution or from extensile opening. In most subsurface rock masses, apertures will be small, probably less than half a millimetre. It will be appreciated, of course, that unlike the examples given in Figure 3.12, the apertures of real discontinuities are likely to vary widely over the extent of the discontinuity. This variation will be difficult, if not impossible, to measure.

Clearly, aperture and its areal variation will have an influence on the shear strength of the discontinuity. Perhaps more important, however, is the influence of aperture on the permeability or hydraulic conductivity of the discontinuity and of the rock mass. For laminar flow, the hydraulic conductivity of a single discontinuity with plane, parallel sides is given by

$$k = \frac{ge^3}{12v} \tag{3.5}$$

where k = hydraulic conductivity (m s^{-1}), g = acceleration due to gravity (m s^{-2}), e = discontinuity aperture (m) and v = kinematic viscosity of the fluid (m^2 s^{-1}) (= 1.01×10^{-6} m^2 s^{-1} for water at 20°C).

If e = 0.05 mm, for example, $k = 1.01 \times 10^{-7}$ m s^{-1} for water at 20°C, but if e is increased to 0.5 mm, k is increased by a factor of 1000 to 1.01×10^{-4} m s^{-1}.

Filling is the term used to describe material separating the adjacent rock walls of discontinuities. Such materials may be calcite, chlorite, clay, silt, fault gouge, breccia, quartz or pyrite, for example. Filling materials will have a major influence on the shear strengths of discontinuities. With the exception of those filled with strong vein materials (calcite, quartz, pyrite), filled discontinuities will generally have lower shear strengths than comparable clean, closed discontinuities. The behaviour of filled discontinuities will depend on a wide range of properties of the filling materials. The following are probably the most important and should be recorded where possible:

(a) mineralogy of the filling material taking care to identify low-friction materials such as chlorite
(b) grading or particle size
(c) water content and permeability
(d) previous shear displacement
(e) wall roughness
(f) width of filling
(g) fracturing, crushing or chemical alteration of wall rock.

58

3.4 Collecting structural data

The task of collecting the data referred to in section 3.3 is usually the responsibility of the mining or engineering geologist, although rock mechanics engineers or mining engineers may sometimes be called on to undertake the necessary fieldwork. In either case, it is essential for the rock mechanics or mining engineer (who will generally initiate a request for the data, and who will use it in mine planning studies) to be familiar with techniques used in collecting the data and with the potential difficulties involved.

The starting point for the development of an engineering understanding of the rock mass structure is a study of the general regional and mine geology as determined during exploration. This will provide some knowledge of the lithologies and of the major structural features (folds, faults and dykes) present in the mining area. Such information provides essential background to rock mechanics studies, but in itself, is inadequate for our purposes. Further studies, involving careful mapping of surface and underground exposures and logging of boreholes drilled for this purpose, are required to obtain the types of data discussed in section 3.3.

The present account of the methods used to collect structural and related geological data is far from exhaustive. Fuller accounts are given by Goodman (1976), ISRM Commission (1978a) and Hoek and Brown (1980).

3.4.1 Mapping exposures

In the early stages of a mining project, it may not be possible to gain access underground. In this case, surface outcrops must be utilised to obtain information on the engineering properties and structure of the rock mass. Measurements may be made on natural outcrops or on faces exposed by surface excavations. In some mining projects, an existing open pit provides an invaluable source of data.

It must be recognised, however, that these surface exposures can be affected by weathering and that the surface rock mass quality may be quite different from that at depth. It is essential, therefore, that any preliminary data obtained from surface exposures are validated by subsequently examining underground exposures. Exploratory openings should be mapped at the earliest possible stage to provide data for the rock mechanics input into mining feasibility studies. As the mining project reaches a more advanced stage, development openings should be mapped to provide information on which stope design can be based.

In all of these instances, there is a basic **sampling** problem to be considered. What proportion of the rock mass should be surveyed to obtain satisfactory results? What degree of confidence can be placed on mean values of discontinuity properties determined using limited amounts of data? There are no complete answers to such questions although the use of statistical techniques, such as those developed by Priest and Hudson (1981) and discussed briefly below, does provide valuable guidance. Even where it is possible to develop a statistical approach to discontinuity mapping, practical considerations, such as a lack of access to the desired underground exposure, can mean that surveyors must use their judgement in interpreting results.

The basic technique used in mapping surface or underground exposures is the **scanline survey** (Figure 3.13). A scanline is a line set on the surface of the rock

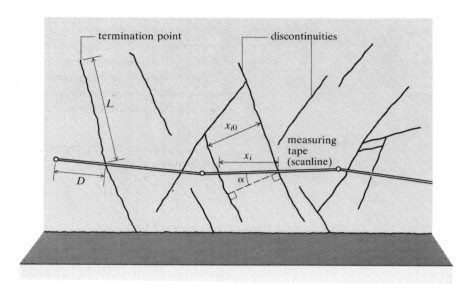

Figure 3.13 Scanline survey.

mass, and the survey consists of recording data for all discontinuities that inter-sect the scanline along its length. An alternative approach is to measure all discontinuities within a defined area on the rock face, but this is more difficult to control and do systematically than are scanline surveys. In underground de-velopment excavations of limited height, it is often possible to combine these two approaches and to record all discontinuities by extrapolating those which do not intersect the scanline to give imaginary intersection points.

In practice, a scanline is a measuring tape fixed to the rock face by short lengths of wire attached to masonry nails hammered into the rock. The nails should be spaced at approximately 3 m intervals along the tape which must be kept as taut and as straight as possible. Where practicable, each scanline location should be photographed with the scanline number or location suitably identified. Once the scanline is established, the location (scanline number and grid co-ordinates), date, rock type, face orientation, scanline orientation and name of the surveyor are recorded on the logging sheet (Figure 3.14). Surveyors should then carefully and systematically work their way along the scanline recording the following features for each discontinuity intersecting the scanline:

(a) D, distance along the scanline to the point at which the discontinuity inter-sects the scanline (Figure 3.13). Fractures obviously caused by blasting are usually not recorded;

(b) L, the length of the discontinuity measured above the scanline (Figure 3.13). In most cases, practical considerations make it necessary to set the scanline at the base of the rock face. This effectively restricts trace length measure-ments to that portion of each trace extending above the scanline. Priest and Hudson (1981) have shown that the mean semi-trace length should be one half the mean complete trace length sampled using a scanline;

(c) T, nature of the termination point (A = at another discontinuity; I = in rock material; O = obscured or extending beyond the extremity of the exposure);

60

(d) dip direction/dip – measured at or near the point of intersection with the scanline using a suitable magnetic compass such as the **Clar compass** (Hoek and Brown, 1980; Hoek and Bray, 1981);

(e) C, curvature or waviness on a numerical scale of, say one to five;

(f) R, roughness of small-scale irregularities on a five-point scale;

(g) comments – particularly on the nature of any infilling present, discontinuity aperture, seepage from the discontinuity and the origin or type of the discontinuity.

Figure 3.14 shows a scanline survey logging sheet completed in this way.

Experience has shown that, for rock mechanics purposes, rock masses can be divided into **homogeneous zones**, or zones within which the rock mass has relatively uniform rock mass structure and geotechnical properties. Where possible, the rock mass should be divided into such zones and at least one scanline survey made in each zone. Clearly, attention should be concentrated on areas in and adjacent to the orebody for mine design studies. However, other sections of the rock mass may also be of interest as sites for permanent underground installations. Figure 3.15 shows the likely locations of scanlines in an

Figure 3.14 Completed scanline survey sheet.

Location SL 11, 9260 N/8850 E			Date 9 July 80			Rock type Agglomerate
Face orientation 048/75						
Scanline orientation 305/04					Recorded by S.D.P.	

D (m)	L (m)	T	Orientation	C	R	Comments
0·00	0·40	A	331/82	2	3	Open 3mm (by blasting?)
0·29	2+	I	124/85	1/2	2	
0·31	0·10	A	155/90	5	3	Contains broken & sheared rock
0·71	0·30	I	305/76	2	2	
0·78	≃3·0	O	308/72	2	4	
0·80–0·98						Obscured by rubble
0·98	1·6	O	304/72	1	2	
1·22	0·70	A	110/76	4	4	
1·50	4·0	O	32/70	1/2	3	Major plane
1·85	0·45	A	128/83	2	3	
1·97	1·0	O	099/82	2	2	Open 15mm (by blasting?)
1·97–2·20						Obscured by rubble
2·70	2·0	A	038/66	3	3	
3·30–3·40	8+	O	138/64	—	—	Shear zone
3·40	1·0	A	126/62	2	3	
3·56	0·30	I	134/60	4	1	

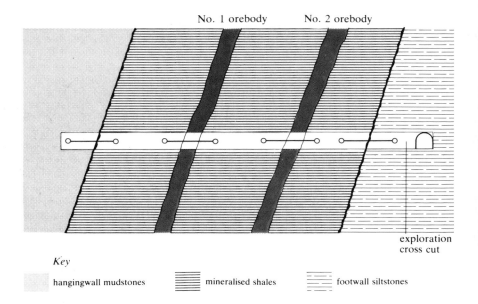

No. 1 orebody   No. 2 orebody

exploration
cross cut

Key

hangingwall mudstones   mineralised shales   footwall siltstones

Figure 3.15 Suggested scanline locations along an exploration cross cut intersecting two orebodies in steeply dipping sedimentary strata.

exploration cross cut driven through a sedimentary sequence containing mineralised shales. These scanlines sample the immediate foot wall and hanging wall of each of the two orebodies, the orebodies themselves and the rocks outside the mineralised zone.

As with other methods of collecting structural data, **bias** may be introduced into scanline survey results by a number of causes. There will be a bias in the observed spacings between discontinuities in a particular set because, generally, the scanline will not be perpendicular to the discontinuity traces. If, as shown in Figure 3.13, the apparent spacing between two discontinuities in a set is x_i and the acute angle between the normal to the discontinuities and the scanline is α, the true spacing in the plane of the face, x_{i0}, can be calculated from

$$x_{i0} = x_i \cos \alpha \qquad (3.6)$$

Only when $\alpha = 0°$, is the true spacing in the plane of the face measured directly. In the extreme case when the discontinuity and scanline are parallel ($\alpha = 90°$), no intersection will be observed. It is necessary, therefore, that scanline surveys of a face be carried out in two orthogonal directions, usually horizontal and vertical. Ideally, equal total horizontal and vertical scanline lengths should be used, but this is often difficult to achieve in practice.

The value x_{i0} given by equation 3.6 will be the true normal spacing of the discontinuities only when the face is normal to the discontinuities. If the scanline intersections N sets of discontinuities, the discontinuity frequency measured along the scanline is given by

$$\lambda = \sum_{i=1}^{N} \lambda_{i0} \cos \alpha_i \qquad (3.7)$$

(a)

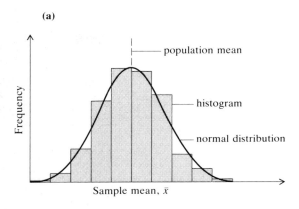

(b)

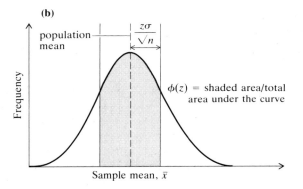

Figure 3.16 Frequency distribution of the sample mean (after Priest and Hudson, 1981).

where λ_{i0} is the frequency of set i measured along the normal to the discontinuities and α_i is the acute angle between the normal and the scanline.

Hudson and Priest (1983) showed that if α_i, β_i are the trend (the azimuth of the vertical plane containing the line) and plunge (the acute angle measured in a vertical plane between the downward directed end of the line and the horizontal) of the normal to the ith discontinuity set and α_s, β_s are the trend and plunge of the scanline, the discontinuity frequency measured along the scanline is

$$\lambda = A \sin \alpha_s \cos \beta_s + B \cos \alpha_s \cos \beta_s + C \sin \beta_s \tag{3.8}$$

where

$$A = \sum_{i=1}^{N} \lambda_{i0} \sin \alpha_i \cos \beta_i$$

$$B = \sum_{i=1}^{N} \lambda_{i0} \cos \alpha_i \cos \beta_i$$

$$C = \sum_{i=1}^{N} \lambda_{i0} \sin \beta_i$$

63

Priest and Hudson (1981) have pointed out that there is also a natural variability in the mean discontinuity spacing $\bar{x}$ computed as

$$\bar{x} = \frac{\sum\limits_{i=1}^{n} x_i}{n} \qquad (3.9)$$

where x_i is the ith discontinuity spacing measurement along a scanline of length L yielding n values. The question arises as to what value n should take in order that the value of $\bar{x}$ can be estimated with acceptable precision. In theory, a plot of the frequency of occurrence of values of $\bar{x}$ determined from several scanline surveys in the one direction with different values of n, should have a normal distribution (Figure 3.16a). It is known that, in this case, a proportion $\phi(z)$ of the different scanlines will yield a mean value within $\pm z\sigma/\sqrt{n}$ of the population mean (Figure 3.16b) where z is the standard normal variable associated with a certain confidence level and σ is the standard deviation of the population of values. Tabulations of values of z and $\phi(z)$ can be found in most statistics textbooks. Selected values are given in Table 3.4.

Table 3.4 Values of $\phi(z)$ for the normal distribution.

z	$\phi(z)$
0.675	0.50
0.842	0.60
1.036	0.70
1.282	0.80
1.645	0.90
1.960	0.95
2.576	0.99

It will be recalled that discontinuity spacings, x, often follow the negative exponential probability density function

$$f(x) = \lambda e^{-\lambda x} \qquad (3.1)$$

It so happens that, for this distribution, the mean and standard deviation of the population are equal. For a sample of size n, the bandwidth of $\phi(z)$ confidence is then $\bar{x} \pm (z\bar{x})/\sqrt{n}$. Alternatively, this bandwidth can be written as $\bar{x} \pm \varepsilon\bar{x}$ where ε is the allowable proportionate error. Hence

$$\varepsilon = \frac{z}{\sqrt{n}} \quad \text{or} \quad n = \left(\frac{z}{\varepsilon}\right)^2 \qquad (3.10)$$

Equation 3.10 can be used to estimate the sample size required to achieve a given error bandwidth to a required confidence level in the estimate of the mean. For example, if the mean spacing is required within an error bandwidth of $\pm 20\%$ at the 80% confidence level, $\varepsilon = 0.2$, $z = 1.282$ and $n = 41$. If, on the other hand, the mean spacing is required to within 10% at the 90% confidence level, $n = 271$.

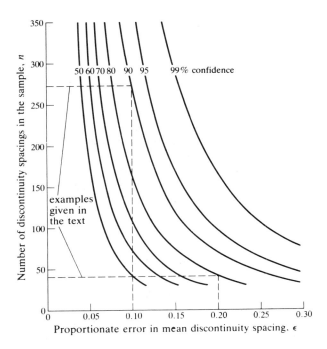

Figure 3.17 Sample number vs. precision of the mean discontinuity spacing estimate for a negative exponential distribution of spacing (after Priest and Hudson, 1981).

Figure 3.17 shows the required number of spacing values versus the error band for various confidence levels. It will be seen that the required sample size increases very rapidly as the allowable error is reduced.

Priest and Hudson (1981) have also analysed the bias in trace lengths measured in scanline surveys, and have developed methods of estimating mean trace lengths, which can serve as measures of discontinuity persistence (Table 3.2), from censored measurements made at exposures of limited extent such as that illustrated in Figure 3.13. A detailed consideration of this analysis is outside the scope of this introductory text. However, the reader should be aware of the uncertainties involved in estimating mean rock mass properties from scanline and other types of discontinuity survey.

3.4.2 Geotechnical drilling and core logging

Core drilling is still the only reliable way of exploring the interior of a rock mass prior to mining. Downhole geophysical or other instruments may be used in drill holes in an attempt to investigate the structure and physical properties of the rock mass, but at their present stage of development, these techniques are much less reliable than sampling the rock by coring.

The aim of geotechnical drilling is to obtain a continuous, correctly oriented sample of the rock mass in as nearly undisturbed a form as possible. Therefore, the standard of the drilling must be considerably higher than that required for normal exploration drilling. In geotechnical drilling, it is necessary to aim for 100% core recovery. Any weak materials such as weathered rock, fault gouge, clay seams or partings in bedding planes, should be recovered, because a knowledge of their presence and properties is essential in predicting the likely behaviour of the rock mass during and following excavation. In normal exploration drilling, these materials are seen to be of little importance and no effort is made to recover them.

Diamond core drilling is expensive, and it is important that the operation be adequately controlled if full value is to be gained from the expenditure. Several factors can influence the quality of the results obtained.

Drilling machine.   A hydraulic feed drilling machine is essential to ensure high core recovery. The independent control of thrust permits the bit to adjust its penetration rate to the properties of the rock being drilled and, in particular, to move rapidly through weathered rock and fault zones before they are eroded away by the drilling fluid.

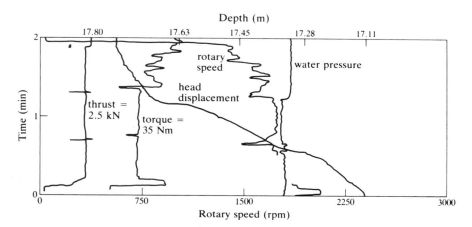

Figure 3.18 Strip chart record of a short instrumented diamond drilling run (after Barr and Brown, 1983).

A range of hydraulic feed machines suitable for geotechnical drilling from surface and underground locations has become available in recent years. The use of lightweight (aluminium) drill rods and hydraulic chucks permits rapid coupling and uncoupling of rods in a one-man operation. Figure 3.18 shows some results obtained with one of these machines, a Craelius Diamec 250, in drilling a horizontal 56 mm diameter hole in an underground limestone quarry. Electronic transducers were used to monitor thrust, rotary speed, penetration, torque and delivery and return water pressures and flows as drilling proceeded. Changes in rock strength were reflected by changes in penetration rate. Open fractures were typified by local steps in the penetration trace and by spikes in the rotary speed and torque traces. Clay- or gouge-filled features also produced irregular torque and rotary speed traces. In the case shown in Figure 3.18, a 12 cm wide clay-filled fissure encountered at a hole depth of 17.61 m, caused the bit to block and the drill to stall. It is suggested that the widespread use of drill-rig instrumentation could greatly improve the quantity and quality of the geotechnical data obtained from coring and non-coring rotary drilling.

Core barrel.   Except in extremely good-quality rock and for the larger core sizes, the objective of recovering a complete, undisturbed core sample of the rock mass can only be achieved if the core passes into an inner tube in the core barrel. When a single-tube core barrel (Figure 3.19a) is used, the core may be damaged

66

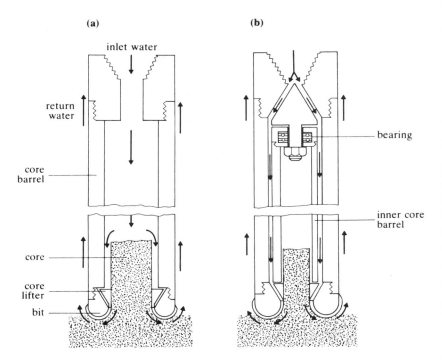

Figure 3.19 Diagrammatic illustration of the distinctive features of (a) single-tube, and (b) double-tube core barrels.

by the rotating barrel and by the circulating water to which it is fully exposed. In a double-tube core barrel (Figure 3.19b), the core is retained in an inner barrel mounted on a bearing assembly. This de-couples the inner barrel from the rotating outer barrel and isolates the core from the drilling water, except at the bit. Some manufacturers also supply triple-tube core barrels which use a split inner tube inside the second barrel. This inner tube is removed from the barrel with the core, further minimising core disturbance, particularly during and after core extraction. Split inner tube wireline core barrels have greatly improved the efficiency and quality of geotechnical drilling in deeper holes.

The more sophisticated core barrels are not required for very good quality, strong rocks or when larger-diameter cores are taken. The usual core diameter for geotechnical drilling is 50–55 mm. With larger core diameters, mechanical breakage of the core and erosion at the bit are less likely to occur, and recovery is correspondingly higher. Unfortunately, drilling costs vary approximately with the square of the core diameter, and so a compromise must be reached between cost and drilling quality. Rosengren (1970) describes successful large-diameter core drilling operations carried out from underground locations at the Mount Isa Mine, Australia. Thin wall bits and up to 6 m long single-tube core barrels were used to take 102 mm and 152 mm diameter cores in hard silica dolomite.

Drilling techniques and contracts. Because the emphasis is on core recovery rather than on depth drilled, the drillers must exercise greater care in geotechnical than in other types of drilling, and must be motivated and rewarded accordingly. It is desirable that geotechnical drilling crews be given special training and that their contracts take account of the specialised nature of their work. The normal method of payment for exploration drilling (fixed rate plus payment per unit length drilled) is generally unworkable for geotechnical drilling. A preferred alternative is to pay drillers on the basis of drilling time, with a bonus for core recovery achieved above a specified value which will vary with the nature of the rock mass.

To obtain good core recovery and avoid excessive breakage of the core, it is essential that the drilling machine be firmly secured to its base, that special care be taken when drilling through weak materials (the readings of instruments monitoring drilling parameters can be invaluable here), and that extreme care be used in transferring the core from the core barrel to the core box and in transporting it to the core shed.

Core orientation. If the fullest possible structural data are to be obtained from the core, it is essential that the core be oriented correctly in space. Not only must the trend and plunge of the borehole axis be measured (the trend and plunge of a line are analogous to the dip direction and dip of a plane), but the orientation of the core around the full 360° of the borehole periphery must be recorded. If this is not done, then the true orientations of discontinuities intersected by the borehole cannot be determined.

Three general approaches may be used to orient the core correctly.

(a) Use the known orientations of geological markers, such as bedding planes, cleavage or an easily identified joint set, to determine the correct orientation of the core and of the other structural features that it contains. Even the most regular geological features do not always have the same attitudes at widely spaced locations within the rock mass and so this approach can be relied upon only in exceptional cases.

(b) Use a device in the core barrel that places orientation marks on the core. Examples of such mechanical devices are the Craelius core orienter which uses a set of lockable prongs to orient the first piece of core in a drilling run using the existing core stub as a guide, and the BHP and Christensen–Hugel core barrels which scribe reference marks on the core in an orientation known from a magnetic borehole survey instrument mounted in the core barrel.

(c) Examine the borehole walls with a suitably oriented downhole tool and relate orientations of features measured on the walls to those found on the core at the corresponding depth. Instruments used for this purpose include cameras, television cameras, periscopes, the seisviewer (an acoustic device) and the borehole impression packer.

A complete discussion of these various techniques is outside the scope of this text. Fuller details are given by Rosengren (1970), Goodman (1976), and Hoek and Brown (1980). Unfortunately, all of these techniques have their own disadvantages and generally only operate successfully under restricted conditions. As with many aspects of mining engineering, there is no universal or simple answer to the question of how to orient core. Quite often, when the standard techniques do not give satisfactory results techniques suited to local conditions can be devised. Rosengren (1970), for example, describes a technique developed for orienting core taken from flat dipping, large-diameter holes at the Mount Isa Mine, Australia. The device consists of a marking pen fitted in a short dummy

68

barrel and attached to a mercury orienting switch. The barrel is lowered into the hole with aluminium rods and when nearly on the face, is rotated until the pen is in a known position, as indicated by the mercury switch. The barrel is then pushed on to the face, and so marks the core stub in a known position.'

Core logging. The final stage in the geotechnical drilling process is the recording of the information obtained from the core. Here again, the value of the entire expensive exercise can be put at risk by the use of poor techniques or insufficient care. Generally, the structural or geotechnical logging of the core is carried out by specially trained operators in a location removed from the drilling site. The log so obtained is additional to the normal driller's or geologist's log.

It must be recognised that the data obtained from geotechnical drilling may not be used in planning studies or detailed mine design until after some time has elapsed. Because of the considerable cost of obtaining the core, measures should be taken to ensure that the fullest amount of useful information is recovered from it. An essential first step in this regard is to take colour photographs of the boxes of core as soon as they become available and before they are disturbed by the logging process or pieces of core are removed for testing or assaying.

The design of the logging sheet and the logging procedures used will vary with the nature of the rock mass and with the project concerned. However, the geotechnical or structural log will usually include information on the size, location and orientation of the borehole, a description of the rock types encountered, together with a strength index (generally the point load index) and/or a weathering index, and, most importantly, data on all discontinuities intersecting the core. These data will include the depth at which the discontinuity is intersected, its nature (joint, bedding plane, drilling break), its orientation, its roughness (generally on a multi-point scale) and the presence and nature of infilling materials. Values of *RQD* or the results of *in situ* testing, such as permeability tests, may be added as required. From the data recorded on the core log and the driller's or geologist's log, composite logs may be prepared for subsequent use by the planning or rock mechanics engineer.

3.5 Presentation of structural data

3.5.1 Major features

The effective utilisation of geological data by a mining or rock mechanics engineer requires that the engineer must first be able to understand and digest the data and to visualise their relation to the proposed mining excavation. It is necessary, therefore, that means be found of presenting the data so that the often complex three-dimensional geometrical relations between excavations and structural features can be determined and portrayed.

Major structural features such as dykes, faults, shear zones and persistent joints may be depicted in a variety of ways. Their traces may be plotted directly on to mine plans with the dips and dip directions marked. Alternatively, structural features may be plotted, level by level, on transparent overlays which can be laid over mine plans so that their influence can be assessed in developing mining layouts.

69

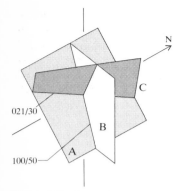

Figure 3.20 Isometric view of the intersection of major discontinuities B (190°/90°) and C (300°/70°) with a stope hanging-wall A (090°/50°).

A complete three-dimensional model of the orebody or of the layout of the excavations is often made from a transparent material such as Perspex or lucite. The intersections of major structural features with excavations, or their traces on particular sections or levels of the mine, can be useful additions to such models. These models are useful for broad planning, demonstration or initial familiarisation purposes, but they are generally unsuitable for use in detailed mine planning.

Isometric drawings may be similarly used to provide three-dimensional visualisations of geological planes and of their intersections with each other and with excavation surfaces. Hoek and Brown (1980) present a series of computer-drawn charts which can be used to construct isometric views of planes with any given dip and dip direction. Figure 3.20 shows a simple example of such a diagram prepared using Hoek and Brown's charts. The immediate hanging wall of a large open stope is represented by plane A which dips at 50° with a dip direction of 090°. A vertical fault B (190/90) and a shear zone C (300/70) intersect in the hanging wall along a line whose trend and plunge are 280° and 79°, respectively. The intersections of B and C with A are the lines 100/50 and 021/23.

3.5.2 Joints and bedding planes

The data for joints and bedding planes differ in two significant respects from the data for major structural features such as faults. First, they are much more numerous giving rise to a distribution of orientations for each set rather than the single orientation used to describe a major feature. Second, their spacings or frequencies are important and must be represented in some way. As illustrated by Figure 3.7, a histogram of spacing values is a convenient way of presenting these data. All discontinuities intersected by a given length of borehole or scanline may be plotted together as in Figure 3.7, or alternatively, the individual discontinuities may be assigned to particular sets which are then plotted separately. Figure 3.21 shows an example in which a distinction is made between bedding plane breaks and other joints in discontinuity frequency plots for two inclined boreholes intersecting two orebodies.

Figures 3.7 and 3.21 show measured spacings along a scanline and measured frequencies along a borehole, respectively. It is also possible to present these data as spacings or frequencies along lines perpendicular to the discontinuity plane at various points along the scanline or borehole.

Figure 3.21 Plots of fracture frequency along the lengths of boreholes (1 ft = 0.305 m) (after Mathews, 1978).

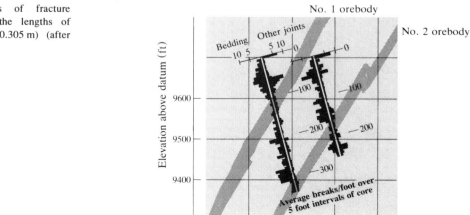

Orientation data are sometimes presented on a **rose diagram** in which the strikes of discontinuities are shown in, say, 5° intervals around a polar diagram and the numbers of observations made for each orientation interval are plotted as radii. Dips and dip directions may be added at the periphery of the circular diagram. This method of presenting discontinuity orientation data is much less useful and versatile than the **hemispherical** or **stereographic projection** which will be discussed in the following section.

3.6 The hemispherical projection

3.6.1 Hemispherical projection of a plane

The hemispherical projection is a method of representing and analysing the three-dimensional relations between planes and lines on a two-dimensional diagram. It has long been a widely used tool in the field of structural geology and has recently found increasing use for solving engineering problems. The basis of the method and its classic geological applications are described by Phillips (1971). Rock engineering applications are described in detail by Goodman (1976), Hoek and Brown (1980), Priest (1980, 1985) and Hoek and Bray (1981). Application of the technique to the problem of structurally controlled failures around underground mining excavations will be discussed in Chapter 9.

Imagine a sphere which is free to move in space so that it can be centred on an inclined plane as illustrated in Figure 3.22. The intersection of the plane and the surface of the sphere is a **great circle**, shown at the perimeter of the shaded

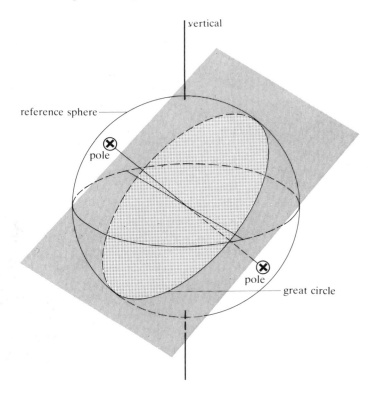

Figure 3.22 The great circle and its poles which define the orientation of a plane.

71

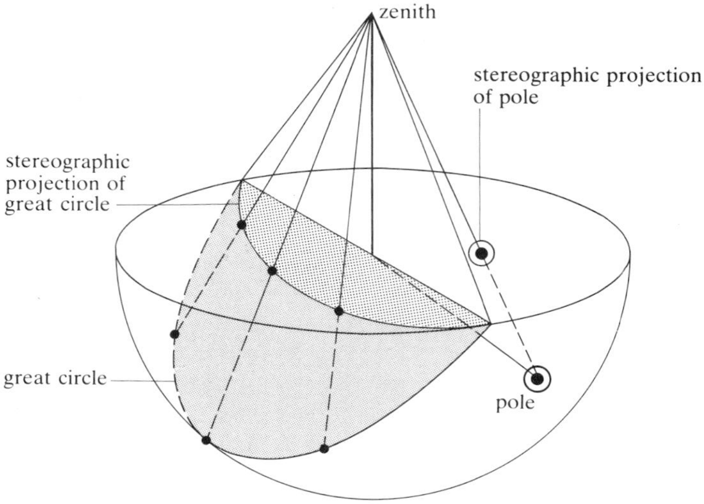

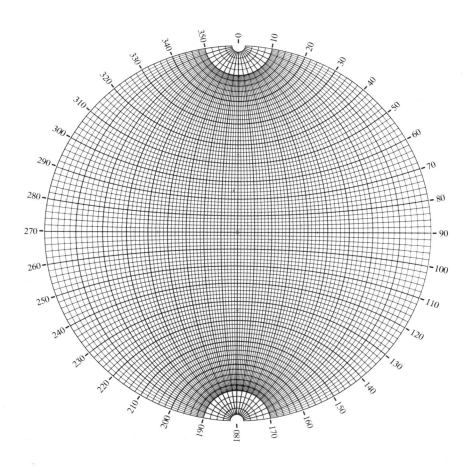

Figure 3.23 Stereographic projection of a great circle and its pole on to the horizontal plane from the lower reference hemisphere.

zenith

stereographic projection of pole

stereographic projection of great circle

great circle

pole

Figure 3.24 Meridional stereographic or equal area net.

area in Figure 3.22. A line perpendicular to the plane and passing through the centre of the sphere intersects the sphere at two diametrically opposite points called the **poles** of the plane.

Because the great circle and the pole representing the plane appear on both the upper and lower parts of the sphere, only one hemisphere need be used to plot and manipulate structural data. In rock mechanics, the **lower-hemisphere projection** is almost always used. The **upper-hemisphere projection** is often used in textbooks on structural geology and can be used for rock mechanics studies if required (for example, Goodman, 1976).

The hemispherical projection provides a means of representing the great circle and pole shown in Figure 3.22 on a horizontal plane. As shown in Figure 3.23, this is achieved by connecting all points on the great circle and the pole with the zenith or point at which a vertical through the centre of the sphere intersects the top of the sphere. The hemispherical projections of the great circle and the pole are then given by the intersections of these projection lines with the horizontal plane.

The projection shown in Figure 3.23 is known as the **stereographic, Wulff**, or **equal-angle projection**. In this projection, any circle on the reference hemisphere projects as a circle on the plane of the projection. This is not the case for an alternative projection known as the **Lambert, Schmidt** or **equal-area projection**. The latter projection is better suited than the equal-angle projection for use in the analysis of discontinuity orientation data, to be discussed in section 3.6.2. The equal-angle projection has an advantage in terms of the solution of some engineering problems and so will be used here. Most of the constructions to be used are the same for both types of projection.

The plotting of planes and their poles is carried out with the aid of a **stereonet** such as that shown in Figure 3.24. The great circles representing planes of constant dip are constructed as circular arcs centred on extensions of the east–west axis of the net. The stereonet also contains a series of **small circles** centred on extensions of the north–south axis. The angle between any two points on a great circle is determined by counting the small circle divisions along the great circle between the two points concerned.

Appendix A sets out the detailed steps required to construct the great circle and pole of a plane using a stereonet such as that shown in Figure 3.24. This procedure involves centring a piece of tracing paper over the stereonet with a drawing pin, marking the north point, marking the dip direction of the discontinuity measured around the periphery of the net from the north point, rotating the tracing paper so that the dip direction coincides with either the east or the west direction on the stereonet, measuring the dip of the discontinuity by counting great circles from the periphery of the net, and drawing in the appropriate great circle. The pole is plotted by counting a further 90° along the east–west axis from the great circle with the tracing paper still in the rotated position. Figure 3.25 shows the great circle and pole of a plane having a dip of 50° and a dip direction of 230°. Appendix A also sets out the steps required to carry out a number of other manipulations. In practice these manipulations are carried out using computer programs. The manual methods are presented here to aid the development of the reader's understanding. Further details are given by Priest (1985, 1992).

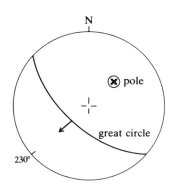

Figure 3.25 Stereographic projection of the great circle and pole of the plane 230°/50°.

3.6.2 Plotting and analysis of discontinuity orientation data

An elementary use of the stereographic projection is the plotting and analysis of field measurements of discontinuity orientation data. If the poles of planes rather than great circles are plotted, the data for large numbers of discontinuities can be rapidly plotted on one diagram and contoured to give the preferred or 'mean' orientations of the dominant discontinuity sets and a measure of the dispersion of orientations about the 'mean'.

Field data may be plotted using a stereonet such as that shown in Figure 3.24 and the method for plotting poles to planes given in Appendix A. However, it is slightly more convenient to use a suitably annotated **polar net** such as that shown in Figure 3.26. Using this net, the tracing paper on which the data are to be plotted does not have to be rotated in the east–west position to plot each pole, as it has to be in the procedure described in Appendix A. A piece of tracing paper is centred over the net using a drawing pin, the north point is marked, and the poles are plotted using the dip directions given in **bold** in Figure 3.26 (the dip direction of the plane ± 180°) and measuring the dips from the centre of the net along the appropriate dip direction lines.

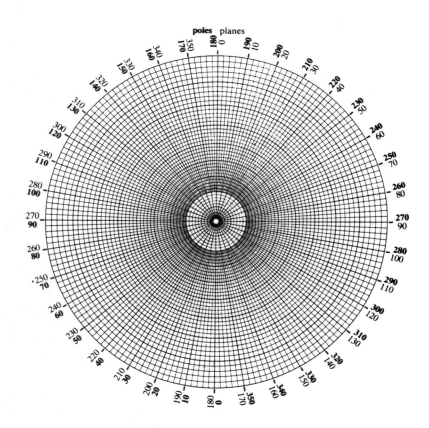

Figure 3.26 Polar stereographic net used for plotting poles of geological planes.

Figure 3.27 shows such a plot of the poles to 351 individual discontinuities whose orientations were measured at a particular field site. Different symbols

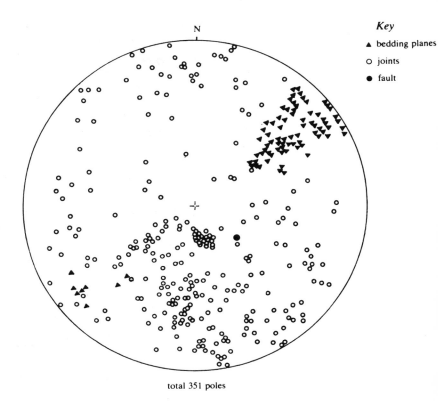

total 351 poles

Figure 3.27 Plots of poles of 351 discontinuities (after Hoek and Brown, 1980).

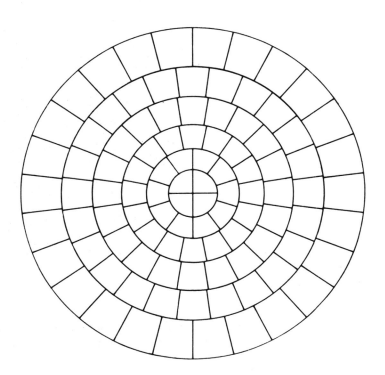

Figure 3.28 Counting net used in conjunction with the polar stereographic net shown in Figure 3.26.

have been used for three different types of discontinuity – joints, bedding planes and a fault. The fault has a dip direction of 307° and a dip of 56°. Contours of pole concentrations may be drawn for the joints and bedding planes to give an indication of the preferred orientations of the various discontinuity sets present. Because of the basic principle of its construction, the equal-area projection is best suited to contouring. However, Hoek and Brown (1980) found that provided a suitable counting net (see below) is used, the equal-angle projection can be used to give results that are almost identical with those given by the equal-area method.

The essential tool required for pole contouring is a **counting net** which divides the surface of the reference hemisphere into a number of equal areas. Figure 3.28 shows a counting net containing 100 equal areas for use with the polar stereographic net shown in Figure 3.26. The most convenient way of using the counting net is to prepare a transparent overlay of it and to centre this overlay on the pole plot by means of a pin through the centre of the net. A piece of tracing paper is mounted on top of the overlay, pierced by the centre pin but fixed by a piece of adhesive tape so that it cannot rotate with respect to the pole plot. Hence one has a sandwich in which the transparent counting net can rotate freely between the pole plot and the piece of tracing paper which are fixed together.

The first step in the analysis is to count all the poles on the net. This is done by keeping the counting net in a fixed position and counting the number of poles falling within each counting cell. These numbers are noted in pencil on the tracing paper at the centre of each cell. In the case of the data shown in Figure 3.27, the pole count will be 350 since the single fault is treated separately and should not be included in the pole population. Once the total pole population has been established, the numbers of poles which make up different percentages of the total are calculated.

The counting net is now rotated to centre the densest pole concentrations in counting cells and so the maximum percentage pole concentrations are determined. By further small rotations of the counting net, the contours of decreasing percentage which surround the maximum pole concentrations can be established. Figure 3.29 shows the contours of pole concentrations determined in this way for the data shown in Figure 3.27. The central orientations of the two major joint sets are 347/22 and 352/83, and that of the bedding planes is 232/81.

The data from which contours of pole concentrations are drawn are usually biased in the manner illustrated by Figure 3.13 and discussed in section 3.4.1. When the data are collected from a single borehole or scanline, it will be necessary to correct the observed pole concentrations to account for the fact that the numbers of discontinuities of a given set intersected will depend on the relative orientations of the set and the borehole or scanline. The correction most widely used is that developed by Terzaghi (1965). In the general case, the number of observations within a given counting area must be weighted by a factor $1/\cos \alpha$ where α is the angle between the borehole axis or the scanline and the normal to the discontinuity. In practice, this correction is generally made only for $|\alpha| \leq 70°$ (Goodman, 1976). For $\alpha > 70°$, the Terzaghi correction is not reliable and so the data should be discarded or an alternative approach used.

Most universities, research establishments, consulting organisations and mining companies now use computers to plot and contour discontinuity orientation

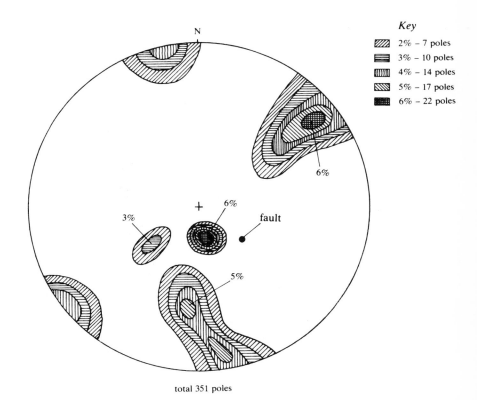

Key

▨	2% – 7 poles
▤	3% – 10 poles
▥	4% – 14 poles
▧	5% – 17 poles
▦	6% – 22 poles

N

3%   6%   fault

+   6%

5%

total 351 poles

Figure 3.29 Contours of pole concentrations for the data plotted in Figure 3.27 (after Hoek and Brown, 1980).

data. For example see Mahtab *et al.* (1972) and Priest (1985, 1992). The Terzaghi correction described above may be applied as a routine step in such programs. When orientation data have been obtained from boreholes or scanlines of similar length oriented in at least three different directions at a site, it is not necessary to apply the Terzaghi correction to a plot of the combined data if discontinuity frequency is not being considered. In this case, it may be necessary to use the procedure given in Appendix A for rotating data from one plane to another if the data are given with respect to the borehole axis rather than as correctly oriented data.

3.7 Rock mass classification

3.7.1 The nature and use of rock mass classification schemes
Whenever possible, it is desirable that mining rock mechanics problems be solved using the analytical tools and engineering mechanics-based approaches discussed in later chapters of this book. However, the processes and interrelations involved in determining the behaviour of the rock surrounding a mining excavation or group of excavations are sometimes so complex that they are not amenable to engineering analysis using existing techniques. In these cases, design decisions may have to take account of previous experience gained in the mine concerned or elsewhere.

In an attempt to quantify this experience so that it may be extrapolated from one site to another, a number of classification schemes for rock masses have been developed. These classification schemes seek to assign numerical values to those properties or features of the rock mass considered likely to influence its behaviour, and to combine these individual values into one overall classification rating for the rock mass. Rating values for the rock masses associated with a number of mining or civil engineering projects are then determined and correlated with observed rock mass behaviour. Aspects of rock mass behaviour that have been studied in this way include the stable spans of unsupported excavations, stand-up times of given unsupported spans, support requirements for various spans, cavability, stable pit slope angles, hangingwall caving angles and fragmentation. A number of these assessments made from geotechnical data collected in the exploration or feasibility study stages of a mining project may provide useful guides to the selection of an appropriate mining method. Some of these applications are outlined in Chapter 12.

Although the use of this approach is superficially attractive, it has a number of serious shortcomings and must be used only with extreme care. The classification scheme approach does not always fully evaluate important aspects of a problem, so that if blindly applied without any supporting analysis of the mechanics of the problem, it can lead to disastrous results. It is particularly important to recognise that the classification schemes give reliable results only for the rock masses and circumstances for which the guide-lines for their application were originally developed. It is for this reason that considerable success has been achieved in using the approach to interpolate experience within one mine or a group of closely related mines, as described by Laubscher (1977), for example.

Goodman (1976) and Hoek and Brown (1980), among others, have reviewed the considerable number of rock mass classification schemes that have been developed for a variety of purposes. Two of these schemes, the NGI tunnelling quality index (Q) developed by Barton et al. (1974) and the CSIR geomechanics or Rock Mass Rating (RMR) scheme developed by Bieniawski (1973, 1976), are currently widely used in civil engineering and, to a lesser extent, in mining practice. Bieniawski's scheme has been used in a number of mining applications for which purpose it has been modified by Laubscher (1977, 1981). The original and modified forms of Bieniawski's scheme will be described in the following sections to illustrate the nature and uses of modern rock mass classification schemes.

3.7.2 Bieniawski's geomechanics classification

Bieniawski (1973, 1976) developed his scheme using data obtained mainly from civil engineering excavations in sedimentary rocks in South Africa. Bieniawski's scheme uses five classification parameters.

1 **Strength of the intact rock material**. The uniaxial compressive strength of the intact rock may be measured on cores as described in section 4.3.2. Alternatively, for all but very low-strength rocks, the point load index (section 4.3.9) may be used.
2 **Rock Quality Designation** (RQD) as described in section 3.3.
3 **Spacing of joints**. In this context, the term joints is used to describe all discontinuities.

Table 3.5 Geomechanics classification of jointed rock masses (after Bieniawski, 1984).

(a) Classification parameters and their ratings

	Parameter		Ranges of values						
1	strength of intact rock material	point–load–strength index (MPa)	> 10	4–10	2–4	1–2	for this low range, uniaxial compression test is preferred		
		uniaxial compressive strength (MPa)	> 250	100–250	50–100	25–50	25–50	1–5	< 1
	rating		15	12	7	4	2	1	0
2	drill core quality *RQD* (1%)		90–100	75–90	50–75	25–50	< 25		
	rating		20	17	13	8	3		
3	joint spacing (m)		> 2	0.6–2	0.2–0.6	0.06–0.2	< 0.06		
	rating		20	15	10	8	5		
4	condition of joints		very rough surfaces, not continuous, no separation, unweathered joint wall rock	slightly rough surfaces, separation < 1 mm, slightly weathered walls	slightly rough surfaces, separation < 1 mm, highly weathered walls	slickensided surfaces or gouge < 5 mm thick or separation 1–5 mm continuous joints	soft gouge > 5 mm thick or separation > 5 mm, continuous joints		
	rating		30	25	20	10	0		
5	groundwater	inflow per 10 m tunnel length (ℓ min^{-1}) or	none	< 10	10–25	25–125	> 125		
		$\dfrac{\text{joint water pressure}}{\text{major principal stress}}$ or	0	0.0–0.1	0.1–0.2	0.2–0.5	> 0.5		
		general conditions	completely dry	damp	wet	dripping	flowing		
	rating		15	10	7	4	0		

(b) Rating adjustment for joint orientations

Strike and dip orientations of joints		very favourable	favourable	fair	unfavourable	very unfavourable
Rating	tunnels	0	– 2	– 5	– 10	– 12
	foundations	0	– 2	– 7	– 15	– 25
	slopes	0	– 5	– 25	– 50	– 60

(c) Rock mass classes determined from total ratings

Ratings	100 ← 81	80 ← 61	60 ← 41	40 ← 21	< 20
Class no.	I	II	III	IV	V
Description	very good rock	good rock	fair rock	poor rock	very poor rock

(d) Meaning of rock classes

Class no.	I	II	III	IV	V
average stand-up time	10 years for 15 m span	6 months for 8 m span	1 week for 5 m span	10 hours for 2.5 m span	30 minutes for 0.5 m span
cohesion of the rock mass (kPa)	> 400	300–400	200–300	100–200	< 100
friction angle of the rock mass	45°	35°–45°	25°–35°	15°–25°	< 15°

4 **Condition of joints**. This parameter accounts for the separation or aperture of discontinuities, their continuity or persistence, their surface roughness, the wall condition (hard or soft) and the nature of any in-filling materials present.

5 **Groundwater conditions**. An attempt is made to account for the influence of groundwater pressure or flow on the stability of underground excavations in terms of the observed rate of flow into the excavation, the ratio of joint water pressure to major principal stress, or by a general qualitative observation of groundwater conditions.

The way in which these parameters are incorporated into Bieniawski's geomechanics classification for jointed rock masses is shown in Part (a) of Table 3.5. For various ranges of each parameter, a rating value is assigned. The allocation of these rating values allows for the fact that all parameters do not necessarily contribute equally to the behaviour of the rock mass. The overall Rock Mass Rating (RMR) is obtained by adding the values of the ratings determined for the individual parameters. This RMR value may be adjusted for the influence of discontinuity orientation by applying the corrections given in Part (b) of Table 3.5. The terms used for this purpose are explained in Table 3.6. (When falling or sliding of blocks of rock from the roof or walls of an excavation is a possibility, this approach should not be relied upon. A wedge analysis of the type described in Chapter 9 should be used.) Part (c) of Table 3.5 sets out the class and description assigned to rock masses with various total ratings. The interpretation of these ratings in terms of stand-up times of underground excavations and rock mass strength parameters is given in Part (d) of Table 3.5. The relations between stand-up time, RMR and unsupported span are further illus-

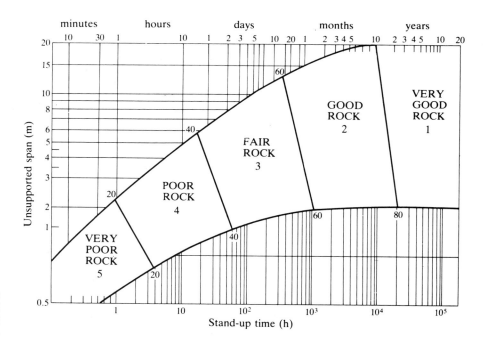

Figure 3.30 Relation between the stand-up times of unsupported underground excavation spans and the CSIR geomechanics classification (after Bieniawski, 1976).

trated in Figure 3.30. The variation with RMR of the *in situ* strengths and deformabilities of jointed rock masses will be discussed in section 4.9.

As an example of the application of Bieniawski's classification, consider a granitic rock mass for which the RMR is determined as shown in Table 3.7.

An adit is to be driven into the granite oriented such that the dominant joint set strikes roughly perpendicular to the adit axis and dips at 35° against the drive direction. From Table 3.6, this situation is described as unfavourable for which a rating adjustment of – 10 is obtained from Part (b) of Table 3.5.

Table 3.6 The effects of joint strike and dip in tunnelling (after Bieniawski, 1984).

Strike perpendicular to tunnel axis				Strike parallel to tunnel axis		
Drive with dip		Drive against dip				Dip 0°–20° irrespective of strike
Dip 45–90°	Dip 20–45°	Dip 45–90°	Dip 20–45°	Dip 45–90°	Dip 20–45°	
very favourable	favourable	fair	unfavourable	very unfavourable	fair	fair

Table 3.7 Determination of rock mass rating.

Parameter	Value or description	Rating
1. strength of intact rock material	150 MPa	12
2. *RQD*	70	13
3. joint spacing	0.5 m	10
4. condition of joints	slightly rough surfaces separation < 1 mm slightly weathered joint wall rock	25
5. groundwater	water dripping	4
	Total RMR	64

Thus the final RMR is reduced to 54 which places the rock mass in Class III with a description of fair. Figure 3.30 gives the stand-up time of an unsupported 3 m wide excavation in the granite as approximately 2 months.

3.7.3 Laubscher's Geomechanics Classification for mining applications

Laubscher (1977, 1984) modified Bieniawski's geomechanics classification on the basis of experience gained in a number of chrysotile asbestos mines in Africa. The classification, set out in Table 3.8, uses the same five classification parameters as Bieniawski's scheme but involves differences in detail. Each of the five classes are divided into subclasses, A and B, new ranges and ratings for intact rock strength (*IRS* in Table 3.8) are used, and the joint spacing and condition of joint parameters are evaluated differently.

The only discontinuities (joints) included in the assessment of RMR are those having trace lengths greater than one excavation diameter or 3 m, and those having trace lengths of less than 3 m that are intersected by other discontinuities to define blocks of rock. True spacings of the three most closely spaced discontinuity sets present in the rock mass are used in conjunction with Figure 3.31 to

81

Table 3.8 Modified geomechanics classification scheme (after Laubscher, 1984).

class	1		2		3		4		5	
rating description	100–81 very good		80–61 good		60–41 fair		40–21 poor		20–0 very poor	
subclasses	A	B	A	B	A	B	A	B	A	B
Item										
1 RQD, %	100–97		96–84	83–71	70–56	55–44	43–31	30–17	16–4	3–0
rating ($RQD\% \times 0.15$)	15		14	12	10	8	6	4	2	0
2 IRS, MPa	> 185	184–165	164–145	144–125	125–104	104–85	84–65	64–45	44–25	24–5
rating ($0.1 \times IRS$)	20	18	16	14	12	10	8	6	4	2
3 joint spacing					refer to Figure 3.31					
rating	25 ←————————————————————→ 0									
4 joint condition including groundwater					refer to Table 3.9					
rating ($40 \times A \times B \times C \times D / 10^8$)	40 ←————————————————————→ 0									

(IRS row also includes a lowest division: 4–0 MPa, rating 0.)

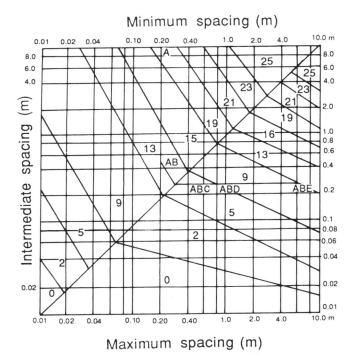

Figure 3.31 Joint spacing ratings for multi-joint systems. In the example, spacings of sets A, B, C, D, and E are 0.2, 0.5, 0.6, 1.0 and 7.0 m, respectively; the combined ratings for AB, ABC, ABD, and ABE are 13, 5, 9 and 13, respectively (after Laubscher, 1984).

obtain a joint spacing rating on a scale of 0 to 25. The way in which the joint condition rating is influenced by a range of factors is set out in Table 3.9. A possible rating of 40 is assigned initially and, depending on the features exhibited by the joints, is then reduced by the percentages given in Table 3.9.

Before the basic rating for the rock mass is applied, it is adjusted to take account of weathering, field and induced stresses, changes in stress due to mining operations, orientations of blocks with exposed bases and blasting effects. For

Table 3.9 Assessment of joint condition – adjustments as accumulative percentages of total possible rating of 40 (after Laubscher, 1984).

Parameter	Description	Dry conditions	Moist conditions	Wet conditions	
				moderate pressure 25–125 $l\,min^{-1}$	severe pressure $>125l\,min^{-1}$
A joint expression (large-scale irregularities)	wavy multi-directional	100	100	95	90
	uni-directional	95–90	95–90	90–85	80–75
	curved	89–80	85–75	80–70	70–60
	straight	79–70	74–65	60	40
B joint expression (small-scale irregularities or roughness)	very rough	100	100	95	90
	striated or rough	99–85	99–85	80	70
	smooth	84–60	80–55	60	50
	polished	59–50	50–40	30	20
C joint wall alteration zone	stronger than wall rock	100	100	100	100
	no alteration	100	100	100	100
	weaker than wall rock	75	70	65	60

Table continued

83

Parameter	Description		Dry conditions	Moist conditions	Wet conditions moderate pressure 25–125 $l\,min^{-1}$	severe pressure $>125l\,min^{-1}$
D joint filling	no fill – surface staining only		100	100	100	100
	non-softening and sheared material (clay or talc free)	coarse sheared medium	95	90	70	50
		sheared fine	90	85	65	45
		sheared	85	80	60	40
	soft sheared material (e.g. talc)	coarse sheared medium	70	65	40	20
		sheared fine	65	60	35	15
		sheared	60	50	30	10
	gouge thickness < amplitude of irregularities		40	30	10	
	gouge thickness > amplitude of irregularities		20	10	flowing material 5	

details of how these adjustments are applied, the reader should consult the papers by Laubscher and Taylor (1976) or Laubscher (1977). The total possible percentage adjustments that can be made to each classification parameter for each of these factors are set out in Table 3.10. Examples of the mining application of Laubscher's classification will be given in subsequent chapters.

Table 3.10 Total possible percentage reductions to ratings for classification parameters (after Laubscher, 1977).

	RQD	IRS	Joint spacing	Condition of joints	Total
weathering	95	96		82	75
field and induced stresses				120–76	120–76
changes in stress				120–60	120–60
strike and dip orientation			70		70
blasting	93			86	80

Problems

1 A scanline survey is to be carried out on the vertical wall of an exploration drive. The rock mass contains two sets of parallel discontinuities whose traces on the wall are mutually inclined at 75° as shown in the diagram. The traces of set A make an angle of 55° with a horizontal scanline. Large numbers of measurements give the apparent mean spacings of sets A and B along the scanline as 0.450 m and 0.800 m, respectively.

(a)   Calculate the mean normal spacing of each set.

84

(b) What is the mean spacing of all discontinuities in the direction of the scanline?

(c) Assuming that the combined discontinuity spacings follow a negative exponential distribution, estimate the *RQD* of the rock mass in the direction of the scanline.

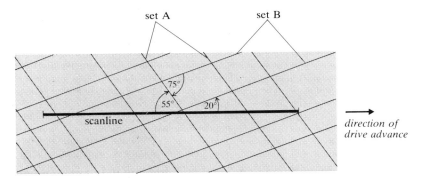

2   A preliminary borehole investigation of a sandstone produced 30 m of core that contained 33 drilling breaks, 283 iron-stained joints and 38 other discontinuities of uncertain origin.

(a) Calculate the bandwidths within which the overall mean discontinuity spacing in the direction of the borehole axis lies at the 80% and 95% confidence levels.

(b) What approximate additional length of borehole is required to provide a value in a bandwidth of ± 8% of the true spacing value at the 95% confidence level?

3   Plot on the stereographic projection the great circle of the plane with the orientation 110/50. What is the apparent dip of this plane in the direction 090°?

4   What are the trend and plunge of the line of intersection of the planes 110/50 and 320/60?

5   Plot the great circles and poles to the planes 156/32 and 304/82. What is the acute angle between these planes? In what plane is it measured?

6   A tunnel of square cross section has planar vertical sidewalls of orientation 230/90. The lineation produced by the intersection of a planar joint with one sidewall plunges at 40° to the north-west. The same joint strikes across the horizontal roof in the direction 005°–185°. What is the orientation of the joint plane?

7   A reference line is scribed on drill core for use in correctly orienting discontinuities intersected by the core. A certain planar discontinuity has the apparent orientation 120/35 measured with the reference line vertical. If the actual trend and plunge of the borehole axis, and of the reference line, are 225° and 60° respectively, determine the true orientation of the discontinuity.

8   Assume that, for the rock mass described in Problem 1, both sets of discontinuities strike perpendicular to the drive axis. The intact rock material has a uniaxial compressive strength of 120 MPa, the joint surfaces are slightly rough with an average separation of 0.2 mm and, although there is water in the joints, the flow into the excavation is quite small.

Determine the basic CSIR geomechanics classification for this rock mass (Table 3.5). How does application of the adjustments for joint orientations for tunnelling given by Tables 3.5 and 3.6 affect this classification?

Is the adjusted RMR value likely to provide a satisfactory guide to roof stability in this case?

4 Rock strength and deformability

4.1 Introduction

The engineering mechanics-based approach to the solution of mining rock mechanics problems used in this book, requires prior definition of the stress–strain behaviour of the rock mass. Important aspects of this behaviour are the constants relating stresses and strains in the elastic range, the stress levels at which yield, fracturing or slip occurs within the rock mass, and the post-peak stress–strain behaviour of the fractured or 'failed' rock.

In some problems, it may be the behaviour of the **intact rock material** that is of concern. This will be the case when considering the excavation of rock by drilling and blasting, or when considering the stability of excavations in good quality, brittle rock which is subject to rockburst conditions. In other instances, the behaviour of **single discontinuities**, or of a small number of discontinuities, will be of paramount importance. Examples of this class of problem include the equilibrium of blocks of rock formed by the intersections of three or more discontinuities and the roof or wall of an excavation, and cases in which slip on a major throughgoing fault must be analysed. A different class of problem is that in which the rock mass must be considered as an assembly of **discrete blocks**. As noted in section 6.7 which describes the distinct element method of numerical analysis, the normal and shear force–displacement relations at block face-to-face and corner-to-face contacts are of central importance in this case. Finally, it is sometimes necessary to consider the global response of a **jointed rock mass** in which the discontinuity spacing is small on the scale of the problem domain. The behaviour of caving masses of rock is an obvious example of this class of problem.

It is important to note that the presence of major discontinuities or of a number of joint sets does not necessarily imply that the rock mass will behave as a discontinuum. In mining settings in which the rock surrounding the excavations is always subject to high compressive stresses, it may be reasonable to treat a jointed rock mass as an equivalent elastic continuum. A simple example of the way in which rock material and discontinuity properties may be combined to obtain the elastic properties of the equivalent continuum is given in section 4.9.2.

Figure 4.1 illustrates the transition from intact rock to a heavily jointed rock mass with increasing sample size in a hypothetical rock mass surrounding an underground excavation. Which model will apply in a given case will depend on the size of the excavation relative to the discontinuity spacing, the imposed stress level, and the orientations and strengths of the discontinuities. Those aspects of the stress–strain behaviour of rocks and rock masses required to solve these various classes of problem, will be discussed in this chapter. Since compressive stresses predominate in geotechnical problems, the emphasis will be on response

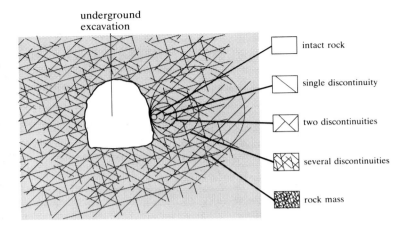

Figure 4.1 Idealised illustration of the transition from intact rock to a heavily jointed rock mass with increasing sample size (after Hoek and Brown, 1980).

to compressive and shear stresses. For the reasons outlined in section 1.2.3, the response to tensile stresses will not be considered in detail.

4.2 Concepts and definitions

Experience has shown that the terminology used in discussions of rock 'strength' and 'failure' can cause confusion. Unfortunately, terms which have precise meanings in engineering science are often used imprecisely in engineering practice. In this text, the following terminology and meanings will be used.

Fracture is the formation of planes of separation in the rock material. It involves the breaking of bonds to form new surfaces. The onset of fracture is not necessarily synonymous with failure or with the attainment of peak strength.

Strength, or **peak strength**, is the maximum stress, usually averaged over a plane, that the rock can sustain under a given set of conditions. It corresponds to point B in Figure 4.2a. After its peak strength has been exceeded, the specimen may still have some load-carrying capacity or strength. The minimum or **residual strength** is reached generally only after considerable post-peak deformation (point C in Figure 4.2a).

Brittle fracture is the process by which sudden loss of strength occurs across a plane following little or no permanent (plastic) deformation. It is usually associated with strain-softening or strain-weakening behaviour of the specimen as illustrated in Figure 4.2a.

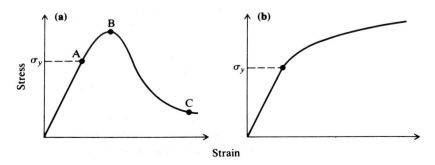

Figure 4.2 (a) Strain-softening; (b) strain-hardening stress–strain curves.

Ductile deformation occurs when the rock can sustain further permanent deformation without losing load-carrying capacity (Figure 4.2b).

Yield occurs when there is a departure from elastic behaviour, i.e. when some of the deformation becomes irrecoverable as at A in Figure 4.2a. The **yield stress** (σ_y in Figure 4.2) is the stress at which permanent deformation first appears.

Failure is often said to occur at the peak strength or be initiated at the peak strength (Jaeger and Cook, 1979). An alternative engineering approach is to say that the rock has failed when it can no longer adequately support the forces applied to it or otherwise fulfil its engineering function. This may involve considerations of factors other than peak strength. In some cases, excessive deformation may be a more appropriate criterion of 'failure' in this sense.

Effective stress is defined, in general terms, as the stress which governs the gross mechanical response of a porous material. The effective stress is a function of the total or applied stress and the pressure of the fluid in the pores of the material, known as the **pore pressure** or **pore-water pressure**. The concept of effective stress was first developed by Karl Terzaghi who used it to provide a rational basis for the understanding of the engineering behaviour of soils. Terzaghi's formulation of the **law of effective stress**, an account of which is given by Skempton (1960), is probably the single most important contribution ever made to the development of geotechnical engineering. For soils and some rocks loaded under particular conditions, the effective stresses, σ'_{ij}, are given by

$$\sigma'_{ij} = \sigma_{ij} - u\delta_{ij} \tag{4.1}$$

where σ_{ij} are the total stresses, u is the pore pressure, and δ_{ij} is the Kronecker delta. This result is so well established for soils that it is often taken to be the definition of effective stress. Experimental evidence and theoretical argument suggest that, over a wide range of material properties and test conditions, the response of rock depends on

$$\sigma'_{ij} = \sigma_{ij} - \alpha u\delta_{ij} \tag{4.2}$$

where $\alpha \leq 1$, and is a constant for a given case (Paterson, 1978).

4.3 Behaviour of isotropic rock material in uniaxial compression

4.3.1 Influence of rock type and condition

Uniaxial compression of cylindrical specimens prepared from drill core, is probably the most widely performed test on rock. It is used to determine the uniaxial or unconfined compressive strength, σ_c, and the elastic constants, Young's modulus, E, and Poisson's ratio, ν, of the rock material. The uniaxial compressive strength of the intact rock is used in rock mass classification schemes (section 3.7), and as a basic parameter in the rock mass strength criterion to be introduced later in this chapter.

Despite its apparent simplicity, great care must be exercised in interpreting results obtained in the test. Obviously, the observed response will depend on the nature and composition of the rock and on the condition of the test specimens.

For similar mineralogy, σ_c will decrease with increasing porosity, increasing degree of weathering and increasing degree of microfissuring. As noted in section 1.2.4, σ_c may also decrease with increasing water content. Data illustrating these various effects are presented by Vutukuri *et al.* (1974).

It must be recognised that, because of these effects, the uniaxial compressive strengths of samples of rock having the same geological name, can vary widely. Thus the uniaxial compressive strength of sandstone will vary with the grain size, the packing density, the nature and extent of cementing between the grains, and the levels of pressure and temperature that the rock has been subjected to throughout its history. However, the geological name of the rock type can give some qualitative indication of its mechanical behaviour. For example, a slate can be expected to exhibit cleavage which will produce anisotropic behaviour, and a quartzite will generally be a strong, brittle rock. Despite the fact that such features are typical of some rock types, it is dangerous to attempt to assign mechanical properties to rock from a particular location on the basis of its geological description alone. There is no substitute for a well-planned and executed programme of testing.

4.3.2 Standard test procedure and interpretation

Suggested techniques for determining the uniaxial compressive strength and deformability of rock material are given by the International Society for Rock Mechanics Commission on Standardization of Laboratory and Field Tests (ISRM Commission, 1979). The essential features of the recommended procedure are:

(a)   The test specimens should be right circular cylinders having a height to diameter ratio of 2.5–3.0 and a diameter preferably of not less than NX core size, approximately 54 mm. The specimen diameter should be at least 10 times the size of the largest grain in the rock.

(b)   The ends of the specimen should be flat to within 0.02 mm and should not depart from perpendicularity to the axis of the specimen by more than 0.001 rad or 0.05 mm in 50 mm.

(c)   The use of capping materials or end surface treatments other than machining is not permitted.

(d)   Specimens should be stored, for no longer than 30 days, in such a way as to preserve the natural water content, as far as possible, and tested in that condition.

(e)   Load should be applied to the specimen at a constant stress rate of $0.5–1.0 \, \text{MPa s}^{-1}$.

(f)   Axial load and axial and radial or circumferential strains or deformations should be recorded throughout each test.

(g)   There should be at least five replications of each test.

Figure 4.3 shows an example of the results obtained in such a test. The axial force recorded throughout the test has been divided by the initial cross-sectional area of the specimen to give the average axial stress, σ_a, which is shown plotted against overall axial strain, ε_a, and against radial strain, ε_r. Where post-peak deformations are recorded (section 4.3.7), the cross-sectional area may change considerably as the specimen progressively breaks up. In this case, it is preferable to present the experimental data as force–displacement curves.

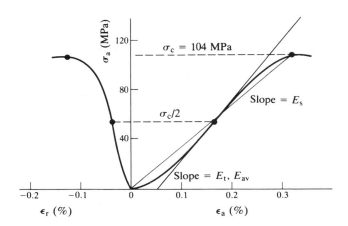

Figure 4.3 Results obtained in a uniaxial compression test on rock.

As shown in Figure 4.3, the axial Young's modulus of the specimen varies throughout the loading history and so is not a uniquely determined constant for the material. It may be calculated in a number of ways, the most common being:

(a) **Tangent Young's modulus**, E_t, is the slope of the axial stress–axial strain curve at some fixed percentage, generally 50%, of the peak strength. For the example shown in Figure 4.3, $E_t = 51.0$ GPa.

(b) **Average Young's modulus**, E_{av}, is the average slope of the more-or-less straight line portion of the axial stress–strain curve. For the example shown in Figure 4.3, $E_{av} = 51.0$ GPa.

(c) **Secant Young's modulus**, E_s, is the slope of a straight line joining the origin of the axial stress–strain curve to a point on the curve at some fixed percentage of the peak strength. In Figure 4.3, the secant modulus at peak strength is $E_s = 32.1$ GPa.

Corresponding to any value of Young's modulus, a value of Poisson's ratio may be calculated as

$$\nu = -\frac{(\Delta \sigma_a / \Delta \varepsilon_a)}{(\Delta \sigma_a / \Delta \varepsilon_r)} \tag{4.3}$$

For the data given in Figure 4.3, the values of ν corresponding to the values of E_t, E_{av}, and E_s calculated above are approximately 0.29, 0.31 and 0.40 respectively.

Because of the axial symmetry of the specimen, the volumetric strain, ε_v, at any stage of the test can be calculated as

$$\varepsilon_v = \varepsilon_a + 2\varepsilon_r \tag{4.4}$$

For example, at a stress level of $\sigma_a = 80$ MPa in Figure 4.3, $\varepsilon_a = 0.220\%$, $\varepsilon_r = -0.055\%$ and $\varepsilon_v = 0.110\%$.

Varying the standard conditions will influence the observed response of the specimen. Some of these effects will be discussed briefly in sections 4.3.3 to

91

4.3.7. More extensive discussions of these effects are given by Hawkes and Mellor (1970), Vutukuri *et al.* (1974) and Paterson (1978).

4.3.3 End effects and the influence of height to diameter ratio

The objective of the test arrangements should be to subject the specimen to uniform boundary conditions with a uniform uniaxial stress and a uniform displacement field being produced throughout the specimen (Figure 4.4a). Due to friction between the specimen ends and the platens and differences between the elastic properties of rock and steel, the specimen will be restrained near its ends and prevented from deforming uniformly. Figure 4.4b illustrates a case in which complete radial restraint occurs at the specimen ends. The result of such restraint is that shear stresses are set up at the specimen–platen contact (Figure 4.4c). This means that the axial stress is not a principal stress and that the stresses within the specimen are not always uniaxial.

As a consequence of these end effects, the stress distribution varies throughout the specimen as a function of specimen geometry. As the height to diameter (H/D) ratio increases, a greater proportion of the sample volume is subjected to an approximately uniform state of uniaxial stress. It is for this essential reason that a H/D ratio of at least 2.0 should be used in laboratory compression testing of rock. Figure 4.5 shows some experimental data which illustrate this effect. When 51 mm diameter specimens of Wombeyan Marble were loaded through 51 mm diameter steel platens, the measured uniaxial compressive strength increased as the H/D ratio was decreased and the shape of the post-peak stress–strain curve became flatter. When the tests were repeated with 'brush' platens (made from an assembly of 3.2 mm square high-tensile steel pins), lateral deformation of the specimens was not inhibited; similar stress–strain curves were obtained for H/D ratios in the range 0.5 to 3.0 However, 'brush' platens were found to be too difficult to prepare and maintain for their use in routine testing to be recommended.

It is tempting to seek to eliminate end effects by treating the specimen–platen interface with a lubricant or by inserting a sheet of soft material between the specimen and the platen. Experience has shown that this can cause lateral tensile stresses to be applied to the specimen by extrusion of the inserts or by fluid pressures set up inside flaws on the specimen ends. For this reason, the ISRM Commission (1979) and other authorities (e.g. Hawkes and Mellor, 1970; Jaeger and Cook, 1979) recommend that treatment of the sample ends, other than by machining, be avoided.

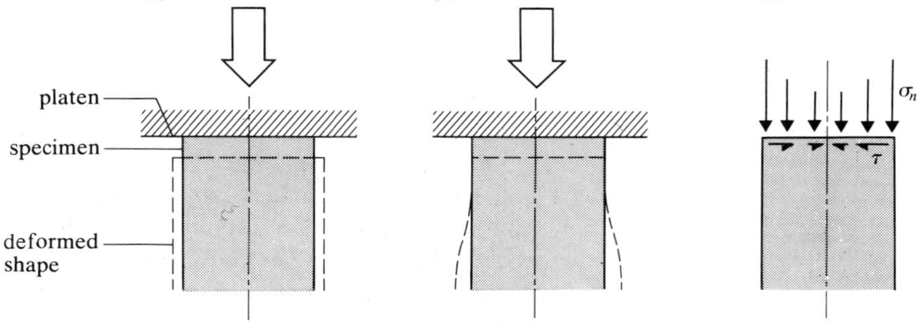

Figure 4.4 Influence of end restraint on stresses and displacements induced in a uniaxial compression test: (a) desired uniform deformation of the specimen; (b) deformation with complete radial restraint at the specimen–platen contact; (c) non-uniform normal stress, σ_n and shear stress, τ, induced at the specimen end as a result of end restraint.

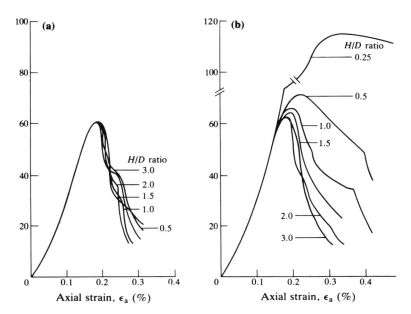

Figure 4.5 Influence of height to diameter (*H/D*) ratio on stress-strain curves obtained in uniaxial compression tests carried out on Wombeyan Marble using (a) brush platens, and (b) solid steel platens (after Brown and Gonano, 1974).

4.3.4 Influence of the standard of end preparation

In Figures 4.3 and 4.5, the axial stress-axial strain curves have initial concave upwards sections before they become sensibly linear. This initial portion of the curve is generally said to be associated with 'bedding-down' effects. However, experience shows that the extent of this portion of the curve can be greatly reduced by paying careful attention to the flatness and parallelism of the ends of the specimen. Analyses of the various ways in which a poor standard of end preparation influence the observed response of the sample have been presented by Hawkes and Mellor (1970).

The ISRM Commission (1979) recommends that in a 50+ mm diameter specimen, the ends should be flat to within 0.02 mm and should not depart from the perpendicular to the specimen axis by more than 0.05 mm. The latter figure implies that the ends could be out of parallel by up to 0.10 mm. Even when spherical seats are provided in the platens, out-of-parallelism of this order can still have a significant influence on the shape of the stress–strain curve, the peak strength and the reproducibility of results. For research investigations, the authors prepare their 50–55 mm diameter specimens with ends flat and parallel to within 0.01 mm.

4.3.5 Influence of specimen volume

It has often been observed experimentally that, for similar specimen geometry, the uniaxial compressive strength of rock material, σ_c, varies with specimen volume. (This is a different phenomenon to that discussed in section 4.1 where the changes in behaviour considered were those due to the presence of varying numbers of geological discontinuities within the sample volume.) Generally, it is observed that σ_c decreases with increasing specimen volume, except at very small specimen sizes where inaccuracy in specimen preparation and surface flaws or contamination may dominate behaviour and cause a strength decrease with

93

decreasing specimen volume. This, coupled with the requirement that the specimen diameter should be at least 10 times the size of the largest grain, provides a reason for using specimen diameters of approximately 50 mm in laboratory compression tests.

Many explanations have been offered for the existence of size effects, but none has gained universal acceptance. A popular approach is to interpret size effects in terms of the distribution of flaws within the material. Much of the data on which conclusions about size effects are based, were obtained using cubical specimens. Brown and Gonano (1975) have shown that in these cases, stress gradients and end effects can greatly influence the results obtained. The most satisfactory explanations of observed size effects in rock and other brittle materials are those in which surface energy is used as the fundamental material property (section 4.5.3).

4.3.6 Influence of strain rate

The ISRM Commission (1979) recommends that a loading rate of 0.5–1.0 $MPa\,s^{-1}$ be used in uniaxial compression tests. This corresponds to a time to the attainment of peak strength in the order of 5–10 min. As the arguments presented below show, it is preferable to regard strain or deformation, rather than axial stress or load, as the controlling variable in the compression testing of rock. For this reason, the following discussion will be in terms of axial strain rate, $\dot{\varepsilon}_a$, rather than axial stress rate.

The times to peak strength recommended by the ISRM Commission (1979) correspond to axial strain rates in the order of 10^{-5}–$10^{-4}\,s^{-1}$. For rocks other than those such as the evaporites which exhibit markedly time-dependent behaviour, departures from the prescribed strain rate by one or two orders of magnitude may produce little discernible effect. For very fast and very slow strain rates, differences in the observed stress–strain behaviour and peak strengths can become quite marked. However, a change in strain rate from $10^{-8}\,s^{-1}$ to $10^{2}\,s^{-1}$ may only increase the measured uniaxial compressive strength by a factor of about two. Generally, the observed behaviour of rock is not significantly influenced by varying the strain rate within the range that it is convenient to use in quasi-static laboratory compression tests.

4.3.7 Influence of testing machine stiffness

Whether or not the post-peak portion of the stress–strain curve can be followed and the associated progressive disintegration of the rock studied, depends on the relative stiffnesses of the specimen and the testing machine. The standard test procedure and interpretation discussed in section 4.3.2 do not consider this post-peak behaviour. However, the subject is important in assessing the likely stability of rock fracture in mining applications including pillar stability and rockburst potential.

Figure 4.6 illustrates the interaction between a specimen and a conventional testing machine. The specimen and machine are regarded as springs loaded in parallel. The machine is represented by a linear elastic spring of constant longitudinal stiffness, k_m, and the specimen by a non-linear spring of varying stiffness, k_s. Compressive forces and displacements of the specimen are taken as positive. Thus as the specimen is compressed, the machine spring extends. (This extension is analogous to that which occurs in the columns of a testing machine during a

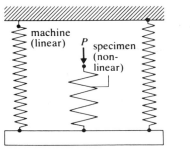

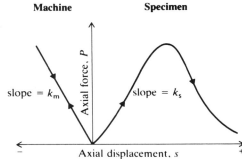

Figure 4.6 Spring analogy illustrating machine–specimen interaction.

compression test.) When the peak strength has been reached in a strain-softening specimen such as that shown in Figure 4.6, the specimen continues to compress, but the load that it can carry progressively reduces. Accordingly, the machine unloads and its extension reduces.

Figure 4.7 shows what will happen if the machine is (a) soft, and (b) stiff, with respect to the specimen. Imagine that the specimen is at peak strength and is compressed by a small amount, Δs. In order to accommodate this displacement, the load on the specimen must reduce from P_A to P_B, so that an amount of energy ΔW_s, given by the area ABED in Figures 4.7 a and b, is absorbed. However, in displacing by Δs from point A, the 'soft' machine only unloads to F and releases stored strain energy ΔW_m, given by the area AFED. In this case $\Delta W_m > \Delta W_s$, and catastrophic failure occurs at, or shortly after, the peak because the energy released by the machine during unloading is greater than that which can be absorbed by the specimen in following the post-peak curve from A to B.

If the machine is stiff with respect to the specimen in the post-peak region, the post-peak curve can be followed. In Figure 4.7b, $\Delta W_m < \Delta W_s$ and energy in excess of that released by the machine as stored strain energy must be supplied in order to deform the specimen along ABC. Note that the behaviour observed up to, and including, the peak, is not influenced by machine stiffness.

For some very brittle rocks, generally those that are fine grained and homogeneous, portions of the post-peak force–displacement or stress–strain curves can be very steep so that it becomes impossible to 'control' post-peak deformation even in the stiffest of testing machines. In these cases, the post-peak curves and the associated mechanisms of fracture may be studied using a judiciously operated **servocontrolled testing machine**.

Figure 4.7 Post-peak unloading using machines that are (a) soft, and (b) stiff, with respect to the specimen.

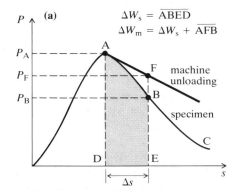

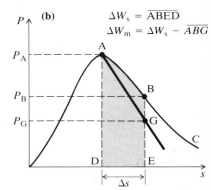

Figure 4.8 Principle of closed-loop control (after Hudson *et al.*, 1972b).

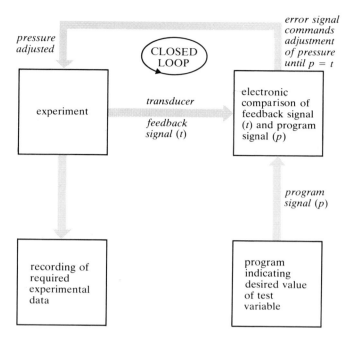

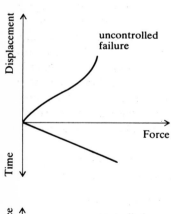

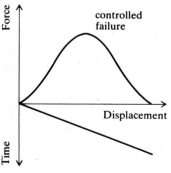

Figure 4.9 Choice between force and displacement as the programmed control variable (after Hudson *et al.*, 1972a).

The essential features of closed-loop servocontrol are illustrated in Figure 4.8. An experimental variable (a force, pressure, displacement or strain component) is programmed to vary in a predetermined manner, generally monotonically increasing with time. The measured and programmed values are compared electronically several thousands of times a second, and a servo valve adjusts the pressure within the actuator to produce the desired equivalence.

Modern servocontrolled testing systems are used to conduct a wide variety of tests in rock mechanics laboratories. The key to the successful use of these systems is the choice of the control variable. The basic choice is between a force (or pressure) and a displacement (or strain) component. Figure 4.9 shows why it is not feasible to obtain the complete uniaxial force–displacement curve for a strain-softening specimen by programming the axial force to increase monotonically with time. When the peak strength of the specimen is reached, the program will attempt to continue to increase the axial force, but the load-carrying capacity can only decrease with further axial displacement. However, the test can be successfully controlled by programming the axial **displacement** to increase monotonically with time.

The post-peak portions of the force–displacement curves obtained in compression tests on some rocks may be steeper than, or not as smooth as, those shown in Figures 4.7 and 4.9. In these cases, better control can be obtained by using the circumferential displacement rather than the axial displacement as the control variable. Figure 4.10 shows the complete axial stress (σ_a)–axial strain (ε_a) and circumferential (or radial) strain (ε_r)–axial strain curves obtained in such a test on a 50 mm diameter by 100 mm long specimen of an oolitic limestone (Portland stone) in which a wrap-around transducer was used to monitor circumferential displacement. Although the possibility of extracting energy from the machine–specimen system offered by this technique is not reproduced in practical mining

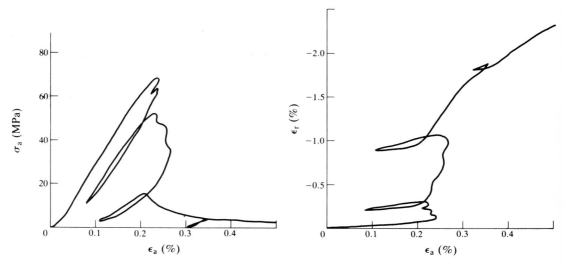

Figure 4.10 Axial stress, σ_a, and radial strain, ϵ_r, vs. axial strain, ϵ_a, curves recorded in a uniaxial compression test on an oolitic limestone (after Elliott, 1982).

problems, this approach does permit progressive post-peak breakdown to be controlled and studied.

Figure 4.11 shows the complete σ_a–ϵ_a curves obtained by Wawersik and Fairhurst (1970) in a series of controlled uniaxial compression tests on a range of rock types. By halting tests on specimens of the same rock at different points on the curve and sectioning and polishing the specimens, Wawersik and Fairhurst were able to study the mechanisms of fracture occurring in the different rock types. They found that the post-peak behaviours of the rocks studied may be divided into two classes (Figure 4.12). For class I behaviour, fracture propagation is stable in the sense that work must be done on the specimen for each incremental decrease in load-carrying ability. For class II behaviour, the fracture process is unstable or self-sustaining; to control fracture, energy must be extracted from the material.

The experiments of Wawersik and Fairhurst and of subsequent investigators, indicate that, in uniaxial compression, two different modes of fracture may occur:

(a) local 'tensile' fracture predominantly parallel to the applied stress;
(b) local and macroscopic shear fracture (faulting).

The relative predominance of these two types of fracture depends on the strength, anisotropy, brittleness and grain size of the crystalline aggregates. However, sub-axial fracturing generally precedes faulting, being initiated at 50-95% of the peak strength.

In very heterogeneous rocks, sub-axial fracturing is often the only fracture mechanism associated with the peaks of the σ_a–ϵ_a curves for both class I and class II behaviour. In such rocks, shear fractures develop at the boundaries and then in the interiors of specimens, well beyond the peak. This observation is at variance with the traditional view that through-going shear fracture occurs *at* the peak. Generally, these shear fractures, observed in 'uncontrolled' tests, are associated with sudden unloading in a soft testing machine.

In homogeneous, fine-grained rocks such as the Solenhofen Limestone (Figure 4.11), the peak compressive strength may be governed by localised faulting.

97

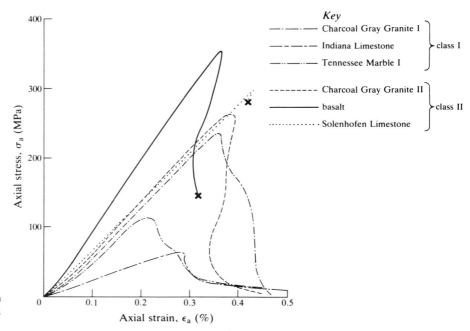

Figure 4.11 Uniaxial stress–strain curves for six rocks (after Wawersik and Fairhurst, 1970).

Because of the internal structural and mechanical homogeneity of these rocks, there is an absence of the local stress concentrations that may produce pre-peak cracking throughout coarser-grained crystalline aggregates. In these homogeneous, fine-grained rocks, fracture initiation and propagation can occur almost simultaneously. If violent post-peak failure of the specimen is to be prevented, the strain energy stored in the unfractured parts of the specimen, and in the testing machine, must be removed rapidly by reversing the sense of platen movement. This produces the artefact of a class II curve.

It is important to recognise that the post-peak portion of the curve does not reflect a true material property. The appearance of localised faulting in laboratory tests on rock and around underground excavations may be explained at a fundamental level by bifurcation or strain localisation analysis. In this approach, it is postulated that the material properties may allow the homogeneous deformation of an initially uniform material to lead to a bifurcation point, at which non-uniform deformation can be incipient in a planar band under conditions of continuing equilibrium and continuing homogeneous deformation outside the zone of localisation (Rudnicki and Rice, 1975). Using a rigorous analysis of this type with the required material properties determined from measured stress–strain and volumetric strain curves, Vardoulakis *et al.* (1988) correctly predicted the axial stress at which a particular limestone failed by faulting in a uniaxial compression test, the orientation of the faults and the Coulomb shear strength parameters (section 4.5.2) of the rock.

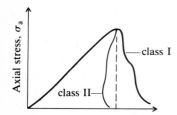

Figure 4.12 Two classes of stress–strain behavior observed in uniaxial compression tests (after Wawersik and Fairhurst, 1970).

4.3.8 Influence of loading and unloading cycles

Figure 4.13 shows the axial force–axial displacement curve obtained by Wawersik and Fairhurst (1970) for a 51 mm diameter by 102 mm long specimen of

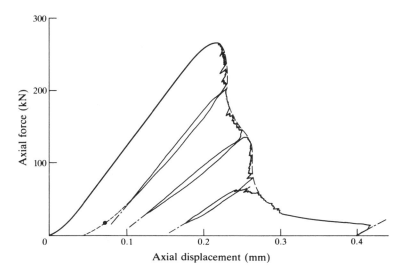

Figure 4.13 Axial force–axial displacement curve obtained for Tennessee Marble with post-peak unloading and reloading (after Wawersik and Fairhurst, 1970).

Tennessee Marble which was unloaded and then reloaded from a number of points in the post-peak range. Several points should be noted about the behaviour observed.

(a) On reloading, the curve eventually joins that for a specimen in which the axial displacement increases monotonically with time.

(b) As displacement continues in the post-peak region, the proportion of the total displacement that is irrecoverable increases.

(c) The unloading–loading loop shows some hysteresis.

(d) The apparent modulus of the rock which can be calculated from the slope of the reloading curve, decreases with post-peak deformation and progressive fragmentation of the specimen.

If rock specimens are subjected to loading and unloading cycles in the pre-peak range, some permanent deformation and hysteresis are generally observed. This is often associated with 'bedding-down' effects, and for this reason, the ISRM Commission (1979) recommends that 'it is sometimes advisable for a few cycles of loading and unloading to be performed'.

4.3.9 The point load test

Sometimes the facilities required to prepare specimens and carry out uniaxial compression tests to the standard described above are not available. In other cases, the number of tests required to determine the properties of the range of rock types encountered on a project may become prohibitive. There may be still further cases, in which the uniaxial compressive strength and the associated stress–strain behaviour need not be studied in detail, with only an approximate measure of peak strength being required. In all of these instances, the point load test may be used to provide an indirect estimate of uniaxial compressive strength.

In this test, a piece of core, generally 50–55 mm in diameter, is loaded between the two hardened steel points of the point load test apparatus (Figure 4.14). The force, P, at which the core breaks is determined from the peak pressure recorded

Figure 4.14 Point load test apparatus (photograph by ELE International Ltd).

on the pressure gauge and the known cross-sectional area of the loading ram. A **point load index** is calculated as

$$I_s = \frac{P}{D^2} \tag{4.5}$$

where D is the diameter of the core.

Broch and Franklin (1972) found that the average of large numbers of determinations of I_s could be correlated with the uniaxial compressive strength of the rock in question. For $D = 50$ mm it was found that

$$\sigma_c \simeq 24I_s \tag{4.6}$$

For other values of D, a size correction is necessary. Bieniawski (1975) suggested the following approximate relation between σ_c, I_s and the core diameter, D, measured in millimetres:

$$\sigma_c = (14 + 0.175D)\,I_s \tag{4.7}$$

Caution must be exercised in carrying out point load tests and interpreting the results. The load should be applied at a point that is at least $0.7D$ from either end of the piece of core if the results are not to be influenced by the proximity of the

100

ends (Bieniawski, 1975). The test is one in which fracture of the rock occurs as a result of induced tension, and it is essential that a consistent mode of failure be produced if the results obtained for different specimens are to be comparable. In this regard, very soft rocks and highly anisotropic rocks or rocks containing marked planes of weakness such as bedding planes, are likely to give spurious results. A high degree of scatter is a general feature of point load test results and large numbers of individual determinations (often in excess of 100) are required to give a reliable estimate of σ_c. Nevertheless, the point load test is a widely used index test for the strength of rock material. It finds particular application in rock mass classification schemes (Tables 3.5 and 3.8).

4.4 Behaviour of isotropic rock material in multiaxial compression

4.4.1 Types of multiaxial compression test

A basic principle of the laboratory testing of rock to obtain data for use in design analyses, is that the boundary conditions applied to the test specimen should simulate those imposed on the rock element *in situ*. This can rarely be achieved. General practice is to study the behaviour of the rock under known uniform applied stress systems.

As was shown in Chapter 2, a general state of three-dimensional stress at a point can be represented by three principal stresses, σ_1, σ_2 and σ_3, acting on mutually orthogonal planes. No shear stresses act on these planes. A plane of particular interest is the boundary of an underground excavation which is a principal plane except in the unusual case in which a shear stress is applied to the boundary surface by the support. The rock surrounding an underground excavation is rarely in a state of uniaxial compression. In the general case, away from the excavation boundary or on the boundary when a normal support stress, σ_3, is applied, there will be a state of **polyaxial** stress ($\sigma_1 \neq \sigma_2 \neq \sigma_3$). The special case in which $\sigma_2 = \sigma_3$ is called **triaxial** stress. It is this form of multiaxial stress that is most commonly used in laboratory testing. On the boundary of an unsupported excavation, $\sigma_3 = 0$, and a state of **biaxial** stress exists. The behaviour of intact, isotropic rock materials under each of these applied stress conditions will be discussed briefly in the following sections.

4.4.2 Biaxial compression ($\sigma_1 \geqslant \sigma_2, \sigma_3 = 0$)

Biaxial compression tests are carried out by applying different normal stresses to two pairs of faces of a cube, plate or rectangular prism of rock. The great difficulty with such tests is that the end effects described in section 4.3.3 exert an even greater influence on the stress distribution within the specimen than in the case of uniaxial compression. For this reason, fluid rather than solid medium loading is preferred. An alternative approach is to generate a biaxial state of stress at the inner surface of a hollow cylinder by loading it axially with a fluid pressure applied to its outer surface (Hoskins, 1969; Jaeger and Cook, 1979) in a triaxial cell (section 4.4.3). However, in this case, the stresses at 'failure' cannot be measured, but must be calculated using the theory of elasticity which may not be applicable at peak stress. The inner boundary of the hollow cylinder is a zone of high stress gradient which could influence the result. For these reasons, it is

101

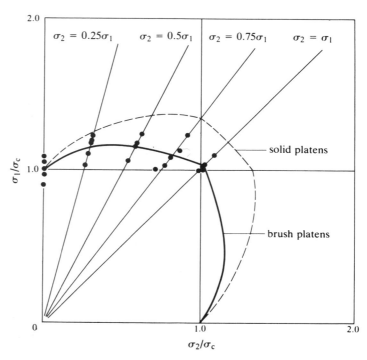

$\sigma_2 = 0.25\sigma_1$ $\sigma_2 = 0.5\sigma_1$ $\sigma_2 = 0.75\sigma_1$ $\sigma_2 = \sigma_1$

solid platens

brush platens

Figure 4.15 Biaxial compression test results for Wombeyan Marble (after Brown, 1974).

recommended that the use of hollow cylinder tests be restricted to the simulation of particular rock mechanics problems such as the behaviour of rock around a shaft, bored raise or borehole.

Brown (1974) carried out a series of biaxial compression tests on 76 mm square by 25 mm thick plates of Wombeyan Marble which were loaded on their smaller faces through (a) 76 mm × 25 mm solid steel platens, and (b) brush platens made from 3.2 mm square steel pins. Figure 4.15 shows the peak strength envelopes obtained in tests carried out at constant σ_2 / σ_1 ratios. The data are normalised with respect to the uniaxial compressive strength of the plates, $\sigma_c = 66$ MPa. The increase in peak strength over σ_c, associated with a given value of σ_2, was greater for the solid platens than for the brush platens. This was attributed to the influence of end effects. When the brush platens were used, the maximum measured increase in peak strength over σ_c was only 15%. For $\sigma_2 = \sigma_1$, no strength increase was observed (i.e. $\sigma_1 = \sigma_c$). The practical consequence of these results is that, for this rock type, the 'strengthening' effect of the intermediate principal stress can be neglected so that the uniaxial compressive strength, σ_c, should be used as the rock material strength whenever $\sigma_3 = 0$. This slightly conservative conclusion is likely to apply to a wide range of rock types.

4.4.3   Triaxial compression $(\sigma_1 > \sigma_2 = \sigma_3)$

This test is carried out on cylindrical specimens prepared in the same manner as those used for uniaxial compression tests. The specimen is placed inside a pressure vessel (Figures 4.16 and 4.17) and a fluid pressure, σ_3, is applied to its surface. A jacket, usually made of a rubber compound, is used to isolate the specimen from the confining fluid which is usually oil. The axial stress, σ_1, is applied to the specimen via a ram passing through a bush in the top of the cell and hardened steel end caps.

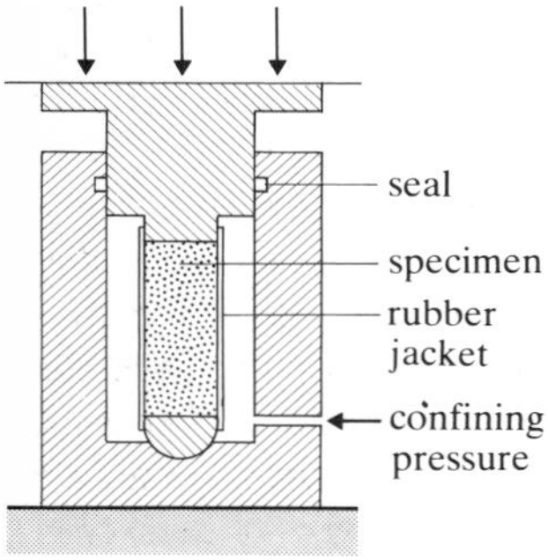

seal

specimen

rubber jacket

confining pressure

Figure 4.16 Elements of a conventional triaxial testing apparatus.

102

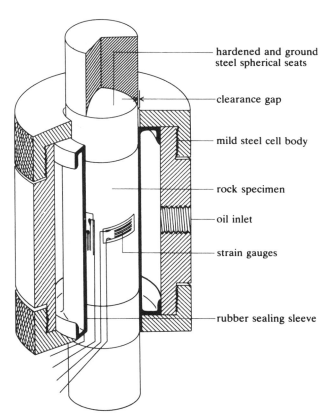

— hardened and ground steel spherical seats

— clearance gap

— mild steel cell body

— rock specimen

— oil inlet

— strain gauges

— rubber sealing sleeve

Figure 4.17 Cut-away view of the triaxial cell designed by Hoek and Franklin (1968). Because this cell does not require drainage between tests, it is well suited to carrying out large numbers of tests quickly.

Pore pressure, u, may be applied or measured through a duct which generally connects with the specimen through the base of the cell. Axial deformation of the specimen may be most conveniently monitored by linear variable differential transformers (LVDTs) mounted inside or outside the cell, but preferably inside. Local axial and circumferential strains may be measured by electric resistance strain gauges attached to the surface of the specimen (Figure 4.17).

It is necessary to have available for use with the triaxial cell a system for generating the confining pressure and keeping it constant throughout the test. If the confining pressure is generated by a screw-driven pressure intensifier, it is possible to use the displacement of the intensifier plunger to measure the volumetric strain of the specimen (Crouch, 1970). Figure 4.18 shows some results obtained using such a system in tests carried out at three different confining pressures on specimens of an oolitic limestone. An important feature of the behaviour of rock material in triaxial compression is illustrated by Figure 4.18. When the specimen is initially loaded it compresses, but a point is soon reached, generally before the peak of the axial stress–axial strain curve, at which the specimen begins to dilate (increase in volume) as a result of internal fracturing. Shortly after the peak strength is reached, the nett volumetric strain of the specimen becomes dilational. Dilation continues in the post-peak range. The amount of dilation decreases with increasing confining pressure. At very high confining pressures, often outside the range of engineering interest, dilation may be totally suppressed with the volumetric strains remaining contractile throughout the test.

103

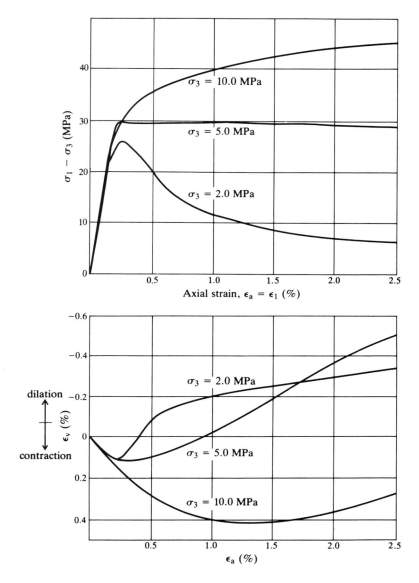

Figure 4.18 Results of triaxial compression tests on an oolitic limestone with volumetric strain measurement (after Elliott, 1982).

Figure 4.19 illustrates a number of other important features of the behaviour of rock in triaxial compression. The axial stress (σ_a)–axial strain (ε_a) data shown were obtained by Wawersik and Fairhurst (1970) for the Tennessee Marble giving the uniaxial stress–strain curve shown in Figure 4.11. These and similar data for other rocks show that, with increasing confining pressure,

(a) the peak strength increases;
(b) there is a transition from typically brittle to fully ductile behaviour with the introduction of plastic mechanisms of deformation including cataclastic flow and grain-sliding effects;
(c) the region incorporating the peak of the σ_a–ε_a curve flattens and widens;

104

Figure 4.19 Complete axial stress –axial strain curves obtained in triaxial compression tests on Tennessee Marble at the confining pressures indicated by the numbers on the curves (after Wawersik and Fairhurst, 1970).

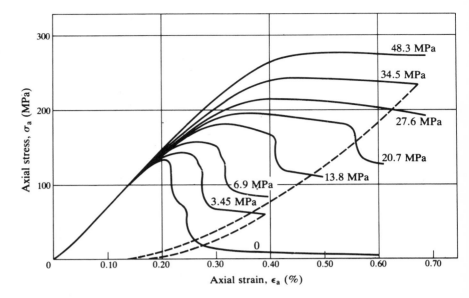

(d) the post-peak drop in stress to the residual strength reduces and disappears at high values of σ_3.

The confining pressure at which the post-peak reduction in strength disappears and the behaviour becomes fully ductile ($\sigma_3 = 48.3$ MPa in Figure 4.19), is known as the **brittle–ductile transition pressure** and varies with rock type. In general, the more siliceous igneous and metamorphic rocks such as granite and quartzite remain brittle at room temperature at confining pressures of up to 1000 MPa or more (Paterson, 1978). In these cases, ductile behaviour will not be of concern in practical mining problems.

The influence of pore-water pressure on the behaviour of porous rock in the triaxial compression test is illustrated by Figure 4.20. A series of triaxial compression tests was carried out on a limestone with a constant value of $\sigma_3 = 69$ MPa, but with various levels of pore pressure in the range $u = 0 - 69$ MPa applied. There is a transition from ductile to brittle behaviour as u is increased from 0 to 69 MPa. In this case, mechanical response is controlled by the effective confining pressure, $\sigma_3' = \sigma_3 - u$, calculated using Terzaghi's classical effective stress law. For less permeable rocks than this limestone, it may appear that the classical effective stress law does not hold. Brace and Martin (1968) conducted triaxial compression tests on a variety of crystalline silicate rocks of low porosity (0.001–0.03) at axial strain rates of $10^{-3} - 10^{-8}$ s^{-1}. They found that the classical effective stress law held only when the strain rate was less than some critical value which depended on the permeability of the rock, the viscosity of the pore fluid and the specimen geometry. At strain rates higher than the critical, static equilibrium could not be achieved throughout the specimen.

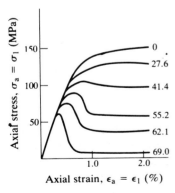

Figure 4.20 Effect of pore pressure (given in MPa by the numbers on the curves) on the stress–strain behaviour of a limestone tested at a constant confining pressure of 69 MPa (after Robinson, 1959).

4.4.4 Polyaxial compression ($\sigma_1 > \sigma_2 > \sigma_3$)

These tests may be carried out on cubes or rectangular prisms of rock with different normal stresses being applied to each pair of opposite faces. The difficulties caused by end effects are even more marked than in the comparable case of biaxial compression (section 4.4.2). By the addition of an internal fluid pressure, the hollow cylinder biaxial compression test may be converted into a polyaxial test. Hoskins (1969) gives a detailed account of such tests. However, the test also suffers from the difficulties noted for the hollow cylinder biaxial compression test.

The results of polyaxial compression tests on prismatic specimens are often conflicting, but generally indicate some influence of the intermediate principal stress, σ_2, on stress–strain behaviour. Generally, the peak strength increases with increasing σ_2 for constant σ_3, but the effect is not as great as that caused by increasing σ_3 by a similar amount (Paterson, 1978). However, doubts must remain about the uniformity of the applied stresses in these tests and the results should be interpreted with great care.

4.4.5 Influence of stress path

In the tests described in the preceding sections, it is usual for two of the principal stresses (σ_2 and σ_3) to be applied and held constant and for the other principal stress (σ_1) to be increased towards the peak strength. This stress path is not necessarily that which an element of rock influenced by an excavation will follow when the excavation is made.

As an example, consider a long excavation of circular cross section made in an elastic rock mass in which the *in situ* principal stresses were p vertically, p horizontally parallel to the axis of the excavation, and $0.5p$ horizontally perpendicular to the axis. Results to be presented in Chapter 7 show that on completion of the excavation, the principal stresses at mid-height on the boundary of the excavation change from $\sigma_1 = p$, $\sigma_2 = p$, $\sigma_3 = 0.5p$, to $\sigma_1 = 2.5p$, $\sigma_2 = (1+v)p$ where v is Poisson's ratio of the rock, and $\sigma_3 = 0$. As a result of excavation, two principal stresses are increased and the other decreased. It is necessary to determine, therefore, whether the behaviour described earlier is stress-path dependent or whether it is simply a function of the final state of stress.

A test of considerable relevance in this regard is the **triaxial extension test** which is carried out in a triaxial cell with the confining pressure, σ_r, greater than the axial stress, σ_a. The test may be commenced at $\sigma_a = \sigma_r$ with σ_a being progressively reduced so that $\sigma_r = \sigma_1 = \sigma_2 > \sigma_a = \sigma_3$. With modern servocontrolled testing machines, almost any desired total or effective stress path can be followed within the limitations imposed by the axisymmetric configuration of the triaxial cell. Swanson and Brown (1971) investigated the effect of stress path on the peak strength of a granite and a quartz diorite. They found that, for both rock types, the peak strengths in all tests fell on the same envelope (Figure 4.21 for Westerly Granite) irrespective of stress path. They also found that the onset of dilatancy, described in section 4.4.3, is stress-path independent. Similarly, Elliott (1982) found the yield locus of a high-porosity, oolitic limestone to be stress-path independent.

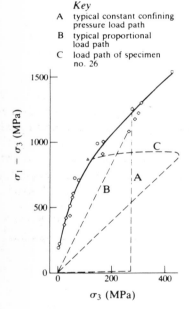

Key

A typical constant confining pressure load path
B typical proportional load path
C load path of specimen no. 26

Figure 4.21 Influence of stress path on the peak strength envelope for Westerly Granite (after Swanson and Brown, 1971).

4.5 Strength criteria for isotropic rock material

4.5.1 Types of strength criterion

A **peak strength criterion** is a relation between stress components which will permit the peak strengths developed under various stress combinations to be predicted. Similarly, a **residual strength criterion** may be used to predict residual strengths under varying stress conditions. In the same way, a **yield criterion** is a relation between stress components which is satisfied at the onset of permanent deformation. Given that effective stresses control the stress–strain behaviour of rocks, strength and yield criteria are best written in **effective stress** form. However, around most mining excavations, the pore-water pressures will be low, if not zero, and so $\sigma'_{ij} \simeq \sigma_{ij}$. For this reason, it is common in mining rock mechanics to use total stresses in the majority of cases and to use effective stress criteria only in special circumstances.

The data presented in the preceding sections indicate that the general form of the peak strength criterion should be

$$\sigma_1 = f(\sigma_2, \sigma_3) \tag{4.8}$$

This is sometimes written in terms of the shear, τ, and normal stresses, σ_n, on a particular plane in the specimen:

$$\tau = f(\sigma_n) \tag{4.9}$$

Because the available data indicate that the intermediate principal stress, σ_2, has less influence on peak strength than the minor principal stress, σ_3, all of the criteria used in practice are reduced to the form

$$\sigma_1 = f(\sigma_3) \tag{4.10}$$

4.5.2 Coulomb's shear strength criterion

In one of the classic papers of engineering science, Coulomb (1776) postulated that the shear strengths of rock and of soil are made up of two parts – a constant cohesion and a normal stress-dependent frictional component. (Actually, Coulomb presented his ideas and calculations in terms of forces; the differential concept of stress that we use today was not introduced until the 1820s.) Thus, the shear strength that can be developed on a plane such as *ab* in Figure 4.22 is

$$s = c + \sigma_n \tan \phi \tag{4.11}$$

where c = cohesion and ϕ = angle of internal friction.

Applying the stress transformation equations to the case shown in Figure 4.22 gives

$$\sigma_n = \tfrac{1}{2}(\sigma_1 + \sigma_3) + \tfrac{1}{2}(\sigma_1 - \sigma_3) \cos 2\beta$$

and

$$\tau = \tfrac{1}{2}(\sigma_1 - \sigma_3) \sin 2\beta$$

Substitution for σ_n and $s = \tau$ in equation 4.11 and rearranging gives the limiting stress condition on any plane defined by β as

$$\sigma_1 = \frac{2c + \sigma_3 \left[\sin 2\beta + \tan \phi \left(1 - \cos 2\beta \right) \right]}{\sin 2\beta - \tan \phi \left(1 + \cos 2\beta \right)} \tag{4.12}$$

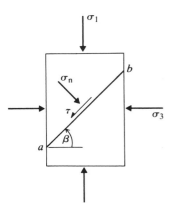

Figure 4.22 Shear failure on plane *ab*.

(a)

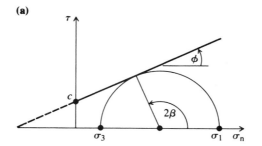

(b)

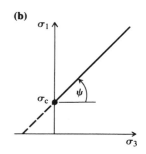

Figure 4.23 Coulomb strength envelopes in terms of (a) shear and normal stresses, and (b) principal stresses.

There will be a critical plane on which the available shear strength will be first reached as σ_1 is increased. The Mohr circle construction of Figure 4.23a gives the orientation of this critical plane as

$$\beta = \frac{\pi}{4} + \frac{\phi}{2} \tag{4.13}$$

This result may also be obtained by putting $d(s - \tau)/d\beta = 0$.

For the critical plane, $\sin 2\beta = \cos\phi$, $\cos 2\beta = -\sin\phi$, and equation 4.12 reduces to

$$\sigma_1 = \frac{2c\cos\phi + \sigma_3(1 + \sin\phi)}{1 - \sin\phi} \tag{4.14}$$

This linear relation between σ_3 and the peak value of σ_1 is shown in Figure 4.23b. Note that the slope of this envelope is related to ϕ by the equation

$$\tan\psi = \frac{1 + \sin\phi}{1 - \sin\phi} \tag{4.15}$$

and that the uniaxial compressive strength is related to c and ϕ by

$$\sigma_c = \frac{2c\cos\phi}{1 - \sin\phi} \tag{4.16}$$

If the Coulomb envelope shown in Figure 4.23b is extrapolated to $\sigma_1 = 0$, it will intersect the σ_3 axis at an apparent value of uniaxial tensile strength of the material given by

$$\sigma_T = \frac{2c\cos\phi}{1 + \sin\phi} \tag{4.17}$$

The measurement of the uniaxial tensile strength of rock is fraught with difficulty. However, when it is satisfactorily measured, it takes values that are generally lower than those predicted by equation 4.17. For this reason, a **tensile cutoff** is usually applied at a selected value of uniaxial tensile stress, T_0, as shown in Figure 4.24. For practical purposes, it is prudent to put $T_0 = 0$.

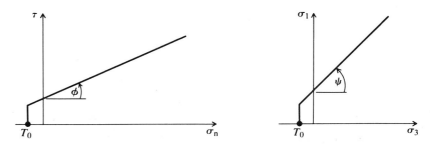

Figure 4.24 Coulomb strength envelopes with a tensile cut-off.

Although it is widely used, Coulomb's criterion is not a particularly satisfactory peak strength criterion for rock material. The reasons for this are:

(a) It implies that a major shear fracture exists at peak strength. Observations such as those made by Wawersik and Fairhurst (1970) show that this is not always the case.
(b) It implies a direction of shear failure which does not always agree with experimental observations.
(c) Experimental peak strength envelopes are generally non-linear. They can be considered linear only over limited ranges of σ_n or σ_3.

For these reasons, other peak strength criteria are preferred for intact rock. However, the Coulomb criterion can provide a good representation of residual strength conditions, and more particularly, of the shear strengths of discontinuities in rock (section 4.7).

4.5.3 Griffith crack theory

In another of the classic papers of engineering science, Griffith (1921) postulated that fracture of brittle materials, such as steel and glass, is initiated at tensile stress concentrations at the tips of minute, thin cracks (now referred to as Griffith cracks) distributed throughout an otherwise isotropic, elastic material. Griffith based his determination of the conditions under which a crack would extend on his **energy instability concept**:

A crack will extend only when the total potential energy of the system of applied forces and material decreases or remains constant with an increase in crack length.

For the case in which the potential energy of the applied forces is taken to be constant throughout, the criterion for crack extension may be written

$$\frac{\partial}{\partial c}(W_d - W_e) \leqslant 0 \tag{4.18}$$

where c is a crack length parameter, W_e is the elastic strain energy stored around the crack and W_d is the surface energy of the crack surfaces.

Griffith (1921) applied this theory to the extension of an elliptical crack of initial length $2c$ that is perpendicular to the direction of loading of a plate of unit thickness subjected to a uniform uniaxial tensile stress, σ. He found that the crack will extend when

109

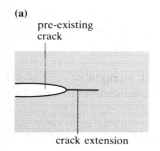

(a) pre-existing crack

crack extension

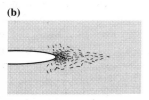

(b)

Figure 4.25 Extension of a pre-existing crack. (a) Griffith's hypothesis, (b) the actual case for rock.

$$\sigma \geqslant \sqrt{\frac{2E\alpha}{\pi c}} \qquad (4.19)$$

where α is the surface energy per unit area of the crack surfaces (associated with the rupturing of atomic bonds when the crack is formed), and E is the Young's modulus of the uncracked material.

It is important to note that it is the surface energy, α, which is the fundamental material property involved here. Experimental studies show that, for rock, a pre-existing crack does not extend as a single pair of crack surfaces, but a fracture zone containing large numbers of very small cracks develops ahead of the propagating crack (Figure 4.25). In this case, it is preferable to treat α as an **apparent surface energy** to distinguish it from the true surface energy which may have a significantly smaller value.

It is difficult, if not impossible, to correlate the results of different types of direct and indirect tensile test on rock using the average tensile stress in the fracture zone as the basic material property. For this reason, measurement of the 'tensile strength' of rock has not been discussed in this chapter. However, Hardy (1973) was able to obtain good correlation between the results of a range of tests involving tensile fracture when the apparent surface energy was used as the unifying material property.

Griffith (1924) extended his theory to the case of applied compressive stresses. Neglecting the influence of friction on the cracks which will close under compression, and assuming that the elliptical crack will propagate from the points of maximum tensile stress concentration (P in Figure 4.26), Griffith obtained the following criterion for crack extension in **plane compression**:

$$(\sigma_1 - \sigma_2)^2 - 8T_0(\sigma_1 + \sigma_2) = 0 \quad \text{if } \sigma_1 + 3\sigma_2 > 0$$

$$\sigma_2 + T_0 = 0 \quad \text{if } \sigma_1 + 3\sigma_2 < 0 \qquad (4.20)$$

where T_0 is the uniaxial tensile strength of the uncracked material (a positive number).

This criterion can also be expressed in terms of the shear stress, τ, and the normal stress, σ_n, acting on the plane containing the major axis of the crack:

$$\tau^2 = 4T_0(\sigma_n + T_0) \qquad (4.21)$$

The envelopes given by equations 4.20 and 4.21 are shown in Figure 4.27. Note that this theory predicts that the uniaxial compressive stress at crack extension will always be eight times the uniaxial tensile strength.

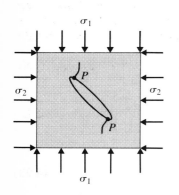

Figure 4.26 Griffith crack model for plane compression.

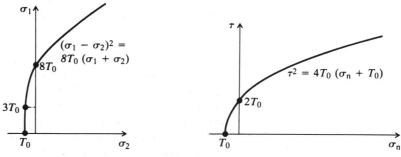

Figure 4.27 Griffith envelopes for crack extension in plane compression.

During the 1960s, a number of attempts were made to apply these results to the peak strength envelopes for rock. Quite often, σ_2 in the plane stress criterion was simply replaced by σ_3 so that the criterion could be applied to triaxial test results. For a number of very good reasons, the classic plane compression Griffith theory did not provide a very good model for the peak strength of rock under multiaxial compression. Accordingly, a number of modifications to Griffith's solution were introduced (see Paterson, 1978 and Jaeger and Cook, 1979 for details). These criteria do not find practical use today. However, Griffith's energy instability concept has formed the basis of the new science of **fracture mechanics** which is being applied increasingly to the study of the fracture of rock. As noted above, the use of this approach with an apparent surface energy taken as the basic material property has been able to explain many observations of apparent size effects and to reconcile the results of different types of indirect tension test on rock. The use of Griffith's essentially microscopic theory to predict the macroscopic behaviour of rock material under a variety of boundary conditions, requires the introduction of a set of Griffith crack size, shape and orientation distribution functions which have not yet been defined.

4.5.4 Empirical criteria

Because the classic strength theories used for other engineering materials have been found not to apply to rock over a wide range of applied compressive stress conditions, a number of empirical strength criteria have been introduced for practical use. These criteria usually take the form of a power law in recognition of the fact that peak σ_1 vs. σ_3 and τ vs. σ_n envelopes for rock material are generally concave downwards (Figures 4.21, 28 and 29). In order to ensure that the parameters used in the power laws are dimensionless, these criteria are best written in normalised form with all stress components being divided by the uniaxial compressive strength of the rock.

Bieniawski (1974) found that the peak triaxial strengths of a range of rock types were well represented by the criterion

$$\frac{\sigma_1}{\sigma_c} = 1 + A\left(\frac{\sigma_3}{\sigma_c}\right)^k \tag{4.22}$$

or

$$\frac{\tau_m}{\sigma_c} = 0.1 + B\left(\frac{\sigma_m}{\sigma_c}\right)^c \tag{4.23}$$

where $\tau_m = \frac{1}{2}(\sigma_1 - \sigma_3)$ and $\sigma_m = \frac{1}{2}(\sigma_1 + \sigma_3)$.

Bieniawski found that, for the range of rock types tested, $k \simeq 0.75$ and $c \simeq 0.90$. The corresponding values of A and B are given in Table 4.1. Note that both A and B take relatively narrow ranges for the rock types tested.

Brady (1977) studied the development of rock fracture around a bored raise in a pillar in mineralised shale in a trial stoping block at the Mount Isa Mine, Australia. Using a boundary element analysis to calculate the elastic stresses induced around the raise as the pillar was progressively mined, he found that fracture of the rock could be accurately modelled using equation 4.22 with $A = 3.0$, $k = 0.75$ and $\sigma_c = 90$ MPa which is approximately half the mean value of 170 MPa measured in laboratory tests.

Table 4.1 Constants in Bieniawski's empirical strength criterion (after Bieniawski, 1974).

Rock type	A	B
norite	5.0	0.8
quartzite	4.5	0.78
sandstone	4.0	0.75
siltstone	3.0	0.70
mudstone	3.0	0.70

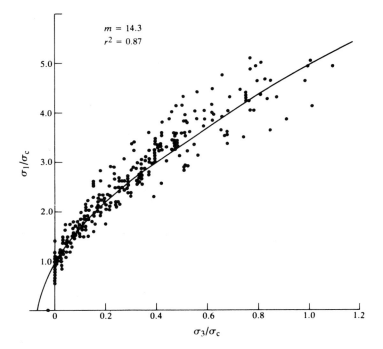

Figure 4.28 Normalised peak strength envelope for sandstones (after Hoek and Brown, 1980).

Hoek and Brown (1980) found that the peak triaxial compressive strengths of a wide range of isotropic rock materials could be described by the equation

$$\frac{\sigma_1}{\sigma_c} = \frac{\sigma_3}{\sigma_c} + \left(m\frac{\sigma_3}{\sigma_c} + 1.0 \right)^{1/2} \tag{4.24}$$

where m varies with rock type. Analysis of published strength data suggests that m increases with rock type in the following general way:

(a) $m \simeq 7$ for carbonate rocks with well developed crystal cleavage (dolomite, limestone, marble);

(b) $m \simeq 10$ for lithified argillaceous rocks (mudstone, siltstone, shale, slate);

(c) $m \simeq 15$ for arenaceous rocks with strong crystals and poorly developed crystal cleavage (sandstone, quartzite);

(d) $m \simeq 17$ for fine-grained polyminerallic igneous crystalline rocks (andesite, dolerite, diabase, rhyolite);

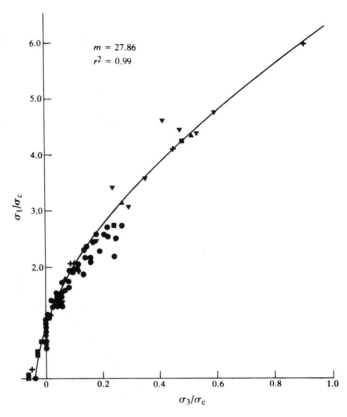

$m = 27.86$
$r^2 = 0.99$

Figure 4.29 Normalised peak strength envelope for granites (after Hoek and Brown, 1980).

(e) $m \simeq 25$ for coarse-grained polyminerallic igneous and metamorphic rocks (amphibolite, gabbro, gneiss, granite, norite, quartz-diorite).

Normalised peak strength envelopes for sandstones and granites are shown in Figures 4.28 and 4.29, respectively. The grouping and analysis of data according to rock type has obvious disadvantages. Detailed studies of rock strength and fracture indicate that factors such as mineral composition, grain size and angularity, grain packing patterns and the nature of cementing materials between grains, all influence the manner in which fracture initiates and propagates. If these factors are relatively uniform within a given rock type, then it might be expected that a single curve would give a good fit to the normalised strength data with a correspondingly high value of the coefficient of determination, r^2. Such a result is shown for granites in Figure 4.29. If, on the other hand, these factors are quite variable from one occurrence of a given rock type to another, then a wider scatter of data points and a poorer fit by a single curve might be anticipated. For sandstones (Figure 4.28) where grain size, porosity and the nature of the cementing material can vary widely, and for limestone which is a name given to a wide variety of carbonate rocks, the values of r^2 are, indeed, quite low.

Despite these difficulties and the sometimes arbitrary allocation of a particular name to a given rock, the results obtained by Hoek and Brown do serve an important practical purpose. By using the approximate value of m found to apply for a particular rock type, it may be possible to carry out preliminary design calculations on the basis of no testing other than a determination of a suitable value of σ_c made using a simple test such as the point load test. A value of σ_c is required as a scaling factor to determine the strength of a particular sample of rock. Thus although the same value of m may apply to granites from different localities, their strengths at different confining pressures may differ by a factor of two or three.

4.5.5 Yield criteria based on plasticity theory

The incremental theory of plasticity (Hill, 1950) is a branch of continuum mechanics that was developed in an attempt to model analytically the plastic deformation or flow of metals. Plastic deformation is permanent or irrecoverable; its onset marks the yield point. Perfectly plastic deformation occurs at constant volume under constant stress. If an increase in stress is required to produce further post-yield deformation, the material is said to be work- or strain-hardening.

As noted in section 4.4.3, plastic or dissipative mechanisms of deformation may occur in rocks under suitable environmental conditions. It would seem reasonable, therefore, to attempt to use plasticity theory to develop yield criteria for rocks. The relevant theory is beyond the scope of this introductory text and only the elements of it will be introduced here.

Because plastic deformation is accompanied by permanent changes in atomic positions, plastic strains cannot be defined uniquely in terms of the current state of stress. Plastic strains depend on loading history, and so plasticity theory must use an incremental loading approach in which incremental deformations are summed to obtain the total plastic deformation. In some engineering problems, the plastic strains are much larger than the elastic strains, which may be neglected. This is not always the case for rock deformation (for example, Elliott and Brown, 1985), and so an elastoplastic analysis may be required.

The total strain increment $\{\dot{\varepsilon}\}$ is the sum of the elastic and plastic strain increments

$$\{\dot{\varepsilon}\} = \{\dot{\varepsilon}^e\} + \{\dot{\varepsilon}^p\} \tag{4.25}$$

A plastic potential function, $Q(\{\sigma\})$, is defined such that

$$\{\dot{\varepsilon}^p\} = \lambda \left\{ \frac{\partial Q}{\partial \sigma} \right\} \tag{4.26}$$

where λ is a non-negative constant of proportionality which may vary throughout the loading history. Thus, from the incremental form of equation 2.38 and equations 4.25 and 4.26

$$\{\dot{\varepsilon}\} = [\mathbf{D}]^{-1} \{\dot{\sigma}\} + \lambda \left\{ \frac{\partial Q}{\partial \sigma} \right\} \tag{4.27}$$

where $[\mathbf{D}]$ is the elasticity matrix.

It is also necessary to be able to define the stress states at which yield will occur and plastic deformation will be initiated. For this purpose, a yield function, $F(\{\sigma\})$, is defined such that $F = 0$ at yield. If $Q = F$, the flow law is said to be

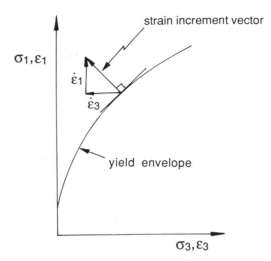

Figure 4.30 The normality condition of the associated flow rule.

associated. In this case, the vectors of $\{\sigma\}$ and $\{\dot{\varepsilon}^p\}$ are orthogonal as illustrated in Figure 4.30. This is known as the normality condition.

For isotropic hardening and associated flow, elastoplastic stress and strain increments may be related by the equation

$$\{\dot{\sigma}\} = [\mathbf{D}^{ep}] \, [\dot{\varepsilon}]$$

where

$$[\mathbf{D}^{ep}] = [\mathbf{D}] - \frac{[\mathbf{D}] \left\{\dfrac{\partial Q}{\partial \sigma}\right\} \left\{\dfrac{\partial F}{\partial \sigma}\right\}^{\mathrm{T}} [\mathbf{D}]}{A + \left\{\dfrac{\partial F}{\partial \sigma}\right\}^{\mathrm{T}} [\mathbf{D}] \left\{\dfrac{\partial Q}{\partial \sigma}\right\}}$$

in which

$$A = -\frac{1}{\lambda} \frac{\partial F}{\partial K} \, \mathrm{d}K$$

where K is a hardening parameter such that yielding occurs when

$$\mathrm{d}F = \left\{\frac{\partial F}{\partial \sigma}\right\}^{\mathrm{T}} \{\dot{\sigma}\} + \frac{\partial F}{\partial K} = 0$$

The concepts of associated plastic flow were developed for perfectly plastic and strain-hardening metals using yield functions such as those of Tresca and von Mises which are independent of the hydrostatic component of stress (Hill, 1950). Although these concepts have been found to apply to some geological materials, it cannot be assumed that they will apply to pressure-sensitive materials such as rocks in which brittle fracture and dilatancy typically occur (Rudnicki and Rice, 1975).

In order to obtain realistic representations of the stresses at yield in rocks and rock masses, it has been necessary to develop yield functions which are more complex than the classical functions introduced for metals. These functions are often of the form $F(I_1, J_2) = 0$ where I_1 is the first invariant of the stress tensor and J_2 is the second invariant of the deviator stress tensor (section 2.4), i.e.

$$J_2 = \tfrac{1}{2}(S_1^2 + S_2^2 + S_3^2)$$

$$= \tfrac{1}{6}[(\sigma_1 - \sigma_2)^2 + (\sigma_2 - \sigma_3)^2 + (\sigma_3 - \sigma_1)^2]$$

More complex functions also include the third invariant of the deviator stress tensor $J_3 = S_1 S_2 S_3$. For example, Desai and Salami (1987) were able to obtain excellent fits to peak strength (assumed synonymous with yield) and stress–strain data for a sandstone, a granite and a dolomite using the yield function

$$F = J_2 - \left(\frac{\alpha}{\alpha_0^{n-2}} I_1^n + I_1^2\right)\left(1 - \beta \frac{J_3^{1/3}}{J_2^{1/2}}\right)^m$$

where α, n, β and m are material parameters and α_0 is one unit of stress.

4.6 Strength of anisotropic rock material in triaxial compression

So far in this chapter, it has been assumed that the mechanical response of rock material is isotropic. However, because of some preferred orientation of the fabric or microstructure, or the presence of bedding or cleavage planes, the behaviour of many rocks is anisotropic. The various categories of anisotropic elasticity were discussed in section 2.10. Because of computational complexity and the difficulty of determining the necessary elastic constants, it is usual for only the simplest form of anisotropy, transverse isotropy, to be used in design analyses. Anisotropic strength criteria are also required for use in the calculations.

The peak strengths developed by transversely isotropic rocks in triaxial compression vary with the orientation of the plane of isotropy, foliation plane or plane of weakness, with respect to the principal stress directions. Figure 4.31 shows some measured variations in peak principal stress difference with the angle of inclination of the major principal stress to the plane of weakness.

Jaeger (1960) introduced an instructive analysis of the case in which the rock contains well-defined, parallel planes of weakness whose normals are inclined at an angle β to the major principal stress direction as shown in Figure 4.32a. Each plane of weakness has a limiting shear strength defined by Coulomb's criterion

$$s = c_w + \sigma_n \tan \phi_w \qquad (4.28)$$

Slip on the plane of weakness (*ab*) will become incipient when the shear stress on the plane, τ, becomes equal to, or greater than, the shear strength, s. The stress transformation equations give the normal and shear stresses on *ab* as

$$\sigma_n = \tfrac{1}{2}(\sigma_1 + \sigma_3) + \tfrac{1}{2}(\sigma_1 - \sigma_3)\cos 2\beta$$

and

$$\tau = \tfrac{1}{2}(\sigma_1 - \sigma_3)\sin 2\beta \qquad (4.29)$$

116

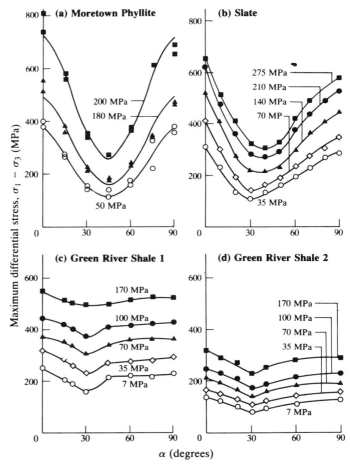

Figure 4.31 Variation of peak principal stress difference with the angle of inclination of the major principal stress to the plane of weakness, for the confining pressures indicated for (a) a phyllite (after Donath, 1972), (b–d) a slate and two shales (after McLamore and Gray, 1967).

Substituting for σ_n in equation 4.28, putting $s = \tau$, and rearranging, gives the criterion for slip on the plane of weakness as

$$(\sigma_1 - \sigma_3)_s = \frac{2(c_w + \sigma_3 \tan \phi_w)}{(1 - \tan \phi_w \cot \beta) \sin 2\beta} \tag{4.30}$$

The principal stress difference required to produce slip tends to infinity as $\beta \to 90°$ and as $\beta \to \phi_w$. Between these values of β, slip on the plane of weakness is possible, and the stress at which slip occurs varies with β according to equation 4.30. By differentiation, it is found that the minimum strength occurs when

$$\tan 2\beta = -\cot \phi_w$$

or when

$$\beta = \frac{\pi}{4} + \phi_w / 2$$

The corresponding value of the principal stress difference is

$$(\sigma_1 - \sigma_3)_{min} = 2(c_w + \mu_w \sigma_3) \left([1 + \mu_w^2]^{1/2} + \mu_w \right)$$

117

where $\mu_w = \tan \phi_w$.

For values of β approaching $90°$ and in the range $0°$ to ϕ_w, slip on the plane of weakness cannot occur, and so the peak strength of the specimen for a given value of σ_3, must be governed by some other mechanism, probably shear fracture through the rock material in a direction not controlled by the plane of weakness. The variation of peak strength with the angle β predicted by this theory is illustrated in Figure 4.31b.

Note that the peak strength curves shown in Figure 4.31, although varying with β and showing pronounced minima, do not take the same shape as Figure 4.32b. (In comparing these two figures note that the abscissa in Figure 4.31 is $\alpha = \pi/2 - \beta$). In particular, the plateau of constant strength at low values of α, or high values of α, predicted by the theory, is not always present in the experimental strength data. This suggests that the two-strength model of Figure 4.32 provides an oversimplified representation of strength variation in anisotropic rocks. Such observations led Jaeger (1960) to propose that the shear strength parameter, c_w, is not constant but is continuously variable with β or α. McLamore and Gray (1967) subsequently proposed that both c_w and $\tan \phi_w$ vary with orientation according to the empirical relations

$$c_w = A - B[\cos 2(\alpha - \alpha_{c0})]^n$$

and

$$\tan \phi_w = C - D[\cos 2(\alpha - \alpha_{\phi 0})]^m$$

where A, B, C, D, m and n are constants, and α_{c0} and $\alpha_{\phi 0}$ are the values of α at which c_w and ϕ_w take minimum values, respectively.

4.7 Shear behaviour of discontinuities

4.7.1 Shear testing

In mining rock mechanics problems other than those involving only fracture of intact rock, the shear behaviour of discontinuities will be important. Conditions for slip on major pervasive features such as faults or for the sliding of individual blocks from the boundaries of excavations are governed by the shear strengths that can be developed by the discontinuities concerned. In addition, the shear and normal stiffnesses of discontinuities can exert a controlling influence on the distribution of stresses and displacements within a discontinuous rock mass. These properties can be measured in the same tests as those used to determine discontinuity shear strengths.

The most commonly used method for the shear testing of discontinuities in rock is the **direct shear test**. As shown in Figure 4.33, the discontinuity surface is aligned parallel to the direction of the applied shear force. The two halves of the specimen are fixed inside the shear box using a suitable encapsulating material, generally an epoxy resin or plaster. This type of test is commonly carried out in the laboratory, but it may also be carried out in the field, using a **portable shear box** to test discontinuities contained in pieces of drill core or as an *in situ* test on samples of larger size. Methods of preparing samples and carrying out

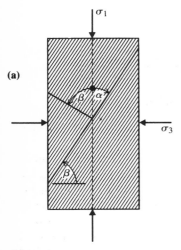

(a)

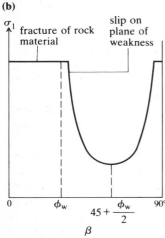

(b)

Figure 4.32 (a) Transversely isotropic specimen in triaxial compression; (b) variation of peak strength at constant confining pressure with the angle of inclination of the normal to the plane of weakness to the compression axis (β).

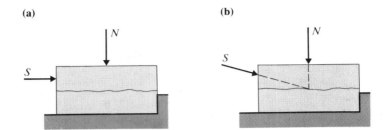

Figure 4.33 Direct shear test configurations with (a) the shear force applied parallel to the discontinuity, (b) an inclined shear force.

these various tests are discussed by the ISRM Commission (1974), Goodman (1976) and Hoek and Bray (1981).

Test arrangements of the type shown in Figure 4.33a can cause a moment to be applied about a lateral axis on the discontinuity surface. This produces relative rotation of the two halves of the specimen and a non-uniform distribution of stress over the discontinuity surface. To minimise these effects, the shear force may be inclined at an angle (usually $10°–15°$) to the shearing direction as shown in Figure 4.33b. This is almost always done in the case of large-scale *in situ* tests. Because the mean normal stress on the shear plane increases with the applied shear force up to peak strength, it is not possible to carry out tests in this configuration at very low normal stresses.

Direct shear tests in the configuration of Figure 4.33a are usually carried out at constant normal force or constant normal stress. Tests are most frequently carried out on dry specimens, but many shear boxes permit specimens to be submerged and drained shear tests to be carried out with excess joint water pressures being assumed to be fully dissipated throughout the test. Undrained testing with the measurement of induced joint water pressures, is generally not practicable using the shear box.

The **triaxial cell** is sometimes used to investigate the shear behaviour of discontinuities. Specimens are prepared from cores containing discontinuities inclined at $25–40°$ to the specimen axis. A specimen is set up in the triaxial cell as shown in Figure 4.32a for the case of anisotropic rocks, and the cell pressure and the axial load are successively applied. The triaxial cell is well suited to testing discontinuities in the presence of water. Tests may be either drained or undrained, preferably with a known level of joint water pressure being imposed and maintained throughout the test.

It is assumed that slip on the discontinuity will occur according to the theory set out in section 4.6. Mohr circle plots are made of the total or effective stresses at slip at a number of values of σ_3, and the points on these circles giving the stresses on the plane of the discontinuity are identified. The required shear strength envelope is then drawn through these points. This requires that a number of tests be carried out on similar discontinuities.

In an attempt to overcome the need to obtain, prepare and set up several specimens containing similar discontinuities, a **stage testing** procedure is sometimes used. A specimen is tested at a low confining pressure as outlined above. When it appears that slip on the discontinuity has just been initiated (represented by a flattening of the axial load–axial displacement curve that must be continuously recorded throughout each test), loading is stopped, the cell pressure is increased to a new value, and loading is recommenced. By repeating this process

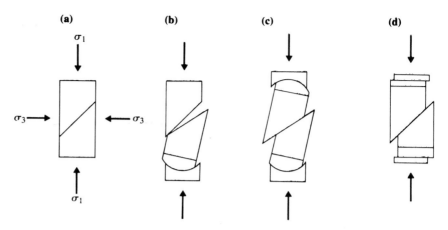

Figure 4.34 Discontinuity shear testing in a triaxial cell (after Jaeger and Rosengren, 1969).

several times, a number of points on the peak strength envelope of the discontinuity can be obtained from the one specimen. However, this approach exacerbates the major difficulty involved in using the triaxial test to determine discontinuity shear strengths, namely the progressive change in the geometry of the cell–specimen system that accompanies shear displacement on the discontinuity.

The problem is illustrated by Figure 4.34. It is clear from Figure 4.34a that, if relative shear displacement of the two parts of the specimen is to occur, there must be lateral as well as axial relative translation. If, as is often the case, one spherical seat is used in the system, axial displacement causes the configuration to change to that of Figure 4.34b, which is clearly unsatisfactory. As shown in Figure 4.34c, the use of two spherical seats allows full contact to be maintained over the sliding surfaces, but the area of contact changes and frictional and lateral forces are introduced at the seats. Figure 4.34d illustrates the most satisfactory method of ensuring that the lateral component of translation can occur freely and that contact of the discontinuity surfaces is maintained. Pairs of hardened steel discs are inserted between the platens and either end of the specimen. No spherical seats are used. The surfaces forming the interfaces between the discs are polished and lubricated with a molybdenum disulphide grease. In this way, the coefficient of friction between the plates can be reduced to the order of 0.005 which allows large amounts of lateral displacement to be accommodated at the interface with little resistance.

This technique was developed by Rosengren (1968) who determined the corrections required to allow for the influence of friction and the change of contact area. His analysis has been re-presented by Goodman (1976) and will not be repeated here. The authors have successfully used this technique in tests on specimens of 150 mm diameter tested at confining pressures of up to 70 MPa.

4.7.2 Influence of surface roughness on shear strength
Shear tests carried out on smooth, clean discontinuity surfaces at constant normal stress generally give shear stress–shear displacement curves of the type shown in Figure 4.35. When a number of such tests are carried out at a range of effective normal stresses, a linear shear strength envelope is obtained (Figure 4.36). Thus the shear strength of smooth, clean discontinuities can be described by the simple Coulomb law

120

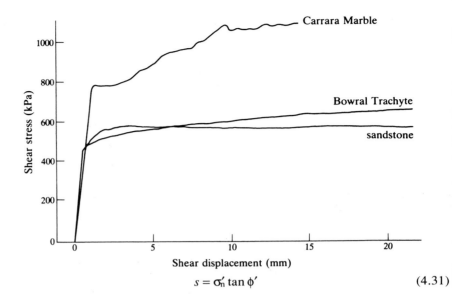

Figure 4.35 Shear stress–shear displacement curves for ground surfaces tested with a constant normal stress of 1.0 MPa (after Jaeger, 1971).

$$s = \sigma_n' \tan \phi' \qquad (4.31)$$

where ϕ' is the effective angle of friction of the discontinuity surfaces. For the case shown in Figure 4.36, $\phi' = 35°$, a typical value for quartz-rich rocks.

Naturally occurring discontinuity surfaces are never as smooth as the artificially prepared surfaces which gave the results shown in Figures 4.35 and 4.36. The shear force–shear displacement curve shown in Figure 4.37a is typical of the results obtained for clean, rough discontinuities. The peak **strength** at constant normal stress is reached after a small shear displacement. With further displacement, the shear resistance falls until the **residual strength** is eventually reached.

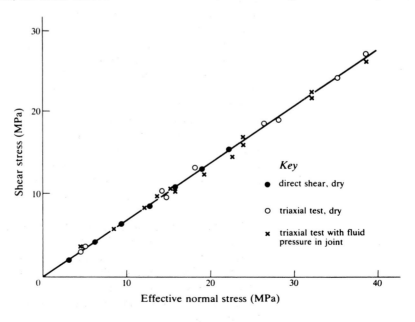

Figure 4.36 Sliding of smooth quartzite surfaces under various conditions (after Jaeger and Rosengren, 1969).

121

Figure 4.37 Results of a direct shear test on a 127 mm × 152 mm graphite-coated joint, carried out at a constant normal force of 28.9 kN. (a) Shear force–shear displacement curves; (b) surface profile contours before testing (mm); (c) relative positions on a particular cross section after 25 mm of sliding (after Jaeger, 1971).

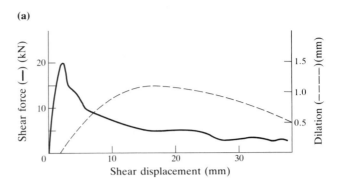

(a)

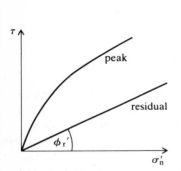

(b)

(c)

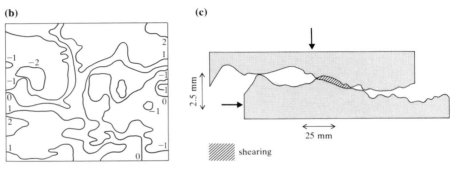

Figure 4.38 Peak and residual effective stress shear strength envelopes.

Tests at a number of normal stresses give peak and residual strength envelopes such as those shown in Figure 4.38.

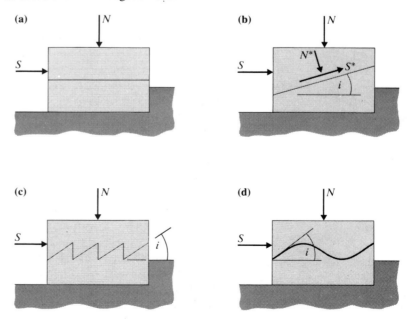

Figure 4.39 Idealised surface roughness models illustrating the roughness angle, i.

122

This behaviour can be explained in terms of surface roughness using a simple model introduced by Patton (1966) (Figure 4.39). A smooth, clean, dry discontinuity surface has a friction angle ϕ, so that at limiting equilibrium for the direct shear test configuration of Figure 4.39a,

$$\frac{S}{N} = \tan \phi$$

If the discontinuity surface is inclined at an angle i to the direction of the shear force, S (Figure 4.39b), then slip will occur when the shear and normal forces on the discontinuity, S^* and N^*, are related by

$$\frac{S^*}{N^*} = \tan \phi \qquad (4.32)$$

Resolving S and N in the direction of the discontinuity surface gives

$$S^* = S \cos i - N \sin i$$

and

$$N^* = N \cos i + S \sin i$$

Substitution of these values in equation 4.32 and simplification gives the condition for slip as

$$\frac{S}{N} = \tan (\phi + i) \qquad (4.33)$$

Thus the inclined discontinuity surface has an apparent friction angle of $(\phi + i)$. Patton extended this model to include the case in which the discontinuity surface contains a number of 'teeth' (Figure 4.39c and d). In a series of model experiments with a variety of surface profiles, he found that, at low values of N, sliding on the inclined surfaces occurred according to equation 4.33. Dilation of the specimens necessarily accompanied this mechanism. As the value of N was increased above some critical value, sliding on the inclined asperity surfaces was inhibited, and a value of S was eventually reached at which shear failure through the asperities occurred. The corresponding values of S and N gave the upper portion of the bilinear shear strength envelope shown in Figure 4.40. Note that, in such cases, the shear strengths that can be developed at low normal loads can be seriously overestimated by extrapolating the upper curve back to $N = 0$ and using a Coulomb shear strength law with a cohesion intercept, c, and a friction angle, ϕ_r.

Natural discontinuities rarely behave in the same way as these idealised models. However, the same two mechanisms – sliding on inclined surfaces at low normal loads and the suppression of dilation and shearing through asperities at higher normal loads – are found to dominate natural discontinuity behaviour. Generally, the two mechanisms are combined in varying proportions with the result that peak shear strength envelopes do not take the idealised bilinear form of Figure 4.40 but are curved. These combined effects are well illustrated by the direct shear test on a graphite-coated joint which gave the results shown in Figure 4.37a. The roughness profile of the initially mating surfaces is shown in Figure 4.37b. The maximum departure from the mean plane over the 127 mm × 152 mm

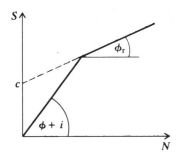

Figure 4.40 Bilinear peak strength envelope obtained in direct shear tests on the models shown in Figure 4.39.

123

surface area was in the order of ± 2.0 mm. After 25 mm of shear displacement at a constant normal force of 28.9 kN, the relative positions of the two parts of the specimen were as shown in Figure 4.37c. Both riding up on asperities and shearing off of some material in the shaded zone took place.

Roughness effects can cause shear strength to be a directional property. Figure 4.41 illustrates a case in which rough discontinuity surfaces were prepared in slate specimens by fracturing them at a constant angle to the cleavage. When the specimens were tested in direct shear with the directions of the ridges on the surfaces parallel to the direction of sliding (test A), the resulting shear strength envelope gave an effective friction angle of 22° which compares with a value of 19.5° obtained for clean, polished surfaces. However, when the shearing direction was normal to the ridges (test B), sliding up the ridges occurred with attendant dilation. A curved shear strength envelope was obtained with a roughness angle of 45.5° at near zero effective normal stress and a roughness angle of 24° at higher values of effective normal stress.

Figure 4.41 Effect of shearing direction on the shear strength of a wet discontinuity in a slate (after Brown *et al.*, 1977).

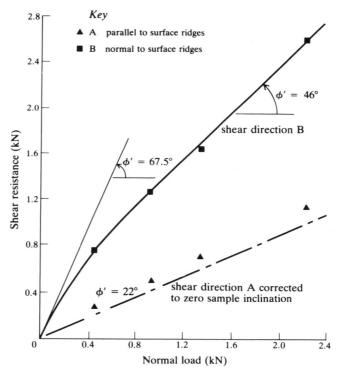

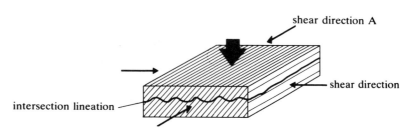

4.7.3 Interrelation between dilatancy and shear strength

All of the test data presented in the previous section were obtained in direct shear tests carried out at constant normal force or stress. Because of the influence of surface roughness, dilatancy accompanies shearing of all but the smoothest discontinuity surfaces in such tests. Goodman (1976) pointed out that although this test may reproduce discontinuity behaviour adequately in the case of sliding of an unconstrained block of rock from a slope (Figure 4.42c), it may not be suited to the determination of the stress–displacement behaviour of discontinuities isolating a block that may potentially slide or fall from the periphery of an underground excavation (Figure 4.42d). In the former case, dilation is permitted to occur freely, but in the latter case, dilation may be inhibited by the surrounding rock and the normal stress may increase with shear displacement.

When laboratory specimens in the configuration of Figure 4.42a are subjected to a shear stress, τ, parallel to the discontinuity, they can undergo shear and normal displacements, u and v, respectively. When a normal compressive stress, σ_n, is applied, the discontinuity will compress. This compressive stress–displacement behaviour is highly non-linear (Figure 4.43a) and at high values of σ_n, becomes asymptotic to a maximum closure, V_{mc}, related to the initial thickness or aperture of the discontinuity.

Suppose that a clean, rough discontinuity is sheared with no normal stress applied. Dilatancy will occur as shown in the upper curve of Figure 4.43b. If the shear resistance is assumed to be solely frictional, the shear stress will be zero throughout. For successively higher values of constant normal stress, A, B, C and D, the initial normal displacement will be a, b, c and d as shown in Figure 4.43a, and the dilatancy–shear displacement and shear stress–shear displacement curves obtained during shearing will be as shown in Figures 4.43b and c. As the normal stress is increased, the amount of dilatancy will decrease because a greater proportion of the asperities will be damaged during shearing.

Figure 4.42 Controlled normal force (a, c) and controlled normal displacement (b, d) shearing modes.

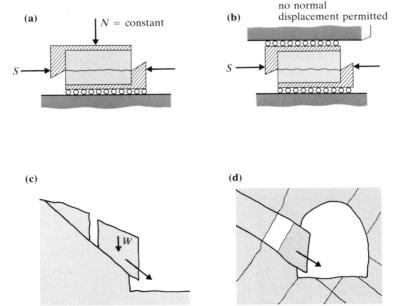

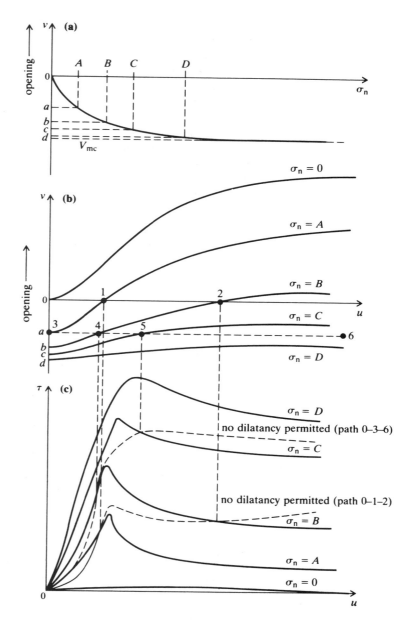

Figure 4.43 Relations between normal stress (σ_n), shear stress (τ), normal displacement (v), and shear displacement (u) in constant displacement shear tests on rough discontinuities (after Goodman, 1980).

Now suppose that a test is carried out on the same specimen with the normal stress initially zero and no dilation permitted during shearing (i.e. $v = 0$ throughout). By the time a shear displacement corresponding to point 1 in Figure 4.43b is reached, a normal stress of $\sigma_n = A$ will have been acquired, and the shear resistance will be that given by the τ–u curve for $\sigma_n = A$ at $u = u_1$. As shearing progresses, the shear stress will increase according to the dashed locus 0–1–2 in Figure 4.43c. If the discontinuity is initially compressed to point 3 in Figure 4.43b by a normal stress $\sigma_n = A$, and shearing then occurs with no further normal displacement being permitted (i.e. $v = a$ throughout), then the τ–u curve followed will be that given by the locus 3–4–5–6 in Figure 4.43c. Note that, in both cases, considerable increases in shear strength accompany shearing without dilatancy, and that the τ–u behaviour is no longer strain softening as it was for constant normal stress tests. This helps explain why limiting dilation on discontinuities by rock bolt, dowel and cable reinforcement (Chapter 11), can stabilise excavations in discontinuous rock.

4.7.4 Influence of scale

As was noted in section 3.3, discontinuity roughness may exist on a number of scales. Figure 3.10 illustrated the different scales of roughness sampled by different scales of shear test. For tests in which dilation is permitted, the roughness angle and, therefore, the apparent friction angle, decrease with increasing scale. For tests in which dilation is inhibited, the influence of scale is less important.

Barton (1973) proposed that the peak shear strengths, of joints, τ, in rock could be represented by the empirical relation

$$\tau = \sigma'_n \tan\left[JRC \log_{10}\left(\frac{JCS}{\sigma'_n}\right) + \phi'_r \right] \qquad (4.34)$$

where σ'_n = effective normal stress, JRC = joint roughness coefficient on a scale of 1 for the smoothest to 20 for the roughest surfaces, JCS = joint wall compressive strength and ϕ'_r = drained, residual friction angle.

Equation 4.34 suggests that there are three components of shear strength – a basic frictional component given by ϕ'_r, a geometrical component controlled by surface roughness (JRC) and an asperity failure component controlled by the ratio (JCS/σ'_n). As Figure 4.44 shows, the asperity failure and geometrical components combine to give the nett roughness component, $i°$. The total frictional resistance is then given by $(\phi'_r + i)°$.

Figure 4.44 Influence of scale on the three components of discontinuity shear strength (after Bandis *et al.*, 1981).

Equation 4.34 and Figure 4.44 show that the shear strength of a rough joint is both scale dependent and stress dependent. As σ'_n increases, the term $\log_{10}(JCS/\sigma'_n)$ decreases, and so the nett apparent friction angle decreases. As the

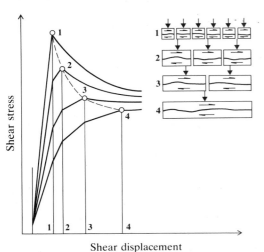

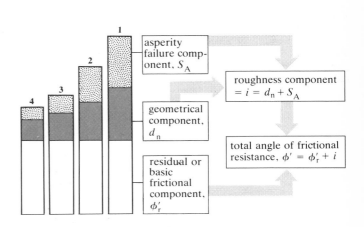

scale increases, the steeper asperities shear off and the inclination of the controlling roughness decreases. Similarly, the asperity failure component of roughness decreases with increasing scale because the material compressive strength, *JCS*, decreases with increasing size as discussed in section 4.3.5.

4.7.5 Infilled discontinuities

The previous discussion referred to 'clean' discontinuities or discontinuities containing no infilling materials. As noted in section 3.3, discontinuities may contain infilling materials such as gouge in faults, silt in bedding planes, low-friction materials such as chlorite, graphite and serpentine in joints, and stronger materials such as quartz or calcite in veins or healed joints. Clearly, the presence of these materials will influence the shear behaviour of discontinuities. The presence of gouge or clay seams can decrease both stiffness and shear strength. Low-friction materials such as chlorite, graphite and serpentine can markedly decrease friction angles, while vein materials such as quartz can serve to increase shear strengths.

Of particular concern is the behaviour of major infilled discontinuities in which the infilling materials are soft and weak, having similar mechanical properties to clays and silts. The shear strengths of these materials are usually described by an effective stress Coulomb law. In a laboratory study of such filled discontinuities, Ladanyi and Archambault (1977) reached the following conclusions:

(a) For most filled discontinuities, the peak strength envelope is located between that for the filling and that for a similar clean discontinuity.
(b) The stiffnesses and shear strength of a filled discontinuity decrease with increasing filling thickness, but always remain higher than those of the filling alone.
(c) The shear stress–displacement curves of filled discontinuities often have two portions, the first reflecting the deformability of the filling materials before rock to rock contact is made, and the second reflecting the deformability and shear failure of rock asperities in contact.
(d) The shear strength of a filled discontinuity does not always depend on the thickness of the filling. If the discontinuity walls are flat and covered with a low-friction material, the shear surface will be located at the filling-rock contact.
(e) Swelling clay is a dangerous filling material because it loses strength on swelling and can develop high swelling pressures if swelling is inhibited.

4.8 Models of discontinuity strength and deformation

In section 4.7, discussion was concentrated on the factors influencing the peak and residual shear strengths of discontinuities. When the responses of discontinuous rock masses are modelled using numerical methods such as joint-element finite element or distinct element methods (Chapter 6) it is also necessary that the shear and normal displacements on discontinuities be considered. The shear and normal stiffnesses of discontinuities can exert controlling influences on the distribution of stresses and displacements within a discontinuous rock mass. Three disconti-

128

nuity strength and deformation models of varying complexity will be discussed here. For simplicity, the discussion is presented in terms of total stresses.

4.8.1 The Coulomb friction, linear deformation model

The simplest coherent model of discontinuity deformation and strength is the Coulomb friction, linear deformation model illustrated in Figure 4.45. Under normal compressive loading, the discontinuity undergoes linear elastic closure up to a limiting value of Δv_m (Figure 4.45a). The discontinuity separates when the normal stress is less than the discontinuity tensile strength, usually taken as zero. For shear loading (Figure 4.45b), shear displacement is linear and reversible up to a limiting shear stress (determined by the value of the normal stress), and then perfectly plastic. Shear load reversal after plastic yield is accompanied by permanent shear displacement and hysteresis. The relation between limiting shear resistance and normal stress is given by equation 4.25.

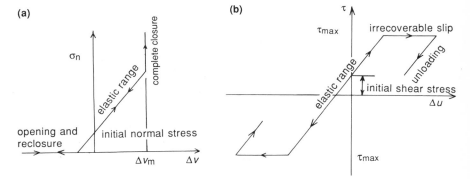

Figure 4.45 Coulomb friction, linear deformation joint model; (a) normal stress (σ_n)–normal closure (Δv) relation; (b) shear deformation (τ)–shear displacement (Δu) relation.

This model may be appropriate for smooth discontinuities such as faults at residual strength, which are non-dilatant in shear. The major value of the model is that it provides a useful and readily implemented reference case for static discontinuity response.

4.8.2 The Barton–Bandis model

The data presented in section 4.7 expressed the non-linear nature of the mechanical responses of rough discontinuities in rock. The effects of surface roughness on discontinuity deformation and strength have been described by Bandis *et al.* (1983, 1985) and Barton *et al.* (1985) in terms of a series of empirical relations between stress and deformation components and the parameters joint roughness coefficient, JRC, and joint wall compressive strength, JCS, introduced in equation 4.34.

The Barton–Bandis discontinuity closure model incorporates hyperbolic loading and unloading curves (Figure 4.46a) in which normal stress and closure, Δv, are related by the empirical expression

$$\sigma_n = \Delta v\,(a - b\,\Delta v) \tag{4.35}$$

where a and b are constants. The initial normal stiffness of the joint, K_{ni}, is equal to the inverse of a and the maximum possible closure, v_m, is defined by the asymptote a/b.

129

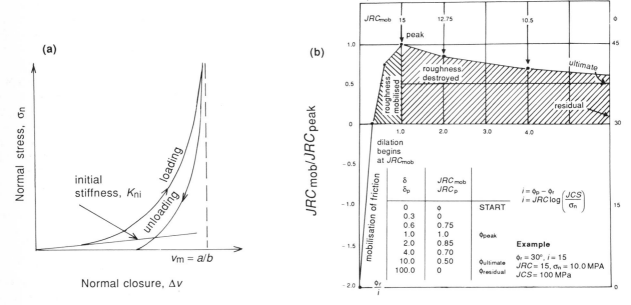

Figure 4.46 The Barton–Bandis model: (a) normal stress–normal closure relation; (b) example of piecewise linear shear deformation simulation (after Barton *et al.*, 1985).

Differentiation of equation 4.35 with respect to Δv yields the expression for normal stiffness

$$K_n = K_{ni} \left[1 - \sigma_n /(v_m K_{ni} + \sigma_n)\right]^{-2}$$

which shows the normal stiffness to be highly dependent on normal stress.

To provide estimates of joint initial stiffness and closure, Bandis *et al.* (1985) present the empirical relations

$$K_{ni} = 0.02 \, (JCS_0/E_0) + 2.0 \, JRC_0 - 10$$

$$v_m = A + B \, (JRC_0) + C \, (JCS_0/E_0)^D$$

where JCS_0 and JRC_0 are laboratory scale values, E_0 is the initial aperture of the discontinuity, and A, B, C and D are constants which depend on the previous stress history.

The peak shear strength is given by equation 4.34. The gradual reduction in shear strength during post-peak shearing is caused by a decline in the effective contribution of roughness due to mismatch and wear. This behaviour is modelled by using different values for the roughness coefficient, JRC_{mob}, that will be mobilised at any given value of shear displacement, u. A set of empirical relations between u, JRC, JCS, σ_n, the mobilised dilation angle and the size of the discontinuity permits the shear stress–shear displacement curve to be modelled in piecewise linear form (Figure 4.46b).

The Barton–Bandis model takes explicit account of more features of discontinuity strength and deformation behaviour than the elementary model discussed in section 4.8.1. However, its practical application may present some difficulties. In particular, the derivation of relations for the mobilisation and degradation of surface roughness from a piecewise linear graphical format rather than from a well-behaved formal expression may lead to some irregularities in numerical simulation of the stress–displacement behaviour.

4.8.3 The continuous-yielding joint model

The continuous-yielding joint model was designed to provide a coherent and unified discontinuity deformation and strength model, taking account of non-linear compression, non-linearity and dilation in shear, and a non-linear limiting shear strength criterion. Details of the formulation of the model are given by Cundall and Lemos (1988).

The key elements of the model are that all shear displacement at a discontinuity has a component of plastic (irreversible) displacement, and all plastic displacement results in progressive reduction in the mobilised friction angle. The displacement relation is

$$\Delta u^{p} = (1 - F)\Delta u$$

where Δu is an increment of shear displacement, Δu^{p} is the irreversible component of shear displacement and F is the fraction that the current shear stress constitutes of the limiting shear stress at the prevailing normal stress.

The progressive reduction in shear stress is represented by

$$\Delta \phi_{m} = -\frac{1}{R}(\phi_{m} - \phi)\,\Delta u^{p}$$

where ϕ_{m} is the prevailing mobilised friction angle, ϕ is the basic friction angle, and R is a parameter with the dimension of length, related to joint roughness.

The response to normal loading is expressed incrementally as

$$\Delta \sigma_{n} = K_{n}\,\Delta v$$

where the normal stiffness K_{n} is given by

$$K_{n} = \alpha_{n}\,\sigma_{n}^{\beta_{n}}$$

in which α_{n} and β_{n} are model parameters.

The shear stress and shear displacement increments are related by

$$\Delta \tau = F K_{s}\,\Delta u$$

where the shear stiffness may also be taken to be a function of normal stress, e.g.

$$K_{s} = \alpha_{s}\,\sigma_{n}^{\beta_{s}}$$

in which α_{s}, β_{s} are further model parameters.

The continuously-yielding joint model has been shown to have the capability to represent satisfactorily single episodes of shear loading and the effects of cyclic loading in a manner consistent with that reported by Brown and Hudson (1974).

131

4.9 Behaviour of discontinuous rock masses

4.9.1 Strength

The determination of the global mechanical properties of a large mass of discontinuous *in situ* rock remains one of the most difficult problems in the field of rock mechanics. Stress–strain properties are required for use in the determination of the displacements induced around mine excavations, and overall strength properties are required in, for example, assessments of pillar strength and the extent of discontinuous subsidence.

A first approach to the determination of the overall strength of a multiply jointed rock mass is to apply Jaeger's single plane of weakness theory (section 4.6) in several parts. Imagine that a rock mass is made up of the material for which the data shown in Figure 4.31b were obtained, but that it contains four sets of discontinuities each identical to the cleavage planes in the original slate. The sets of discontinuities are mutually inclined at 45° as shown in the sketches in Figure 4.47. A curve showing the variation of the peak principal stress difference with the orientation angle, α, may be constructed for a given value of σ_3 by superimposing four times the appropriate curve in Figure 4.31b with each curve displaced from its neighbour by 45° on the α axis. Figure 4.47 shows the resulting rock mass strength characteristics for three values of σ_3. In this case, failure always takes place by slip on one of the discontinuities. Note that, to a very good approximation, the strength of this hypothetical rock mass may be assumed to be

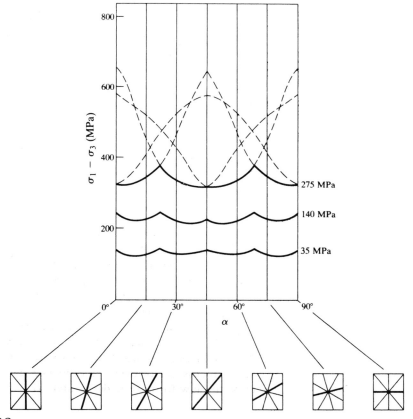

Figure 4.47 Composite peak strength characteristics for a hypothetical rock mass containing four sets of discontinuities each with the properties of the cleavage in the slate for which the data shown in Figure 4.31b were obtained.

132

isotropic. This would not be the case if one of the discontinuity sets had a substantially lower shear strength than the other sets.

Because of the difficulty of determining the overall strength of a rock mass by measurement, empirical approaches are generally used. As discussed in section 4.5.4, Brady (1977) found that the power law of equation 4.22 could be applied to the mineralised shale at the Mount Isa Mine. An attempt to allow for the influence of rock quality on rock mass strength was made by Bieniawski (1976) who assigned Coulomb shear strength parameters, c and φ, to the various rock mass classes in his geomechanics classification (Table 3.5).

The most completely developed of these empirical approaches is that introduced by Hoek and Brown (1980) who proposed the empirical rock mass strength criterion

$$\sigma_{1s} = \sigma_3 + (m\sigma_c \, \sigma_3 + s\sigma^2)^{1/2} \qquad (4.36)$$

where σ_{1s} = major principal stress at peak strength, σ_3 = minor principal stress, m and s = constants that depend on the properties of the rock and the extent to which it had been broken before being subjected to the failure stresses, and σ_c = uniaxial compressive strength of the intact rock material.

For intact rock material, $s = 1.0$, and for a completely granulated specimen or a rock aggregate, $s = 0$. The uniaxial compressive strength of the rock mass is given by $C_0 = \sigma_c \, s^{1/2}$ and is zero when $s = 0$. The principal stress form of a peak strength criterion is particularly appropriate for use in underground excavation design. However, Hoek and Brown (1980) showed that their criterion could also be expressed in terms of shear strength, τ, and normal stress, σ_n, as

$$\tau_N = A \, (\sigma_N - \sigma_{tN})^B \qquad (4.37)$$

where

$$\tau_N = \tau/\sigma_c$$

$$\sigma_N = \sigma_n/\sigma_c$$

$$\sigma_{tN} = \left[\tfrac{1}{2} \, m - (m^2 + 4s)^{1/2} \right]$$

and A and B are constants depending upon the value of m.

By evaluating such experimental data as were available and using a good measure of judgement, Hoek and Brown (1980) estimated the way in which the parameters varied with rock type and rock mass quality, presenting the values in tabular form. Because of the lack of more rigorous methods of estimating the strength of rock masses, this table has been used widely in engineering practice. Experience gained from these applications showed that the estimated rock mass strengths were reasonable when used for slope stability or caving studies in which the rock mass is disturbed and loosened by the excavation and subsequent relaxation processes. However, the estimated rock mass strengths generally appeared to be too low for most underground excavations where, presumably because of the greater confinement, loosening and loss of interlock does not occur to the same extent as in a slope.

On the basis of this experience, Hoek and Brown (1988) proposed a revised set of relations between Bieniawski's rock mass rating (RMR) and the parameters m and s:

Disturbed rock masses

$$\frac{m}{m_i} = \exp\left(\frac{RMR - 100}{14}\right) \tag{4.38}$$

$$s = \exp\left(\frac{RMR - 100}{6}\right) \tag{4.39}$$

Undisturbed or interlocking rock masses

$$\frac{m}{m_i} = \exp\left(\frac{RMR - 100}{28}\right) \tag{4.40}$$

$$s = \exp\left(\frac{RMR - 100}{9}\right) \tag{4.41}$$

where m_i is the value of m for the intact rock as given in section 4.5.4.

Experience has also shown that neither the RMR nor the NGI tunnelling quality index (Q) is ideally suited for use as an index for estimating rock mass strengths. A new purpose-designed classification scheme is required. For this and other

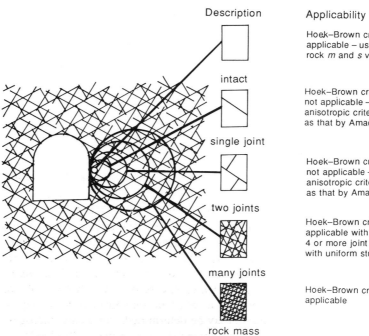

Figure 4.48 Applicability of the Hoek–Brown empirical rock mass strength criterion at different scales (after Hoek and Brown, 1988).

reasons the Hoek–Brown empirical rock mass strength criterion must be used with extreme care. Hoek and Brown (1988) have discussed the limitations of their strength criterion and given guidance on its use. It is important to recognise that the criterion does not take account of anisotropy or of the influence of the intermediate principal stress, both of which can have considerable significance in jointed rock masses. Figure 4.48 summarises the applicability of the Hoek–Brown criterion in terms of the scale of the rock mass.

Figure 4.49 shows an example of the application of Hoek and Brown's criterion to a sandstone rock mass at the site of a proposed underground excavation. Triaxial compression tests on samples of intact sandstone gave the upper envelope with $m_i = 15$. Logging of the cores showed that the rock mass rating varied from 65 (good) to 44 (fair). For a sandstone rock mass in the undisturbed or interlocked condition, equations 4.40 and 4.41 give $m = 4.298$, $s = 0.0205$ for RMR = 65 and $m = 2.030$, $s = 0.0020$ for RMR = 44. These values, used in conjunction with a mean value of $\sigma_c = 35$ MPa, give the estimated bounds to the *in situ* rock mass strength shown in Figure 4.49.

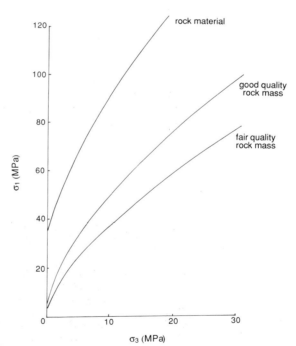

Figure 4.49 Hoek–Brown rock mass strength criterion applied to a sandstone rock mass with $\sigma_c = 35$ MPa.

4.9.2 Deformability

The study of the complete stress–strain behaviour of jointed rock masses involving post-yield plastic deformation, is beyond the scope of this introductory text. However, it is of interest to consider the pre-peak behaviour with a view to determining equivalent overall elastic constants for use in design analyses.

In the simplest case of a rock mass containing a single set of parallel discontinuities, a set of elastic constants for an equivalent transversely isotropic continuum may be determined. For a case analogous to that shown in Figure 2.10, let the rock material be isotropic with elastic constants E and ν, let the discontinuities have normal and shear stiffnesses K_n and K_s as defined in section 4.7.5, and let the mean discontinuity spacing be S. By considering the deformations resulting from the application of unit shear and normal stresses with respect to the x,y plane in Figure 2.10, it is found that the equivalent elastic constants required for use in equation 2.42 are given by

$$E_1 = E$$

$$\frac{1}{E_2} = \frac{1}{E} + \frac{1}{K_n S}$$

$$v_1 = v$$

$$v_2 = \frac{E_2}{E} v$$

$$\frac{1}{G_2} = \frac{1}{G} + \frac{1}{K_s S}$$

If, for example, $E = 10$ GPa, $v = 0.20$, $K_n = 5$ GPa m^{-1}, $K_s = 0.1$ GPa m^{-1} and $S = 0.5$ m, then $G = 4.17$ GPa, $E_1 = 10$ GPa, $E_2 = 2.0$ GPa, $v_1 = 0.20$, $v_2 = 0.04$ and $G_2 = 49.4$ MPa.

Similar solutions for cases involving more than one set of discontinuities are given by Amadei and Goodman (1981) and by Gerrard (1982). It is often found in practice that the data required to apply these models are not available or that the rock mass structure is less regular than that assumed in developing the analytical solutions. In these cases, it is common to determine E as the modulus of deformation or slope of the force–displacement curve obtained in an *in situ* compression test. There are many types of *in situ* compression test including uniaxial compression, plate bearing, flatjack, pressure chamber, borehole jacking and dilatometer tests.

The results of such tests must be interpreted with care particularly when tests are conducted under deviatoric stress conditions on samples containing discontinuities that are favourably oriented for slip. Under these conditions, initial loading may produce slip as well as reflecting the elastic properties of the rock material and the elastic deformabilities of the joints. Using a simple analytical model, Brady *et al.* (1985) have demonstrated that, in this case:

(a) the loading–unloading cycle must be accompanied by hysteresis; and
(b) it is only in the initial stage of unloading (Figure 4.50) that inelastic response is suppressed and the true elastic response of the rock mass is observed.

Bieniawski (1978) compiled values of *in situ* modulus of deformation determined using a range of test methods at 15 different locations throughout the world. He found that for values of rock mass rating, RMR, greater than about 55, the mean deformation modulus, E_M, measured in GPa, could be approximated by the empirical equation

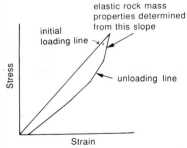

Figure 4.50 Determination of the Young's modulus of a rock mass from the response on initial unloading in a cyclic loading test (after Brady *et al.*, 1985).

$$E_M = 2(RMR) - 100 \qquad (4.42)$$

Serafim and Pereira (1983) found that an improved fit to their own and to Bieniawski's data, particularly in the range of E_M between 1 and 10 GPa, is given by the relation

$$E_M = 10^{\frac{RMR - 10}{40}} \qquad (4.43)$$

It is important to recognise that equations 4.42 and 4.43 empirically relate RMR values to mean values of deformation modulus determined from sets of data that show considerable scatter. Barton (1983) has shown that, in a crude sense, the dispersion of deformation modulus values arising from inherent variations in the rock mass, may be of similar magnitude to the dispersion associated with different methods of data interpretation. Deformation modulus may also be a highly anisotropic rock mass property.

Because of the high costs involved in carrying out *in situ* tests, geophysical methods are sometimes used to estimate *in situ* moduli. These methods generally involve studying the transmission of elastic compression and shear waves through the rock mass. Using a hammer seismograph technique known as the *petite sismique*, Bieniawski (1978) found that the measured static modulus of deformation, E, could be empirically related to the shear wave frequency, f, generated by a hammer blow and received at distances of up to 30 m on the rock surface, by the equation

$$E = 0.054f - 9.2$$

where E is in GPa and f is in Hz.

Problems

1 From the data given in Figure 4.18, calculate the tangent modulus and Poisson's ratio for the initial elastic behaviour of the limestone with $\sigma_3 = 2.0$ MPa.

2 A porous sandstone has a uniaxial compressive strength of $\sigma_c = 75$ MPa. The results of a series of triaxial compression tests plotted on shear stress–normal stress axes give a linear Coulomb peak strength envelope having a slope of 45°.

Determine the axial stress at peak strength of a jacketed specimen subjected to a confining pressure of $\sigma_3 = 10$ MPa. If the jacket had been punctured during the test, and the pore pressure had built up to a value equal to the confining pressure, what would the peak axial stress have been?

3(a) Establish an approximate peak strength envelope for the marble for which the data shown in Figure 4.19 were obtained.
3(b) In what ways might the observed stress–strain behaviour of the specimens have differed had the tests been carried out in a conventional testing machine having a longitudinal stiffness of 2.0 GN m^{-1}? Assume that all specimens were 50 mm in diameter and 100 mm long.

4 A series of laboratory tests on intact specimens of quartzite gave the following mean peak strengths. The units of stress are MPa, and compression is taken as positive.

	$\frac{1}{2}(\sigma_1 + \sigma_3)$		$\frac{1}{2}(\sigma_1 + \sigma_3)$
	100		100
	135		130
triaxial	160		150
compression	200		180
$\sigma_2 = \sigma_3$	298		248
	435		335
	σ_1	σ_2	σ_3
biaxial	0	0	−13.5
tension/	0	−13	−13
compression	218	50	0
	225	100	0
	228	150	0
	210	210	0

Develop a peak strength criterion for the quartzite for use in underground excavation design. Experience has shown that the *in situ* uniaxial compressive strength of the quartzite is one-half the laboratory value.

5 A series of triaxial compression tests on specimens of a slate gave the following results:

Confining pressure σ_3 (MPa)	Peak axial stress σ_1 (MPa)	Angle between cleavage and σ_1 $\alpha°$
2.0	62.0	40
5.0	62.5	32
10.0	80.0	37
15.0	95.0	39
20.0	104.0	27

In each test, failure occurred by shear along the cleavage. Determine the shear strength criterion for the cleavage planes.

6 In a further series of tests on the slate for which the data of Problem 5 were obtained, it was found that, when failure occurred in directions other than along the cleavage, the peak strength of the rock material was given by

$$\sigma_1 = 150 + 2.8\sigma_3$$

where σ_1 and σ_3 are in MPa.

Construct a graph showing the expected variation of peak axial stress at a confining pressure of 10 MPa, as the angle between the cleavage and the specimen axis varies from 0° to 90°.

7 The following results were obtained in a series of direct shear tests carried out on 100 mm square specimens of granite containing clean, rough, dry joints.

Normal stress	Peak shear strength	Residual shear strength	Displacement at peak shear strength	
			Normal	Shear
σ_n (MPa)	τ_p (MPa)	τ_r (MPa)	v(mm)	u(mm)
0.25	0.25	0.15	0.54	2.00
0.50	0.50	0.30	0.67	2.50
1.00	1.00	0.60	0.65	3.20
2.00	1.55	1.15	0.45	3.60
3.00	2.15	1.70	0.30	4.00
4.00	2.60	–	0.15	4.20

(a) Determine the basic friction angle and the initial roughness angle for the joint surfaces.

(b) Establish a peak shear strength criterion for the joints, suitable for use in the range of normal stresses, 0–4 MPa.

(c) Assuming linear shear stress–shear displacement relations to peak shear strength, investigate the influence of normal stress on the shear stiffness of the joints.

8 A triaxial compression test is to be carried out on a specimen of the granite referred to in Problem 7 with the joint plane inclined at $35°$ to the specimen axis. A confining pressure of $\sigma_3 = 1.5$ MPa and an axial stress of $\sigma_1 = 3.3$ MPa are to be applied. Then a joint water pressure will be introduced and gradually increased with σ_1 and σ_2 held constant. At what joint water pressure is slip on the joint expected to occur? Repeat the calculation for a similar test in which $\sigma_1 = 47$ MPa and $\sigma_3 = 1.5$ MPa.

9 In the plane of the cross section of an excavation, a rock mass contains four sets of discontinuities mutually inclined at $45°$. The shear strengths of all discontinuities are given by a linear Coulomb criterion with $c' = 100$ kPa and $\phi' = 30°$.

Develop an isotropic strength criterion for the rock mass that approximates the strength obtained by applying Jaeger's single plane of weakness theory in several parts.

10 A fair quality andesite in the hangingwall of a steeply dipping orebody has a rock mass rating of 44. The uniaxial compressive strength of the intact rock material, $\sigma_c = 75$ MPa. The rock mass strength may be described by Hoek and Brown's non-linear criterion using the parameters given in Table 4.2.

An analysis of progressive hangingwall caving of the type described in Chapter 16 requires the use of equivalent Coulomb shear strength parameters for the rock mass for various values of the normal stress on the failure plane. Write down expressions for these values, c_i and ϕ_i, in terms of the Hoek–Brown parameters A and B in equation 4.37. Using these expressions, prepare a graph showing the variation of c_i and ϕ_i with normal stress for normal stresses in the range 0–10 MPa.

11 A certain slate can be treated as a transversely isotropic elastic material. Block samples of the slate are available from which cores may be prepared with the cleavage at chosen angles to the specimen axes.

139

Nominate a set of tests that could be used to determine the five independent elastic constants in equation 2.42 required to characterise the stress–strain behaviour of the slate in uniaxial compression. What measurements should be taken in each of these tests?

5 Pre-mining state of stress

5.1 Specification of the pre-mining state of stress

The design of an underground structure in rock differs from other types of structural design in the nature of the loads operating in the system. In conventional surface structures, the geometry of the structure and its operating duty define the loads imposed on the system. For an underground rock structure, the rock medium is subject to initial stress prior to excavation. The final, post-excavation state of stress in the structure is the resultant of the initial state of stress and stresses induced by excavation. Since induced stresses are directly related to the initial stresses, it is clear that specification and determination of the pre-mining state of stress is a necessary precursor to any design analysis.

The method of specifying the *in situ* state of stress at a point in a rock mass, relative to a set of reference axes, is demonstrated in Figure 5.1. A convenient set of Cartesian global reference axes is established by orienting the x axis towards mine north, y towards mine east, and z vertically downwards. The ambient stress components expressed relative to these axes are denoted $p_{xx}, p_{yy}, p_{zz}, p_{xy}, p_{yz}, p_{zx}$. Using the methods established in Chapter 2, it is possible to determine, from these components, the magnitudes of the field principal stresses $p_i \, (i = 1, 2, 3)$, and the respective vectors of direction cosines $(\lambda_{xi}, \lambda_{yi}, \lambda_{zi})$ for the three principal axes. The corresponding direction angles yield a dip angle, α_i, and a bearing, or dip azimuth, β_i, for each principal axis. The specification of the pre-mining state of stress is completed by defining the ratio of the principal stresses in the form $p_1 : p_2 : p_3 = 1.0 : q : r$ where both q and r are less than unity.

The assumption made in this discussion is that it is possible to determine the *in situ* state of stress in a way which yields representative magnitudes of the components of the field stress tensor throughout a problem domain. The state of stress in the rock mass is inferred to be spatially quite variable, due to the presence of structural features such as faults or local variation in rock material properties. Spatial variation in the field stress tensor may be sometimes observed

Figure 5.1 Method of specifying the *in-situ* state of stress relative to a set of global reference axes.

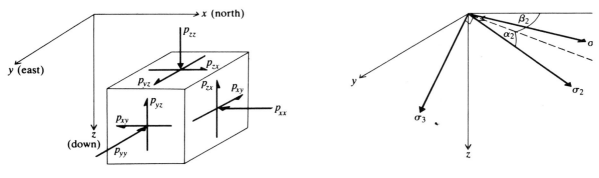

141

as an apparent violation of the equation of equilibrium for the global z (vertical) direction. Since the ground surface is always traction-free, simple statics requires that the vertical normal stress component at a sub-surface point be given by

$$p_{zz} = \gamma z \qquad (5.1)$$

where γ is the rock unit weight, and z is the depth below ground surface.

Failure to satisfy this equilibrium condition (equation 5.1) in any field determination of the pre-mining state of stress may be a valid indication of heterogeneity of the stress field. For example, the vertical normal stress component might be expected to be less than the value calculated from equation 5.1, for observations made in the axial plane of an anticlinal fold.

A common but unjustified assumption in the estimation of the *in situ* state of stress is a condition of uniaxial strain ('complete lateral restraint') during development of gravitational loading of a formation by superincumbent rock. For elastic rock mass behaviour, horizontal normal stress components are then given by

$$p_{xx} = p_{yy} = \left(\frac{\nu}{1-\nu}\right) p_{zz} \qquad (5.2)$$

where ν is Poisson's ratio for the rock mass.

If it is also assumed that the shear stress components p_{xy}, p_{yz}, p_{zx} are zero, the normal stresses defined by equations 5.1 and 5.2 are principal stresses.

Reports and summaries of field observations (Hooker *et al.*, 1972; Brown and Hoek, 1978) indicate that for depths of stress determinations of mining engineering interest, equation 5.2 is rarely satisfied, and the vertical direction is rarely a principal stress direction. This state of affairs arises from the complex load path and geologic history to which an element of rock is typically subjected in reaching its current equilibrium state.

5.2 Factors influencing the *in situ* state of stress

The ambient state of stress in an element of rock in the ground subsurface is determined by both the current loading conditions in the rock mass, and the stress path defined by its geologic history. Stress path in this case is a more complex notion than that involved merely in changes in surface and body forces in a medium. Changes in the state of stress in a rock mass may be related to temperature changes and thermal stress, and chemical and physicochemical processes such as leaching, precipitation and recrystallisation of constituent minerals. Mechanical processes such as fracture generation, slip on fracture surfaces and viscoplastic flow throughout the medium, can be expected to produce both complex and heterogeneous states of stress. Consequently, it is possible to describe, in only semi-quantitative terms, the ways in which the current observed state of a rock mass, or inferred processes in its geologic evolution, may determine the current ambient state of stress in the medium. The following discussion is intended to illustrate the role of common and readily comprehensible factors on pre-mining stresses.

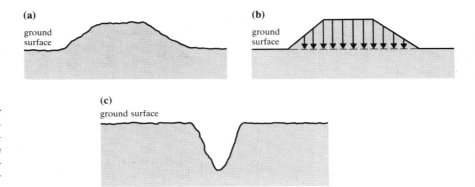

Figure 5.2 The effect of irregular surface topography (a) on the subsurface state of stress may be estimated from a linearised surface profile (b). A V-notch valley (c) represents a limiting case of surface linearisation.

5.2.1 Surface topography

Previous discussion has indicated that, for a flat ground surface, the average vertical stress component should approach the depth stress (i.e. $p_{zz} = \gamma z$). For irregular surface topography, such as that shown in Figure 5.2a, the state of stress at any point might be considered as the resultant of the depth stress and stress components associated with the irregular distribution of surface surcharge load. An estimate of the latter effect can be obtained by linearising the surface profile, as indicated in Figure 5.2b. Expressions for uniform and linearly varying strip loads on an elastic half-space can be readily obtained by integration of the solution for a line load on a half-space (Boussinesq, 1883). From these expressions, it is possible to evaluate the state of stress in such locations as the vicinity of the subsurface of the base of a V-notch valley (Figure 5.2c). Such a surface configuration would be expected to produce a high horizontal stress component, relative to the vertical component at this location. In all cases, it is to be expected that the effect of irregular surface topography on the state of stress at a point will decrease rapidly as the distance of the point below ground surface increases. These general notions appear to be confirmed by field observations (Endersbee and Hofto, 1963).

5.2.2 Erosion and isostasy

Erosion of a ground surface, either hydraulically or by glaciation, reduces the depth of rock cover for any point in the ground subsurface. It can be reasonably assumed that the rock mass is in a lithologically stable state prior to erosion, and thus that isostasy occurs under conditions of uniaxial strain in the vertical direction. Erosion and isostasy reduce the vertical stress according to equation 5.1, and the horizontal stresses by a lesser amount according to equation 5.2. It may be readily appreciated that these geologic processes can lead to ambient states of stress characterised by high horizontal/vertical ratios, particularly at shallow depths in the rock mass. Analysis of this problem also indicates that the horizontal/vertical stress ratio decreases as the depth increases, approaching the pre-erosion value when the current depth to the point of interest is significantly greater than the depth of overburden removed.

5.2.3 Residual stress

Residual stresses exist in a finite body when its interior is subject to a state of stress in the absence of applied surface tractions. The phenomenon has long been

143

recognised in the mechanics of materials. For example, Love (1944) describes the generation of residual stresses in a cast-iron body on cooling, due to the exterior cooling more rapidly than the interior. Timoshenko and Goodier (1970) also discuss the development of residual (or initial) stresses in common engineering materials. In general, residual stresses may be related to physical or chemical processes occurring non-homogeneously in restricted volumes of material. For example, non-uniform cooling of a rock mass, or the presence in a rock mass subject to uniform cooling, of contiguous lithological units with different coefficients of thermal expansion, will produce states of stress which are locally 'locked-in'.

Processes other than cooling that produce residual stresses may involve local mineralogical changes in the rock medium. Local recrystallisation in a rock mass may be accompanied by volumetric strain. Changes in the water content of a mineral aggregation, by absorption or exudation and elimination of chemically or physically associated water, can result in strains and residual stresses similar in principle to those associated with spatially non-uniform cooling.

A comprehensive understanding of the thermal history and subtle geologic evolution of the members of a rock formation is not considered a practical possibility. The problem of residual stresses therefore remains an inhibiting factor in predicting the ambient state of stress in a rock mass, from either basic mechanics or detailed geological investigations. The inverse process may be a more tractable proposition; i.e. anomalous or non-homogeneous states of stress in a formation may be related to the features or properties of the rock mass which reflect the spatial non-uniformity of its thermal, chemical or petrological history.

5.2.4 Inclusions

Inclusions in a rock mass are lithological units that post-date the formation of the host rock mass. Common inclusions are extrusive features such as dykes and sills, and veins of such minerals as quartz and fluorspar. The existence of a vertical, subplanar inclusion in a rock mass may have influenced the current *in-situ* state of stress in two ways. First, if the inclusion were emplaced under pressure against the horizontal passive resistance of the surrounding rock, a high-stress component would operate perpendicular to the plane of inclusion. A second possible influence of an inclusion is related to the relative values of the deformation moduli of the inclusion and the surrounding rock. Any loading of the system, for example by change of effective stress in the host rock mass or imposed displacements in the medium by tectonic activity, will generate relatively high or low stresses in the inclusion, compared with those in the host rock mass. A relatively stiff inclusion will be subject to relatively high states of stress, and conversely. An associated consequence of the difference in elastic moduli of host rock and inclusion is the existence of high-stress gradients in the host rock in the vicinity of the inclusion. In contrast, the inclusion itself will be subject to a relatively homogeneous state of stress (Savin, 1961).

An example of the effect of an inclusion on the ambient state of stress is provided by studies of conditions in and adjacent to dykes in the Witwatersrand Quartzite. The high elastic modulus of dolerite, compared with that of the host quartzite, should lead to a relatively high state of stress in the dyke, and a locally high stress gradient in the dyke margins. These effects appear to be confirmed in practice (Gay, 1975).

5.2.5 Tectonic stress

The state of stress in a rock mass may be derived from a pervasive force field imposed by tectonic activity. Stresses associated with this form of loading operate on a regional scale, and may be correlated with such structural features as thrust faulting and folding in the domain. Active tectonism need not imply that an area be seismically active, since elements of the rock mass may respond viscoplastically to the imposed state of stress. However, the stronger units of a tectonically stressed mass should be characterised by the occurrence of one subhorizontal stress component significantly greater than both the overburden stress and the other horizontal component. It is probable also that this effect should persist at depth. The latter factor may therefore allow distinction between near-surface effects, related to erosion, and latent tectonic activity in the medium.

5.2.6 Fracture sets and discontinuities

The existence of fractures in a rock mass, either as sets of joints of limited continuity, or as major, persistent features transgressing the formation, constrains the equilibrium state of stress in the medium. Thus vertical fractures in an uplifted or elevated rock mass, such as a ridge, can be taken to be associated with low horizontal stress components. Sets of fractures whose orientations, conformation and surface features are compatible with compressive failure in the rock mass, can be related to the properties of the stress field inducing fracture development (Price, 1966). In particular, a set of conjugate faults is taken to indicate that the direction of the major principal field stress prior to faulting coincides with the acute bisector of the faults' dihedral angle, the minor principal stress axis with the obtuse bisector, and the intermediate principal stress axis with the line of intersection of the faults (Figure 5.3). This assertion is based on a simple analogy with the behaviour of a rock specimen in true triaxial compression. Such an

Figure 5.3 Relation between fault geometry and the field stresses causing faulting.

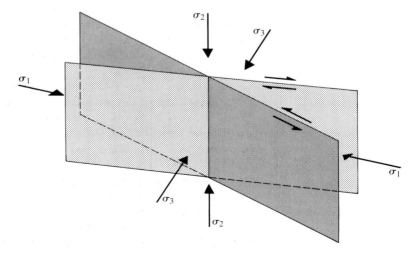

interpretation of the orientation of the field principal stresses does not apply to the state of stress prevailing following the episode of fracture. In fact, the process of rock mass fracture is intrinsically an energy dissipative and stress redistributive event.

The implication of the stress redistribution during any clastic episode is that the ambient state of stress may be determined by the need to maintain equilibrium conditions on the fracture surfaces. It may bear little relation to the pre-fracture state of stress. A further conclusion, from considerations of the properties of fractured rock, and of significance in site investigation, relates to the problem of spatial variability of the field stress tensor. A fracture field in a rock mass is usually composed of members which are variably oriented. It is inferred that a mechanically compatible stress field may also be locally variable, in both magnitudes and orientations of the principal stresses. A heterogeneous stress field is thus a natural consequence of an episode of faulting, shearing or extensive slip, such as occurs between beds in parallel folding. Successive episodes of fracturing, where, for example, one fault set transgresses an earlier set, may be postulated to lead to increasing complexity in the stress distribution throughout the medium.

It is clear from this brief discussion that the ambient, subsurface state of stress in a rock mass presents prohibitive difficulty in estimation *ab initio*. Its direct determination experimentally also presents some difficulty. In particular, the spatial variability of the stress tensor suggests that any single experimental determination may bear little relation to volume averages of the tensor components. In the design of a mine excavation or mine structure, it is the average state of stress in the zone of influence of the structure which exerts a primary control on the post-excavation stress distribution in the excavation near-field rock. The requirements for successful definition of the *in situ* state of stress are a technique for a local determination of the stress tensor, and a strategy for integration of a set of observations to derive a representative solution for the field stress tensor throughout the sampled volume.

5.3 Methods of *in situ* stress determination

5.3.1 General procedures

The need for reliable estimates of the pre-mining state of stress has resulted in the expenditure of considerable effort in the development of stress measurement devices and procedures. Methods developed to date exploit two separate and distinct principles in the measurement methodology, although most methods use a borehole to gain access to the measurement site. The most common set of procedures is based on determination of strains in the wall of a borehole, or other deformations of the borehole, induced by overcoring that part of the hole containing the measurement device. If sufficient strain or deformation measurements are made during this stress-relief operation, the six components of the field stress tensor can be obtained directly from the experimental observations using solution procedures developed from elastic theory. The second type of procedure, represented by flatjack measurements and hydraulic fracturing (Haimson, 1978), determines a circumferential normal stress component at particular locations in

146

(a)

the wall of a borehole. At each location, the normal stress component is obtained by the pressure, exerted in a slot or fissure, which is in balance with the local normal stress component acting perpendicular to the measurement slot. The circumferential stress at each measurement location may be related directly to the state of stress at the measurement site, preceding boring of the access hole. If sufficient boundary stress determinations are made in the hole periphery, the local value of the field stress tensor can be determined directly.

5.3.2 Triaxial strain cell

The range of devices for direct and indirect determination of *in situ* stresses includes photoelastic gauges, USBM borehole deformation gauges, and biaxial and triaxial strain cells. The soft inclusion cell, as described by Leeman and Hayes (1966) and Worotnicki and Walton (1976) is the most convenient of these devices, since it allows determination of all components of the field stress tensor

(b)

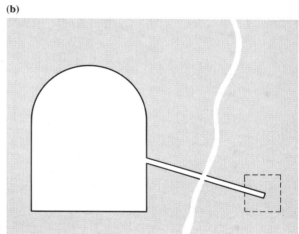

Figure 5.4 (a) A triaxial strain cell (of CSIRO design), and (b), (c), (d) its method of application.

(c)

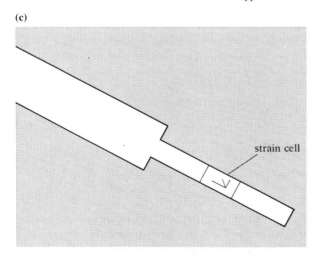

strain cell

(d)

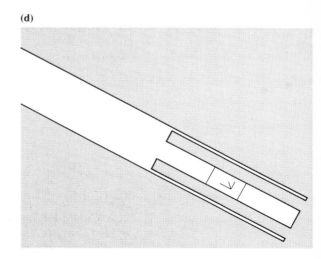

(a)   **(b)**   **(c)**

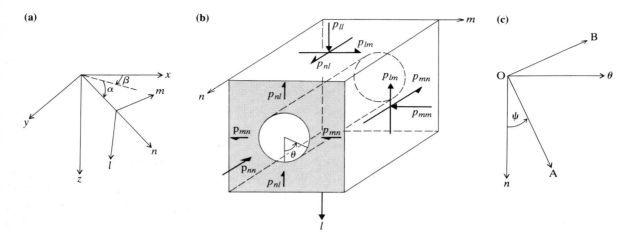

Figure 5.5 (a) Definition of hole local axes; (b) field stress components relative to hole local axes and position co-ordinate angle, θ; (c) reference axes on hole wall.

in a single stress relief operation. Such a strain cell, as shown in Figure 5.4a, consists of at least three strain rosettes, mounted on a deformable base or shell. The method of operation is indicated in Figures 5.4b, c and d. The cell is bonded to the borehole wall using a suitable epoxy or polyester resin. Stress relief in the vicinity of the strain cell induces strains in the gauges of the strain rosettes, equal in magnitude but opposite in sign to those originally existing in the borehole wall. It is therefore a simple matter to establish, from measured strains in the rosettes, the state of strain in the wall of the borehole prior to stress relief. These borehole strain observations are used to deduce the local state of stress in the rock, prior to drilling the borehole, from the elastic properties of the rock and the expressions for stress concentration around a circular hole.

The method of determination of components of the field stress tensor from borehole strain observations is derived from the solution (Leeman and Hayes, 1966) for the stress distribution around a circular hole in a body subject to a general triaxial state of stress. Figure 5.5a shows the orientation of a stress measurement hole, defined by dip, α, and dip direction, β, relative to a set of global axes x, y, z. Relative to these axes, the ambient field stress components (prior to drilling the hole) are $p_{xx}, p_{yy}, P_{zz}, p_{xy}, p_{yz}, p_{zx}$. A convenient set of local axes, l, m, n, for the borehole is also shown in Figure 5.5a, with the n axis directed parallel to the hole axis, and the m axis lying in the horizontal (x,y) plane. The field stress components expressed relative to the hole local axes, i.e. p_{ll}, p_{ln}, etc., are readily related to the global components p_{xx}, p_{xz}, etc., through the stress transformation equation and a rotation matrix defined by

$$[\mathbf{R}] = \begin{bmatrix} \lambda_{xl} & \lambda_{xm} & \lambda_{xn} \\ \lambda_{yl} & \lambda_{ym} & \lambda_{yn} \\ \lambda_{zl} & \lambda_{zm} & \lambda_{zn} \end{bmatrix} = \begin{bmatrix} -\sin\alpha\cos\beta & \sin\beta & \cos\alpha\cos\beta \\ -\sin\alpha\sin\beta & -\cos\beta & \cos\alpha\sin\beta \\ \cos\alpha & 0 & \sin\alpha \end{bmatrix}$$

In Figure 5.5b, the location of a point on the wall of the borehole is defined by the angle θ measured clockwise in the l, m plane. Boundary stresses at the point are related to the local field stresses, for an isotropic elastic medium, by the expression

148

$$\sigma_{rr} = \sigma_{r\theta} = \sigma_{rn} = 0$$

$$\sigma_{\theta\theta} = p_{ll}(1 - 2\cos 2\theta) + p_{mm}(1 + 2\cos 2\theta) - 4p_{lm}\sin 2\theta$$

$$\sigma_{nn} = p_{nn} + 2v(-p_{ll}\cos 2\theta + p_{mm}\cos 2\theta - 2p_{lm}\sin 2\theta)$$

$$\sigma_{\theta n} = 2p_{mn}\cos\theta - 2p_{nl}\sin\theta$$

(5.3)

Equations 5.3 define the non-zero boundary stress components, $\sigma_{\theta\theta}$, $\sigma_{\theta n}$, σ_{nn}, relative to the n, θ axes, aligned respectively with the hole axis, and the orthogonal direction which is tangential, in the l, m plane, to the hole boundary. Another set of right Cartesian axes, OA, OB, may be embedded in the hole boundary, as shown in Figure 5.5c, where the angle ψ defines the rotation angle from the n, θ axes to the OA, OB axes. The normal components of the boundary stress, in the directions OA, OB are given by

$$\sigma_A = \tfrac{1}{2}(\sigma_{nn} + \sigma_{\theta\theta}) + \tfrac{1}{2}(\sigma_{nn} - \sigma_{\theta\theta})\cos 2\psi + \sigma_{\theta n}\sin 2\psi$$

$$\sigma_B = \tfrac{1}{2}(\sigma_{nn} + \sigma_{\theta\theta}) - \tfrac{1}{2}(\sigma_{nn} - \sigma_{\theta\theta})\cos 2\psi - \sigma_{\theta n}\sin 2\psi$$

(5.4)

Suppose the direction OA in Figure 5.5c coincides with the orientation and location of a strain gauge used to measure the state of strain in the hole wall. Since plane stress conditions operate at the hole boundary during the stress relief process, the measured normal strain component is related to the local boundary stress components by

$$\varepsilon_A = \frac{1}{E}\left(\sigma_A - v\sigma_B\right)$$

or

$$E\varepsilon_A = \sigma_A - v\sigma_B$$

(5.5)

Substituting the expressions for σ_A, σ_B (equations 5.4) in equation 5.5, and then substituting equations 5.3 in the resulting expression, gives the required relation between the local state of strain in the hole wall and the field stresses as

$$E\varepsilon_A = p_{ll}\left\{\tfrac{1}{2}[(1-v) - (1+v)\cos 2\psi] - (1-v^2)(1-\cos 2\psi)\cos 2\theta\right\}$$

$$+ p_{mm}\left\{\tfrac{1}{2}[(1-v) - (1+v)\cos 2\psi] + (1-v^2)(1-\cos 2\psi)\cos 2\theta\right\}$$

$$+ p_{nn}\tfrac{1}{2}[(1-v) + (1+v)\cos 2\psi]$$

$$- p_{lm}2(1-v^2)(1-\cos 2\psi)\sin 2\theta$$

$$+ P_{mn}2(1+v)\sin 2\psi\cos\theta$$

$$- p_{nl}2(1+v)\sin 2\psi\sin\theta$$

(5.6a)

or

$$a_1\,p_{ll} + a_2\,p_{mm} + a_3\,p_{nn} + a_4\,p_{lm} + a_5\,p_{mn} + a_6\,p_{nl} = b \qquad (5.6\text{b})$$

Equations 5.6a and 5.6b indicate that the state of strain in the wall of a borehole, at a defined position and in a particular orientation, specified by the angles θ and ψ, is determined linearly by the field stress components. In equation 5.6b, the coefficients a_i $(i = 1 - 6)$ can be calculated directly from the position and orientation angles for the measurement location and Poisson's ratio for the rock. Thus if six independent observations are made of the state of strain in six positions/orientations on the hole wall, six independent simultaneous equations may be established. These may be written in the form

$$[\mathbf{A}]\,[\mathbf{p}] = [\mathbf{b}] \qquad (5.7)$$

where $[\mathbf{p}]$ represents a column vector formed from the stress components $p_{ll}, p_{mm}, p_{nn}, p_{lm}, p_{mn}, p_{nl}$. Provided the positions/orientations of the strain observations are selected to ensure a well conditioned coefficient matrix $[\mathbf{A}]$, equation 5.7 can be solved directly for the field stress components p_{ll}, p_{lm}, etc.. A Gaussian elimination routine, similar to that given by Fenner (1974), presents a satisfactory method of solving the set of equations.

The practical design of a triaxial strain cell usually provides more than the minimum number of six independent strain observations. The redundant observations may be used to obtain large numbers of equally valid solutions for the field stress tensor (Brady et al., 1976). These may be used to determine a locally averaged solution for the ambient state of stress in the zone of influence of the stress determination. Confidence limits for the various parameters defining the field stress tensor may also be attached to the measured state of stress.

5.3.3 Flatjack measurements

Stress measurement using strain gauge devices is usually performed in small-diameter holes, such that the volume of rock whose state of stress is sampled is about 10^{-3} m^3. Larger volumes of rock can be examined if a larger diameter opening is used as the measurement site. For openings allowing human access, it may be more convenient to measure directly the state of stress in the excavation wall, rather than the state of strain. This eliminates the need to determine or estimate a deformation modulus for the rock mass. The flatjack method presents a particularly attractive procedure for determination of the boundary stresses in an opening, as it is a null method, i.e. the measurement system seeks to restore the original, post-excavation local state of stress at the experiment site. Such methods are intrinsically more accurate than those relying on positive disturbance of the initial condition whose state is to be determined.

Three prerequisites must be satisfied for a successful *in situ* stress determination using flatjacks. These are:

(a) a relatively undisturbed surface of the opening constituting the test site;
(b) an opening geometry for which closed-form solutions exist, relating the far-field stresses and the boundary stresses; and

150

(c) a rock mass which behaves elastically, in that displacements are recoverable when the stress increments inducing them are reversed.

The first and third requirements virtually eliminate the use as a test site of an excavation developed by conventional drilling and blasting. Cracking associated with blasting, and other transient effects, may cause extensive disturbance of the elastic stress distribution in the rock and may give rise to non-elastic displacements in the rock during the measurement process. The second requirement restricts suitable opening geometry to simple shapes. An opening with circular cross section is by far the most convenient.

The practical use of a flatjack is illustrated in Figure 5.6. The jack consists of a pair of parallel plates, about 300 mm square, welded along the edges. A tubular non-return connection is provided to a hydraulic pump. A measurement site is established by installing measurement pins, suitable for use with a DEMEC or similar deformation gauge, in a rock surface and perpendicular to the axis of the proposed measurement slot. The distance d_0 between the pins is measured, and the slot is cut, using, for example, a series of overlapping core-drilled holes. Closure occurs between the displacement measuring stations. The flatjack is grouted in the slot, and the jack pressurised to restore the original distance d_0 between the displacement monitoring pins. The displacement cancellation pressure corresponds closely to the normal stress component directed perpendicular to the slot axis prior to slot cutting.

Determination of the field stresses from boundary stresses using flatjacks follows a procedure similar to that using strain observations. Suppose a flatjack is used to measure the normal stress component in the direction OA in Figure 5.5c, i.e. the plane of the flatjack slot is perpendicular to the axis OA. If σ_A is the jack cancellation pressure, substitution of the expressions for $\sigma_{\theta\theta}$, σ_{nn}, $\sigma_{\theta n}$ (equations 5.3) into the expression for σ_A (equation 5.4) yields

Figure 5.6 (a) Core drilling a slot for a flatjack test and (b) slot pressurisation procedure.

(a)   **(b)**

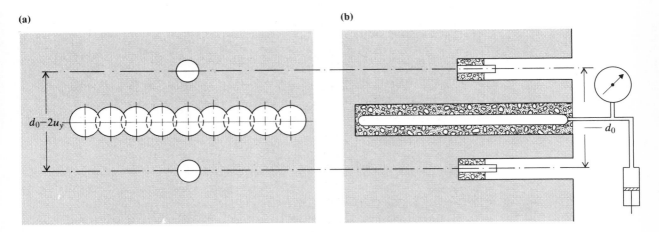

$d_0 - 2u_y$

d_0

151

$$\sigma_A = (p_{ll}) \frac{1}{2} \{ (1 - \cos 2\psi) - 2 \cos 2\theta \, [(1 + \nu) - (1 - \nu) \cos 2\psi] \}$$

$$+ (p_{mm}) \frac{1}{2} \{ (1 - \cos 2\psi) + 2 \cos 2\theta \, [(1 + \nu) - (1 - \nu) \cos 2\psi] \}$$

$$+ (p_{nn}) \frac{1}{2} (1 + \cos 2\psi) - p_{lm} \, 2 \sin 2\theta \, \{ (1 + \nu) - (1 - \nu) \cos 2\psi \}$$

$$+ p_{mn} \, 2 \sin 2\psi \cos \theta - p_{nl} \, 2 \sin 2\psi \sin \theta \tag{5.8}$$

or

$$C_1 \, p_{ll} + C_2 \, p_{mm} + C_3 \, p_{nn} + C_4 \, p_{lm} + C_5 \, p_{mn} + C_6 \, p_{nl} = \sigma_A \tag{5.9}$$

Equation 5.9 confirms that the state of stress in any position/orientation in the hole boundary is linearly related to the field stress components, by a set of coefficients which are simply determined from the geometry of the measurement system. Thus if six independent observations of σ_A are made at various locations defined by angles θ and ψ, a set of six simultaneous equations is established:

$$[C] \, [p] = [\sigma] \tag{5.10}$$

The terms of the coefficient matrix $[C]$ in this equation are determined from θ and ψ for each boundary stress observation, using equation 5.8.

In the design of a measurement programme, the boundary stress measurement positions and orientations must be selected carefully, to ensure that equations 5.10 are both linearly independent and well conditioned. The criterion for a poorly conditioned set of equations is that the determinant of the $[C]$ matrix is numerically small compared with any individual term in the matrix.

5.3.4 Hydraulic fracturing

A shortcoming of the methods of stress measurement described previously is that close access to the measurement site is required for operating personnel. For example, hole depths of about 10 m or less are required for effective use of most triaxial strain cells. Virtually the only method which permits remote determination of the state of stress is the hydraulic fracturing technique, by which stress measurements can be conducted in deep boreholes such as exploration holes drilled from the surface.

The principles of the technique are illustrated in Figure 5.7. A section of a borehole is isolated between inflatable packers, and the section is pressurised with water, as shown in Figure 5.7a. When the pressure is increased, the state of stress around the borehole boundary due to the field stresses is modified by superposition of hydraulically-induced stresses. If the field principal stresses in the plane perpendicular to the hole axis are not equal, application of sufficient pressure induces tensile circumferential stress over limited sectors of the boundary. When the tensile stress exceeds the rock material tensile strength, fractures initiate and propagate perpendicular to the hole boundary and parallel to the major principal stress, as indicated in Figure 5.7b. Simultaneously, the fluid pressure falls in the test section. After relaxation of the pressure and its sub-

152

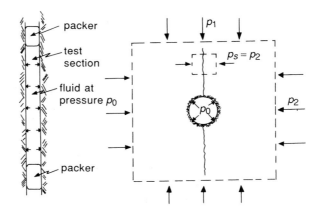

Figure 5.7 Principles of stress measurement by hydraulic fracturing: (a) packed-off test section; (b) cross section of hole, and fracture orientation relative to plane principal stresses.

sequent re-application, the peak borehole pressure achieved is less than the initial boundary fracturing pressure by an amount corresponding to the tensile strength of the rock material.

A record of borehole pressure during an hydraulic fracturing experiment, in which typically several cycles of pressure application and decline are examined, is shown in Figure 5.8. Two key parameters defined on the borehole pressure record are the instantaneous shut-in pressure p_s and the crack re-opening pressure p_r. The shut-in pressure or fracture closure pressure defines the field principal stress component perpendicular to the plane of the fracture. As suggested by Figure 5.7b, this corresponds to the minor principal stress p_2 acting in the plane of section. The crack re-opening pressure is the borehole pressure sufficient to separate the fracture surfaces under the state of stress existing at the hole boundary. The crack re-opening pressure, the shut-in pressure and the pore pressure, u, at the test horizon may be used to estimate the major principal stress in the following way.

The minimum boundary stress, σ_{min}, around a circular hole in rock subject to biaxial stress, with field stresses of magnitudes p_1 and p_2, is given by

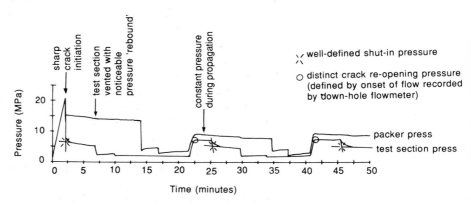

Figure 5.8 Pressure vs. time record for a hydraulic fracturing experiment (after Enever and Chopra, 1986).

$$\sigma_{min} = 3p_2 - p_1 \tag{5.11}$$

When a pressure p_0 is applied to the interior of the borehole, the induced tangential stress $\sigma_{\theta\theta}$ at the hole wall is

$$\sigma_{\theta\theta} = -p_0$$

The minimum tangential boundary stress is obtained by superimposing this stress on that given by equation 5.11, i.e.

$$\sigma_{min} = 3p_2 - p_1 - p_0 \tag{5.12}$$

and the minimum effective boundary stress is

$$\sigma'_{min} = 3p_2 - p_1 - p_0 - u \tag{5.13}$$

The crack re-opening pressure p_r corresponds to the state of borehole pressure p_0 where the minimum effective boundary stress is zero, i.e. introducing p_r in equation 5.13

$$3p_2 - p_1 - p_r - u = 0$$

or

$$p_1 = 3p_2 - p_r - u \tag{5.14}$$

Because $p_2 = p_s$, equation 5.14 confirms that the magnitudes of the major and minor plane principal stresses p_1 and p_2 can be determined from measurements of shut-in pressure p_s and crack re-opening pressure p_r. The orientation of the principal stress axes may be deduced from the position of the boundary fractures, obtained using a device such as an impression packer. The azimuth of the major principal stress axis is defined by the hole diameter joining the trace of fractures on opposing boundaries of the hole.

Although hydraulic fracturing is a simple and apparently attractive stress measurement technique, it is worth recalling the assumptions implicit in the method. First, it is assumed that the rock mass is continuous and elastic, at least in the zone of influence of the hole and the hydraulically induced fractures. Second, the hole axis is assumed to be parallel to a field principal stress axis. Third, the induced fracture plane is assumed to include the hole axis. If any of these assumptions is not satisfied, an invalid solution to the field stresses will be obtained. A further limitation is that it provides only plane principal stresses, and no information on the other components of the triaxial stress field. The usual assumption is that the vertical normal stress component is a principal stress, and that it is equal to the depth stress.

5.4 Presentation of *in situ* stress measurement results

The product of a stress measurement exercise is the set of six components of the field stress tensor, usually expressed relative to a set of local axes for the meas-

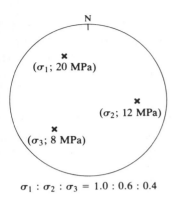

$\sigma_1 : \sigma_2 : \sigma_3 = 1.0 : 0.6 : 0.4$

Figure 5.9 Presentation of principal stress data on lower hemisphere stereographic projection.

urement hole. These yield the stress components expressed relative to the global axes by a simple transformation. The principal stress magnitudes and orientations are then determined from these quantities using the methods described in Chapter 2. If a single determination is made of the field stress tensor, the orientations of the principal stress axes can be plotted directly on to a stereonet overlay as shown in Figure 5.9. The value of this procedure is that the required mutual orthogonality of the principal stress directions can be determined by direct measurement. It therefore provides a useful check on the validity and consistency of the solution for the principal stresses. Other points of confirmation of the correctness of the solution are readily established by considering the magnitudes of the various stress invariants, calculated from the several sets of components (expressed relative to different sets of reference axes) of the field stress tensor.

In the case where redundant experimental observations have been collected, many independent solutions are possible for the *in situ* stress tensor. The methods suggested by Friday and Alexander (Brady *et al.*, 1976) can then be used to establish mean values of the components and principal directions of the stress tensor. An example of the way in which an over-determination of experimental parameters can be used is illustrated in Figure 5.10. More than 1000 independent solutions for the field stresses allowed construction of histograms of principal stress magnitudes, and contour plots of principal stress directions. Presumably, greater reliability can be attached to the mean values of principal stress magnitudes and orientations obtained from these plots, than to any single solution for the field stresses. The usual tests for consistency can be applied to the mean solution in the manner described earlier.

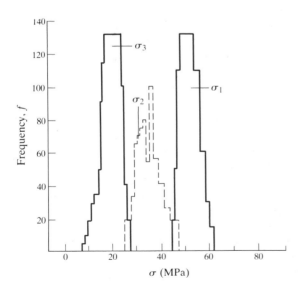

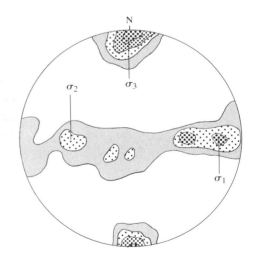

Figure 5.10 Histogram plots of principal stress frequencies and contour plots and principal stress orientations obtained from redundant strain observations.

A number of other assessments, in addition to those to test the internal consistency of a solution for the field stresses, can be performed to take account of specific site conditions. A primary requirement is that the ambient state of stress cannot violate the *in situ* failure criterion for the rock mass. As was noted in Chapter 4, establishing a suitable rock mass failure criterion is not a simple procedure process, but an essential proposition is that the field stresses should not violate the failure criterion for the intact rock material. The latter rock property may be established from standard laboratory tests on small specimens. Since the *in situ* strength of rock is typically much less than the strength measured on small specimens, the proposed test may not be a sensitive discriminant of the acceptability of a field stress determination. However, it ensures that widely inaccurate results are identified and re-examined.

A second determinant of the mechanical acceptability of an *in situ* stress measurement is derived from the requirement for conditions of static equilibrium on pervasive planes of weakness in the rock mass. Application of this criterion is best illustrated by example. Suppose a fault plane has the orientation 295°/50° (dip direction/dip), and that the measured *in situ* stress field is defined by:

σ_1, magnitude 15 MPa, dips 35° towards 085°;

σ_2, magnitude 10 MPa, dips 43° towards 217°;

σ_3, magnitude 8 MPa, dips 27° towards 335°.

The groundwater pressure at the measurement horizon is 2.8 MPa, and the angle of friction for the fault surface 25°. These data can be used to determine the normal and resultant shear stress components acting on the fault, and thus to calculate the angle of friction on the fault surface required to maintain equilibrium.

The given data are applied in the following way. When plotted on a stereonet, the direction angles (α, β, γ) between the principal axes and the pole to the fault plane are measured directly from the stereonet as (24°, 71°, 104°). These yield direction cosines (l_1, l_2, l_3) of (0.914, 0.326, $-$0.242) of the fault normal relative to the principal stress axes. The effective principal stresses are given by

$$\sigma_1' = 12.2 \text{ MPa}$$

$$\sigma_2' = 7.2 \text{ MPa}$$

$$\sigma_3' = 5.2 \text{ MPa}$$

Working now in terms of effective stresses, the resultant stress is given by

$$R = (l_1^2 \, \sigma_1'^2 + l_2^2 \, \sigma_2'^2 + l_3^2 \, \sigma_3'^2)^{1/2} = 11.46 \text{ MPa}$$

and the normal stress by

$$\sigma_n' = l_1^2 \, \sigma_1' + l_2^2 \, \sigma_2' + l_3^2 \, \sigma_3' = 11.26 \text{ MPa}$$

The resultant shear stress is

$$\tau = (R^2 - \sigma_n'^2)^{1/2} = 2.08 \text{ MPa}$$

The angle of friction mobilised by the given state of stress on the plane is

$$\phi_{mob} = \tan^{-1}(2.08/11.26) = 10.5°$$

Since ϕ_{mob} is less than the measured angle of friction for the fault, it is concluded that the *in situ* state of stress is compatible with the orientation and strength properties of the fault. It is to be noted also that a similar conclusion could be reached by some simple constructions on the stereonet. Clearly, the same procedures would be followed for any major structural feature transgressing the rock mass.

For the example considered, the measured state of stress was consistent with static equilibrium on the plane of weakness. In cases where the field stresses apparently violate the equilibrium condition, it is necessary to consider carefully all data related to the problem. These include such factors as the possibility of a true cohesive component of discontinuity strength and the possible dilatant properties of the discontinuity in shear. Questions to be considered concerning stress measurement results include the probable error in the determination of both principal stress magnitudes and directions, and the proximity of the stress measurement site to the discontinuity. Thus the closer the measurement site to the discontinuity, the more significance to be attached to the no-slip criterion. Only when these sorts of issues have been considered in detail should the inadmissibility of a solution for the field stress tensor be decided.

5.5 Results of *in situ* stress measurements

Figure 5.11 Variation with depth below surface of (a) measured values of *in situ* vertical stress, p_{zz}, and (b) ratio of average measured horizontal stresses to the vertical stress (after Brown and Hoek, 1978).

A comprehensive collation of the results of measurement of the pre-mining state of stress, at locations of various major mining and civil engineering operations, has been prepared by Brown and Hoek (1978). The results are summarised in Figure 5.11. The first observation from this figure is that the measurements of p_{zz} (in MPa) are scattered about the trend line

$$p_{zz} = 0.027z$$

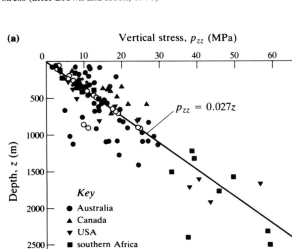

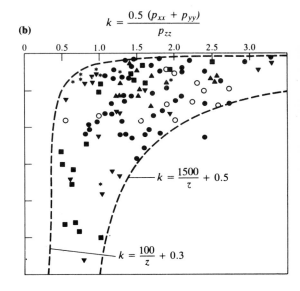

157

where z (in m) is the depth below ground surface. Since 27 kN m^{-3} represents a reasonable average unit weight for most rocks, it appears that the vertical component of stress is closely related to depth stress. A further observation concerns the variation with depth of the ratio of the average of the horizontal stresses to the vertical stress, i.e. [$0.5 (p_{xx} + p_{yy})$] / p_{zz}. At shallow depth, values of the ratio are extremely variable, and frequently greater than unity. For increasing depth, variability of the ratio decreases, and approaches unity. Some of the variability in the stress ratio at shallow depths and low stress levels may be due to experimental error. However, the convergence of the ratio to a value of unity at depth is consistent with the principle of time-dependent elimination of shear stress in rock masses. The postulate of regression to a lithostatic state by viscoplastic flow is commonly referred to as Heim's Rule (Talobre, 1957).

The final observation arising from inspection of Figures 5.11a and b is a confirmation of the assertion made at the beginning of this chapter. The virgin state of stress in a rock mass is not amenable to calculation by any known method, but must be determined experimentally. In jointed and fractured rock masses, a highly variable stress distribution is to be expected, and indeed has been confirmed by several investigations of the state of stress in such settings. For example, Bock (1986) described the effect of horizontal jointing on the state in a granite, confirming that each extensive joint defined a boundary of a distinct stress domain. In the analysis of results from a jointed block test, Brown *et al.* (1986) found that large variations in state of stress occurred in the different domains of the block generated by the joints transgressing it. Richardson *et al.* (1986) reported a high degree of spatial variation of the stress tensor in foliated gneiss, related to rock fabric, and proposed methods of deriving a representative solution of the field stress tensor from the individual point observations. In some investigations in Swedish bedrock granite, Carlsson and Christiansson (1986) observed that the local state of stress is clearly related to the locally dominant geological structure.

These observations of the influence of rock structure on rock stress suggest that a satisfactory determination of a representative solution of the *in situ* state of stress is probably not possible with a small number of random stress measurements. The solution is to develop a site-specific strategy to sample the stress tensor at a number of points in the mass, taking account of the rock structure. It may then be necessary to average the results obtained, in a way consistent with the distribution of measurements, to obtain a site representative value.

Problems

1 At a particular site, the surface topography can be represented in a vertical cross section by a vertical cliff separating horizontal, planar ground surfaces, as shown in the figure (a) below. The upper and lower surfaces AB and CD can be taken to extend infinite distances horizontally from the toe of the 100 m high cliff. The effect of the ground above the elevation D′CD can be treated as a wide surcharge load on a half-space.

(a) The stress components due to a line load of magnitude P on a half space are given by

$$\sigma_{rr} = \frac{2P \sin \theta}{\pi r}$$

$$\sigma_{\theta\theta} = \sigma_{r\theta} = 0$$

where r and θ are defined in figure (b).

Relative to x, z reference axes, show that the stress components due to the strip load defined in figure (c) are given by

$$\sigma_{xx} = \frac{p}{2\pi}\left[2(\theta_2 - \theta_1) + (\sin 2\theta_2 - \sin 2\theta_1)\right]$$

$$\sigma_{zz} = \frac{p}{2\pi}\left[2(\theta_2 - \theta_1) - (\sin 2\theta_2 - \sin 2\theta_1)\right]$$

$$\sigma_{zx} = \frac{p}{2\pi}\left[\cos 2\theta_1 - \cos 2\theta_2\right]$$

(b) If the unit weight of the rock is 27 kN m^{-3}, calculate the stress components induced at a point P, 80 m vertically below the toe of the cliff, by the surcharge load.

(c) If the state of stress at a point Q remote from the toe of the cliff and on the same elevation as P, is given by $\sigma_{xx} = 2.16$ MPa, $\sigma_{zz} = 3.24$ MPa, $\sigma_{zx} = 0$, estimate the magnitudes and orientations of the plane principal stresses at P.

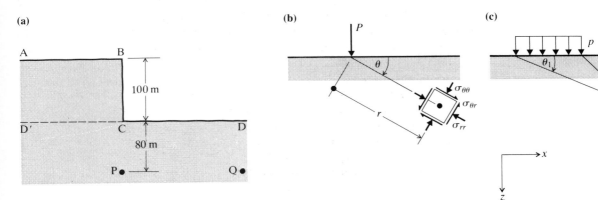

(a)

A   B

100 m

D′   C   D

80 m

P●   Q●

(b)

P

θ

r

$\sigma_{\theta\theta}$

$\sigma_{\theta r}$

σ_{rr}

(c)

p

θ_1 θ_2

x

z

2 An element of rock 800 m below ground surface is transgressed by a set of parallel, smooth continuous joints, dipping as shown in the figure below. The fissures are water filled below an elevation 100 m below the ground surface. The vertical stress component p_{zz} is a principal stress, and equal to the depth stress. From the calculated depth stress, p_{zz}, calculate the range of possible magnitudes of the horizontal stress component, p_{xx}. The unit weight of the rock mass is 26 kN m^{-3}, and the unit weight of water 9.8 kN m^{-3}. The resistance to slip on the joints is purely frictional, with an angle of friction ϕ' of 20°.

State any assumptions used in deriving the solution.

159

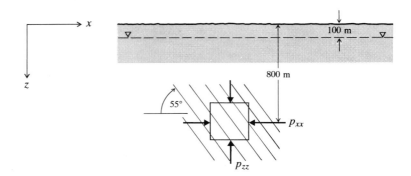

3 In an underground mine, flatjacks were used to measure the state of stress in the walls and crown of a long horizontal tunnel of circular cross section. Independent observations indicated that the long axis of the tunnel was parallel to a field principal stress. For slots cut parallel to the tunnel axis, the cancellation pressures were 45 MPa in both side walls, at the midheight of the tunnel, and 25 MPa in the crown. For slots cut in each side wall, perpendicular to the tunnel axis and at the tunnel midheight, the cancellation pressures were 14.5 MPa.

(a) By considering the symmetry properties of equations 5.3 defining boundary stresses around a circular hole in a triaxial stress field, deduce the principal stress directions.
(b) Calculate the magnitudes of the field principal stresses, assuming $v = 0.25$.

4 A CSIRO hollow inclusion strain cell was used in an overcoring experiment to determine the state of strain in the walls of a borehole. The borehole was oriented 300°/70°. Using the angular co-ordinates and orientations defined in Figure 5.5, the measured states of strain (expressed in microstrains) in the wall of the hole were, for various θ and ψ:

$\theta°$ \ $\psi°$	0°	45°	90°	135°
0°	—	213.67	934.41	821.11
120°	96.36	—	349.15	131.45
240°	96.36	—	560.76	116.15

Young's modulus of the rock was 40 GPa, and Poisson's ratio, 0.25.

(a) Set up the set of nine equations relating measured strain and gauge location. (Note that, for $\psi = 0°$, identical equations are obtained, independent of θ.)
(b) Select the best conditioned set of six equations, and solve for the field stresses, expressed relative to the hole local axes.
(c) Transform the field stresses determined in (b) to the mine global axes (x – north, y – east, z – down).
(d) Determine the magnitudes of the field principal stresses, and their orientations relative to the mine global axes.

160

(Note: This problem is most conveniently handled with a calculator capable of solving six simultaneous equations.)

5 Measurement of the state of stress in a rock mass produced the following results:

σ_1, of magnitude 25 MPa, is oriented 109°/40°;

σ_2, of magnitude 18 MPa, is oriented 221°/25°;

σ_3, of magnitude 12 MPa, is oriented 334°/40°.

The groundwater pressure at the measurement site is 8 MPa. A fault, oriented 190°/60°, is cohesionless, and has an estimated angle of friction of ϕ'°.

Comment on the consistency of this set of observations, and describe any other subsequent investigations you might consider necessary.

6 Methods of stress analysis

6.1 Analytical methods for mine design

Basic issues to be considered in the development of a mine layout include the location and design of the access and service openings, and the definition of stoping procedures for ore extraction. These issues are not mutually independent. However, geomechanics questions concerning stoping activity may be more pervasive than those related to the siting and design of permanent openings, since the former persist throughout the life of the mine, and possibly after the completion of mining.

The scope of the problems which arise in designing and planning the extraction of an orebody can be appreciated by considering the implementation of a method such as room-and-pillar mining. It is necessary to establish parameters such as stope dimensions, pillar dimensions, pillar layout, stope mining sequence, pillar extraction sequence, type and timing of placement of backfill, and the overall direction of mining advance. These geomechanics aspects of design and planning must also be integrated with other organisational functions in the planning process. It is not certain that this integration is always achieved, or that economic and geomechanics aspects of mine planning and design are always compatible. However, it is clear that sound mining rock mechanics practice requires effective techniques for predicting rock mass response to mining activity. A particular need is for methods which allow parameter studies to be undertaken quickly and efficiently, so that a number of operationally feasible mining options can be evaluated for their geomechanical soundness. Alternatively, parameter studies may be used to identify and explore geomechanically appropriate mining strategies and layouts, which can then be used to develop detailed ore production schemes.

The earliest attempts to develop a predictive capacity for application in mine design involved studies of physical models of mine structures. Their general objective was to identify conditions which might cause extensive failure in the prototype. The difficulty in this procedure is maintaining similitude in the material properties and the loads applied to model and prototype. These problems can be overcome by loading a model in a centrifuge. However, such facilities are expensive to construct and operate, and their use is more suited to basic research than to routine design applications. An additional and major disadvantage of any physical modelling concerns the expense and time to design, construct and test models which represent the prototype in sufficient detail to resolve specific mine design questions. The general conclusion is that physical models are inherently limited in their potential application as a predictive tool in mine design. Base friction modelling provides an exception to this statement. If it is possible to deal with a two-dimensional model of a mine structure, and to examine discrete sections of the complete mine layout, the procedure described by Bray and Goodman (1981) provides a useful and inexpensive method for design evaluation. The method is particularly appropriate where structural features exercise a dominant role in rock mass response.

A conventional physical model of a structure yields little or no information on stresses and displacements in the interior of the medium. The earliest method for quantitative experimental determination of the internal state of stress in a body subject to applied load was the photoelastic method. The principle exploited in the method is that, in two dimensions, and for isotropic elasticity, the stress distribution is independent of the elastic properties of the material, and is the same for plane stress and plane strain. In its original application, a two-dimensional model of a structure was prepared from a transparent material such as glass or plastic, and mounted in a beam of monochromatic, polarised light. Application of loads to the model, and passage of the light beam through an analyser onto a screen, produced a series of bands, or fringes, of light extinction and enhancement. Generation of the fringes is due to dependence of the propagation velocity of light through the medium on the local principal stress components. A fringe, also called an isochromatic, represents a contour line of constant principal stress difference. Thus a fringe pattern produced by a photoelastic model represents a mapping of contours of maximum shear stress throughout the medium. Calibration of the system allows the shear stress magnitude of any contour level to be determined. For excavation design in rock, it is necessary to establish the distribution of principal stresses throughout the medium. Thus in addition to the maximum shear stress distribution, it is necessary to establish contour plots of the first stress invariant. Since, as is shown later, this quantity satisfies the Laplace equation, various analogues can be used to define its spatial variation in terms of a set of isopachs, or contour plots of $(\sigma_1 + \sigma_3)$. Taken together with the photoelastic data, these plots allow the development of contour plots of the principal stresses throughout the problem domain.

It is clear from this brief discussion that the photoelastic method of stress analysis is a rather tedious way of predicting the stress distribution in a mine structure. It is therefore rarely used in design practice. However, the method is a useful research technique, for examining such problems as blocky media (Gaziev and Erlikmann, 1971) and three-dimensional structures (Timoshenko and Goodier, 1970) using the frozen-stress method.

A major detraction from the use of physical models of any sort for prediction of the rock mass response to mining is their high cost in time and effort. Since many mine design exercises involve parameter studies to identify an optimum mining strategy, construction and testing of models is inherently unsuited to the demands of the design process. Their use can be justified only for a single, confirmatory study of a proposed extraction strategy, to verify key aspects of the mine structural design.

6.2 Principles of classical stress analysis

A comprehensive description of the fundamentals of stress analysis is beyond the scope of this book. Texts such as those by Timoshenko and Goodier (1970) and Prager (1959) may be consulted as general discourses on engineering elasticity and plasticity, and related methods of the analysis of stress. The intention here is to identify key elements in the analytical determination of the stress and displacement fields in a body under applied load. The particular concern is to ensure that the conditions to be satisfied in any closed-form solution for the stress distribution in a body are appreciated. Techniques can then be established to verify the accuracy of any solution to a particular problem, such as the stress distribution around an underground excavation with a defined shape. This procedure is important, since there exist many analytical solutions, such as those collated by Poulos and Davis (1974), which can be used in excavation design. Use of any solution in a design exercise could not be justified unless suitable tests were applied to establish its validity.

The following discussion considers as an example a long, horizontal opening of regular cross section excavated in an elastic medium. A representative section of the problem geometry is shown in Figure 6.1. The far-field stresses are $p_{yy}\,(=p)$, $p_{xx}\,(=Kp)$, and p_{zz}, and other field stress components are zero, i.e. the long axis of the excavation is parallel to a pre-mining principal stress axis. The problem is thus one of simple plane strain. It should be noted that in dealing with excavations in a stressed medium, it is possible to consider two approaches in the analysis. In the first case, analysis proceeds in terms of displacements, strains and stresses induced by excavation in a stressed medium, and the final state of stress is obtained by superposition of the field stresses. Alternatively, the analysis proceeds by determining the displacements, strains and stresses obtained by applying the field stresses to a medium containing the excavation. Clearly, in the two cases, the equilibrium states of stress are identical, but the displacements are not. In this discussion, the first method of analysis is used.

The conditions to be satisfied in any solution for the stress and displacement distributions for a particular problem geometry and loading conditions are:

Figure 6.1 An opening in a medium subject to initial stresses, for which is required the distribution of total stresses and excavation-induced displacements.

(a) the boundary conditions for the problem;
(b) the differential equations of equilibrium;
(c) the constitutive equations for the material;
(d) the strain compatibility equations.

For the types of problem considered here, the boundary conditions are defined by the imposed state of traction or displacement at the excavation surface and

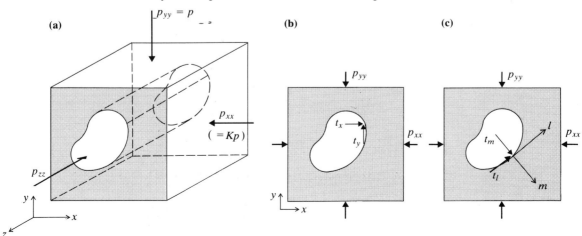

164

the far-field stresses. For example, an excavation surface is typically traction free, so that, in Figures 6.1b and c, t_x and t_y, or t_l and t_m, are zero over the complete surface of the opening. The other conditions are generally combined analytically to establish a governing equation, or field equation, for the medium under consideration. The objective then is to find the particular function which satisfies both the field equation for the system and the boundary conditions for the problem.

It is instructive to follow the procedure developed by Airy (1862) and described by Timoshenko and Goodier (1970), in establishing a particular form of the field equation for isotropic elasticity and plane strain. The differential equations of equilibrium in two dimensions for zero body forces are

$$\frac{\partial \sigma_{xx}}{\partial x} + \frac{\partial \sigma_{xy}}{\partial y} = 0 \qquad (6.1)$$

$$\frac{\partial \sigma_{xy}}{\partial x} + \frac{\partial \sigma_{yy}}{\partial y} = 0$$

or

$$\frac{\partial^2 \sigma_{xy}}{\partial x \, \partial y} = -\frac{\partial^2 \sigma_{xx}}{\partial x^2} = -\frac{\partial^2 \sigma_{yy}}{\partial y^2} \qquad (6.2)$$

For plane strain conditions and isotropic elasticity, strains are defined by

$$\varepsilon_{xx} = \frac{1}{E'} \, (\sigma_{xx} - v' \, \sigma_{yy})$$

$$\varepsilon_{yy} = \frac{1}{E'} \, (\sigma_{yy} - v' \, \sigma_{xx}) \qquad (6.3)$$

$$\gamma_{xy} = \frac{1}{G} \, \sigma_{xy}$$

$$= \frac{2(1 + v')}{E'} \, \sigma_{xy}$$

where

$$E' = \frac{E}{1 - v^2}$$

$$v' = \frac{v}{1 - v}$$

The strain compatibility equation in two dimensions is given by

$$\frac{\partial^2 \varepsilon_{yy}}{\partial x^2} + \frac{\partial^2 \varepsilon_{xx}}{\partial y^2} = \frac{\partial^2 \gamma_{xy}}{\partial x \, \partial y} \qquad (6.4)$$

165

Substituting the expressions for the strain components, (equations 6.3) in equation 6.4, and then equations 6.2 in the resultant expression yields

$$\frac{1}{E'}\left(\frac{\partial^2\sigma_{yy}}{\partial x^2} - v'\frac{\partial^2\sigma_{xx}}{\partial x^2}\right) + \frac{1}{E'}\left(\frac{\partial^2\sigma_{xx}}{\partial y^2} - v'\frac{\partial^2\sigma_{yy}}{\partial y^2}\right) = 2\frac{(1+v')}{E'}\frac{\partial^2\sigma_{xy}}{\partial x\,\partial y}$$

$$= -\frac{(1+v')}{E'}\left(\frac{\partial^2\sigma_{xx}}{\partial x^2} + \frac{\partial^2\sigma_{yy}}{\partial y^2}\right)$$

which becomes, on simplification,

$$\frac{\partial^2\sigma_{xx}}{\partial x^2} + \frac{\partial^2\sigma_{xx}}{\partial y^2} + \frac{\partial^2\sigma_{yy}}{\partial x^2} + \frac{\partial^2\sigma_{yy}}{\partial y^2} = 0$$

or

$$\left(\frac{\partial^2}{\partial x^2} + \frac{\partial^2}{\partial y^2}\right)(\sigma_{xx} + \sigma_{yy}) = 0 \tag{6.5}$$

Equation 6.5 demonstrates that the two-dimensional stress distribution for isotropic elasticity is independent of the elastic properties of the medium, and that the stress distribution is the same for plane strain as for plane stress. The latter point validates the use of photoelastic plane-stress models in estimating the stress distribution in bodies subject to loading in plane strain. Also, as noted in section 6.1, equation 6.5 demonstrates that the sum of the plane normal stresses, $\sigma_{xx} + \sigma_{yy}$, satisfies the Laplace equation.

The problem is to solve equations 6.1 and 6.5, subject to the imposed boundary conditions. The method suggested by Airy introduces a new function $U(x, y)$, in terms of which the stress components are defined by

$$\sigma_{xx} = \frac{\partial^2 U}{\partial y^2}$$

$$\sigma_{yy} = \frac{\partial^2 U}{\partial x^2} \tag{6.6}$$

$$\sigma_{xy} = \frac{\partial^2 U}{\partial x\,\partial y}$$

These expressions for the stress components satisfy the equilibrium equations, 6.1, identically. Introducing them in equation 6.5 gives

$$\nabla^4 U = 0$$

where

$$\nabla^2 = \frac{\partial^2}{\partial x^2} + \frac{\partial^2}{\partial y^2} \tag{6.7}$$

Equation 6.7 is called the biharmonic equation.

166

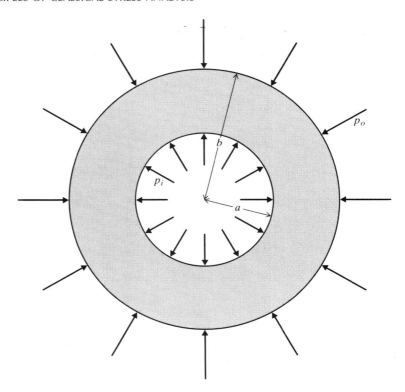

Figure 6.2 A thick-walled cylinder of elastic material, subject to interior pressure, p_i, and exterior pressure, p_o.

Several methods may be used to obtain solutions to particular problems in terms of an Airy stress function. Timoshenko and Goodier (1970) transform equations 6.5 and 6.6 to cylindrical polar co-ordinates, and illustrate a solution procedure by reference to a thick-walled cylinder subject to internal and external pressure, as shown in Figure 6.2. For this axisymmetric problem, the biharmonic equation assumes the form

$$\frac{\mathrm{d}^4 U}{\mathrm{d}r^4} + \frac{2}{r}\frac{\mathrm{d}^3 U}{\mathrm{d}r^3} - \frac{1}{r^2}\frac{\mathrm{d}^2 U}{\mathrm{d}r^2} + \frac{1}{r^3}\frac{\mathrm{d} U}{\mathrm{d}r} = 0$$

for which a general solution for U is given by

$$U = A \ln r + B r^2 \ln r + C r^2 + D$$

In this expression, the constants A, B, C, D are determined by considering both the requirement for uniqueness of displacements and the pressure boundary conditions for the problem. It can be shown that uniqueness of displacements requires $B = 0$, and that the stress components are then given by

$$\sigma_{rr} = \frac{A}{r^2} + 2C$$

$$\sigma_{\theta\theta} = -\frac{A}{r^2} + 2C$$

$$\sigma_{r\theta} = 0 \tag{6.8}$$

where

$$A = \frac{a^2 b^2 (p_i - p_o)}{(b^2 - a^2)}$$

$$2C = \frac{p_o b^2 - p_i a^2}{(b^2 - a^2)}$$

In these expressions, a and b are the inner and outer radii of the cylinder, and p_i and p_o are the pressures applied to its inner and outer surfaces.

For problems involving two-dimensional geometry and biaxial stress, the most elegant solution of the biharmonic equation is obtained in terms of complex variable theory. The topic is discussed in some detail by Jaeger and Cook (1979). Briefly, it is shown that the Airy stress function may be expressed as the real part of two analytic functions ϕ and χ of a complex variable z, in the form

$$U = R\left[\bar{z}\phi(z) + \chi(z)\right]$$

$$= \tfrac{1}{2}\left[\bar{z}\phi(z) + \overline{z\phi(z)} + \chi(z) + \overline{\chi(z)}\right] \tag{6.9}$$

Expressions for the stress components may then be established from U (equation 6.9) by successive differentiation. The displacements are obtained by setting up explicit expressions for the normal strain components ε_{xx} and ε_{yy} in terms of the stress components, and integrating. It is then found that stresses and displacements are given by

$$\sigma_{xx} + \sigma_{yy} = 4R\left[\phi'(z)\right]$$

$$-\sigma_{xx} + \sigma_{yy} + 2i\sigma_{xy} = 2\left[\bar{z}\phi''(z) + \psi'(z)\right] \tag{6.10}$$

$$2G(u_x + iu_y) = -\left[\mu\phi(z) - z\overline{\phi'(z)} - \overline{\psi(z)}\right]$$

where

$$\psi(z) = \chi'(z)$$

and

$$\mu = 3 - 4\nu \text{ for plane strain}$$

In applying these results, it is often useful to invoke the transformation between the rectangular Cartesian and cylindrical polar co-ordinates, given in complex variable form by

$$\sigma_{rr} + \sigma_{\theta\theta} = \sigma_{xx} + \sigma_{yy}$$

$$\tag{6.11}$$

$$-\sigma_{rr} + \sigma_{\theta\theta} + 2i\sigma_{r\theta} = \left[-\sigma_{xx} + \sigma_{yy} + 2i\sigma_{xy}\right] e^{i2\theta}$$

The solution to particular problems in two dimensions involves selection of suitable forms of the analytic functions $\phi(z)$ and $\chi(z)$. Many useful solutions involve polynomials in z or z^{-1}. For example, one may take

$$\phi(z) = 2cz, \qquad \psi(z) = \frac{d}{z} \tag{6.12}$$

where c and d are real.

Using the relations 6.9 and 6.10, equation 6.11 yields

$$\sigma_{rr} + \sigma_{\theta\theta} = 2c$$

$$-\sigma_{rr} + \sigma_{\theta\theta} + 2i\sigma_{r\theta} = -\frac{2d}{r^2}$$

so that

$$\sigma_{rr} = c + \frac{d}{r^2} \tag{6.13}$$

$$\sigma_{\theta\theta} = c - \frac{d}{r^2}$$

$$\sigma_{r\theta} = 0$$

For the axisymmetric problem defined by Figure 6.2, introducing the boundary conditions $\sigma_{rr} = p_i$ when $r = a$, and $\sigma_{rr} = p_o$ when $r = b$, into equations 6.13 yields

$$c = \frac{p_o b^2 - p_i a^2}{(b^2 - a^2)}$$

$$\tag{6.14}$$

$$d = \frac{a^2 b^2 (p_i - p_o)}{(b^2 - a^2)}$$

Equations 6.13 and 6.14 together are identical in form to equations 6.8. Thus the choice of the analytic functions in the form given by equation 6.12 is sufficient to represent conditions in a thick-walled cylinder subject to internal and external pressure. Expressions for the displacements induced in the cylinder by application of the internal and external pressures are obtained directly from $\phi(z)$, $\psi(z)$ and the third of equations 6.10.

It is clear that expressions for the stress and displacement distributions around openings of various shapes may be obtained by an heuristic selection of the forms of the analytic source functions. For example, for a circular hole with a traction-free surface, in a medium subject to a uniaxial stress p_{xx} at infinity, the source functions are

$$\phi(z) = \frac{1}{4} p_{xx} \left(z + \frac{A}{z} \right), \qquad \psi(z) = -\frac{1}{2} p_{xx} \left(z + \frac{B}{z} + \frac{C}{z^3} \right) \tag{6.15}$$

The real constants A, B, C are then selected to satisfy the known boundary conditions. These conditions are that, for all θ, $\sigma_{rr} = \sigma_{r\theta} = 0$ at $r = a$ (the hole boundary), and $\sigma_{rr} \rightarrow p_{xx}$ for $\theta = 0$ and $r \rightarrow \infty$. The resulting equations yield

$$A = 2a^2, \quad B = a^2, \quad C = -a^4$$

and the stress components are given by

$$\sigma_{rr} = \frac{1}{2} p_{xx} \left(1 - \frac{a^2}{r^2} \right) + \frac{1}{2} p_{xx} \left(1 - \frac{4a^2}{r^2} + \frac{3a^4}{r^4} \right) \cos 2\theta$$

$$\sigma_{\theta\theta} = \frac{1}{2} p_{xx} \left(1 + \frac{a^2}{r^2} \right) - \frac{1}{2} p_{xx} \left(1 + \frac{3a^4}{r^4} \right) \cos 2\theta \qquad (6.16)$$

$$\sigma_{r\theta} = -\frac{1}{2} p_{xx} \left(1 + \frac{2a^2}{r^2} - \frac{3a^4}{r^4} \right) \sin 2\theta$$

In spite of the apparent elegance of this procedure, it appears that seeking source analytic functions to suit particular problem geometries may be a tedious process. However, the power of the complex variable method is enhanced considerably by working in terms of a set of curvilinear co-ordinates, or through a technique called conformal mapping. There is considerable similarity between the two approaches, which are described in detail by Muskhelishvili (1963) and Timoshenko and Goodier (1970).

A curvilinear co-ordinate system is most conveniently invoked to match the shape of a relatively simple excavation cross section. For example, for an excavation of elliptical cross section, an orthogonal elliptical (ξ, η) co-ordinate system in the z plane can be readily established from an orthogonal Cartesian system in the ζ plane through the transformation

$$z = x + iy = c \cosh \zeta = c \cosh (\xi + i\eta)$$

so that

$$x = c \cosh \xi \cos \eta, \quad y = c \sinh \xi \sin \eta \qquad (6.17)$$

Equations 6.17 are also the parametric equations for an ellipse of major and minor axes $c \cosh \xi$, $c \sinh \xi$. The analysis of the stress distribution around the opening proceeds by expressing boundary conditions, source analytic functions and stress components in terms of the elliptic co-ordinates. The detail of the method then matches that described for a circular opening.

The conformal mapping method involves finding a transformation which will map a chosen geometric shape in the z plane into a circle of unit radius in the ζ plane. The problem boundary conditions are simultaneously transformed to an appropriate form for the ζ plane. The problem is solved in the ζ plane and the resulting expressions for stress and displacement distributions then inverted to obtain those for the real problem in the z plane. Problem geometries which have been analysed with this method include a square with rounded corners, an equilateral triangle and a circular hole with a concentric annular inclusion.

6.3 Closed-form solutions for simple excavation shapes

The preceding discussion has established the analytical basis for determining the stress and displacement distributions around openings with two-dimensional geometry. In rock mechanics practice, there is no need for an engineer to undertake the analysis for particular problem configurations. It has been noted already that comprehensive collections of solutions exist for the analytically tractable problems. The collection by Poulos and Davis (1974) is the most thorough. The practical requirement is to be able to verify any published solution which is to be applied to a design problem. This is achieved by systematic checking to determine if the solution satisfies the governing equations and the specified far-field and boundary conditions. The verification tests to be undertaken therefore match the sets of conditions employed in the development of the solution to a problem, as defined in section 6.1, i.e. imposed boundary conditions, differential equations of equilibrium, strain compatibility equations and constitutive equations. The method of verification can be best demonstrated by the following example which considers particular features of the stress distribution around a circular opening.

6.3.1 Circular excavation

Figure 6.3a shows the circular cross section of a long excavation in a medium subject to biaxial stress, defined by $p_{yy} = p$, and $p_{xx} = Kp$. The stress distribution around the opening may be readily obtained from equations 6.6, by superimposing the induced stresses associated with each of the field stresses p and Kp. The complete solutions for stress and displacement distributions around the circular opening, originally due to Kirsch (1898), are

Figure 6.3 Problem geometry, co-ordinate system and nomenclature for specifying the stress and displacement distribution around a circular excavation in a biaxial stress field.

$$\sigma_{rr} = \frac{p}{2}\left[(1+K)\left(1-\frac{a^2}{r^2}\right) - (1-K)\left(1-4\frac{a^2}{r^2}+\frac{3a^4}{r^4}\right)\cos 2\theta\right]$$

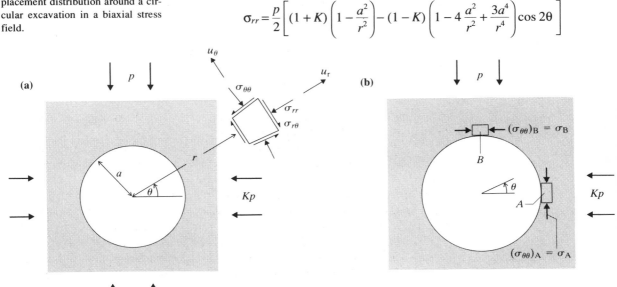

$$\sigma_{\theta\theta} = \frac{p}{2}\left[(1+K)\left(1+\frac{a^2}{r^2}\right)+(1-K)\left(1+\frac{3a^4}{r^4}\right)\cos 2\theta\right]$$

$$\sigma_{r\theta} = \frac{p}{2}\left[(1-K)\left(1+\frac{2a^2}{r^2}-\frac{3a^4}{r^4}\right)\sin 2\theta\right] \qquad (6.18)$$

$$u_r = -\frac{pa_2}{4Gr}\left[(1+K)-(1-K)\left\{4(1-\nu)-\frac{a^2}{r^2}\right\}\cos 2\theta\right]$$

$$u_\theta = -\frac{pa_2}{4Gr}\left[(1-K)\left\{2(1-2\nu)+\frac{a^2}{r^2}\right\}\sin 2\theta\right]$$

In these expressions u_r, u_θ are displacements induced by excavation, while σ_{rr}, $\sigma_{\theta\theta}$, $\sigma_{r\theta}$ are total stresses after generation of the opening.

By putting $r = a$ in equation 6.18, the stresses on the excavation boundary are given as

$$\sigma_{\theta\theta} = p\,[(1+K)+2(1-K)\cos 2\theta]$$

$$\sigma_{rr} = 0 \qquad (6.19)$$

$$\sigma_{r\theta} = 0$$

Equations 6.19 confirm that the solutions satisfy the imposed condition that the excavation boundary is traction free. Similarly, for $\theta = 0$, and r large, the stress components are given by

$$\sigma_{rr} = Kp, \quad \sigma_{\theta\theta} = p, \quad \sigma_{r\theta} = 0$$

so that the far-field stresses recovered from the solutions correspond to the imposed field stresses. With regard to the equilibrium requirements, the differential equation of equilibrium in two dimensions for the tangential direction and no body forces, is

$$\frac{\partial \sigma_{r\theta}}{\partial r}+\frac{1}{r}\frac{\partial \sigma_{\theta\theta}}{\partial \theta}+\frac{2\sigma_{r\theta}}{r}=0$$

Evaluating each of the terms of this equation from the expressions for $\sigma_{\theta\theta}$ and $\sigma_{r\theta}$ leads to

$$\frac{\partial \sigma_{r\theta}}{\partial r}=p(1-K)\left(-\frac{2a^2}{r^3}+\frac{6a^4}{r^5}\right)\sin 2\theta$$

$$\frac{1}{r}\frac{\partial \sigma_{\theta\theta}}{\partial \theta}=-p(1-K)\left(\frac{1}{r}+\frac{3a^4}{r^5}\right)\sin 2\theta$$

$$\frac{2\sigma_{r\theta}}{r}=p(1-K)\left(\frac{1}{r}+\frac{2a^2}{r^3}-\frac{3a^4}{r^5}\right)\sin 2\theta$$

172

Inspection indicates that the equilibrium equation for the tangential direction is satisfied by these expressions. It is obviously an elementary exercise to confirm that the stress components satisfy the other two-dimensional (i.e. radial) equilibrium equation. Similarly, the strain components can be determined directly, by differentiation of the solutions for displacements, and the expressions for stress components derived by employing the stress–strain relationships. Such a test can be used to confirm the mutual consistency of the solutions for stress and displacement components.

Boundary stresses.   Equations 6.19 define the state of stress on the boundary of a circular excavation in terms of the co-ordinate angle θ. Clearly, since the surface is traction free, the only non-zero stress component is the circumferential component $\sigma_{\theta\theta}$. For $K < 1.0$, the maximum and minimum boundary stresses occur in the side wall ($\theta = 0$) and crown ($\theta = \pi/2$) of the excavation. Referring to Figure 6.3b, these stresses are defined by the following:

$$\text{at point A:} \quad \theta = 0, \quad (\sigma_{\theta\theta})_A = \sigma_A = p(3 - K)$$

$$\text{at point B:} \quad \theta = \frac{\pi}{2}, \quad (\sigma_{\theta\theta})_B = \sigma_B = p(3K - 1)$$

These expressions indicate that, for the case when $K = 0$, i.e. a uniaxial field directed parallel to the y axis, the maximum and minimum boundary stresses are

$$\sigma_A = 3p, \quad \sigma_B = -p$$

These values represent upper and lower limits for stress concentration at the boundary. That is, for any value of $K > 0$, the sidewall stress is less than $3p$, and the crown stress is greater than $-p$. The existence of tensile boundary stresses in a compressive stress field is also noteworthy.

In the case of a hydrostatic stress field ($K = 1$), equation 6.19 becomes

$$\sigma_{\theta\theta} = 2p$$

i.e. the boundary stress takes the value $2p$, independent of the co-ordinate angle θ. This represents the optimum distribution of local stress, since the boundary is uniformly compressed over the complete excavation periphery.

Equations 6.18 are considerably simplified for a hydrostatic stress field, taking the form

$$\sigma_{rr} = p\left(1 - \frac{a^2}{r^2}\right)$$

$$\sigma_{\theta\theta} = p\left(1 + \frac{a^2}{r^2}\right) \tag{6.20}$$

$$\sigma_{r\theta} = 0$$

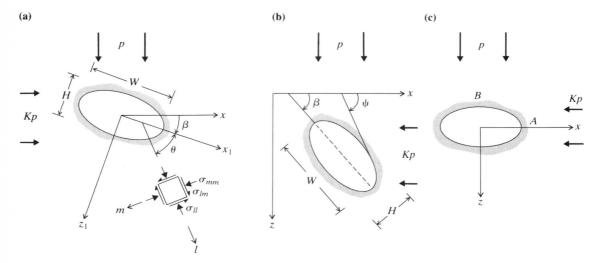

Figure 6.4 Problem geometry, co-ordinate axes and nomenclature for specifying the stress distribution around an elliptical excavation in a biaxial stress field.

The independence of the stress distribution of the co-ordinate angle θ, and the fact that $\sigma_{r\theta}$ is everywhere zero, indicates that the stress distribution is axisymmetric.

6.3.2 Elliptical excavation

Solutions for the stress distribution in this case are quoted by Poulos and Davis (1974) and Jaeger and Cook (1979). In both cases, the solutions are expressed in terms of elliptical curvilinear co-ordinates. Their practical use is somewhat cumbersome. Bray (1977) produced a set of formulae which results in considerable simplification of the calculation of the state of stress at points in the medium surrounding an elliptical opening. The problem geometry is defined in Figure 6.4a, with the global x axis parallel to the field stress component Kp, and with an axis of the ellipse defining the local x_1 axis for the opening. The width, W, of the ellipse is measured in the direction of the x_1 axis, and the height, H, in the direction of the local z_1 axis. The attitude of the ellipse in the biaxial stress field is described by the angle β between the global x and local x_1 axes. The position of any point in the medium is defined by its Cartesian co-ordinates (x_1, z_1) relative to the local x_1, z_1 axes.

Bray's solution specifies the state of stress at a point in the medium in terms of a set of geometrical parameters, and relative to a set of local axes, denoted l and m, centred on the point of interest. The various geometric parameters are defined as follows:

$$e_0 = \frac{(W + H)}{(W - H)}$$

$$b = \frac{4\,(x_1^2 + z_1^2)}{(W^2 - H^2)}$$

$$d = \frac{8(x_1^2 - z_1^2)}{(W^2 - H^2)} - 1$$

$$u = b + \frac{e_0}{|e_0|}(b^2 - d)^{1/2}$$

$$e = u + \frac{e_0}{|e_0|}(u^2 - 1)^{1/2}$$

$$\psi = \arctan\left[\left(\frac{e+1}{e-1}\right)\frac{z_1}{x_1}\right]$$

$$\theta = \arctan\left[\left(\frac{e+1}{e-1}\right)^2\frac{z_1}{x_1}\right]$$

$$C = 1 - ee_0$$

$$J = 1 + e^2 - 2e\cos 2\psi$$

The stress components are given by

$$\sigma_{ll} = \frac{p(e_0 - e)}{J^2}\left\{(1+K)(e^2-1)\frac{C}{2e_0}\right.$$

$$\left. + (1-K)\left[\left[\frac{J}{2}(e-e_0) + Ce\right]\cos 2(\psi + \beta) - C\cos 2\beta\right]\right\}$$

$$\sigma_{mm} = \frac{p}{J}\left\{(1+K)(e^2-1) + 2(1-K)e_0[e\cos 2(\psi + \beta) - \cos 2\beta]\right\} - \sigma_{ll}$$

$$\sigma_{lm} = \frac{p(e_0 - e)}{J^2}\left\{(1+K)\frac{Ce}{e_0}\sin 2\psi + (1-K)\left[e(e_0 + e)\sin 2\beta\right.\right.$$

$$\left.\left. + e\sin 2(\psi - \beta) - \left[\frac{J}{2}(e_0 + e) + e^2 e_0\right]\sin 2(\psi + \beta)\right]\right\} \qquad (6.21)$$

In applying these formulae, it should be noted that the angle θ, defining the orientation of the local reference axes l, m relative to the ellipse local axes x_1, z_1, is not selected arbitrarily. It is defined uniquely in terms of the ellipse shape and the point's position co-ordinates.

The boundary stresses around an elliptical opening with axes inclined to the field stress directions are obtained by selecting values of x_1, z_1 which fall on the boundary contour. For this case, $e = e_0$, $\sigma_{ll} = \sigma_{lm} = 0$, and the l axis is directed normal to the boundary. For the problem geometry defined in Figure 6.4b, the boundary stress is given by

$$\sigma = \frac{p}{2q}\left\{(1+K)[(1+q^2) + (1-q^2)\cos 2(\psi - \beta)]\right.$$

$$\left. - (1-K)[(1+q)^2\cos 2\psi + (1-q^2)\cos 2\beta]\right\} \qquad (6.22)$$

When the axes of the ellipse are oriented parallel to the field stress directions, equation 6.22 reduces to

$$\sigma = \frac{p}{2q} \{(1 + K) [(1 + q^2) + (1 - q^2) \cos 2\psi]$$

$$- (1 - K) [(1 + q)^2 \cos 2\psi + (1 - q^2)] \} \qquad (6.23)$$

In assessing the state of stress in the side wall and crown of an elliptical excavation, i.e. for points A and B ($\psi = \pi/2$, $\psi = 0$) in Figure 6.4c, it is useful to introduce the effect of the local boundary curvature on boundary stress. For an ellipse of major and minor axes 2a and 2b, the radius of curvature at points A and B, ρ_A and ρ_B, is found from simple analytical geometry (Lamb, 1956) to be

$$\rho_A = \frac{b^2}{a}, \quad \rho_B = \frac{a^2}{b}$$

Since $q = W/H = a/b$, it follows that

$$q = \sqrt{\frac{W}{2\rho_A}}, \quad \frac{1}{q} = \sqrt{\frac{H}{2\rho_B}}$$

Sidewall and crown stresses in the ellipse boundary, for the problem defined in Figure 6.4c, may then be expressed as

$$\sigma_A = p(1 - K + 2q) = p\left(1 - K + \sqrt{\frac{2W}{\rho_A}}\right)$$

$$(6.24)$$

$$\sigma_B = p\left(K - 1 + \frac{2K}{q}\right) = p\left(1 - K + \sqrt{\frac{2H}{\rho_B}}\right)$$

It will be shown later that the formulae for stress distribution about ideal excavation shapes, such as a circle and an ellipse, can be used to establish useful working ideas of the state of stress around regular excavation shapes.

6.4 Computational methods of stress analysis

The preceding discussion indicated that even for a simple two-dimensional excavation geometry, such as an elliptical opening, quite complicated expressions are obtained for the stress and displacement distributions. Many design problems in rock mechanics practice involve more complex geometry. Although insight into the stress distributions around complex excavation shapes may be obtained from the closed form solutions for approximating simple shapes, it is sometimes necessary to seek a detailed understanding of stress distribution for more complicated configurations. Other conditions which arise which may require more powerful analytical tools include non-homogeneity of the rock mass in the problem domain and non-linear constitutive behaviour of the medium. These conditions generally present difficulties which are not amenable to solution by conventional analysis.

Solutions to the more complex excavation design problems may usually be obtained by use of computational procedures. The use of these techniques is now firmly embedded in rock mechanics practice. Their application will undoubtedly increase in the future, as computing power increases and its cost decreases. In particular, it can be expected that engineers will have immediate access to design-dedicated desk-top computers. The following discussion is intended to indicate the potential of various computational methods of analysis in excavation design. The description is limited to the formulation of solution procedures for plane geometric problems, with the implication that, conceptually at least, there is no difficulty in extending a particular procedure to three-dimensional geometry.

Computational methods of stress analysis fall into two categories – differential methods and integral methods. In differential methods, the problem domain is divided (discretised) into a set of subdomains or elements. A solution procedure may then be based on numerical approximations of the governing equations, i.e. the differential equations of equilibrium, the strain–displacement relations and the stress–strain equations, as in classical finite difference methods. Alternatively, the procedure may exploit approximations to the connectivity of elements, and continuity of displacements and stresses between elements, as in the finite element method.

The characteristic of integral methods of stress analysis is that a problem is specified and solved in terms of surface values of the field variables of traction and displacement. Since the problem boundary only is defined and discretised, the so-called boundary element methods of analysis effectively provide a unit reduction in the dimensional order of a problem. The implication is a significant advantage in computational efficiency, compared with the differential methods.

The differences in problem formulation between differential and integral methods of analysis lead to various fundamental and operational advantages and disadvantages for each. For a method such as the finite element method, non-linear and heterogeneous material properties may be readily accommodated, but the outer boundary of the problem domain is defined arbitrarily, and discretisation errors occur throughout the domain. On the other hand, boundary element methods model far-field boundary conditions correctly, restrict discretisation errors to the problem boundary, and ensure fully continuous variation of stress and displacement throughout the medium. However, these methods are best suited to linear material behaviour and homogeneous material properties; non-linear behaviour and medium heterogeneity negate the intrinsic simplicity of a boundary element solution procedure.

In describing various computational procedures, the intention is not to provide a comprehensive account of the methods. Instead, the aim is to identify the essential principles of each method.

6.5 The boundary element method

Attention in the following discussion is confined to the case of a long excavation of uniform cross section, developed in an infinite elastic body subject to initial

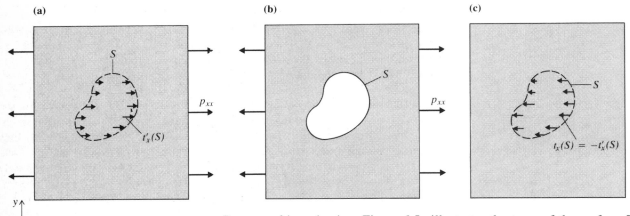

(a)

(b)

(c)

Figure 6.5 Superposition scheme demonstrating that generation of an excavation is mechanically equivalent to introducing a set of tractions on a surface in a continuum.

stress. By way of introduction, Figure 6.5a illustrates the trace of the surface S of an excavation to be generated in a medium subject to uniaxial stress, p_{xx}, in the x direction. At any point on the surface, the pre-excavation load condition is defined by a traction $t'_x(S)$. After excavation of the material within S, the surface of the opening is to be traction free, as shown in Figure 6.5b. This condition is achieved if a distribution of surface traction, $t_x(S)$, equal in magnitude, but opposite in sense to that shown in Figure 6.5a, is induced in a medium that is stress free at infinity. The required induced traction distribution is shown in Figure 6.5c. Superposition of Figures 6.5a and c confirms that their resultant is a stressed medium with an internal traction-free surface S. It is concluded from this that if a procedure is established for solving the problem illustrated in Figure 6.5c, the solution to the real problem (Figure 6.5b) is immediately available. Thus the following discussion deals with excavation-induced tractions, displacements and stresses, and the method of achieving particular induced traction conditions on a surface in a continuum.

For a medium subject to general biaxial stress, the problem posed involves distributions of induced tractions, $t_x(S), t_y(S)$, at any point on the surface S, as illustrated in Figure 6.6a. In setting up the boundary element solution procedure, the requirements are to discretise and describe algebraically the surface S, and to find a method of satisfying the imposed induced traction conditions on S.

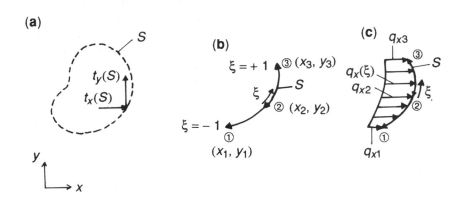

Figure 6.6 Surface, element and load distribution description for development of a quadratic, indirect boundary element formulation.

The geometry of the problem surface S is described conveniently in terms of the position co-ordinates, relative to global x, y axes, of a set of nodes, or collocation points, disposed around S. Three adjacent nodes, forming a representative boundary element of the surface S, are shown in Figure 6.6b. The complete geometry of this element of the surface may be approximated by a suitable interpolation between the position co-ordinates of the nodes. In Figure 6.6b an element intrinsic co-ordinate ξ is defined, with the property that $-1 \leq \xi \leq 1$ over the range of the element. Considering nodes 1, 2, 3 of the element, suppose that a set of functions is defined by

$$N_1 = -\tfrac{1}{2}\xi(1-\xi)$$

$$N_2 = (1-\xi^2) \tag{6.25}$$

$$N_3 = \tfrac{1}{2}\xi(1+\xi)$$

The property of these functions is that each takes the value of unity at a particular node, and zero at the other two nodes. They are therefore useful for interpolation of element geometry from the nodal co-ordinates in the form

$$x(\xi) = x_1 N_1 + x_2 N_2 + x_3 N_3 = \Sigma\, x_\alpha N_\alpha$$

$$y(\xi) = \Sigma\, y_\alpha N_\alpha \tag{6.26}$$

It is seen that the properties of the interpolation functions 6.25 ensure that equations 6.26 return the position co-ordinates of the nodes, 1, 2, 3, for $\xi = -1, 0, 1$ respectively. Also, equations 6.25 and 6.26 can be interpreted to define a transformation from the element local co-ordinate ξ to the global x, y system.

In seeking a solution to the boundary value problem posed in Figure 6.6a, it is known that the stress and displacement distribution in the medium exterior to S is uniquely determined by the conditions on the surface S (Love, 1944). Thus if some method can be established for inducing a traction distribution on S identical to the known, imposed distribution, the problem is effectively solved. Suppose, for example, continuous distributions $q_x(S)$, $q_y(S)$, of x- and y-directed line load singularities are disposed over the surface S in the continuum. Using the solutions for stress components due to unit line loads (Appendix B), and the known tangent to S at any point i, the x and y component tractions T_{xi}^x, T_{yi}^x and T_{xi}^y, T_{yi}^y induced by the distributions of x- and y-directed line loads, can be determined. When point i is a node of the surface, the condition to be achieved to realise the known condition on S is

$$\int_S [q_x(S)T_{xi}^x + q_y(S)T_{xi}^y]\,\mathrm{d}S = t_{xi}$$

$$\tag{6.27}$$

$$\int_S [q_x(S)T_{yi}^x + q_y(S)T_{yi}^y]\,\mathrm{d}S = t_{yi}$$

Discretisation of equation 6.27 requires that the surface distributions of fictitious load, $q_x(S)$ and $q_y(S)$, be expressed in terms of the nodal values of these quantities. Suppose that, for any element, the interpolation functions 6.25 are also used to define fictitious load distributions with respect to the element intrinsic co-ordinate ξ, i.e.

$$q_x(\xi) = q_{x1}N_1 + q_{x2}N_2 + q_{x3}N_3 = \Sigma\, q_{x\alpha}N_\alpha$$

$$q_y(\xi) = \Sigma\, q_{y\alpha}N_\alpha$$

(6.28)

The discretised form of equation 6.27 can be written

$$\sum_{j=1}^{n} \int_{S_e} [q_x(S)\, T_{xi}^x + q_y(S)\, T_{xi}^y]\, dS = t_{xi} \tag{6.29}$$

$$\sum_{j=1}^{n} \int_{S_e} [q_x(S)\, T_{yi}^x + q_y(S)\, T_{yi}^y]\, dS = t_{yi} \tag{6.30}$$

where n is the number of boundary elements, and each surface integral is evaluated over the range S_e^j of each boundary element j. Considering a particular element, one of the surface integrals can be expressed by

$$\int_{S_e} [q_x(S)\, T_{xi}^x + q_y(S)\, T_{xi}^y]\, dS =$$

$$\Sigma\, q_{x\alpha} \int_{-1}^{1} N_\alpha(\xi)\, T_{xi}^x(\xi)\, \frac{dS}{d\xi}\, d\xi + \Sigma\, q_{y\alpha} \int_{-1}^{1} N_\alpha(\xi)\, T_{xi}^y(\xi)\, \frac{dS}{d\xi}\, d\xi \tag{6.31}$$

The integrals of the interpolation function (N)–kernel (T) products defined in equation 6.31 can be evaluated readily by standard Gaussian quadrature methods. When all components of equations 6.29 and 6.30 have been calculated using the procedure defined in equation 6.31, it is found that for the m boundary nodes

$$\sum_{j=1}^{m} (q_{xj}\, T_x^{x*} + q_{yj}\, T_x^{y*}) = t_{xi}$$

$$\sum_{j=1}^{m} (q_{xj}\, T_x^{x*} + q_{yj}\, T_y^{y*}) = t_{yi}$$

(6.32)

where T_x^{x*}, etc., are the results of the various interpolation function–kernel integrations and, for the end nodes of each element, a summation with the appropriate integral for the adjacent element. When equations similar to 6.32 have been established for each of the m boundary nodes, they may be recast in the form

$$[\mathbf{T}^*]\,[\mathbf{q}] = [\mathbf{t}] \tag{6.33}$$

Equation 6.33 represents a set of $2m$ simultaneous equations in $2m$ unknowns, which are the nodal values of fictitious boundary load intensity.

Once equation 6.33 has been solved for the vector $[\mathbf{q}]$ of nodal load intensities, all other problem unknowns can be calculated readily. For example, nodal displacements, or displacements at an internal point i in the medium, can be determined from

$$u_{xi} = \int_S [q_x(S)\,U_{xi}^x + q_y(S)\,U_{xi}^y]\,\mathrm{d}S$$

$$\tag{6.34}$$

$$u_{yi} = \int_S [q_x(S)\,U_{yi}^x + q_y(S)\,U_{yi}^y]\,\mathrm{d}S$$

Similarly, stress components at an internal point i in the medium are given by

$$\sigma_{xxi} = \int_S [q_x(S)\,\sigma_{xxi}^x + q_y(S)\,\sigma_{xxi}^y]\,\mathrm{d}S$$

$$\sigma_{yyi} = \int_S [q_x(S)\,\sigma_{yyi}^x + q_y(S)\,\sigma_{yyi}^y]\,\mathrm{d}S \tag{6.35}$$

$$\sigma_{xyi} = \int_S [q_x(S)\,\sigma_{xyi}^x + q_y(S)\,\sigma_{xyi}^y]\,\mathrm{d}S$$

In equations 6.34 and 6.35, U_{xi}^x, σ_{xxi}^x, U_{xi}^y, σ_{xxi}^y, etc., are displacements and stresses induced by x- and y-directed unit line loads, given by the expressions in Appendix B. Equations 6.34 and 6.35 may be discretised using the methods defined by equations 6.29–6.31, and all the resulting integrals can be evaluated using standard quadrature formulae.

In setting up equations 6.27, 6.34 and 6.35, the principle of superposition is exploited implicitly. Thus the method is applicable to linear elastic, or at least piece-wise linear elastic, behaviour of the medium. Also, since both element geometry and fictitious load variation are described in terms of quadratic interpolation functions (Equations 6.25), the method may be described as a quadratic, isoparametric formulation of the boundary element method.

The introduction of fictitious load distributions, $q_x(S)$, $q_y(S)$, in the solution procedure to satisfy the imposed boundary conditions results in this approach being called an indirect formulation of the boundary element method. In the alternative direct formulation, the algorithm is developed from a relation between nodal displacements $[u]$ and tractions $[t]$, based on the Betti Reciprocal Work Theorem (Love, 1944). These formulations are also isoparametric, with element geometry, surface tractions and displacements following imposed quadratic variation with respect to the element intrinsic co-ordinate.

181

6.6 The finite element method

The basis of the finite element method is the definition of a problem domain surrounding an excavation, and division of the domain into an assembly of discrete, interacting elements. Figure 6.7a illustrates the cross section of an underground opening generated in an infinite body subject to initial stresses p_{xx}, p_{yy}, p_{xy}. In Figure 6.7b, the selected boundary of the problem domain is indicated, and appropriate supports and conditions are prescribed at the arbitrary outer boundary to render the problem statically determinate. The domain has been divided into a set of triangular elements. A representative element of the set is illustrated in Figure 6.7c, with the points i, j, k defining the nodes of the element. The problem is to determine the state of total stress, and the excavation-induced displacements, throughout the assembly of finite elements. The following description of the solution procedure is based on that by Zienkiewicz (1977).

In the displacement formulation of the finite element method considered here, the initial step is to choose a set of functions which define the displacement components at any point within a finite element, in terms of the nodal displacements. The various steps of the solution procedure then develop from the imposed displacement field. Thus, since strain components are defined uniquely in terms of various derivatives of the displacements, the imposed displacement variation defines the state of strain throughout an element. These induced strains and the elastic properties of the medium together determine the induced stresses in an element. Superposition of the initial and the induced stresses yields total stresses in the element.

The assumption in the finite element method is that transmission of internal forces between the edges of adjacent elements can be represented by interactions at the nodes of the elements. It is therefore necessary to establish expressions for nodal forces which are statically equivalent to the forces acting between elements along the respective edges. Thus the procedure seeks to analyse the continuum problem (Figure 6.7a) in terms of sets of nodal forces and displacements for the discretised domain (Figure 6.7b). The solution procedure described here, for

Figure 6.7 Development of a finite element model of a continuum problem, and specification of element geometry and loading for a constant strain, triangular finite element.

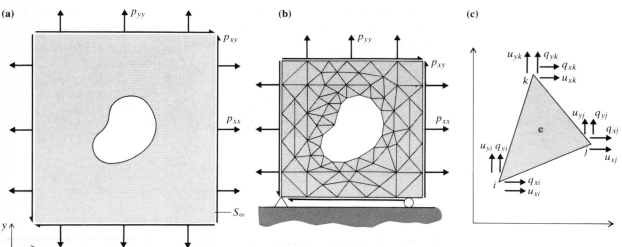

purposes of illustration, considers triangular element geometry, linear variation of displacement with respect to element intrinsic co-ordinates, and resultant constant stress within an element.

6.6.1 Displacement variation

In Figure 6.7c, induced nodal displacements are u_{xi}, u_{yi}, etc., and displacements $[\mathbf{u}]$ at any point within the element are to be obtained by suitable interpolation from the nodal values. Introducing a matrix of interpolation functions, $[\mathbf{N}]$, a suitable interpolation formula is

$$[\mathbf{u}] = \begin{bmatrix} u_x \\ u_y \end{bmatrix} = \Sigma\,[\mathbf{N}_i]\,[\mathbf{u}_i] = [\overline{\mathbf{N}}_i, \overline{\mathbf{N}}_j, \overline{\mathbf{N}}_k] \begin{bmatrix} \mathbf{u}_i \\ \mathbf{u}_j \\ \mathbf{u}_k \end{bmatrix} \qquad (6.36)$$

$$= [\mathbf{N}]\,[\mathbf{u}^e]$$

where

$$[\mathbf{u}_i] = \begin{bmatrix} u_{xi} \\ u_{yi} \end{bmatrix} \qquad [\mathbf{N}_i] = \overline{\mathbf{N}}_i = \begin{bmatrix} N_i & 0 \\ 0 & N_i \end{bmatrix}$$

the components of $[\mathbf{N}]$, i.e. the terms N_i, are prescribed functions of position, and $[\mathbf{u}^e]$ is a column vector listing the nodal displacements $u_{xi}, u_{yi}, u_{xj} \ldots$ etc.

The interpolation functions which constitute the elements of $[\mathbf{N}]$ must be chosen to return the nodal displacements at each of the nodes. This requires that

$$[\mathbf{N}_i]_{xi,\,yi} = [\mathbf{I}]$$

$$[\mathbf{N}_i]_{xi,\,yj} = [\mathbf{0}], \text{ etc.}$$

where $[\mathbf{I}]$ and $[\mathbf{0}]$ are the identity and null matrices respectively. Also, since both components of displacement at a point are to be interpolated in the same way, it is clear that

$$[\mathbf{N}_i] = N_i\,[\mathbf{I}]$$

where N_i is a scalar function of position within the element.

A simple development of a linear interpolation function is demonstrated by representing the displacements in terms of linear functions of position, i.e.

$$u_x = \alpha_1 + \alpha_2 x + \alpha_3 y$$

$$u_y = \alpha_4 + \alpha_5 x + \alpha_6 y \qquad (6.37)$$

The six interpolation constants are determined by ensuring that the displacements u_x, u_y assume the nodal values when nodal co-ordinates are inserted in

183

equation 6.37. Thus α_1, α_2, α_3 are determined by solving the simultaneous equations

$$u_{xi} = \alpha_1 + \alpha_2 x_i + \alpha_3 y_i$$

$$u_{xj} = \alpha_1 + \alpha_2 x_j + \alpha_3 y_j$$

$$u_{xk} = \alpha_1 + \alpha_2 x_k + \alpha_3 y_k$$

Solution for α_1, α_2, α_3 and some rearrangement, produces

$$u_x = \frac{1}{2\Delta} [(a_i + b_i x + c_i y)u_{xi} + (a_j + b_j x + c_j y)u_{xj}$$

$$+ (a_k + b_k x + c_k y)u_{xk}] \qquad (6.38)$$

where
$$a_i = x_j y_k - x_k y_j$$

$$b_i = y_j - y_k$$

$$c_i = x_k - x_j$$

with cyclic permutation of i, j, k to obtain a_j, etc., and

$$2\Delta = 2 \times \text{area of the triangular element}$$

$$= 2 \begin{bmatrix} 1 & x_i & y_i \\ 1 & x_j & y_j \\ 1 & x_k & y_k \end{bmatrix}$$

Solution for α_4, α_5, α_6 yields an interpolation function for u_y identical to equation 6.38, with u_{yi} replacing u_{xi}, etc. The variation of displacements throughout an element is therefore described by

$$[\boldsymbol{u}] = \begin{bmatrix} u_x \\ u_y \end{bmatrix} = [\mathbf{N}] [\mathbf{u}^e] = [N_i \mathbf{I}, N_j \mathbf{I}, N_k \mathbf{I}] [u^e] \qquad (6.39)$$

where $N_i = (a_i + b_i x + c_i y)/2\Delta$, with similar expressions N_j, N_k and $\mathbf{I}$ is a 2×2 identity matrix.

By defining the displacement field in an element in terms of the nodal displacements, the interpolation procedure ensures continuity of the displacements both across an element interface with an adjacent element, and within the element itself.

Once the displacement field in an element is defined, the state of strain can be established from the strain–displacement relations. For plane strain problems, a strain vector may be defined by

$$[\varepsilon] = \begin{bmatrix} \varepsilon_{xx} \\ \varepsilon_{yy} \\ \gamma_{xy} \end{bmatrix} = \begin{bmatrix} \dfrac{\partial u_x}{\partial x} \\ \dfrac{\partial u_y}{\partial y} \\ \dfrac{\partial u_x}{\partial y} + \dfrac{\partial u_y}{\partial x} \end{bmatrix} = \begin{bmatrix} \dfrac{\partial}{\partial x} & 0 \\ 0 & \dfrac{\partial}{\partial y} \\ \dfrac{\partial}{\partial y} & \dfrac{\partial}{\partial x} \end{bmatrix} \begin{bmatrix} u_x \\ u_y \end{bmatrix}$$

or

$$[\varepsilon] = [L][u] \tag{6.40}$$

Since displacements are specified by equation 6.39, equation 6.40 becomes

$$[\varepsilon] = [L][N][u^e] = [B][u^e]$$

where

$$[B] = \begin{bmatrix} \dfrac{\partial N_i}{\partial x} & 0 & \dfrac{\partial N_j}{\partial x} & 0 & \dfrac{\partial N_k}{\partial x} & 0 \\ 0 & \dfrac{\partial N_i}{\partial y} & 0 & \dfrac{\partial N_j}{\partial y} & 0 & \dfrac{\partial N_k}{\partial y} \\ \dfrac{\partial N_i}{\partial y} & \dfrac{\partial N_i}{\partial x} & \dfrac{\partial N_j}{\partial y} & \dfrac{\partial N_j}{\partial x} & \dfrac{\partial N_k}{\partial y} & \dfrac{\partial N_k}{\partial x} \end{bmatrix}$$

For the case of linear displacement variation, the terms $\partial N_i/\partial x$, etc., of the **B** matrix are constant, and thus the strain components are invariant over the element.

6.6.2 Stresses within an element

The state of total stress within an element is the sum of the induced stresses and the initial stresses. Ignoring any thermal strains, total stresses, for conditions of plane strain, are given by

$$\begin{bmatrix} \sigma_{xx} \\ \sigma_{yy} \\ \sigma_{xy} \end{bmatrix} = \dfrac{E(1-\nu)}{(1+\nu)(1-\nu)} \begin{bmatrix} 1 & \nu/(1-\nu) & 0 \\ \nu/(1-\nu) & 1 & 0 \\ 0 & 0 & (1-2\nu)/2(1-\nu) \end{bmatrix} \begin{bmatrix} \varepsilon_{xx} \\ \varepsilon_{yy} \\ \gamma_{xy} \end{bmatrix} + \begin{bmatrix} \sigma_{xx}^0 \\ \sigma_{yy}^0 \\ \sigma_{xy}^0 \end{bmatrix}$$

or

$$[\sigma] = [D][\varepsilon] + [\sigma^0]$$

$$= [D][B][u^e] + [\sigma^0] \tag{6.41}$$

185

where $[\boldsymbol{\sigma}]$ is the vector of total stresses, $[\mathbf{D}]$ is the elasticity matrix, and $[\boldsymbol{\sigma}^0]$ is the vector of initial stresses.

6.6.3 Equivalent nodal forces

The objective in the finite element method is to establish nodal forces q_{xi}, q_{yi} etc., equivalent to the internal forces operating between the edges of elements, and the body force

$$[\mathbf{b}] = \left[\begin{array}{c} b_x \\ b_y \end{array}\right]$$

operating per unit volume of the element. The internal nodal forces are determined by imposing a set of virtual displacements $[\delta\mathbf{u}^e]$ at the nodes, and equating the internal and external work done by the various forces in the displacement field. For imposed nodal displacements $[\delta\mathbf{u}^e]$, displacements and strains within an element are

$$[\delta\mathbf{u}] = [\mathbf{N}] [\delta\mathbf{u}^e], \quad [\delta\boldsymbol{\varepsilon}] = [\mathbf{B}] [\delta\mathbf{u}^e]$$

The external work done by the nodal forces $[\mathbf{q}^e]$ acting through the virtual displacements is

$$\Delta W^e = [\delta\mathbf{u}^e]^T [\mathbf{q}^e]$$

and the internal work per unit volume, by virtue of the virtual work theorem for a continuum (Charlton, 1959) is given by

$$\Delta W^i = [\delta\boldsymbol{\varepsilon}]^T [\boldsymbol{\sigma}] - [\delta\mathbf{u}]^T [\mathbf{b}]$$

$$= ([\mathbf{B}] [\delta\mathbf{u}^e])^T [\boldsymbol{\sigma}] - ([\mathbf{N}] [\delta\mathbf{u}^e])^T [\mathbf{b}]$$

$$= [\delta\mathbf{u}^e]^T ([\mathbf{B}]^T [\boldsymbol{\sigma}] - [\mathbf{N}]^T [\mathbf{b}])$$

Integrating the internal work over the volume V_e of the element, and equating it with the external work, gives

$$[\mathbf{q}^e] = \int_{V_e} [\mathbf{B}]^T [\boldsymbol{\sigma}] \, dV - \int_{V_e} [\mathbf{N}]^T [\mathbf{b}] \, dV$$

Introducing equation 6.41,

$$[\mathbf{q}^e] = \int_{V_e} [\mathbf{B}]^T [\mathbf{D}] [\mathbf{B}] [\mathbf{u}^e] \, dV + \int_{V_e} [\mathbf{B}]^T [\boldsymbol{\sigma}^0] \, dV - \int_{V_e} [\mathbf{N}]^T [\mathbf{b}] \, dV$$

$$(6.42)$$

Examining each component of the r.h.s. of this expression, for a triangular element the term $\int_{V_e} [\mathbf{B}]^T [\mathbf{D}] [\mathbf{B}] \mathbf{u} \, dV$ yields a 6×6 matrix of functions which must be integrated over the volume of the element, the term $\int_{V_e} [\mathbf{B}]^T [\sigma^0] \, dV$ a 6×1 matrix, and $\int_{V_e} [\mathbf{N}]^T [\mathbf{b}] \, dV$ a 6×1 matrix. In general, the integrations may be carried out using standard quadrature theory. For a constant strain triangular element, of volume V_e, the elements of $[\mathbf{B}]$ and $[\mathbf{N}]$ are constant over the element, and equation 6.42 becomes

$$[\mathbf{q}^e] = V_e [\mathbf{B}]^T [\mathbf{D}] [\mathbf{B}] [\mathbf{u}^e] + V_e [\mathbf{B}]^T [\sigma^0] - V_e [\mathbf{N}]^T [\mathbf{b}]$$

In all cases, equation 6.42 may be written

$$[\mathbf{q}^e] = [\mathbf{K}^e] [\mathbf{u}^e] + [\mathbf{f}^e] \tag{6.43}$$

In this equation, equivalent internal nodal forces $[\mathbf{q}^e]$ are related to nodal displacements $[\mathbf{u}^e]$ through the element stiffness matrix $[\mathbf{K}^e]$ and an initial internal load vector $[\mathbf{f}^e]$. The elements of $[\mathbf{K}^e]$ and $[\mathbf{f}^e]$ can be calculated directly from the element geometry, the initial state of stress and the body forces.

6.6.4 Solution for nodal displacements

The computational implementation of the finite element method involves a set of routines which generate the stiffness matrix $[\mathbf{K}^e]$ and initial load vector $[\mathbf{f}^e]$ for all elements. These data, and applied external loads and boundary conditions, provide sufficient information to determine the nodal displacements for the complete element assembly. The procedure is illustrated, for simplicity, by reference to the two-element assembly shown in Figure 6.8.

Suppose the applied external forces at the nodes are defined by

$$[\mathbf{r}]^T = [r_{x1} \ r_{y1} \ r_{x2} \ r_{y2} \ r_{x3} \ r_{y3} \ r_{x4} \ r_{y4}]$$

Figure 6.8 A simple finite element structure to illustrate the relation between nodal connectivity and construction of the global stiffness matrix.

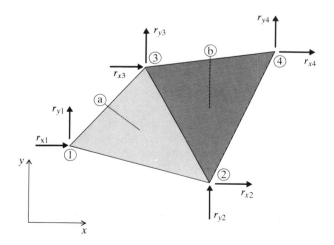

187

Equilibrium at any node requires that the applied external force at the node be in balance with the resultant internal equivalent nodal force. Suppose the internal equivalent nodal force vector for elements a and b is given by

$$[\mathbf{q}^a]^T = [q_{x1}^a\ q_{y1}^a\ q_{x2}^a\ q_{y2}^a\ q_{x3}^a\ q_{y3}^a]$$

$$[\mathbf{q}^b]^T = [q_{x2}^b\ q_{y2}^b\ q_{x4}^b\ q_{y4}^b\ q_{x3}^b\ q_{y3}^b]$$

The nodal equilibrium condition requires

$$\text{for node 1: } r_{x1} = q_{x1}^a, \quad r_{y1} = q_{y1}^a$$

$$\text{for node 2: } r_{x2} = q_{x2}^a + q_{x2}^b, \quad r_{y2} = q_{y2}^a + q_{y2}^b$$

with similar conditions for the other nodes. The external force–nodal displacement equation for the assembly then becomes

$$
\begin{bmatrix} r_{x1} \\ r_{y1} \\ r_{x2} \\ r_{y2} \\ r_{x3} \\ r_{y3} \\ r_{x4} \\ r_{y4} \end{bmatrix}
=
\begin{bmatrix}
K^a & & 0 & 0 \\
 & & 0 & 0 \\
 & K^a + K^b & & \\
 & & & \\
0 & 0 & & K^b \\
0 & 0 & &
\end{bmatrix}
\begin{bmatrix} u_{x1} \\ u_{y1} \\ u_{x2} \\ u_{y2} \\ u_{x3} \\ u_{y3} \\ u_{x4} \\ u_{y4} \end{bmatrix}
+
\begin{bmatrix} f_{x1} \\ f_{y1} \\ f_{x2} \\ f_{y2} \\ f_{x3} \\ f_{y3} \\ f_{x4} \\ f_{y4} \end{bmatrix}
$$

where appropriate elements of the stiffness matrices $[\mathbf{K}^a]$ and $[\mathbf{K}^b]$ are added at the common nodes. Thus assembly of the global stiffness matrix $[\mathbf{K}]$ proceeds simply by taking account of the connectivity of the various elements, to yield the global equation for the assembly

$$[\mathbf{K}]\,[\mathbf{u}^g] = [\mathbf{r}^g] - [\mathbf{f}^g] \tag{6.44}$$

Solution of the global equation 6.44 returns the vector $[\mathbf{u}^g]$ of nodal displacements. The state of stress in each element can then be calculated directly from the appropriate nodal displacements, using equation 6.41.

In practice, special attention is required to render $[\mathbf{K}]$ non-singular, and account must be taken of any applied tractions on the edges of elements. Also, most finite element codes used in design practice are based on curvilinear quadrilateral elements and higher-order displacement variation with respect to the element intrinsic co-ordinates. For example, a quadratic isoparametric formulation imposes quadratic variation of displacements and quadratic description of element shape. Apart from some added complexity in the evaluation of the element stiffness matrix and the initial load vector, the solution procedure is essentially identical to that described here.

188

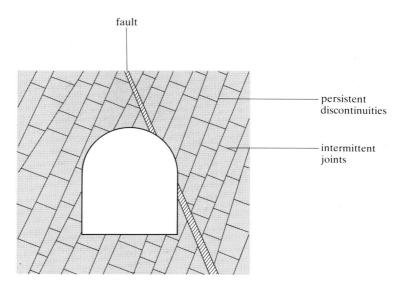

fault

persistent
discontinuities

intermittent
joints

Figure 6.9 A schematic representation of a rock mass, in which the behaviour of the excavation periphery is controlled by discrete rock blocks.

6.7 The distinct element method

Both the boundary element method and the finite element method are used extensively for analysis of underground excavation design problems. Both methods can be modified to accommodate discontinuities such as faults, shear zones, etc., transgressing the rock mass. However, any inelastic displacements are limited to elastic orders of magnitude by the analytical principles exploited in developing the solution procedures. At some sites, the performance of a rock mass in the periphery of a mine excavation may be dominated by the properties of pervasive discontinuities, as shown in Figure 6.9. This is the case since discontinuity stiffness (i.e. the force/displacement characteristic) may be much lower than that of the intact rock. In this situation, the elasticity of the blocks may be neglected, and they may be ascribed rigid behaviour. The distinct element method described by Cundall (1971) was the first to treat a discontinuous rock mass as an assembly of quasi-rigid blocks interacting through deformable joints of definable stiffness. It is the method discussed here. The technique evolved from the conventional relaxation method described by Southwell (1940) and the dynamic relaxation method described by Otter *et al.* (1966). In the distinct element approach, the algorithm is based on a force-displacement law specifying the interaction between the quasi-rigid rock units, and a law of motion which determines displacements induced in the blocks by out-of-balance forces.

6.7.1 Force–displacement laws

The blocks which constitute the jointed assemblage are taken to be rigid, meaning that block geometry is unaffected by the contact forces between blocks. The deformability of the assemblage is conferred by the deformability of the joints, and it is this property of the system which renders the assemblage statically determinate under an equilibrating load system. It is also noted that, intuitively, the deformability of joints in shear is likely to be much greater than their normal deformability.

189

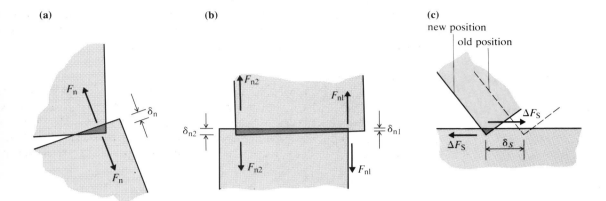

Figure 6.10 Normal and shear modes of interaction between distinct elements.

In defining the normal force mobilised by contact between blocks, a notional overlap δ_n is assumed to develop at the block boundaries, as shown in Figure 6.10a. The normal contact force is then computed assuming a linear force–displacement law, i.e.

$$F_n = K_n \delta_n \tag{6.45}$$

where K_n is the joint normal stiffness. When the faces of two blocks are aligned in a subparallel attitude, as shown in Figure 6.10b, interaction is assumed to occur at two point contacts, for each of which equation 6.45 is taken to define the contact force. While the realism of this two-point mode of interaction might be questioned, it is likely that when rock blocks are disturbed from their initial, topographically matched equilibrium condition, the number of contacts will be small. It is improbable that the location of these contacts will affect the shear deformability of a joint.

Equation 6.45 indicates that the normal contact force between blocks is determined uniquely by the relative spatial positions of the blocks. However, the shear contact force at any stage depends on the deformation path to which the contact has been subjected. Thus it is necessary to compute the progressive shear displacements of blocks, which are then used to determine the incremental shear force operating between two blocks. For an increment of shear displacement δ_s, as shown in Figure 6.10c, the increment of shear force ΔF_s is given by

$$\Delta F_s = K_s \delta_s \tag{6.46}$$

where K_s is the joint shear stiffness.

The deformation relations defined by equations 6.45 and 6.46 are elastic, in that they describe non-dissipative, reversible processes. Under some circumstances, these relations will not apply. For example, when separation occurs at a joint, normal and shear forces at the block surfaces vanish. If, at some stage, the computed shear force, F_s, at a contact exceeds the maximum frictional resistance ($F_n \tan \phi$, for a cohesionless surface), slip occurs, and the shear force assumes

190

the limiting value $F_n \tan \phi$. Consequently, in any algorithmic treatment, after each increment of normal and shear displacement, the total shear force must be evaluated. If the shear force is less than the limiting frictional resistance, elastic deformability conditions are re-established at the joints.

6.7.2 Law of motion

Equations 6.45 and 6.46 indicate how a set of forces acting on a block can be determined from the position of a block relative to its neighbours. For each block, these forces may be combined to determine the resultant force, and a moment. Using Newton's Second Law of motion, it is possible to determine the translation of the block centroid and the rotation of the block about the centroid; i.e. for the x direction

$$\ddot{u}_x = \frac{F_x}{m} \qquad (6.47)$$

where $\ddot{u}_x$ is the acceleration of the block centroid in the x direction, F_x is the x component of the resultant force on the block, and m is the mass of the block.

The translation of the block centroid can be determined from equation 6.47, by numerical integration. Suppose a time increment Δt is selected, over which it is intended to determine the block translation. Block velocity and translation are approximated by

$$\dot{u}_x(t_1) = \dot{u}(t_0) + \ddot{u}_x \Delta t$$

$$u_x(t_1) = u_x(t_0) + \dot{u}_x \Delta t$$

Similar expressions are readily established for block translation in the y direction, and for block rotation.

6.7.3 Computational scheme

The distinct element method is conceptually and algorithmically the simplest of the methods of analysis considered here. In its computational implementation, precautions and some effort are required to achieve satisfactory performance. First, the time step Δt in the integration of the law of motion cannot be chosen arbitrarily, and an excessively large value of Δt results in numerical instability. Second, for an assemblage of blocks which is mechanically stable, the dynamic relaxation method described above provides no mechanism for dissipation of energy in the system. Computationally, this is expressed as continued oscillation of the blocks as the integration proceeds in the time domain. It is therefore necessary to introduce a damping mechanism to remove elastic strain energy as the blocks displace to an equilibrium position. Viscous damping is used in practice.

The computational scheme proceeds by following the motion of blocks through a series of increments of displacements controlled by a time-stepping iteration. Iteration through several thousand time steps may be necessary to achieve equilibrium in the block assemblage.

191

6.8 Linked computational schemes

The preceding discussion has noted the advantages and limitations of the various computational methods. In many cases, the nature of mine excavation design problems means that the boundary element method can be used for design analyses, particularly if the intention is to carry out a parameter study in assessing various options. In those cases where non-linear material or discrete, rigid-block displacements are to be modelled, the scale of a mining problem frequently precludes the effective or economical use of finite element or distinct element codes. The solution is to develop linked schemes, where the far-field rock is modelled with boundary elements, and the more complex constitutive behaviour is modelled with the appropriate differential method of analysis. A domain of complex behaviour is then embedded in an infinite elastic continuum. The advantages of this approach include, first, elimination of uncertainties associated with the assumption of an outer boundary for the problem domain, as required by the differential methods. Second, far-field and elastic material behaviour is represented in a computationally economical and mechanically appropriate way with boundary elements. Finally, zones of complex constitutive behaviour in a mine structure are frequently small and localised, so that only these zones may require the versatility conferred by a differential method. The implied reduction in the size of zones to be modelled with a differential method again favours computational efficiency. An example of the development of a linked method and its application in excavation design is given by Lorig and Brady (1982).

The principles used in the development of a boundary element (b.e.) – distinct element (d.e.) linkage algorithm are illustrated in Figure 6.11. The b.e. and d.e. domains are isolated as separate problems, and during a computational cycle in the d.e. routine, continuity conditions for displacement and traction are enforced at the interface between the domains. This is achieved in the following way. As shown by Brady and Wassyng (1981), the boundary constraint equation developed in a direct b.e. formulation can be manipulated to yield a stiffness matrix $[\mathbf{K}_i]$ for the interior surface of the domain. In each computational cycle of the d.e. iteration, displacements $[\mathbf{u}_i]$ of interface d.e. nodes calculated in a previous cycle are used to determine the reactions $[\mathbf{r}_i]$ developed at the nodes from the expression

Figure 6.11 Resolution of a coupled distinct element–boundary element problem into component problems.

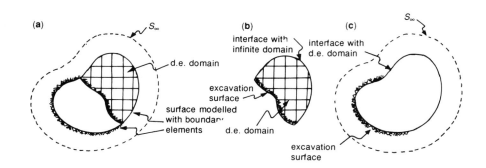

192

$$[\mathbf{r}_i] = [\mathbf{K}_i] [\mathbf{u}_i] \qquad (6.48)$$

Then the forces applied to the d.e. interface nodes are equal and opposite to the reactions developed on the b.e. interface; i.e.

$$[\mathbf{q}_i] = - [\mathbf{r}_i] \qquad (6.49)$$

Equation 6.48 represents formal satisfaction of the requirement for continuity of displacement at the interface, while equation 6.49 implies satisfaction of the condition for force equilibrium.

In the d.e. routine, nodal forces determined from equation 6.49 are introduced in the equation of motion (equation 6.47) for each block in contact with the interface. In practice, several iterative cycles may elapse before it is necessary to update the interface nodal forces. When equilibrium is achieved in the d.e. assembly, the interface nodal displacements and tractions (derived from the nodal reactions) are used to determine stresses and displacements at interior points in the d.e. domain.

7 Excavation design in massive elastic rock

7.1 General design methodology

Mining excavations are of two types – service openings and production openings. Service openings include mine accesses, ore haulage drives, airways, crusher chambers and underground workshop space. They are characterised by a duty life approaching the mining life of the orebody. It is therefore necessary to design these openings so that their operational rôle can be maintained at low cost over a relatively long operational life.

Mine production openings are distinguished by having a temporary rôle in the operation of the mine. These openings include the ore sources, or stopes, and related excavations such as drill headings, stope accesses, and ore extraction and service ways. In these cases, it is necessary to assure control of the rock in the excavation periphery only for the life of the stope, which may be as little as a few months.

The issue considered in this chapter is the design of service openings and production openings subject to entry by personnel. (The design of non-entry excavations is considered later in conjunction with the related mining methods.) A logical methodology is established for the design of these openings. Although it is proposed for general application in the design of permanent mine openings, it can also be regarded as a basis for the evolution of designs for specific temporary openings. Such an evolutionary process could be associated with the implementation of an observational principle of excavation design.

Excavation design in massive elastic rock represents the simplest design problem posed in mining rock mechanics. The case considered in this chapter involves a single excavation, which is taken to imply that the opening will not be mined in the zone of influence of any existing opening. The rock medium is considered massive if the volume of rock to be mined to generate the opening is traversed by only one or two persistent structural features. The rock mass strength properties are assumed to be defined by a compressive failure criterion of the type defined in Chapter 4, with a related uniaxial compressive strength C_0. The tensile strength T_0 of the rock mass is usually taken as zero.

In the design of any mine opening, two main factors are to be borne in mind. First, the existence of an extensive zone of failed rock around the periphery of an opening is common in mining practice. Second, the basic mining problem is not necessarily to prevent rock mass failure, but to ensure that large, uncontrolled displacement of excavation peripheral rock cannot occur. This may be achieved by attention to excavation shape, excavation practice, and possibly by the application of one or several rock support and reinforcement procedures. The general problem of mine excavation design devolves into questions of excavation location and geometry, and, frequently, development of an excavation sequence and a support specification. Problems related to location and geometry are considered in this chapter.

The design of a mine excavation proceeds from an initial configuration which satisfies its duty requirements, such as minimum dimensions required for operating equipment, or airway resistance to achieve some ventilation capacity. Siting and orientation are also determined by the duty requirements and by the need for integration with other elements of the mining layout. The suitability of the selected design is then assessed by following the methodology defined in Figure 7.1. It is observed that a key step in the design process is the determination of the stress distribution around the excavation. This can be achieved using any of the methods described in Chapter 6. The logical path then involves comparison of boundary stresses around the excavation with the uniaxial compressive and tensile strengths of the rock mass. If no boundary failure is predicted, it remains to examine the effect of any major discontinuities which will transgress the excavation. This requires consideration of both the general effect of the structural features on boundary stresses and local stability problems in the vicinity of the discontinuity/boundary intersection. These considerations may educe design changes to achieve simultaneous satisfaction of local and general stability conditions for the excavation perimeter.

Figure 7.1 A logical methodology for mine excavation design in massive rock.

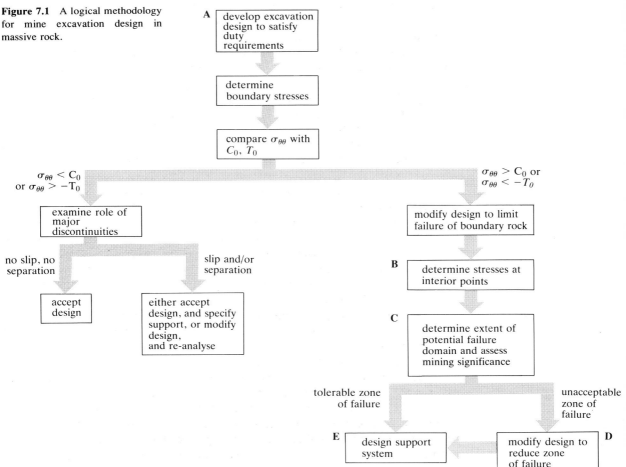

Excavation design for the case where rock mass strength is low, or field stresses are high, proceeds using the path defined by the right-hand branch of Figure 7.1. These conditions imply that fracturing of the rock on the excavation periphery will occur. Iteration on the design variables proceeds to restrict the extent of boundary failure and the zone of failure in the excavation near field, and to mitigate any difficulties arising from either the presence of major planes of weakness or their interaction with zones of induced rock fracture. The final phase of the design is the specification of support and reinforcement measures to control the performance of fractured peripheral rock. Referring to Figure 7.1, the design logic implies iteration over the steps A, B, C, D, E, using practically feasible shape, orientation and location parameters, until a geomechanically sound and operationally acceptable design is attained.

In the discussion that follows, elastic analyses will be used to illustrate some important design issues and principles. In some cases they will be employed to predict the extent of non-linear processes, such as slip and separation on discontinuities, or rock mass failure. In these cases, the analyses produce only a first-order estimate of the range of these processes, but this is adequate for most mine excavation design problems. In addition, the following examples deal with simple excavation configurations, for which the stress fields are amenable to description by simple algebraic expressions. In practice, for general excavation shapes, the methods of stress analysis described in Chapter 6 may be employed, and the results used in the manner described in the following discussion.

7.2 Zone of influence of an excavation

The concept of a zone of influence is important in mine design, since it may provide considerable simplification of a design problem. The essential idea of a zone of influence is that it defines a domain of significant disturbance of the pre-mining stress field by an excavation. It differentiates between the near field and far field of an opening. The extent of an opening's effective near-field domain can be explained by the following example.

The stress distribution around a circular hole in a hydrostatic stress field, of magnitude p, is given by equations 6.20 as

$$\sigma_{rr} = p \left(1 - \frac{a^2}{r^2} \right)$$

$$\sigma_{\theta\theta} = p \left(1 + \frac{a^2}{r^2} \right) \qquad (7.1)$$

$$\sigma_{r\theta} = 0$$

Equations 7.1 indicate that the stress distribution is axisymmetric, and this is illustrated in Figure 7.2a. Using equations 7.1, it is readily calculated that for $r = 5a$, $\sigma_{\theta\theta} = 1.04p$ and $\sigma_{rr} = 0.96p$, i.e. on the surface defined by $r = 5a$, the state

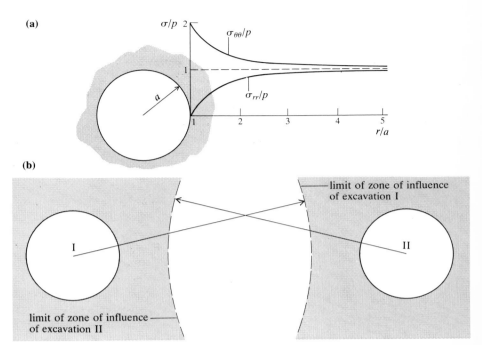

Figure 7.2 (a) Axisymmetric stress distribution around a circular opening in a hydrostatic stress field; (b) circular openings in a hydrostatic stress field, effectively isolated by virtue of their exclusion from each other's zone of influence.

of stress is not significantly different (within ±5%) from the field stresses. If a second excavation (II) were generated outside the surface described by $r = 5a$ for the excavation I, as shown in Figure 7.2b, the pre-mining stress field would not be significantly different from the virgin stress field. The boundary stresses for excavation II are thus those for an isolated excavation. Similarly, if excavation I is outside the zone of influence of excavation II, the boundary stresses around excavation I are effectively those for an isolated opening. The general rule is that openings lying outside one another's zones of influence can be designed by ignoring the presence of all others. For example, for circular openings of the same radius, a, in a hydrostatic stress field, the mechanical interaction between the openings is insignificant if the distance $D_{I, II}$ between their centres is

$$D_{I, II} \geq 6a$$

It is important to note that, in general, the zone of influence of an opening is related to both excavation shape and pre-mining stresses.

Other issues related to the notion of zone of influence include the state of stress in a medium containing a number of excavations, and interaction between different-sized excavations. Figure 7.2b illustrates the overlap of the zones of influence of two circular openings. In the overlap region, the state of stress is produced by the pre-mining stresses and the stress increments induced by each of the excavations I and II. In the other sections of each zone of influence, the state of stress is that due to the particular excavation.

Figure 7.3 illustrates a large-diameter opening (I) with a small-diameter opening (II) in its zone of influence. Since excavation I is outside the zone of influence

197

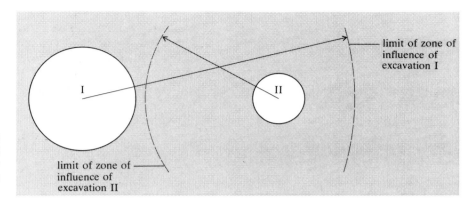

Figure 7.3 Illustration of the effect of contiguous openings of different dimensions. The zone of influence of excavation I includes excavation II, but the converse does not apply.

of excavation II, a fair estimate of the boundary stresses around I is obtained from the stress distribution for a single opening. For excavation II, the field stresses are those due to the presence of excavation I. An engineering estimate of the boundary stresses around II can be obtained by calculating the state of stress at the centre of II, prior to its excavation. This can be introduced as the far-field stresses in the Kirsch equations (Equations 6.18) to yield the required boundary stresses for the smaller excavation.

Problems related to zone of influence arise frequently in metalliferous mining. Haulages, access and service openings must frequently be located, for reasons of economy and practicality, in the zone of influence of the major production openings. An example is shown in Figure 7.4a, with access openings on the footwall side of an inclined orebody. In this case, a zone of influence could be defined for an ellipse inscribed in the stope cross section, for any particular stage of up-dip advance of mining. Suppose the stope is outside the zone of influence of each access drive. Then, reasonable estimates of the access opening boundary stresses could be obtained from the local stresses due to the pseudo-elliptical stope and the boundary stress concentrations due to the shape of the access drive.

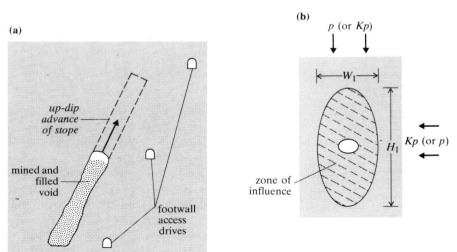

Figure 7.4 (a) A practical problem involving semi-coupling between a large excavation (a cut-and-fill stope) and smaller access openings; (b) nomenclature for definition of the zone of influence of an elliptical opening.

198

The preceding discussion suggests that it is useful to consider the zone of influence of an elliptical excavation in the course of a design exercise. It is therefore appropriate to formalise its definition. The general case of a zone around an elliptical excavation in which the stresses depart from the maximum *in situ* stress (p or Kp) by more than $c\%$ has been considered by Bray (1986). From this analysis, the zone may be approximated by an ellipse with axes W_1 and H_1 equal to the greater of each of the following sets of values:

$$W_1 = H[A\alpha \mid q\,(q+2) - K(3+2q)\mid]^{1/2}$$

or

$$W_1 = H[\alpha\,\{A(K+q^2) + Kq^2\}]^{1/2}$$

$$H_1 = H[A\alpha \mid K(1+2q) - q\,(3q+2)\mid]^{1/2}$$

or

$$H_1 = H[\alpha\,\{A(K+q^2) + 1\}]^{1/2}$$

where W and H are the width and height of the elliptical excavation, $q = W/H$, $A = 100/2c$ and $\alpha = 1$, if $K < 1$, and $\alpha = 1/K$, if $K > 1$.

For the special case of a zone of influence defined by a 5% departure from the field stresses, A is set equal to 10 in the preceding expressions.

7.3 Effect of planes of weakness on elastic stress distribution

In excavation design problems where major discontinuities penetrate the prospective location of the opening, questions arise concerning the validity of elastic analysis in the design process and the potential effect of the discontinuity on the behaviour of the excavation periphery. It is now shown that, in some cases, an elastic analysis presents a perfectly valid basis for design in a discontinuous rock mass, and in others, provides a basis for judgement of the engineering significance of a discontinuity. The following discussion takes account of the low shear strengths of discontinuities compared with that of the intact rock. It assumes that a discontinuity has zero tensile strength, and is non-dilatant in shear, with a shear strength defined by

$$\tau = \sigma_n \tan \phi \tag{7.2}$$

As observed earlier, although the following discussion is based on a circular opening, for purposes of illustration, the principles apply to an opening of arbitrary shape. In the latter case, a computational method of stress analysis would be used to determine the stress distribution around the opening.

Case 1. (Figure 7.5) From the Kirsch equations (equations 6.18), for $\theta = 0$, the shear stress component $\sigma_{r\theta} = 0$, for all r. Thus σ_{rr}, $\sigma_{\theta\theta}$ are the principal stresses σ_{xx}, σ_{yy} and σ_{xy} is zero. The shear stress on the plane of weakness is zero, and there is no tendency for slip on it. The plane of weakness therefore has no effect on the elastic stress distribution.

199

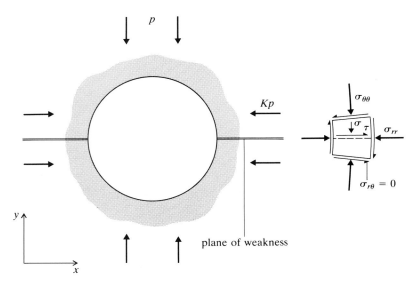

Figure 7.5 A plane of weakness, oriented perpendicular to the major principal stress, intersecting a circular opening along the horizontal diameter.

Case 2. (Figure 7.6a) Equations 6.18, with $\theta = \pi/2$, indicate that no shear stress is mobilised on the plane of weakness, and thus the elastic stress distribution is not modified by slip. The possibility of separation on the plane of weakness arises if tensile stress can develop in the crown of the opening, i.e. if $K < \frac{1}{3}$. If $K \geq \frac{1}{3}$, the elastic stress distribution is unaltered by either slip or separation.

For the case where $K < \frac{1}{3}$, separation on the plane of weakness leads to de-stressing of a region in the crown of the opening (and also in the floor, although this is of no engineering consequence). A reasonable estimate of the extent of the de-stressed zone, for purposes of support design, for example, can be obtained by considering the circumscribed ellipse, illustrated in Figure 7.6b. Separation

Figure 7.6 A plane of weakness intersecting a circular opening and oriented parallel to the major principal stress, showing development of a de-stressed zone for $K < \frac{1}{3}$.

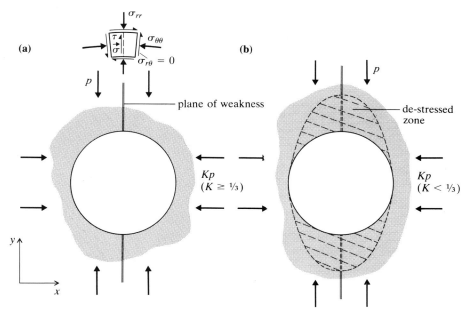

on the plane of weakness is prevented when $\sigma_B = 0$; from equation 6.24, this occurs when

$$\sigma_B = p\left(K - 1 + \frac{2K}{q}\right) = 0$$

or

$$q = \frac{2K}{1 - K}$$

Since $q = W/H = 2a/H$, it is readily shown that the height Δh of the de-stressed zone above the crown of the opening is given by

$$\Delta h = a\left(\frac{1 - 3K}{2K}\right)$$

Case 3. (Figure 7.7) A flat-lying feature whose trace on the excavation boundary is located at an angle θ above the horizontal diameter is shown in Figure 7.7a. Considering the small element of the boundary, shown in Figure 7.7b, the normal and shear stress components on the plane of weakness are given by

$$\sigma_n = \sigma_{\theta\theta} \cos^2\theta$$

(7.3)

$$\tau = \sigma_{\theta\theta} \sin\theta\cos\theta$$

The limiting condition for slip under this state of stress is

$$\tau = \sigma_n \tan\phi$$

or, introducing equations 7.3,

Figure 7.7 A flat-lying plane of weakness intersecting a circular excavation non-diametrically.

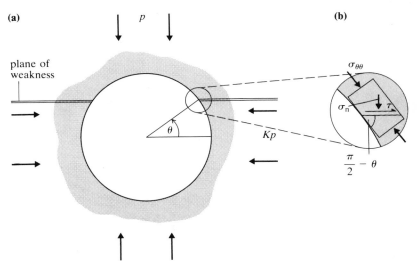

$$\sigma_{\theta\theta} \sin\theta \cos\theta = \sigma_{\theta\theta} \cos^2\theta \tan\phi \qquad (7.4)$$

or

$$\tan\theta = \tan\phi$$

Thus if $\theta = \phi$, the condition for slip is satisfied on the plane of weakness. (This conclusion could have been established by noting that the resultant stress, in this case $\sigma_{\theta\theta}$, is constrained to act within the angle ϕ to the normal to the plane.) It is observed that the sense of slip, defined by the sense of the shear stress, involves outward displacement of the upper (hanging wall) surface of the fault relative to the lower surface. This implies boundary stresses lower than the elastic values in the crown of the opening. A prudent design response would anticipate the generation of subvertical tension fractures in the crown.

The equilibrium state of stress at the boundary-plane of weakness intersection can be established from equation 7.4, which may be rewritten in the form

$$\sigma_{\theta\theta} \frac{\sin(\theta - \phi)}{\cos\theta} = 0$$

For $\theta > \phi$, this condition can be satisfied only if $\sigma_{\theta\theta} = 0$. Thus the regions near the intersection of the opening and the plane of weakness are either de-stressed, or at low confining stress. They may be expected to be areas from which loosening of rock may commence, and therefore deserve special attention in support design.

Case 4. (Figure 7.8) The problem illustrated in Figure 7.8a is introduced as a simple example of an arbitrarily inclined plane of weakness intersecting an opening. The far-field stresses are defined by components p (vertical) and $0.5p$ (horizontal). For a feature inclined at an angle of $45°$, the normal and shear stress components are obtained by substitution in the Kirsch equations (equations 6.18), and are given by

$$\sigma_n = \sigma_{\theta\theta} = \frac{p}{2} * 1.5\left(1 + \frac{a^2}{r^2}\right)$$

$$\tau = \sigma_{r\theta} = \frac{p}{2} * 0.5\left(1 + \frac{2a^2}{r^2} - \frac{3a^4}{r^4}\right)$$

The variation of the ratio τ/σ_n is plotted in Figure 7.8b. The maximum value of the ratio, 0.357 at $r/a = 2.5$, corresponds to a mobilised angle of friction of $19.6°$. The far-field value of the ratio of the stresses corresponds to a mobilised angle of friction of $18.5°$. If the rock mass were in a state of limiting equilibrium under the field stresses, the analysis indicates that mining the excavation could develop an extensive zone of slip along the plane of weakness. On the other hand, an angle of friction for the plane of weakness exceeding $19.6°$ would be sufficient to preclude slip anywhere in the medium.

202

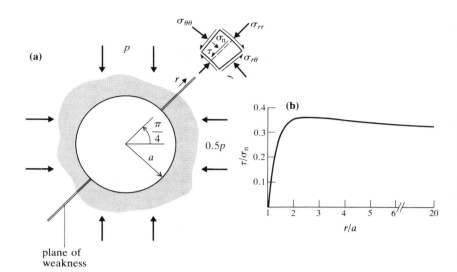

Figure 7.8 An inclined, radially oriented plane of weakness intersecting a circular excavation.

Case 5. (Figure 7.9) The design problem shown in Figure 7.9a involves a circular opening to be excavated close to, but not intersecting, a plane of weakness. For purposes of illustration, the stress field is taken as hydrostatic. From the geometry given in Figure 7.9a, equations 7.1 for the stress distribution around a circular hole in a hydrostatic stress field, and the transformation equations, the normal and shear stresses on the plane are given by

$$\sigma_n = \tfrac{1}{2}(\sigma_{rr} + \sigma_{\theta\theta}) + \tfrac{1}{2}(\sigma_{rr} - \sigma_{\theta\theta})\cos 2\alpha$$

$$= p\left(1 - \frac{a^2}{r^2}\cos 2\alpha\right)$$

Figure 7.9 Shear stress/normal stress ratio on a plane of weakness close to, but not intersecting, a circular excavation.

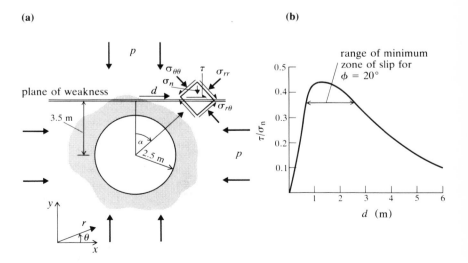

$$\tau = \sigma_{r\theta} \cos 2\alpha - \tfrac{1}{2}(\sigma_{rr} - \sigma_{\theta\theta}) \sin 2\alpha$$

$$= p \frac{a^2}{r^2} \sin 2\alpha$$

The value of the ratio τ/σ_n determined from these expressions is plotted for various points along the plane of weakness in Figure 7.9b.

The peak value of the shear stress/normal stress ratio corresponds to a mobilised angle of friction of about 24°. If the angle of friction for the plane of weakness exceeds 24°, no slip is predicted on the plane, and the elastic stress distribution can be maintained.

For a plane of weakness with an angle of friction of 20°, the extent of the predicted zone of slip is shown in Figure 7.9b. Clearly a zone of slip is also predicted for the reflection of the depicted zone about the vertical centreline of the excavation. For both zones, the sense of slip produces inward displacement of rock on the underside of the plane of weakness. This would be expressed as increased boundary stresses in the segment between the fault and the excavation. The effect of the fault is to deflect and concentrate the stress trajectories in the region between the excavation and the fault.

The following comments are offered to establish some practical guidelines for the type of analysis described above. First, the procedures indicate whether inelastic effects such as separation and slip on planes of weakness are likely to be significant in the performance of an excavation. If the zones of inelastic response are small relative to the dimensions of the excavation, their effect on the stress distribution around the excavation may reasonably be ignored. If the zones are relatively large, the stress distribution around the opening can be determined only by comprehensive analysis using, for example, a finite element package. However, even in this case, some useful engineering insights into the behaviour of excavation peripheral rock can be established by exploiting quite simple conceptual models of the effects of inelastic deformation. Finally, the procedures allow quick and inexpensive exploration of the effects of varying the principal design options, i.e. excavation location, orientation, shape and excavation sequence. In fact, in a design exercise, the types of analysis discussed above should usually precede a more sophisticated analysis which might be needed to model inelastic behaviour of discontinuities in the rock mass.

7.4 Excavation shape and boundary stresses

The previous discussion has indicated that useful information on boundary stresses around a mine opening can be established from the elastic solution for the particular problem geometry even in the presence of discontinuities. It is now shown that simple, closed form solutions have greater engineering value than might be apparent from a first inspection.

Figure 7.10 illustrates a long opening of elliptical cross section, with axes parallel to the pre-mining stresses. For the particular cases of $\beta = 0$, $\psi = 0$, and $\beta = 0$, $\psi = \pi/2$, equation 6.21 reduces to

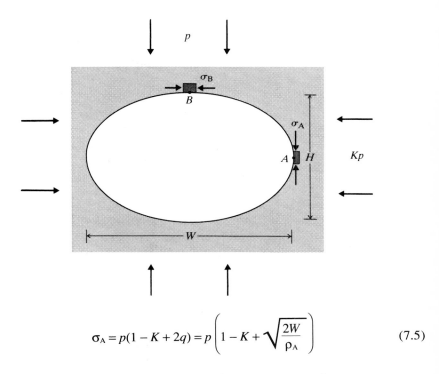

Figure 7.10 Definition of nomenclature for an elliptical excavation with axes parallel to the field stresses.

$$\sigma_A = p(1 - K + 2q) = p\left(1 - K + \sqrt{\frac{2W}{\rho_A}}\right) \qquad (7.5)$$

$$\sigma_B = p\left(K - 1 + \frac{2k}{q}\right) = p\left(K - 1 + K\sqrt{\frac{2H}{\rho_B}}\right) \qquad (7.6)$$

where σ_A and σ_B are boundary circumferential stresses in the side wall (A) and crown (B) of the excavation, and ρ_A and ρ_B are the radii of curvature at points A and B. Equation 7.5 indicates that if ρ_A is small, σ_A is large. Equation 7.6 defines a similar relation between ρ_B and σ_B. A generalisation drawn from these results is that high boundary curvature (i.e. $1/\rho$) leads to high boundary stresses, and that boundary curvature can be used in a semi-quantitative way to predict boundary stresses.

Figure 7.11 shows an ovaloidal opening oriented with its major axis perpendicular to the pre-mining principal stress. The width/height ratio for the opening is three, and the radius of curvature for the side wall is $H/2$. For a ratio of 0.5 of the horizontal and vertical field principal stresses, the sidewall boundary stress is given, by substitution in equation 7.5, as

$$\sigma_A = p\left(1 - 0.5 + \sqrt{\frac{2 \times 3\,H}{H/2}}\right)$$

$$= 3.96\,p$$

An independent boundary element analysis of this problem yielded a sidewall boundary stress of $3.60p$, which is sufficiently close for practical design purposes. Although the radius of curvature, for the ovaloid, is infinite at point B in

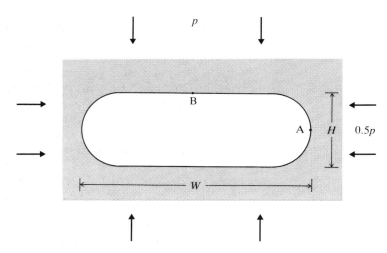

Figure 7.11 Ovaloidal opening in
a medium subject to biaxial stress.

the centre of the crown of the excavation, it is useful to consider the state of
stress at the centre of the crown of an ellipse inscribed in the ovaloid. This
predicts a value of σ_B, from equation 7.6, of $-0.17\,p$, while the boundary element
analysis for the ovaloid produces a value of σ_B of $-0.15\,p$. This suggests that
excavation aspect ratio (say W/H), as well as boundary curvature, can be used to
develop a reasonably accurate picture of the state of stress around an opening.

A square hole with rounded corners, each with radius of curvature $\rho = 0.2B$, is
shown in Figure 7.12a. For a hydrostatic stress field, the problem shown in Figure
7.12b is mechanically equivalent to that shown in Figure 7.12a. The inscribed
ovaloid has a width of $2B[2^{1/2} - 0.4(2^{1/2} - 1)]$, from the simple geometry. The
boundary stress at the rounded corner is estimated from equation 7.5 as

$$\sigma_A = p\left\{1 - 1 + \left[\frac{2B(2^{1/2} - 0.4(2^{1/2} - 1))}{0.2B}\right]^{1/2}\right\}$$

$$= 3.53\,p$$

Figure 7.12 Square opening with
rounded corners, in a medium subject to hydrostatic stress.

(a)   **(b)**

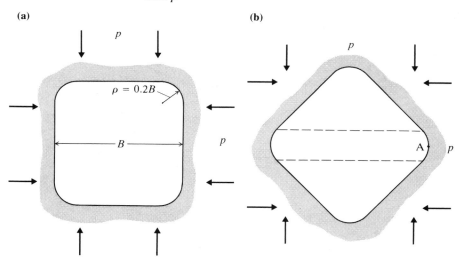

The corresponding boundary element solution is 3.14p.

The effect of boundary curvature on boundary stress appears to be a particular consequence of St Venant's Principle, in that the boundary state of stress is dominated by the local geometry, provided the excavation surface contour is relatively smooth. An extension of this idea demonstrates an important design concept, namely that changing the shape of an opening presents a most effective method of controlling boundary stresses. This is illustrated by the problem shown in Figure 7.13a. The arched opening has a width of 4.0 m and a height of 4.5 m, which are not unusual dimensions for a mine haulage. For a field stress ratio K of 0.3, an inscribed ellipse indicates approximate sidewall stresses of 2.5p, using equation 7.5. If the observed performance of the opening involved crushing of the sidewalls, its redesign should aim to reduce stresses in these areas. Inspection of equation 7.5 indicates this can be achieved by reducing the excavation width/height ratio. For example, if the width/height ratio is reduced to 0.5, the peak sidewall stress is calculated to be 1.7p. While the practicality of mining an opening to this shape is not certain, the general principle is clear, that the maximum boundary stress can be reduced if the opening dimension is increased in the direction of the major principal stress. For this case, a practical solution could be achieved as shown in Figure 7.13b, by mining an opening with a low width/height ratio, and leaving a bed of mullock in the base of the excavation.

Having taken the steps noted above to minimise the maximum boundary stress, failure of boundary rock may be unavoidable under the local conditions of field stresses and rock mass strength. In that case, orienting the major axis of the excavation parallel to the major principal field stress cannot be expected to provide the optimal solution. The extent of rock mass failure may be greater than for other orientations, including that in which the long axis is perpendicular to the major principal field stress (Ewy *et al.*, 1987). Indeed, it has been proposed that, for an excavation subject to an extremely high vertical principal field stress and extensive sidewall failure, an elliptical excavation with the long axis horizontal may be the preferred excavation shape for the prevailing conditions of rock mass rupture and local stability (Ortlepp and Gay, 1984).

Figure 7.13 Effect of changing the relative dimensions of a mine haulage drive, to mitigate sidewall failure.

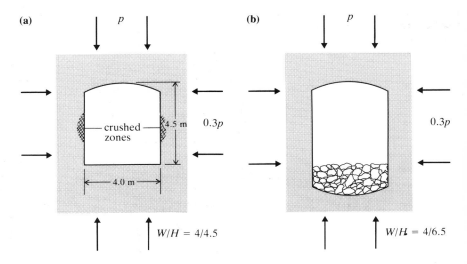

207

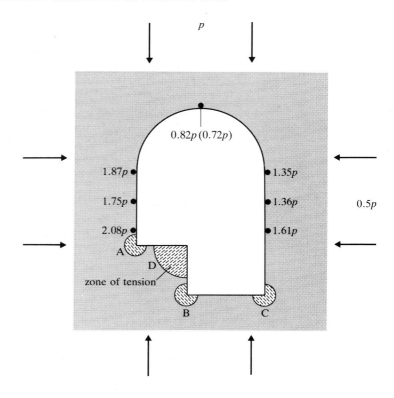

Figure 7.14 States of stress at selected points on the boundary of an excavation with a moderately irregular cross section.

A fairly general excavation cross section is shown in Figure 7.14. Such an excavation geometry might be used in a crusher station, battery charging station or machine workshop, where a bench is retained for equipment installation. Using the general notions developed above, the opening geometry (width/height ratio = 2/3) and pre-mining stress ratio ($K = 0.5$), the following information concerning boundary stresses can be deduced:

(a) The zones A, B, C are likely to be highly stressed, since the boundary curvature at these locations is high. Local cracking is to be expected in these zones, but this would compromise neither the integrity of the excavation nor the validity of the stress analysis.

(b) The bench area D is likely to be at a low state of stress, due to the notionally negative curvature of the prominence forming the bench.

(c) The boundary stress at the centre of the crown would be approximately $0.72p$, estimated from equation 7.6. (The boundary element solution is $0.82p$.)

(d) An estimate of the sidewall boundary stress, obtained by considering an inscribed ellipse and applying equation 7.5, yields $\sigma_A = 1.83p$. For the sidewall locations in the left wall shown in Figure 7.14, boundary element analysis gives values of $1.87p$, $1.75p$ and $2.08p$. For the locations in the right wall, the σ_A values are $1.35p$, $1.36p$ and $1.61p$. The average of these six values is $1.67p$. Boundary element analysis also confirms conclusions (a) and (b). The extent of the zone of tensile stress determined by the boundary element analysis is shown in Figure 7.14.

The demonstration, in an elastic analysis, of a zone of tensile stress, such as in the bench of the current excavation design, has significant engineering implications. Since a rock mass must be assumed to have zero tensile strength, stress redistribution must occur in the vicinity of the bench. This implies the development of a de-stressed zone in the bench and some loss of control over the behaviour of rock in this region. The important point is that a rock mass in compression may behave as a stable continuum. In a de-stressed state, small imposed or gravitational loads can cause large displacements of component rock units.

The conclusion from these studies is that a useful appreciation of the state of stress at key sections of an excavation boundary can be established from simple, closed-form solutions. Inscription of a simple excavation shape in the design cross section, and determination of boundary curvature, are simple techniques allowing the key features of the boundary stress distribution for an excavation to be defined. More comprehensive definition of the boundary stress distribution would be required if studies, such as those described, identified zones of mechanically unacceptable states of stress around the excavation periphery.

7.5 Delineation of zones of rock failure

In assessing the performance of excavations and rock structures, it is useful to distinguish between failure of the structure, and failure or fracture of the rock mass. Failure of a structure implies that it is unable to fulfil the designed duty requirement. Failure of a rock structure in massive rock is synonymous with extensive rock fracture, since the stable performance of the structure under these conditions cannot be assured. In a mine structure, control of displacements in a fractured rock mass may require the installation of designed support elements, or implementation of a mining sequence which limits the adverse consequences of an extensive fracture domain. On the other hand, limited fractured rock zones may pose no mining problem, and a structure or opening may completely satisfy the design duty requirements. A simple method of estimating the extent of fracture zones provides a basis for the prediction of rock mass performance, modification of excavation designs, or assessing support and reinforcement requirements.

In Chapter 4, it was observed that a compressive failure criterion for a rock mass may be expressed in the form

$$\sigma_1^{(f)} = F(\sigma_3) \tag{7.7}$$

This indicates that for any value of the minimum principal stress at a point, a major principal stress value can be determined which, if reached, leads to local failure of the rock. The uniaxial compressive strength, C_0, corresponds to $\sigma_3 = 0$. The tensile strength of a rock mass is taken to be universally zero.

Prediction of the extent of boundary failure may be illustrated by reference to a circular excavation in a biaxial stress field, as shown in Figure 7.15. Boundary stresses are given by the expression

$$\sigma_{\theta\theta} = p[1 + K + 2(1 - K)\cos 2\theta]$$

209

For a rock mass uniaxial compressive strength of 16 MPa, the data of Figure 7.15 indicate that compressive failure of boundary rock occurs over intervals defined by

$$7.5\,[1.3 + 1.4 \cos 2\theta] \geq 16$$

i.e. for θ given by

$$-26° \leq \theta \leq 26° \quad \text{or} \quad 154° \leq \theta \leq 206°$$

Similarly, boundary tensile failure occurs over intervals satisfying the condition

$$7.5\,[\,1.3 + 1.4 \cos 2\theta] \leq 0$$

i.e.

$$79° \leq \theta \leq 101° \quad \text{or} \quad 259° \leq \theta \leq 281°$$

These intervals are illustrated in Figure 7.15. The extent of these zones, relative to the excavation perimeter, is sufficient to imply that the opening, as designed, could not perform its duty rôle. The proper design response would be to increase the height of the opening.

For arbitrarily shaped openings, assessment of boundary failure involves comparison of computed boundary stresses and the uniaxial strength parameters.

Determination of the extent of failure zones in the interior of the rock mass may be accomplished by considering the detail of the stress distribution around an opening. This may be represented conveniently by contour maps of

Figure 7.15 Prediction of the extent of boundary failure around a circular excavation, using the rock mass failure criterion and the elastic stress distribution.

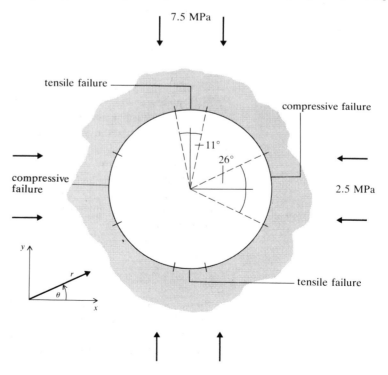

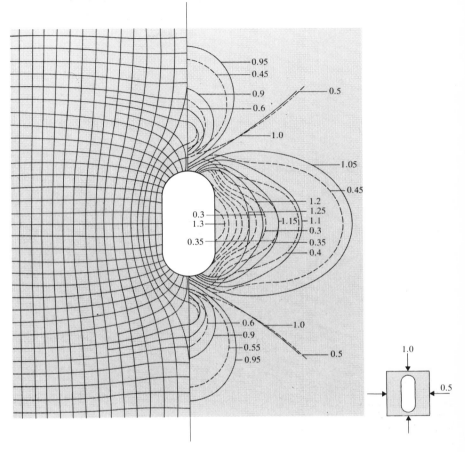

Figure 7.16 Contour plots of principal stress, and stress trajectories, around an ovaloidal opening in a biaxial stress field (after Elissa, 1980).

the principal stresses. An example of a set of contour plots for an ovaloidal excavation is shown in Figure 7.16, in which stress trajectories are also presented. For general conditions of opening geometry and pre-mining stresses, stress contour maps are readily generated and displayed using computational techniques.

Mapping of the zone of failure uses the rock mass failure criterion and the stress contour maps, in the following way. Various values of σ_3 ($0.05p$, $0.1p$, etc.) are used to calculate corresponding, limiting values of $\sigma_1^{(f)}$ from the failure criterion. The contour plots of σ_1 and σ_3 are superimposed. Selecting a particular σ_3 isobar, the intersection is found with the σ_1 isobar which satisfies the failure criterion. Repetition of this process for the various values of σ_3 generates a set of points which together define the boundary of the failure domain.

When dealing computationally with failure domain delineation, explicit generation of principal stress contour maps is not required. At any interior point, the computed state of stress is inserted directly into the failure criterion to determine local rock behaviour. The condition at a large number of locations throughout the rock mass can be displayed symbolically on a computer graphics terminal, for visual identification of the failure zone.

211

Failure domains determined in the manner discussed are not described accurately. This can be appreciated from the observation that any rock failure causes a change in the geometry of the elastic domain, which can cause further failure. That is, the problem is a non-linear one, and the solution procedure suggested here examines only the initial, linear component of the problem. For mining engineering purposes, the suggested procedure is usually adequate.

7.6 Support and reinforcement of massive rock

Mining activity frequently takes place under conditions sufficient to induce extensive failure around mine access and production openings. An understanding of the mechanics of the techniques exploited to control the performance of fractured rock is therefore basic to effective mining practice under these conditions. In this section, some basic principles of rock support and reinforcement are introduced. This discussion is extended, and a discussion of practical methods and procedures is given in Chapter 11.

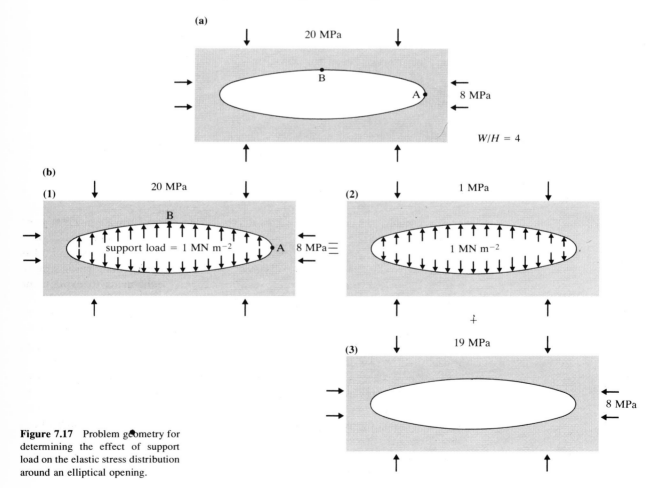

Figure 7.17 Problem geometry for determining the effect of support load on the elastic stress distribution around an elliptical opening.

In evaluating the possible mechanical roles of a support system, it is instructive to consider the effect of support on the elastic stress distribution in the rock medium. One function of support is taken to be the application or development of a support load at the excavation surface. It is assumed initially that this is uniformly distributed over the boundary.

Figure 7.17a shows an elliptical opening with width/height ratio of 4, in a stress field with vertical stress 20 MPa and horizontal stress 8 MPa. Boundary stresses at points A and B in the side wall and crown of the excavation may be calculated directly from equations 7.5 and 7.6, i.e.

$$\sigma_A = 172.0 \, \text{MPa}, \quad \sigma_B = -8.0 \, \text{MPa}$$

If a set of vertical supports is installed, sufficient to generate a vertical load of 1 MNm^{-2} uniformly distributed over the excavation surface, the boundary stresses around the supported excavation can be determined from the superposition scheme shown in Figure 7.17b. Thus

$$\sigma_{A1} = \sigma_{A2} + \sigma_{A3}$$

$$= 1 + 19 \left(1 - \frac{8}{19} + 8 \right)$$

$$= 161.0 \, \text{MPa}$$

$$\sigma_{B1} = \sigma_{B2} + \sigma_{B3}$$

$$= 0 + 19 \left(\frac{8}{19} - 1 + \frac{2 \times 8}{19} \times \frac{1}{4} \right)$$

$$= -7.0 \, \text{MPa}$$

From these results, it is concluded that support pressure does not modify the elastic stress distribution around an underground opening significantly. If failure of the rock mass is possible in the absence of support, installation of support is unlikely to modify the stress distribution sufficiently to preclude development of failure. It is therefore necessary to consider other mechanisms to explain the mode of action of support and reinforcement systems.

The following example is intended to illustrate the main features of the function of surface support on excavation peripheral rock, and is based on an analysis by Ladanyi (1974). It is not a basis for comprehensive description of the interaction between installed support and tunnel peripheral rock, which is a more complex, statically indeterminate problem. The problem geometry is shown in Figure 7.18a, with a circular opening excavated in a medium subject to hydrostatic stress. The field stresses and rock mass strength are such that an annulus of failed rock is generated in the excavation periphery. The main questions are the relation between the radius, r_e, of the failed zone, the applied support presure, p_i, and the stress distribution in the fractured rock and elastic domains. It is assumed that the strength of the rock mass is described by a Mohr–Coulomb criterion, i.e.

213

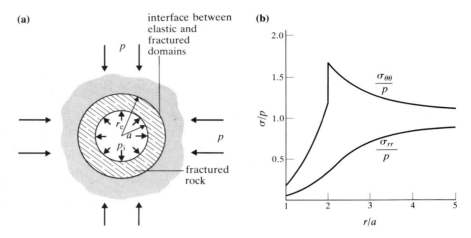

Figure 7.18 Stress distribution around a circular opening in a hydrostatic stress field, due to development of a fracture zone.

$$\sigma_1 = \sigma_3 \frac{(1 + \sin \phi)}{(1 - \sin \phi)} + \frac{2c \cos \phi}{1 - \sin \phi}$$

or

$$\sigma_1 = b\sigma_3 + C_0 \tag{7.8}$$

The strength of fractured rock is taken to be purely frictional, with the limiting state of stress within the fractured rock mass defined by

$$\sigma_1 = \sigma_3 \frac{(1 + \sin \phi^f)}{(1 - \sin \phi^f)}$$

or

$$\sigma_1 = d\sigma_3 \tag{7.9}$$

where ϕ^f is the angle of friction for fractured rock.

Since the problem is axisymmetric, there is only one differential equation of equilibrium,

$$\frac{d\sigma_{rr}}{dr} = \frac{\sigma_{\theta\theta} - \sigma_{rr}}{r} \tag{7.10}$$

This condition is to be satisfied throughout the problem domain. In the fractured rock, $\sigma_{\theta\theta}$ and σ_{rr} are related through equation 7.9, since limiting friction is assumed in the fractured medium. Introducing equations 7.9 into equation 7.10, gives

$$\frac{d\sigma_{rr}}{dr} = (d - 1) \frac{\sigma_{rr}}{r}$$

Integrating this expression, and introducing the boundary condition, $\sigma_{rr} = p_i$ when $r = a$, yields the stress distribution relations

$$\sigma_{rr} = p_i \left(\frac{r}{a} \right)^{d-1}$$

$$\sigma_{\theta\theta} = dp_i \left(\frac{r}{a} \right)^{d-1}$$

(7.11)

Equations 7.11 are satisfied throughout the fractured domain and on its boundaries. At the outer limit of the fractured annulus, fractured rock is in equilibrium with intact, elastic rock. If p_1 is the equilibrium radial stress at the annulus outer boundary, r_e,

$$p_1 = p_i \left(\frac{r_e}{a} \right)^{d-1}$$

or

$$r_e = a \left(\frac{p_1}{p_i} \right)^{1/(d-1)}$$

(7.12)

Simple superposition indicates that the stress distribution in the elastic zone is defined by

$$\sigma_{\theta\theta} = p \left(1 + \frac{r_e^2}{r^2} \right) - p_1 \frac{r_e^2}{r^2}$$

$$\sigma_{rr} = p \left(1 - \frac{r_e^2}{r^2} \right) + p_1 \frac{r_e^2}{r^2}$$

(7.13)

At the inner boundary of the elastic zone, the state of stress is defined by

$$\sigma_{\theta\theta} = 2p - p_1, \quad \sigma_{rr} = p_1$$

This state of stress must represent the limiting state for intact rock, i.e. substituting for $\sigma_{\theta\theta}(\sigma_1)$ and $\sigma_{rr}(\sigma_3)$ in equation 7.8 gives

$$2p - p_1 = bp_1 + C_0$$

or

$$p_1 = \frac{2p - C_0}{1 + b}$$

(7.14)

Substitution of equation 7.14 in equation 7.12 yields

$$r_e = a \left[\frac{2p - C_0}{(1 + b) \, p_i} \right]^{1/(d-1)}$$

(7.15)

Equations 7.11, 7.13, and 7.15, together with the support pressure, field stresses and rock properties, completely define the stress distribution and fracture domain in the periphery of the opening.

215

A numerical example provides some insight into the operational function of installed support. Choosing particular values of ϕ and ϕ^f of $35°$, $p_i = 0.05p$ and $C_0 = 0.5p$, leads to $r_e = 1.99a$. The stress distribution around the opening is shown in Figure 7.18b. The main features of the stress distribution are, first, the high and increasing gradient in the radial variation of $\sigma_{\theta\theta}$, both in an absolute sense and compared with that for σ_{rr}; and secondly, the significant step increase in $\sigma_{\theta\theta}$ at the interface between the fractured and intact domains. These results suggest that the primary role of installed support in massive rock subject to peripheral failure is to maintain radial continuity of contact between rock fragments. It also serves to generate a radial confining stress at the excavation boundary. The mode of action of the support is to generate and maintain a high triaxial state of stress in the fractured domain, by mobilising friction between the surfaces of the rock fragments. The significance of frictional action in the fractured rock can be readily appreciated from equation 7.15. This indicates that the radius of the fractured zone is proportional to some power of the friction parameter.

A significant issue neglected in this analysis is the dilatancy of undisturbed, fractured rock. The inclusion of a dilatancy term would result in a significant increase in the effective angle of friction of the fractured material, and a consequent marked increase in effectiveness of the installed support.

Problems

1   A long opening of circular cross section is located 1000 m below ground surface. In the plane perpendicular to the tunnel axis, the field principal stresses are vertical and horizontal. The vertical stress p is equal to the depth stress, and the horizontal stress is defined by $0.28p$. The unit weight of the rock mass is 27 kN m^{-3}, the compressive strength is defined by a Coulomb criterion with $c = 20$ MPa, $\phi = 25°$, and the tensile strength by $T_0 = 0$.

(a) Predict the response of the excavation peripheral rock to the given conditions.
(b) Propose an alternative design for the excavation.

2   The figure on the left represents a cross section through a long opening. The magnitudes of the plane components of the field stresses are $p_{xx} = 13.75$ MPa, $p_{yy} = 19.25$ MPa, $p_{xy} = 4.76$ MPa, expressed relative to the reference axes shown.

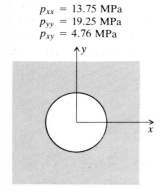

$p_{xx} = 13.75$ MPa
$p_{yy} = 19.25$ MPa
$p_{xy} = 4.76$ MPa

(a) Calculate the maximum and minimum boundary stresses in the excavation perimeter, defining the locations of the relevant points.
(b) If the strength of the rock mass is defined by a maximum shear strength criterion, and the shear strength is 20 MPa, estimate the extent of boundary failure, in terms of the angular range over the perimeter.
(c) Comment on the significance of this result for any mining operations in the opening.

3   The figure overleaf shows the locations of two vertical, parallel shafts, each 4 m in diameter. The pre-mining stress field is defined by $p_{xx} = p_{yy} = P_{zz} = 20$ MPa. Estimate approximate values for the boundary stresses around each opening, and calculate the principal stresses at point A, and their orientations.

216

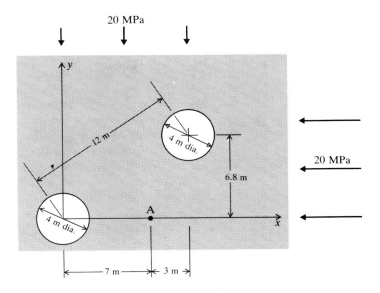

4   In the development of a haulage level in a mine, a horizontal opening of horseshoe cross section, 4 m wide and 4 m high, is to be mined parallel to an existing haulageway of the same cross section and on the same horizon. The field principal stresses are p(vertical) and $0.5p$ (horizontal). Ignoring any boundary loosening due to blasting effects, propose a minimum distance between the centrelines of the haulageways such that, during development of the second opening, no support problem or local instability is produced in the existing opening.

5   The figure below shows a horizontal section through the midheight of a vertical, lenticular orebody. The orebody is mined with long vertical blast holes in such a way that a slot approximately elliptical in a horizontal section is extended along the strike of the orebody. The field principal stresses in the horizontal plane are of magnitudes 25 MPa and 10 MPa, and oriented as shown. For the case where the length (L)/breadth (B) ratio of the mined excavation is 2.8, determine the magnitudes and corresponding locations of the maximum and minimum boundary stresses. If the rock mass tensile strength is zero, and compressive failure is defined by a Coulomb law with $c = 30$ MPa, $\phi = 30°$, determine

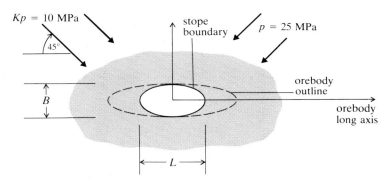

the likelihood of boundary failure. Assess the consequences of the results for mining activity.

6   The figure below represents a plan view of a vertical shaft, with a clay-filled fault located near the shaft boundary. The state of stress at the particular horizon is hydrostatic, of magnitude 8 MPa. The resistance to slip on the fault is purely cohesive, and of magnitude 1.5 MPa.

   Determine the maximum shear stress generated on the fault after development of the shaft, and determine if the elastic stress distribution can be maintained.

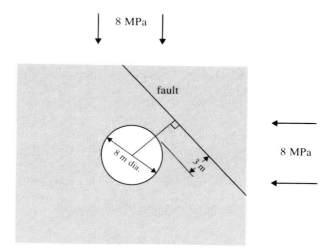

8 Excavation design in stratified rock

8.1 Design factors

A stratified host rock mass for an ore deposit is not uncommon in mining practice. An orebody in a sedimentary setting is frequently conformable with the surrounding rock, in which the stratification is conferred by bedding planes or related depositional features. The main geometric characteristics of these features are their planar geometry and their persistence. They can be assumed to be continuous over plan areas greater than that of any excavation created during mining. There are two principal engineering properties of bedding planes which are significant in an underground mining context. The first is the low or zero tensile strength in the direction perpendicular to the bedding plane. The second is the relatively low shear strength of the surfaces, compared with that of the intact rock. Both these properties introduce specific modes of rock mass response to mining, which must be considered in the excavation design procedure. An associated issue is that, for flat-lying stratiform orebodies, the typical mining method involves entry of personnel into the mined void. The performance of the bed of rock spanning the excavation, i.e. the immediate roof, then assumes particular importance in assuring maintenance of a geomechanically sound, local mining environment.

Excavations in a stratified rock mass are usually mined to a cross-sectional geometry in which the immediate roof and floor of the excavation coincide with bedding planes, as illustrated in Figure 8.1. Factors to be considered in the design of such an excavation include:

(a) the state of stress at the excavation boundary and in the interior of the rock medium, compared with the strength of the anisotropic rock mass;
(b) the stability of the immediate roof;
(c) floor heave in the excavation.

Figure 8.1 An excavation in a stratified rock mass, with geometry conforming with dominant rock structure.

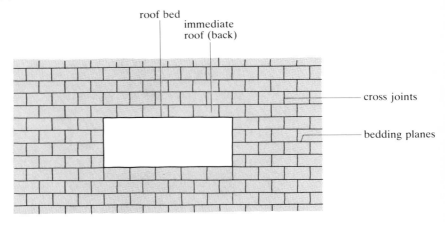

roof bed

immediate roof (back)

cross joints

bedding planes

In addressing these three different design factors, it is necessary to consider different deformation modes of the excavation near-field rock. For example, the first factor reflects concern with surface spalling and internal fracture in the rock medium, in the post-excavation stress field. The second factor involves the problem of detachment of the immediate roof from the host medium, and its loading and deflection into the mined void under gravity loading. The third problem, floor heave, is an issue where the floor rocks are relatively weak and the material yields under the stresses operating beneath the excavation side walls. It is a problem more frequently encountered in a room-and-pillar mining layout, rather than in designing a single excavation.

Since an excavation design must satisfy different rock mass performance criteria for the various modes of rock response, it is clear that a number of different analytical methods are to be employed in the design process. It also implies that it may be necessary to iterate in the design process, to satisfy the various performance requirements simultaneously.

8.2 Rock mass response to mining

Adverse performance of the rock mass in the post-excavation stress field may be caused by either failure of the anisotropic medium or slip on the pervasive weakness planes. The initial phase of the design process involves determining the elastic stress distribution in the medium around the selected excavation configuration. Following the procedure proposed for an excavation in massive elastic rock, one can then define any zones of tensile stress, or compressive stress exceeding the strength of the rock mass. The excavation shape may be modified to eliminate or restrict these zones, or alternatively, the extent of domains requiring support and reinforcement may be defined. Concurrently, it is necessary to determine the extent of the zone around the excavation in which slip can occur on bedding planes.

The criterion for slip on bedding planes is obtained from the shear strength of the surfaces. For the reference axes illustrated in Figure 8.2, interbed slip is possible if

$$|\sigma_{zx}| \geq \sigma_{zz} \tan \phi + c \qquad (8.1)$$

Figure 8.2 Slip-prone zones around an excavation in stratified rock.

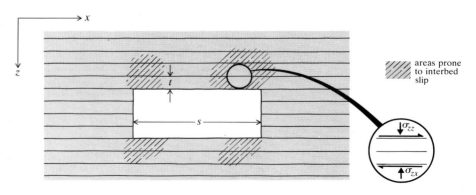

Hence, evaluation of the extent of slip requires that the stress components σ_{zz} and σ_{zx} be determined, from the results of the elastic stress analysis, at points coinciding with the locations of bedding planes. The particular zones to be examined are the roof and haunches of the excavation. It is then a relatively simple matter to map the domains where the elastic stresses satisfy the criterion for slip. Of course, the mapped zones do not indicate the complete extent of potential slip, but they can be used to obtain a reasonable impression of the mining significance of the problem. The design process seeks to limit the extent of the slip domain, while simultaneously restricting the extent of any other adverse rock mass response.

Potential slip on bedding planes is a general problem in design in a stratified rock mass. Its extent is clearly related to the pre-mining stress field and the planned shape of the excavation. As a general rule, a problem configuration in which the span/bed thickness ratio (s/t) is low will be subject to slip only in the haunch area. This may be expressed in the rock mass as the opening of cracks subperpendicular to bedding, perhaps coincident with any cross joints in the medium, as illustrated in Figure 8.3a. For a configuration in which the s/t ratio is high (i.e. beds relatively thin compared with excavation span), the zone of slip may include virtually the complete span of the immediate roof. Since the sense of slip on bedding is such as to cause inward displacement towards the span centreline of beds, the tendency is for isolation of the lower bed, at its centre, from the one immediately above it. The process of separation of a roof bed from its uppermost neighbour is of great engineering consequence, since it implies that load sharing with the upper beds is eliminated. The mechanics of the problem can be appreciated by reference to Figure 8.3b. Prior to decoupling of the roof layer, its gravitational load is carried in the more extensive body of rock with which it is integrated. Detachment of the bed from its uppermost neighbour means that the bed itself must support its full gravitational load.

Reference to Figure 8.3 gives some indication of the types of problem presented by design of roof spans in strata-bound excavations. For thick roof strata, any slip and cracking over the haunches would appear to introduce the

Figure 8.3 The effects of slip and separation on excavation peripheral rock.

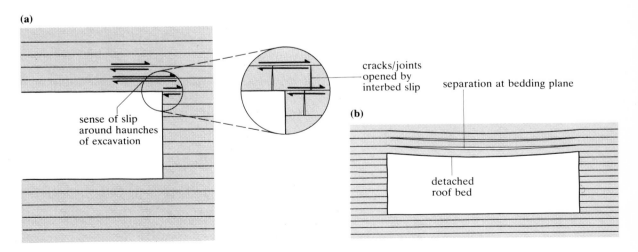

(a)

sense of slip
around haunches
of excavation

cracks/joints
opened by
interbed slip

separation at bedding plane

(b)

detached
roof bed

possibility of failure by shear displacement of the roof bed past the abutment. For thin roof strata, the implied problem is one of stability of the roof bed under the deflection and lateral thrust associated with detachment and gravity loading.

8.3 Roof bed deformation mechanics

Prediction and control of the deformation behaviour of the immediate roof of an opening has been the subject of formal engineering investigation for more than a century. Fayol (1885) published the results of investigations of the behaviour of stacks of beams spanning a simple support system, simulating the bedding sequence of a roof span. By noting the deflection of the lowest beam as successive beams were loaded onto the stack, Fayol demonstrated that, at a certain stage, none of the added load of an upper beam was carried by the lowest member. The gravitational load of the uppermost beams was clearly being transferred laterally to the supports, rather than vertically, as transverse loading of the lower members. The process of lateral load distribution, associated with friction mobilised between the surfaces of the upper beams, was described as **arching**. The basic concept proposed by Fayol was that a rock arch was generated above a mine opening in a stratified mass, with the rock beds between the excavation roof and the rock arch constituting an effectively decoupled immediate roof for the opening. Fayol proceeded to apply his results to surface subsidence phenomena, rather than to underground excavation design.

The first rigorous analysis of roof bed performance was attempted by Jones and Llewellyn-Davies (1929). They mapped the morphology of roof failures, and sought to explain the localisation of failure in terms of arching principles. Bucky and Taborelli (1938) studied physical models of the creation and extension of wide roof spans. They used initially intact beams of rock-like material, and found that, at a particular span, a vertical tension fracture was induced at the centre of the lower beam. Increase in the mined span produced a new central fracture, and closed the earlier fracture. This suggested that the central fracture is the dominant transverse discontinuity in the roof bed.

Recognising the relationship between vertical deflection, lateral thrust and stability of a naturally or artificially fractured roof bed, Evans (1941) undertook a seminal set of investigations of roof deformation mechanics at the Royal School of Mines. This work established the notion of a 'voussoir beam' spanning an excavation, using the analogy with the voussoir arch considered in masonry structures. Evans also developed an analytical procedure for assessing roof beam stability, but an error in statics and failure to handle the basic indeterminacy of the problem limited its practical application.

Significant investigations of roof bed mechanics subsequent to those by Evans have been reported by Adler and Sun (1968), Barker and Hatt (1972), Wright (1974) and Sterling (1980). The studies by Sterling capture the salient features of the work by other researchers and provide a coherent picture of the deformation and failure modes of roof rock.

The experimental arrangement used by Sterling is illustrated in Figure 8.4. A rock beam, of typical dimensions 660 mm × 75 mm 75 × mm, was constrained between steel end plates linked by strain-gauged tie rods. The beam was loaded

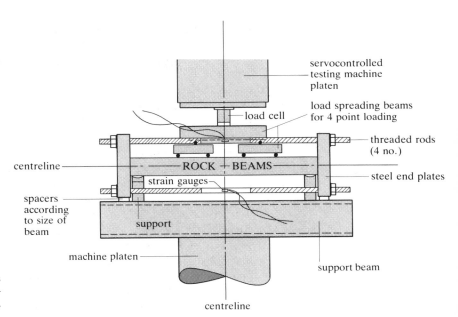

Figure 8.4 Experimental apparatus for testing the load–deflection behaviour of roof beams (after Sterling, 1980).

transversely by a servocontrolled testing machine and a load spreading system. The experiment design provided data on applied transverse load, induced beam deflection, induced lateral thrust, and eccentricity of the lateral thrust. The typical response of an initially intact limestone beam is given in Figure 8.5. The load–deflection plot, shown in Figure 8.5a, shows an initial elastic range (0–1). At this stage a transverse, central crack developed in the beam, accompanied, in the test rig, by a relaxation of the applied load, at constant transverse deflection. Increase in transverse load produced a linear load–deflection plot (2–7). Loading and unloading in the range 2–7 was reversible, and the downward extension of the plot is observed to pass through the origin. Load increase (7–10) produced a pronounced non-linear response, accompanied physically by local crushing at either the top centre of the beam or lower edges of the beam ends. Subsequent loading showed decreased load capacity at increased deflection (10–17), accom-

Figure 8.5 (a) Load–deflection and (b) induced lateral thrust–transverse load plots, for laterally constrained rock beams (after Sterling, 1980).

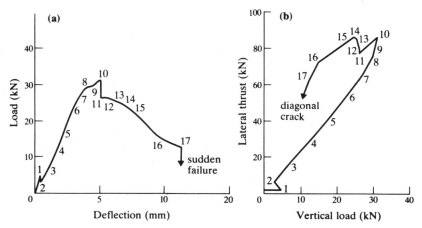

panied by spalling at the upper centre or lower ends of the specimen, and finally localised specimen disintegration.

The main features of the load–deflection plot are confirmed by the plot of lateral thrust and vertical load (Figure 8.5b). From the small original thrust corresponding to lateral prestress, the initial response (0–1) is flat, corresponding to continuous, elastic behaviour of the beam. Central vertical cracking of the beam (1–2) with increase in lateral thrust, reflects the significance of induced thrust in determining the subsequent performance of the voussoir beam. The linear range of response (2–7) was reversible, and extrapolates downwards to the original loading conditions. Past the peak load capacity of the beam (10), the reducing lateral thrust caused by local spalling results in reduced vertical load capacity for the beam.

This and other tests conducted by Sterling allow formulation of the following principles concerning roof rock behaviour over mined spans:

(a) roof beds cannot be simulated by continuous, elastic beams or plates, since their behaviour is dominated by the blocks (voussoirs) generated by natural cross joints or induced transverse fractures;

(b) roof bed behaviour is determined by the lateral thrusts generated by deflection, under gravity loading, of the voussoir beam against the confinement of the abutting rock;

(c) a voussoir beam behaves elastically (i.e. the lateral thrust – vertical deflection plot is linear and reversible) over the range of its satisfactory performance, the upper limit of which approaches the peak transverse load capacity;

(d) for a voussoir beam with low span/thickness ratio, the most likely failure mode is shear failure at the abutments;

(e) for a roof with high span/thickness ratio, roof span stability is limited by the possibility of buckling of the beam, with no significant spalling of central or abutment voussoirs;

(f) a roof with low rock material strength or moderate span/thickness ratio may fail by crushing or spalling of central or abutment voussoirs.

An alternative study of the performance of excavations in bedded and jointed rock led to conclusions consistent with the model developed from experimental

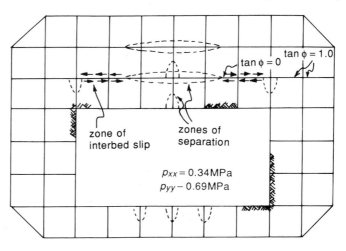

Figure 8.6 Results of linked d.e.–b.e. analysis of an excavation in stratified rock.

observations. Lorig and Brady (1983) describe application of a linked boundary element–distinct element (b.e.–d.e.) computational scheme to analysis of roof deformation mechanics. The key results of the analysis are indicated in Figure 8.6. Slip is observed over the abutments of the excavation, the immediate roof bed detaches from the overlying strata, and tension cracks open in the centre of the roof span. The distributions of normal stress and shear stress in the roof bed were generally consistent with the voussoir beam model proposed by Evans (1941), and considered below. One notable difference was observed between the Evans model and the results of the numerical study. This was that bed separation, proposed by Evans to include the complete excavation span, was indicated over only the centre of the span in the computational analysis.

8.4 Roof design procedure for plane strain

The design procedure proposed here is based on that developed by Evans (1941), and modified and extended by Beer and Meek (1982). The voussoir beam model for a roof bed is illustrated in Figure 8.7a, and the forces operating in the system are defined in Figure 8.7b. The essential idea conveyed in the figures is that, in the equilibrium condition, the lateral thrust is not transmitted either uniformly or axially through the beam cross section. The section of the beam transmitting lateral load is assumed to be approximated by the parabolic arch traced on the beam span. Since various experimental investigations support the intuitive idea that the central transverse crack determines the deformational behaviour of the discontinuous beam, the problem may be analysed in terms of the problem geometry illustrated in Figure 8.7b. The three possible failure modes discussed earlier can be readily appreciated from the problem configuration. These are: shear at the abutment when the limiting shear resistance, $T \tan \phi$, is less than the required abutment vertical reaction $V(= \frac{1}{2} W)$; crushing at the hinges formed in the beam crown and lower abutment contacts; buckling of the roof beam with increasing eccentricity of lateral thrust, and a consequent tendency to form a 'snap-through' mechanism. Each of these modes is examined in the following analysis.

In analysing the performance of a roof bed in terms of a voussoir beam, it is observed that the problem is fundamentally indeterminate. A solution is obtained in the current analysis only by making particular assumptions concerning the load distribution in the system and the line of action of resultant forces. The assumption of a parabolically shaped thrust line has already been noted. The other assumption is that triangular load distributions operate over the abutment surfaces of the beam and at the central section, as shown in Figure 8.7b. These distributions are reasonable since roof beam deformation mechanics suggest that elastic hinges form in these positions, by opening or formation of transverse cracks.

The roof beam illustrated in Figure 8.7b, of span, s, and thickness, t, supports its own weight, W, by vertical deflection and induced lateral compression. The triangular end load distributions each operate over a depth nt of the beam, and unit thickness of the system is considered in the out-of-plane direction. Noting that the line of action of the resultant of each distributed load acts horizontally

225

(a)

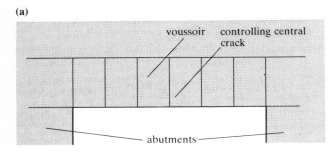

(b)

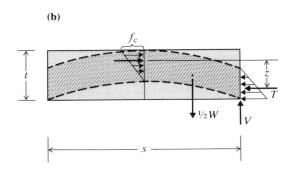

(c)

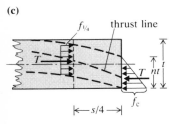

Figure 8.7 Voussoir beam geometry and load specification for roof beam analysis.

through the centroid of each distribution, the initial moment arm of the couple is given by

$$z_0 = t(1 - \frac{2}{3}n) \tag{8.2}$$

Moment equilibrium of the body illustrated in Figure 8.7b requires that the couple associated with the gravitational load and the abutment vertical reaction, V, be balanced by the moment of the distributed end loads, i.e.

$$\frac{\gamma}{8} ts^2 = \frac{f_c}{2} ntz \tag{8.3}$$

or

$$f_c = \frac{1}{4} \frac{\gamma s^2}{nz} \tag{8.4}$$

When the beam deflects, the compression arch is shortened, and the lateral thrust finally mobilised is that which allows equation 8.4 to be satisfied. The relation between mobilised thrust and beam deformation exploits the assumption of the shape of the arch (or initial thrust line) operating longitudinally in the beam. Suppose the arch profile is described by the parabolic expression

$$s'^2 = 4az$$

where s' is the beam half span and a is chosen to satisfy the known geometry. Elementary analytic geometry and the theory of hyperbolic functions then allows the arc length, L, of the arch to be expressed in terms of its height, z, and span, s, by

$$L = s + \frac{16}{3} \frac{z^2}{s} \tag{8.5}$$

Thus if the arch of height z_0 is shortened in compression by an increment ΔL, the corresponding lever arm, z, in the deformed arch may be determined directly from equation 8.5 as

$$z = \left[\frac{3s}{16} \left(\frac{16 z_0^2}{3s} - \Delta L \right) \right]^{1/2} \tag{8.6}$$

226

where z is the height of the incrementally deformed arch.

The net incremental shortening of the arch results from elastic compression of the roof beam and elastic deformation or localised spreading of the abutments. Each abutment is subject to the same distributed load as the ends of the beam, but the loaded area is small compared with the beam span. Therefore, elastic compression of the abutments will be small compared with the beam compression, and may be neglected. Arch shortening in compression is then accounted for completely by beam compression.

Determination of arch shortening is based on the average value of the longitudinal stress operating in the beam. The method of estimation of this stress is illustrated in Figure 8.7c. At the quarter-span of the beam, the equivalent, internal lateral load is taken to be uniformly distributed over the arch depth. Over each quarter of the beam length, and therefore for the complete beam, the average longitudinal stress is then given by

$$f_{av} = \frac{1}{2} f_c \left(\frac{2}{3} + \frac{n}{2} \right) \tag{8.7}$$

The incremental elastic shortening, ΔL, of the arch is defined by

$$\Delta L = \frac{f_{av}}{E'} L \tag{8.8}$$

Shortening of the arch accompanying beam deflection changes the internal load distribution in the beam. Since the effective height of the arch after deflection is defined explicitly by equation 8.6, it is then possible to calculate the load/depth ratio, n, for the distribution of linearly varying internal end loads. Generalising equation 8.2, the load/depth ratio is given by

$$n = \frac{3}{2} \left(1 - \frac{z}{t} \right) \tag{8.9}$$

It will be clear from this discussion that a completely explicit solution for the final deformation and state of loading in a roof beam is not possible. However, the procedure now proposed is a useful introduction to the relaxation technique for solving these types of problems. In this case, the technique involves an assumption for the internal load distribution, and iteration to determine the equilibrium state. The beam is assumed initially to be free of lateral thrust, and subject only to gravity loading. From these specified loads, the relaxation process aims to find the final state of lateral thrust. The operational response of the beam can then be determined.

The relaxation procedure starts from an assumption of the load/depth ratio, n, from which can be calculated an initial lever arm for the abutment and half-span internal forces, using equation 8.2. The various roof beam load and deformation parameters can then be calculated directly, from their respective formulae, i.e.

maximum longitudinal stress:
$$f_c = \frac{1}{4} \frac{\gamma s^2}{nz} \tag{a}$$

227

average longitudinal stress:
$$f_{av} = \frac{1}{2} f_c \left(\frac{2}{3} + \frac{n}{2} \right)$$
(b)

arc length of the arch:
$$L = s + \frac{16}{3} \frac{z^2}{s}$$
(c)

(8.10)

elastic shortening of the arch:
$$\Delta L = \frac{f_{av}}{E'} L$$
(d)

arch height and lever arm:
$$z = \left[\frac{3s}{16} \left(\frac{16 z_c^2}{3s} - \Delta L \right) \right]^{1/2}$$
(e)

where z_c is the value of z from the previous computation cycle

lateral load depth ratio:
$$n = \frac{3}{2} \left(1 - \frac{z}{l} \right)$$
(f)

The solution procedure involves sequential calculation of $f_c, f_{av}, L, \Delta L, z$ and n from the formulae (a)–(f) of equation 8.10. Each sequence of computation produces a new value of n, which is then introduced into formula (a) to recommence the solution cycle. Iterations of the solution sequence are continued until stable values of the maximum longitudinal stress and the lateral depth ratio are obtained.

The computational sequence implied in initial use of equation 8.2, and the iterative application of the formulae of equation 8.10, are readily implemented on a programmable calculator. The relaxation procedure produces fairly rapid convergence to a solution. However, a simple test needs to be introduced in the algorithm to prevent drift away from the equilibrium position if computation cycles are continued near the solution, which corresponds to a slowly and monotonically decreasing value of f_c and a slowly increasing value of n. A number of simple checks have been conducted by the authors to assess the stability of the solution obtained by the relaxation procedure. For example, it has been found that the solution is insensitive to the choice of initial value of the load/depth ratio.

The performance of the roof bed is evaluated from the voussoir beam analysis by considering each of the possible modes of roof failure. For the case of compressive failure of the rock in the arch, the maximum longitudinal stress, f_c, is to be compared with the uniaxial compressive strength of the rock, σ_c. The latter quantity needs to be considered carefully, not only because of questions related to the scale of the rock volume involved in a roof beam. In many sedimentary rocks, the rock fabric is characterised by a foliation parallel to the bedding. The theory for the strength of rock containing a single set of uniformly oriented planes of weakness is described in Chapter 4. It indicates that a small inclination of the plane of weakness to the major principal stress direction results in a significant reduction in the compressive strength. Thus a prudent evaluation of the potential for compressive failure in the roof beam should be based on the minimum value of σ_c for the transversely isotropic lithological unit.

The possibility of shear failure of the roof beam by slip at the abutments is assessable directly from the roof analysis. Referring to Figure 8.7b, the lateral thrust, T, at the abutment must mobilise a frictional resistance to slip sufficient to provide the abutment shear force, V. The maximum frictional resistance that can be mobilised is

$$F = T \tan \phi = \tfrac{1}{2} f_c nt \tan \phi$$

and the abutment shear force is given by

$$V = \tfrac{1}{2} \gamma st$$

The factor of safety (F of S) against abutment shear failure is then defined by

$$\text{F of S} = \frac{f_c n}{\gamma s} \tan \phi$$

The third mode of roof failure to be evaluated is buckling of the span, to form a snap-through mechanism. From Figure 8.7b it is observed that buckling is synonymous with the lever arm of the arch, z, becoming negative. Thus in the relaxation solution procedure, a test is included to evaluate the behaviour of z as the iteration routine proceeds. Of course, z cannot be evaluated when the term within the brackets of formula (e) of equation 8.10 becomes negative. This quantity therefore serves as an indicator of buckling as the mode of roof failure. Formulae (d) and (e) of equation 8.10 also imply that the buckling mode is favoured in wide roof spans with low rock mass deformation modulus. Both these relations satisfy conventional ideas of simple buckling phenomena, such as in columns and plates. The important consequence of the relation between roof buckling and rock deformation modulus is that it is necessary to measure or estimate the *in situ* deformation modulus of the roof bed, in the direction parallel to bedding. In particular, explicit account must be taken of the presence and compressibility of cross joints in estimating the elastic modulus.

Example. As an example of the application of this plane strain design analysis, consider the case of a long excavation to be made in a horizontally bedded limestone having a bed thickness of $t = 1.0$ m. The *in situ* mechanical properties of the limestone have been estimated as $\gamma = 0.026\,\text{MN m}^{-3}$, $E' = 1.5\,\text{GPa}$, $\sigma_c = 20\,\text{MPa}$ and $\phi = 30°$. For operational reasons a span of 10 m has been proposed. A check on the stability of the roof beam is required.

Using an initial value of load/depth ratio of $n = 0.50$, the iterative solution presented above gives final values of $n = 0.75$, $z = 0.50$ and $f_c = 1.73$ MPa. The corresponding factors of safety against compression failure and against slip at the abutments are 11.5 and 2.9, respectively. Since z is not negative, buckling of the roof beam is not predicted.

Factors of safety against all modes of beam failure are acceptable and so the span is confirmed as being permissible. The span could be increased if required, but this would decrease the factor of safety against compressive failure, increase the factor of safety against shear failure on cross joints at the abutments, and increase the likelihood of buckling failure.

8.5 Roof design for square and rectangular excavations

A key concept in the plane strain analysis of roof bed behaviour is that a rock mass behaves as a no-tension medium. A crack is assumed to develop at the centre of the roof span. This corresponds, in a roof bed capable of sustaining tensile stress, to the location of the greatest longitudinal tensile stress. For cases where the mine excavations have span and length dimensions not compatible with the plane strain condition, it is necessary to account for the development of cracks in the roof plate. These occur in preferred directions related to orientations of lines of maximum principal tensile stress in the lower surface of the roof plate. The cracks are directly analogous to the yield lines postulated in the behaviour of reinforced concrete slabs. In a square roof plate, elastic analysis indicates that the maximum principal tensile stresses are oriented perpendicular to the plate diagonals. This causes symmetric, diagonal cracking, as indicated in Figure 8.8. For rectangular excavations, the crack orientations are more complex. Whatever the crack pattern, the ideas for the voussoir beam can be extended to accommodate the more general geometry of the slabs of the immediate roof. The following discussion is based on the development proposed by Beer and Meek (1982). The suggested solution procedure is still the relaxation method, but some modifications are introduced in the load–deformation formulae of equation 8.10.

8.5.1 Square plate
Using the problem geometry defined in Figure 8.8, the weight of a segment of the roof is given by

$$w = \frac{\gamma s^2 t}{4}$$

Since the weight acts through the centroid of the triangular segment, the moment arm is $s/6$, and the moment equilibrium equation becomes

$$\frac{\gamma s^3 t}{24} = f_c \frac{nt}{2} sz$$

or

$$f_c = \frac{1}{12} \frac{\gamma s^2}{nz} \tag{8.11}$$

The elastic shortening of the arch is calculated from the average state of stress in the plane of the plate. Each of the average stresses is defined by equation 8.7, and the elastic shortening is then given by

$$\Delta L = f_{av} \frac{(1-v)}{E} L \tag{8.12}$$

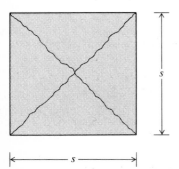

Figure 8.8 Yield lines for a square roof plate (after Beer and Meek, 1982).

The solution procedure for a square plate is identical to that for the roof beam, with equations 8.11 and 8.12 replacing formulae (a) and (d) of equation 8.10. The potential for buckling or compressive failure of the roof plate is determined using the methods described previously. The possibility of shear failure at the abutment is assessed from the appropriate factor of safety, given by

$$\text{F of S} = \frac{2 f_c n \tan \phi}{\gamma s} \qquad (8.13)$$

where f_c is the equilibrium value of the maximum compressive stress.

8.5.2 Rectangular plate

The lines of tensile cracking for a rectangular roof plate can be readily related to the plate symmetry. Referring to Figure 8.9, Beer and Meek suggest that the pattern of cracking is described by the generation of triangular and trapezoidal segments, with the shape of the triangular segments given by

$$y = \frac{a}{2} [(k^2 + 3)^{1/2} - k]$$

where $k = a/b$, and a and b are the short and long spans of the plate.

Both the weight and moment arms of a trapezoidal segment are greater than those for a triangular segment. Thus it is the performance of the arching developed between the trapezoidal segments which controls the behaviour of the roof. The weight of a trapezoidal segment is given by

$$w = \gamma \frac{a}{2} (b - y)t$$

and the centroid is located a distance $\bar{x}$ from the plate edge, where

$$\bar{x} = \frac{ab}{(b-y)} \left(\frac{1}{4} - \frac{yk}{3a} \right)$$

Thus the moment equilibrium equation for the rectangular plate becomes

$$\gamma a^2 bt \left(\frac{1}{4} - \frac{yk}{3a} \right) = f_c \frac{ntbz}{2}$$

or

$$f_c = \gamma a^2 \left(\frac{1}{4} - \frac{yk}{3a} \right) / nz \qquad (8.14)$$

Calculation of the elastic shortening of the short span requires the average stresses acting in the roof plate, parallel to the x and y axes shown in Figure 8.9. For the x direction, f_{av} can be calculated directly from equation 8.7. For the y direction, the average stress is approximated by

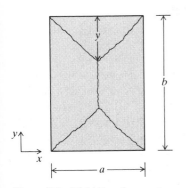

Figure 8.9 Yield lines for a rectangular roof plate (after Beer and Meek, 1982).

$$f_{av}^y = \frac{7}{12} k f_c \qquad (8.15)$$

This expression is selected since it returns the appropriate average antiplane stresses for both a square roof plate ($k = 1$) and a roof beam ($k = 0$). Using equations 8.7 and 8.15, the elastic shortening of the arch is given by

$$\Delta L = f_{av} \frac{(1 - kv)}{E} L \qquad (8.16)$$

In setting up the relaxation routine for a rectangular plate, formulae (a) and (d) of equation 8.10 are replaced by equations 8.14 and 8.16. Other aspects of the solution procedure are identical to those proposed previously. The factor of safety against shear failure at the abutments is calculated from the equilibrium value of f_c, and is given by

$$F \text{ of } S = \frac{f_c n b}{\gamma a (b - y)} \tan \phi \qquad (8.17)$$

Design of roof beams and roof plates is to be based on the determination of factors of safety against roof compressive failure or abutment shear failure, or by assuring that roof buckling cannot occur. The degree of uncertainty inherent in the analysis, due to the static indeterminacy of the problem, has already been noted. The required factor of safety for a design should reflect this uncertainty. In the absence of any significant volume of field observations of roof strata performance, it is suggested that a factor of safety exceeding two is necessary to assure satisfactory operational performance. If the field and rock mass conditions suggest that buckling may be an operative failure mode, the deformation modulus introduced in equation 8.10 should be less than one-half the measured or estimated value.

8.6 Improved design procedures

It is clear that there is ample scope for improvement on the roof design procedures proposed here. The relaxation method represents an appropriate approach for handling the type of problem posed in mine roof design. However, the simple relaxation scheme which is presented makes a number of mechanically reasonable, but unsubstantiated, assumptions to achieve a solution. These should lead to a rather conservative application of the results of design analyses undertaken using the scheme.

The roof design problem is not mechanically intractable. For example, there is no difficulty in devising a solution procedure based on interfaced boundary element and distinct element methods of analysis. The distinct element domain represents satisfactorily the excavation near-field, i.e. the joint and bedding-defined block assemblage close to the excavation. The boundary element region can represent the rest of the rock mass interacting with the block assemblage, and behaving as an elastic continuum. The individual rock blocks can be modelled as elastic distinct elements, and the joints ascribed appropriate shear and normal stiffnesses in the usual way. The initial, pre-excavation state of a rock mass may be established by consolidating the block assemblage under the field stresses, and then using the peripheral nodes of the distinct element set to define, within the computational scheme, the nodes of the boundary element domain. This approach renders the problem statically determinate. The roof problem is then analysed by removing blocks to represent the excavated material. A relaxation scheme may then be used to study the displacement of the blocks and the boundary element nodes, under the out-of-balance forces caused by excavation.

9 Excavation design in jointed rock

9.1 Design factors

A jointed rock mass presents a more complex design problem than considered previously. The complexity arises from either the number (greater than two) of joint sets which define the degree of discontinuity of the medium, or the presence of a discrete structural feature transgressing such a simply jointed system as a cross-jointed, stratified mass. The problem which arises in these types of media is the generation of distinct rock blocks, of various geometric configurations defined by the natural fracture surfaces and the excavation surface, as illustrated in Figure 9.1. Since the blocks exist in the immediate periphery of an excavation whose surface has been subject to the removal of support forces by the mining operation, the possibility arises of collapse of the block assembly in the prevailing gravitational and local stress fields.

The issues to consider in the design of an opening in a jointed medium are a natural extension of those proposed for the structurally simpler media considered previously. That is, it is necessary initially to determine the likelihood of induced fracture in the rock mass, in the total stress field after mining. For continuous features, such as faults or bedding planes which persist over dimensions exceeding those of the excavation, it is necessary to examine the possibility and consequences of slip under excessive shear stress. Also, since joints have effectively zero tensile strength, a jointed rock mass is unequivocally a no-tension medium. Any part of a jointed medium which is notionally subject to tensile stress will, in fact, de-stress. The process of de-stressing a discontinuum implies loss of control, and possibly local collapse of the medium. Stable, continuous behaviour

Figure 9.1 Generation of a discrete prism in the crown of an excavation by the surfaces of defined geological features and the excavation boundary.

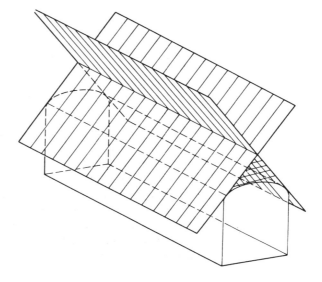

234

of a jointed or granular medium exploits frictional resistance to shear stress, and this resistance is mobilised by compressive normal stress. Thus generation and maintenance of a state of mechanically sustainable compressive stress in the excavation peripheral rock is a basic principle of design in this type of medium.

In addition to considering the quasi-continuous behaviour of a jointed medium in compression, it is also necessary to take account of its explicitly discontinuous properties. Since the rock mass prior to mining consists of an assembly of topographically matched blocks delineated by the joint surfaces, it is necessary to predict the behaviour of the individual blocks when an adjacent surface is developed. For blocks defined in the crown and side walls of an excavation, the requirement is to examine the potential for displacement of each block under the influence of the surface tractions arising from the local stress field, the fissure water pressure and the gravitational load.

9.2 Identification of potential failure modes

The initial phase of the design of an opening in jointed rock follows the methodology defined in Figure 7.1, for design in a continuous medium. This procedure seeks to limit or control the consequences of induced fracture in the rock mass. In the case of jointed rock masses, overall rock mass strength criteria of the type discussed in section 4.8 should be used. Examples of such applications are given by Hoek and Brown (1980). Subsequent analyses to assess effects such as slip and separation on major planes of weakness also employ the general principles established in Chapter 7. In low-stress settings, of course, the strength of the rock mass may obviate any need for the detailed stress analysis proposed in the general design strategy.

In dealing with the specific problems posed by discontinuities in the rock mass, there are two basic analytical requirements. The first is to identify the modes of potential collapse in the block assemblage. The second is to determine the state of equilibrium of the rock blocks and prisms which are involved in the identified collapse modes.

Potential collapse modes are defined by detailed examination of the configuration of blocks identified from the known geometric properties of the discontinuities in the side walls and crown of the excavation. The simplest procedure is to present the known structural data on a stereonet, and to infer the shapes of the blocks generated by the discontinuities and a particular surface of the excavation. By comparing the orientations of the various lines of intersection of the planes which define a block geometry with the inferred direction of the resultant force on the block, kinematically possible directions of displacement of the block can be established. In carrying out this geometric analysis, particular attention must be paid to the crown of the opening, since an obvious mode of collapse is simple detachment of rock prisms in this zone, under gravity load. However, subtle and consequential modes of collapse can also arise in the side walls of an excavation. The intention here is to indicate how unfavourable block geometry and kinematically feasible displacement paths can be recognised, using a stereographic presentation of the appropriate geometric data representing an excavation surface and the structural features.

235

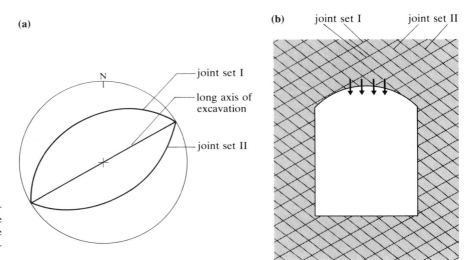

Figure 9.2 Stereographic projection of joints striking parallel to the long axis of an excavation, and the corresponding cross section (looking north-east) of the excavation.

9.2.1 Excavation roof

The simplest structural condition admitting a kinematically possible collapse mode is represented by two features, or sets of features, which strike parallel or subparallel to the long axis of the proposed excavation. The condition is represented by the stereographic projection shown in Figure 9.2a, where it is observed that the line of intersection of the joints is parallel to the equatorial plane, which represents the roof of the excavation. Similarly, the joints' intersections with the roof also strike parallel to the excavation axis. The two conditions illustrated in Figure 9.2a correspond to the possibilities of either free vertical displacement of the triangular prism under gravity load, or slip of the prism on the joint surface with the shallower dip. Cross sections relating to the respective stereographic projections are shown in Figure 9.2b, from which the kinematics of block collapses can be readily appreciated.

Three features or sets of features in a rock mass increase the number of distinct block configurations which can be identified in the excavation periphery. The joint attitudes defined in Figure 9.3a generate a tetrahedral wedge. It is seen that a vertical line through the centre of the net (corresponding to the direction of the weight vector) lies within the trihedral angle defined by the planes of weakness,

Figure 9.3 Stereographic projections of joint sets forming tetrahedral wedges in the crown of an excavation, corresponding to collapse modes of (a) free vertical displacement, and (b) slip on a base plane.

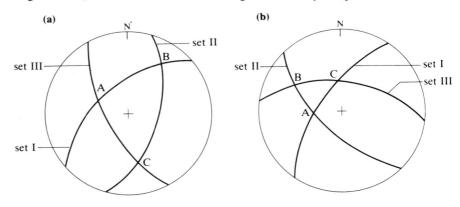

i.e. the three lines of intersection, OA, OB, OC, of the planes of weakness are disposed about the weight vector. This structural condition therefore results in generation of a tetrahedral wedge which is kinematically free to fall vertically into the underlying excavation.

Figure 9.3b shows the case in which a tetrahedral wedge is defined by the planes of weakness, but the weight vector does not act within the wedge trihedral angle. This is indicated by the fact that point O, which represents a vertical line through the centre of the net, lies outside the triangle ABC, which represents the trace of the joints' intersections on the crown of the opening. This precludes free vertical displacement of the wedge under its own weight. Instead, kinematically possible displacement of the wedge involves slip on the plane of weakness I.

It is to be noted that many other block configurations are generated by each of the rock structural conditions illustrated here. Merely by considering the stereonet as an upper hemisphere rather than lower hemisphere projection, mirror images of the tetrahedral wedges already discussed are identified. By considering it as upper and lower hemisphere projections simultaneously, the lines of intersection of the features are seen to define wedges which are topologically complementary to the wedges explicitly identified as collapse prone. Thus the procedure discussed defines only the key block in the jointed assemblage, whose removal may provoke displacement of the contiguous rock units.

9.2.2 Excavation side walls

The vertical side walls of an excavation project on the stereonet as a straight line in the equatorial plane. Since the parallel side walls of an excavation are represented stereographically by the same line, it is necessary to note in any geometric analysis which side of the projection line represents the solid rock. This idea is represented in the example illustrated in Figure 9.4, where three joint sets exist in a rock mass in which a long excavation, striking north-east–south-west, is to be mined. In Figure 9.4a, the planes of weakness I, II, III, define a tetrahedral wedge. If AB represents the southeastern wall of the excavation, the joint

Figure 9.4 (a) Stereographic projection; (b) cross section (looking north-east), showing collapse modes of tetrahedral wedges in the side walls of an excavation.

(a)

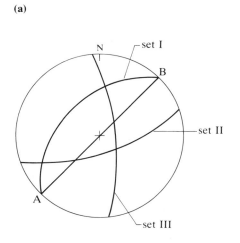

(b)

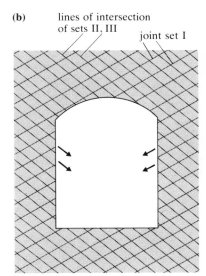

geometry allows slip of the tetrahedral wedge, on the base plane I, into the excavation. If AB represents the northwestern wall of the excavation, the feasible mode of wall collapse involves block slip in a direction parallel to the lines of intersection of planes II and III. These collapse modes are represented schematically in Figure 9.4b.

In using the stereonet to identify kinematically feasible, rigid-body displacement modes from the jointed mass, no attempt has been made to give a comprehensive description of the conformational relations between the rock blocks and the excavation surface. This topic is the subject of current research, and involves consideration of the basic principles of topology (Gen Hua Shi and Goodman, 1981; Warburton, 1981).

9.3 Symmetric triangular roof prism

Having identified the feasible block collapse modes associated with joint attitudes and excavation surface geometry, it is necessary to determine the potential for block displacement under the ambient conditions which exist in the post-excavation state of the opening periphery. In the following discussion, attention is confined to analysis of crown stability problems. Methods for analysis of sidewall problems may be established by some minor variations in the proposed procedures.

A rock block in the crown of an excavation is subject to its own weight, W, surface forces associated with the prevailing state of stress, and possibly fissure water pressure and some support load. Assume, for the moment, that fissure water pressure is absent, that the block surface forces can be determined by some independent analytical procedure, and that the block weight can be determined from the joint orientations and the excavation geometry.

Figure 9.5a represents the cross section of a long, uniform, triangular prism generated in the crown of an excavation by symmetrically inclined joints. The semi-apical angle of the prism is α. Considering unit length of the problem geometry in the antiplane direction, the block is acted on by its weight W, a support force R, and normal and shear forces N and S on its superficial contacts with the adjacent country rock. The magnitude of the resultant of W and R is P. To assess the stability of the prism under the imposed forces, replace the force P by a force P_ℓ, as shown in Figure 9.5b, and find the magnitude of P_ℓ required to establish a state of limiting equilibrium of the block.

Figure 9.5 Free-body diagrams of a crown prism (a) subject to surface forces (N and S), its own weight (*W*) and a support force (*R*), and (b) in a state of limiting equilibrium.

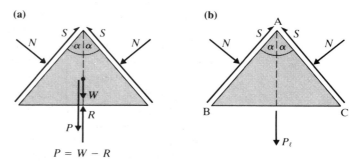

From Figure 9.5b, the equation of static equilibrium for the vertical direction is satisfied if

$$P_\ell = 2(S \cos \alpha - N \sin \alpha) \tag{9.1}$$

If the resistance to slip on the surfaces AB, AC is purely frictional, in the limiting equilibrium condition,

$$S = N \tan\phi \tag{9.2}$$

and equation 9.1 becomes

$$P_\ell = 2N(\tan \phi \cos \alpha - \sin \alpha) \tag{9.3}$$

$$= 2N \sec \phi \sin(\phi - \alpha)$$

Thus for $N > 0$, the condition $P_\ell > 0$ can be satisfied only if $\alpha < \phi$. Thus if $\alpha > \phi$, $P_\ell < 0$, and even in the absence of its own weight, the prism would be displaced from the crown under the influence of the joint surface forces. For the case $\alpha < \phi$, the prism is potentially stable, but stability can only be assured by a more extensive analysis.

The following analysis is intended to establish the key factors affecting the stability of a symmetric roof prism, for the case $\alpha < \phi$. It is an example of a relaxation method of analysis, proposed originally by Bray (1977). The procedure takes explicit account of the deformation properties of the joints defining the crown prism. Initially, the joint normal and shear stiffnesses K_n and K_s are assumed to be sufficiently high for the presence of the joints to be ignored. It is then possible to determine the stress distribution around the opening assuming the rock behaves as an elastic continuum. Since no body forces are induced in the medium by the process of excavating the opening, the elastic analysis takes account implicitly of the weight of the medium. Such an analysis allows the state of stress to be calculated at points in the rock mass coinciding with the surfaces of the prism. It is then a simple matter to estimate the magnitudes of the surface forces acting on the prism from the magnitudes of the stress components and the area and orientation of each surface.

The relaxation method proceeds by introducing the joint stiffnesses K_n and K_s, and examining the displacements subsequently experienced by the block caused by joint deformation. Since real joint stiffnesses are low compared with the elasticity of the rock material, the deformability of the prism can be neglected in this process. As defined previously, the block is subject to its own weight, W, and some support force, R, whose resultant is $P = W - R$, as well as the joint surface forces. The analysis proceeds through the displacements of the body under the influence of the internal surface forces and the vertical force producing limiting equilibrium, P_ℓ, defined by equation 9.3. The state of stability of the prism is then assessed through the factor of safety against roof failure, defined by

239

$$\text{F of S} = \frac{P_\ell}{P} \tag{9.4}$$

Before the relaxation process (i.e. before applying the limiting force P_ℓ and reducing the joint stiffnesses), the state of loading of the prism is as shown in Figure 9.6a. In this case, the surface forces N_0, S_0 account completely for the static equilibrium of the prism. These original surface forces are related to the internal horizontal force H_0 by

$$N_0 = H_0 \cos \alpha$$

$$\tag{9.5}$$

$$S_0 = H_0 \sin \alpha$$

When the resultant force P_ℓ is applied, the wedge is displaced vertically through a distance u_y. Displacement u_s and u_n, with the directions indicated in Figure 9.6b, occur at the joint surface, and the normal and shear forces are incrementally perturbed, changing to the new equilibrium values N and S. Since the prism is not deformed in the joint relaxation, joint deformations, u_s and u_n, are readily related to the vertical, rigid-body displacement, u_y, of the prism. From Figure 9.6b

$$u_s = u_y \cos \alpha$$

$$\tag{9.6}$$

$$u_n = u_y \sin \alpha$$

Noting that the block moves away from the surrounding rock during joint relaxation, increments of surface force are related to displacement increments by

$$N_0 - N = K_n u_n = K_n u_y \sin \alpha$$

$$\tag{9.7}$$

$$S - S_0 = K_s u_s = K_s u_y \cos \alpha$$

Also, the equation of static equilibrium of the prism for the x direction requires

$$H = N \cos + S \sin \alpha \tag{9.8}$$

Figure 9.6 Free-body diagrams of a crown prism (a) subject to surface forces corresponding to elastic stresses, and (b) in a state of limiting equilibrium after external load application and joint relaxation.

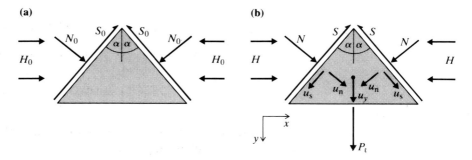

Substitution of N_0, S_0, from equation 9.5, in equation 9.7, yields

$$N = H_0 \cos \alpha - K_n u_y \sin \alpha$$

$$S = H_0 \sin \alpha + K_s u_y \cos \alpha$$

(9.9)

Since the problem being considered involves the limiting equilibrium state of the prism, introduction of the limiting friction defined by equation 9.2 in equation 9.9 yields

$$H_0 \sin \alpha + K_s u_y \cos \alpha = (H_0 \cos \alpha - K_n u_y \sin \alpha) \tan \phi$$

Simple trigonometric substitutions and rearrangement of this expression produce

$$H_0 \sin (\phi - \alpha) = u_y(K_s \cos \alpha \cos \phi + K_n \sin \alpha \sin \phi)$$

or

$$u_y = \frac{H_0 \sin (\phi - \alpha)}{(K_s \cos \alpha \cos \phi + K_n \sin \alpha \sin \phi)}$$

Inserting this expression for u_y in equation 9.9 gives

$$N = H_0 \left[\cos \alpha - \frac{K_n \sin \alpha \sin (\phi - \alpha)}{K_s \cos \alpha \cos \phi + K_n \sin \alpha \sin \phi} \right]$$

$$S = H_0 \left[\sin \alpha - \frac{K_n \cos \alpha \sin (\phi - \alpha)}{K_s \cos \alpha \cos \phi + K_n \sin \alpha \sin \phi} \right]$$

which, on rearrangement, yields

$$N = \frac{H_0}{D} (K_s \cos^2 \alpha + K_n \sin^2 \alpha) \cos \phi$$

$$S = \frac{H_0}{D} (K_s \cos^2 \alpha + K_n \sin^2 \alpha) \sin \phi$$

(9.10)

where

$$D = K_s \cos \alpha \cos \phi + K_n \sin \alpha \sin \phi$$

Introducing these expressions for N and S into equation 9.8, and simplifying,

$$H = \frac{H_0}{D} (K_s \cos^2 \alpha + K_n \sin^2 \alpha) \cos (\phi - \alpha)$$

(9.11)

The expression for N in equation 9.10, when substituted in equation 9.3 for the limiting vertical force, yields

241

$$P_\ell = \frac{2H_0}{D}(K_s \cos^2 \alpha + K_n \sin^2 \alpha)\sin(\phi - \alpha) \tag{9.12}$$

For the case where $K_n \gg K_s$, which is the usual condition in practice, equation 9.12 becomes

$$P_\ell = \frac{2H_0 \sin \alpha \sin(\phi - \alpha)}{\sin \phi} \tag{9.13}$$

This analysis indicates that when the elastic state of stress has been determined, the net vertical external load required to produce a state of limiting equilibrium can be estimated, from equation 9.13, using the known prism geometry and joint frictional properties. The factor of safety against roof collapse can then be calculated from equation 9.4. In particular, if the result of the determination of P_ℓ is that $P_\ell > W$, the weight of the prism, the analysis suggests that the wedge is stable in the absence of any installed support. If $P_\ell < W$, stability of the prism can be assured only by the application of a positive support load.

It is instructive to examine the relation between the limiting vertical load, P_ℓ, and the horizontal force components on the prism surfaces. Introducing equation 9.2 in equation 9.8, rearranging, and then substituting the resulting expression for N in equation 9.3, gives

$$P_\ell = 2H \tan(\phi - \alpha) \tag{9.14}$$

Equations 9.13 and 9.14 emphasise the important role of the horizontal force components acting on the prism surfaces. It is clear that any process which acts to reduce these surface forces, applied by the adjacent rock, will reduce the limiting vertical load. This is equivalent, of course, to increasing the tendency for collapse of the prism from the roof. Excavation development procedures may affect these internal forces. For example, uncontrolled blasting practice near the excavation periphery will inject high-pressure gases directly into the joints, promote vertical displacement greater than that associated with elastic joint deformability, and thereby reduce the final horizontal force responsible for block retention in the crown of the opening.

9.4 Asymmetric triangular roof prism

The problem of the symmetric prism in the crown of an excavation was rendered analytically determinate by virtue of the symmetry of loading and through the joint deformability exploited in the relaxation process. In the case of an asymmetric prism, a further condition must be introduced to obtain a tractable problem. The relaxation analysis is identical in principle to that described for the symmetric prism. It commences from an elastic analysis, ignoring joint deformability, to obtain the normal and shear forces operating on the surfaces defining the prism interfaces with the surrounding rock. These results can be used directly to make an initial assessment of block stability. Since the relaxation process

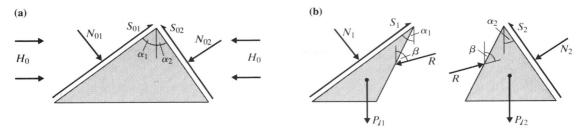

Figure 9.7 (a) Free-body diagram of an asymmetric crown prism subject to surface forces corresponding to elastic stresses, and (b) constituent free bodies for relaxation analysis.

results in a reduction of the normal and shear forces operating at the joints, the elastic analysis provides a basis for a first assessment of the potential stability of a roof prism. Referring to the problem geometry defined in Figure 9.7a, if

$$S_{01} \cos \alpha_1 - N_{01} \sin \alpha_1 + S_{02} \cos \alpha_2 - N_{02} \sin \alpha_2 > W$$

the prism is potentially stable under its own weight. If the opposite condition applies, the prism must be regarded as liable to collapse in the force field existing after joint relaxation.

In seeking to apply the relaxation method to the analysis of the problem illustrated in Figure 9.7a, it is observed that a uniform, rigid-body displacement of the prism would result in incompatible normal and shear displacements at the joint surfaces. To allow each surface to express its own deformation properties in achieving an equilibrium state, it is necessary to postulate an arbitrarily oriented slip surface within the body of the prism, and passing through the apex. In a jointed medium, the existence of such a surface, on which incremental slip can occur, presents no particular conceptual difficulty.

The relaxation routine proceeds by considering the limiting equilibrium of the two problems shown in Figure 9.7b. In its final state of loading, each block will be acted on by its own weight, W_1 or W_2, respectively, and support forces, R_1 or R_2, such that

$$P_1 = W_1 - R_1$$

$$P_2 = W_2 - R_2$$

Thus for the system as a whole

$$P = P_1 + P_2 \tag{9.15}$$

As before, the stability of the system is examined by finding the limiting vertical forces, $P_{\ell 1}$ and $P_{\ell 2}$, which, when replacing P_1 and P_2, bring blocks 1 and 2 to a state of limiting equilibrium. Writing the equations of static equilibrium for the vertical direction for each block yields

$$P_{\ell 1} = S_1 \cos \alpha_1 - N_1 \sin \alpha_1 - R \cos \beta$$

$$P_{\ell 2} = S_2 \cos \alpha_2 - N_2 \sin \alpha_2 + R \cos \beta$$

243

where, for limiting friction,

$$S_1 = N_1 \tan \phi_1$$

$$S_2 = N_2 \tan \phi_2$$

Therefore

$$P_\ell = P_{\ell 1} + P_{\ell 2} = N_1 \sec \phi_1 \sin (\phi_1 - \alpha_1) + N_2 \sec \phi_2 \sin (\phi_2 - \alpha_2) \qquad (9.16)$$

Considering the displacements and changes in normal and shear forces associated with joint relaxation, in a way analogous to that used for the symmetric prism, leads to the result

$$P_\ell = \frac{H_0}{D_1} (K_{s1} \cos^2 \alpha_1 + K_{n1} \sin^2 \alpha_1) \sin (\phi_1 - \alpha_1)$$

$$+ \frac{H_0}{D_2} (K_{s2} \cos^2 \alpha_2 + K_{n2} \sin^2 \alpha_2) \sin (\phi_2 - \alpha_2) \qquad (9.17)$$

where

$$D_1 = K_{s1} \cos \alpha_1 \cos \phi_1 + K_{n1} \sin \alpha_1 \sin \phi_1$$

with a similar expression for D_2.

When $K_{n1} \gg K_{s1}, K_{n2} \gg K_{s2}$,

$$P_\ell = \frac{H_0 \sin \alpha_1 \sin (\phi_1 - \alpha_1)}{\sin \phi_1} + \frac{H_0 \sin \alpha_2 \sin (\phi_2 - \alpha_2)}{\sin \phi_2} \qquad (9.18)$$

In applying the magnitude of P_ℓ (calculated from either equation 9.17 or 18) in equation 9.4 to determine the factor of safety against roof collapse, particular care needs to be exercised in the case in which one term of each of the right-hand sides of these equations is negative. Such a circumstance implies that conditions on a single joint surface cannot be sustained by the post-excavation stress field. When this condition applies, the operational performance of the crown of the opening may be assured by timely emplacement of active support.

The relation between the post-relaxation horizontal forces and the limiting vertical load, corresponding to equation 9.14 for a symmetric wedge, is readily shown to be

$$P_\ell = H_1 \tan (\phi_1 - \alpha_1) + H_2 \tan (\phi_2 - \alpha_2) \qquad (9.19)$$

9.5 Roof stability analysis for a tetrahedral wedge

A comprehensive relaxation analysis for a non-regular tetrahedral wedge in the crown of an excavation presents some conceptual difficulties. These arise from the extra number of degrees of freedom to be accommodated in the analysis. For example, on any face of the tetrahedron it is necessary to consider two components of mutually orthogonal shear displacement as well as a normal displacement component. Maintenance of statical determinacy during the relaxation process would require that the wedge be almost isotropically deformable internally. For this reason, a thorough analysis of the stability of a tetrahedral wedge in the crown is not handled conveniently by the relaxation method presented earlier. A computational method which takes explicit account of the deformation properties of the rock mass, and particular joints and joint systems, presents the most opportune basis for a mechanically appropriate analysis.

In some circumstances, it may be necessary to assess the stability of a roof wedge in the absence of adequate computational tools. In this case, it is possible to make a first estimate of wedge stability from an elastic analysis of the problem and the frictional properties of the joints. Suppose the orientation of the dip vector **OA** of a joint surface, which is also a face of a tetrahedral wedge, is defined by the dip angle α, and dip direction β, measured relative to the global (x, y, z) reference axes, illustrated in Figure 9.8a. The direction cosines of the outward normal to the plane are given by

$$\mathbf{n} = (n_x, n_y, n_z) = (\sin \alpha \cos \beta, \sin \alpha \sin \beta, -\cos \alpha)$$

The normal component of traction at any point on the joint surface can be estimated from the elastic stress components and the direction cosines by substitution in the equation

$$t_n = n_x^2 \sigma_{xx} + n_y^2 \sigma_{yy} + n_z^2 \sigma_{zz} + 2(n_x n_y \sigma_{xy} + n_y n_z \sigma_{yz} + n_z n_x \sigma_{zx})$$

If the normal traction t_n is determined at a sufficient number of points on the joint surface, its average value and the area of the surface can be used to estimate the total normal force N. Thus for each of the three confined faces of the tetrahedron,

Figure 9.8 (a) Geometry for determination of the unit normal vector to a plane; (b) lines of action of mobilised shear forces on the face of a tetrahedral wedge.

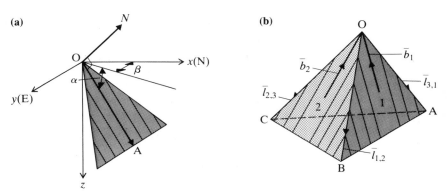

the respective normal forces N_1, N_2, N_3 can be calculated directly from the joint surface geometry and the elastic stress distribution.

In determining the stability of a wedge under surface and gravitational forces, it is necessary to take account of the directions of the shear resistances mobilised by the joint normal forces. Suppose the outward normals to faces 1, 2, 3 of the tetrahedron OABC shown in Figure 9.8b are given by

$$\mathbf{n}_1 = (n_{x1}, n_{y1}, n_{z1}) \quad \text{etc.}$$

and that the faces are numbered in a sense compatible with the right-handed system of reference axes. The lines of intersection of the faces are then given by cross products of the normals to the faces, i.e.

$$\mathbf{l}_{1,2} = \mathbf{n}_1 \times \mathbf{n}_2 \quad \text{etc.}$$

The bisector of an apical angle of a face of the tetrahedron, and directed towards the apex, as shown in Figure 9.8b, is obtained from the orientations of the adjacent lines of intersection which define the face, i.e.

$$\mathbf{B} = -\tfrac{1}{2}(\mathbf{l}_{1,2} + \mathbf{l}_{3,1})$$

From this, one can readily establish the unit vector parallel to the bisector,

$$\mathbf{b}_1 = (b_{x1}, b_{y1}, b_{z1}) \tag{9.20}$$

It can be reasonably assumed that, in the case where the crown trihedral angle of the tetrahedron includes the z axis, the mobilised shear resistance on any face is directed parallel to the bisector of the face apical angle. Also, the inward unit normal to any face, defining the line of action of the normal component of the surface force, is given by

$$\mathbf{a} = (-\sin\alpha\cos\beta, -\sin\alpha\sin\beta, \cos\alpha)$$

The magnitudes of the maximum shear forces that can be mobilised on the various faces are given by

$$S_1 = N_1 \tan\phi_1 \quad \text{etc.}$$

and the x, y, z components of the shear resistance on any face can be determined directly from its magnitude and the components of the appropriate unit vector for the face, defined by equation 9.20. Taking account of all applied normal forces and mobilised shear resistances, the net vertical force associated with the internal surface forces is

$$F_z = \sum_{i=1}^{3} N_i (b_{zi} \tan\phi_i + a_{zi}) \tag{9.21}$$

246

Introducing the weight of the wedge, if the resultant vertical force satisfies the condition

$$F_z + W < 0 \qquad (9.22)$$

the wedge is potentially stable under the set of surface and body forces. An added condition to be satisfied in this assessment is that the sum of each pair of terms on the right-hand side of equation 9.21, i.e. $(b_z \tan \phi + a_z)$, must be negative. If the sum of any such pair of terms is positive, this implies that the particular surface will be subject to slip under the prevailing state of stress. In such a case, the initiation of slip must be anticipated to lead to expansion of the area of slip over the other block surfaces, and subsequent detachment of the wedge from the crown of the opening.

The case considered above concerned potential displacement of the wedge in the vertical direction. For particular joint attitudes, the kinematically possible displacement may be parallel to the dip vector of a plane of weakness, or parallel to the line of intersection of two planes. In these cases, some simple modifications are required to the above analysis. Since, in all cases, the lines of action of the maximum shear resistances are subparallel to the direction of displacement, equations 9.21 and 9.22 should be developed by considering the direction of the feasible displacement as the reference direction. This merely involves dot products of the various operating forces with a unit vector in the reference direction.

9.6 Pragmatic design in jointed rock

In the course of considering the behaviour of rock prisms and wedges in the periphery of underground excavations, it was seen that, once a kinematically feasible collapse mode exists, the stability of the system depends on:

(a) the tractions on the joint-defined surfaces of the block, and therefore on the final state of stress around the excavation and the attitudes of the joints;
(b) the frictional properties of the joints;
(c) the weight of the prism, i.e. its volume and unit weight.

Effective excavation design in a jointed rock mass requires a general understanding of the engineering significance of each of these factors.

It has already been observed that it is the normal component of traction on a joint surface which is responsible for mobilising friction to prevent displacement of a rock prism. Thus if at any stage in the mining life of an opening in jointed rock, the peripheral rock is de-stressed, wedge collapses will occur from the excavation crown and side walls. De-stressing may be due to such effects as mining adjacent stopes, local fracture of rock and its subsequent non-transmission of stress, blasting practice causing local stress relief, and local stress relaxation due to time-dependent effects. In initially low-stress environments, of course, the internal forces available to prevent block failures will always be low, and pervasive peripheral failures are to be expected. In all cases, the design of the excavation should take account of the near-field state of stress to be expected throughout its projected mining life.

247

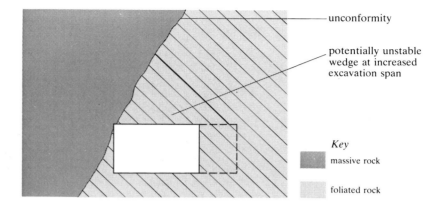

Figure 9.9 Problem geometry demonstrating how an increase in excavation span increases the volume of a roof prism, without comparable increase in the restraining force.

The rôle of friction in controlling peripheral rock performance was discussed briefly in Chapter 7. In the current context, it is noted that, in their initial, topographically matched state, joints are almost universally dilatant in shear. Thus the effective angle of friction exceeds the value which might be determined by a shear test on a disturbed specimen of a joint surface. Any mining activity which disturbs the initial, interlocked state of a joint surface automatically reduces the capacity of the rock mass to support its constituent blocks at the excavation periphery. The chief sources of joint disturbance are local blasting effects, transient effects due to the impulsive nature of the excavation process, and large scale, far-field blasting.

The effect of wedge size on the possibility of boundary collapse may appear to be obvious, on a superficial examination. However, there are some subtle considerations which may have serious practical consequences if ignored. An example is illustrated in Figure 9.9, in which an opening has been developed in a rock mass in such a way that a rock prism has been generated in the crown of the opening. If it were decided to widen the opening, a stage would inevitably be reached where the roof prism would collapse. This is so since the wedge increases in weight with the square of the span, while the mobilised support force, to a first approximation, increases only linearly with span. For a three-dimensional problem, the same conclusion applies, since the wedge weight always increases by a power of the linear dimension higher by unity than does the surface area. The important principle demonstrated by this example is that a marginal increase in the span of an opening in jointed rock can cause a significant reduction in the stability of the system, through a marked increase in the disturbing force (the block weight) relative to the mobilised resisting force.

Since mining engineering suffers from few of the cosmetic requirements of civil engineering, mine openings can be excavated to shapes that are more appropriate and effective geomechanically than the latter types. In mining practice, the general rule is that an opening should be mined to a shape conformable with the dominant structural features in the rock mass. Although such an excavation shape might not be aesthetically satisfying, it would represent the optimum design for the particular setting, in terms of peripheral stability and support and maintenance costs. An example is illustrated in Figure 9.10a, representing the cross section of a long excavation developed in a rock mass with a steeply

(a)   **(b)**

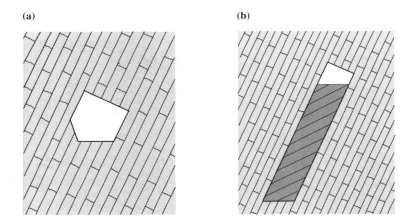

Figure 9.10 Maintenance of excavation boundary stability in jointed rock by mining to conformity with the rock structure, for (a) a mine drive, and (b) a cut-and-fill stope.

Figure 9.11 Cut-and-fill stope mined to conformity with rock structure at the Mount Isa Mine, Australia (after Mathews and Edwards, 1969).

inclined set of continuous joints, and an orthogonal, flat-dipping set. The crown of the excavation has been mined so that segments of the boundary coincide with a member of each joint set, to eliminate the potential roof prism. The right-hand side of the excavation has been mined to coincide with a member of the continuous joint, to eliminate the side wall prism. The prism defined on the lower side of the left-hand wall presents no problem of potential instability. For the excavation shown, virtually all boundary stress transmission occurs across joints

which are oriented perpendicular to the excavation surface. There is thus no tendency for local slip and stress relief on these features. Provided the excavation periphery is maintained in a state of compression, this design ensures that there will be no source of instability in the excavation crown and side walls.

The design principle illustrated in Figure 9.10 is of particular value in mining methods such as shrink stoping and cut-and-fill stoping. In these cases, miners work beneath the subhorizontal rock face exposed by the subvertical advance of mining. Effective control of the stope crown, with the added requirement of limited support emplacement, is achieved by a stope shape mined to conformability with the dominant rock structure. Figure 9.11 shows the industrial implementation at the Mount Isa Mine, Australia, of the design principle illustrated in Figure 9.10b.

In the design of a permanent mine excavation, such as a crusher station or an underground workshop, some scope usually exists for orienting and shaping the opening to produce an economic design. The general rule is that no major permanent opening should be located and oriented so that its long axis is parallel to the strike of a significant geological feature, such as a fault or shear zone. If it is impossible to avoid the zone containing the feature, the principle should be to orient the excavation axis as near perpendicular as possible to the strike of the major discontinuity. The objective in this case is to limit the size of wedges formed in the crown of the excavation, and to restrict the area of the excavation periphery subject to potential de-stabilisation by the feature.

10 Energy, mine stability and rockbursts

10.1 Mechanical relevance of energy changes

The discussion in preceding chapters was concerned with the design of single, or mechanically isolated, excavations in different types of rock media. In all cases, a design objective was to achieve a static stress distribution, or a set of static forces on joint-defined internal surfaces, which could be sustained by the constituent elements of the rock mass. This approach would be completely satisfactory if rockbursts and similar dynamic events did not need to be considered in underground excavation design, or if the stress concentrations which occur around openings were achieved in a pseudo-static way. Rockbursts are a pervasive problem in mines which operate at high extraction ratio, and involve release and transmission of seismic energy from the zone of influence of mining. Furthermore, in metalliferous mining, the development of mine excavations, for both access and ore production, frequently involves near-instantaneous generation of segments of the excavation surface. As observed in Chapter 1, the development of an underground opening is mechanically equivalent to application of a set of tractions over a surface representing the excavation boundary. Thus, typical excavation development practice is represented mechanically by the impulsive application of these surface forces in the rock medium.

Rockbursts arise from unstable energy changes in the host rock mass for mining. Also, in conventional mechanics, it is well known that impulsive loading of a structural member results in transient stresses greater than the final, static stresses in the system, and that the most effective method of determining transient stresses and deformations under impulsive loading is by consideration of the energy changes in the system. In fact, the amount of energy that a particular member can store or dissipate is frequently an important criterion in mechanical design. A component which is operationally subject to rapid loading must be constructed to a specification which reflects its duty as a transient energy absorber. In the mining context, it is reasonable to propose that rock around mine excavations will be subject, during development, to transient stresses exceeding the equilibrium static stresses, due to rapid application of surface tractions. It is thus inferred that both rockbursts and these transient effects may be best studied through methods which account for energy changes in the system.

As observed later, energy changes in a mine domain arise from generation and displacement of excavation surfaces and energy redistribution accompanying seismic events. Because mine development practice frequently employs drill and blast methods for rock removal to generate a new excavation surface, chemical explosives are also a source of energy input to the rock mass. Energy from that source is not the concern in this chapter. Attention will be restricted to the strain energy changes which arise from the way in which surface forces are applied to parts of a mine structure.

251

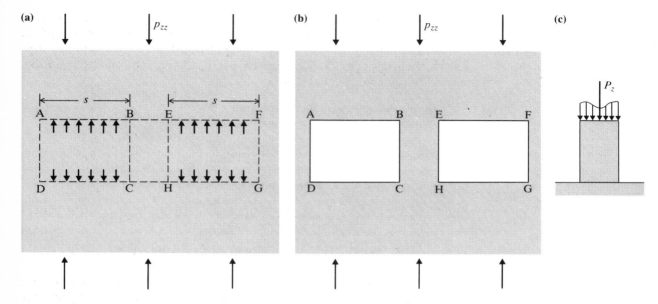

Figure 10.1 (a) Pre-mining state of loading around rooms in a stratiform orebody subject to uniaxial stress; (b) post-excavation state; (c) free-body diagram showing induced axial pillar load.

An instructive insight into the problem posed by sudden generation of a rock surface is presented by the example illustrated in Figure 10.1. Consider unit thickness of the problem geometry in the antiplane direction. Figure 10.1a shows the pre-mining state of a body of rock, in which two rooms are to be mined to isolate a central pillar. The post-excavation state is shown in Figure 10.1b. For the sake of simplicity, it is assumed that the pre-mining stress field is uniaxial, vertical and of magnitude p_{zz}. After excavation, the support loads previously applied to the surrounding medium by the rock within the surfaces ABCD and EFGH are taken to be shared equally between the pillar and the abutments. Since the span of each excavation is s, the induced axial pillar load, P_z, is given by

$$P_z = s p_{zz}$$

Figure 10.2 (a) Static loading of a pillar; (b) induced load–deflection diagram for the pillar; (c) load–deflection diagram for boundary unloading.

The induced load is assumed to be uniformly distributed over the pillar plan area, and the closure is assumed uniform over the pillar and the adjacent excavations. The response to mining of the pillar and the adjacent rock can then be represented by the rock strut on a rigid base, shown in Figure 10.1c.

Consider first the case of gradual mining of the excavations ABCD and EFGH, corresponding to gradual removal of the surface support forces and gradual

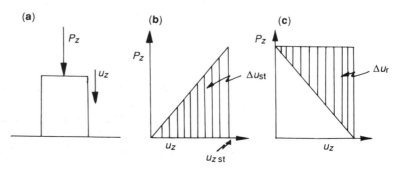

application of the mining-induced load to the pillar. The response, as illustrated in Figure 10.2, consists of an axial, static compression $u\ (= u_{z\,\text{st}})$ of the pillar, and a static decompression of the mined adjacent rock, both displacements varying linearly with change in the operating load P_z. During the excavation process, energy conservation requires that

Work done by the applied load

$$= \int_0^{u_z} P_z \, du_z$$

$$= \frac{1}{2} P_z u_z \qquad (10.1)$$

$$= \text{Increase in strain energy in pillar } (\Delta U_{\text{st}})$$

The complementary work (i.e. the work done by the excavation surfaces against the reducing support force) may be identified with energy U_r released by the rock mass by the process of excavation. From Figure 10.2b, the released energy is given by

$$\text{Work done by excavation surface loads} = \frac{1}{2} P_z u_z \qquad (10.2)$$

It is observed, from equations 10.1 and 10.2, that the released energy is equal to the increase in pillar strain energy, consistent with energy conservation.

After the pillar load is applied, induced static stress σ_{zz} and axial compression u_z are related through the elasticity of the rock; i.e.

$$\sigma_{zz} = P_z/A$$

or

$$P_z = \sigma_{zz} A \qquad (10.3)$$

and

$$u_z = \varepsilon_{zz} L = (\sigma_{zz}/E)\,L \qquad (10.4)$$

where A and L are pillar plan area and length.
From equations 10.1, 10.3, and 10.4

$$\Delta U_{\text{st}} = \left(\frac{1}{2}\,\sigma_{zz}^2/E\right) \times A \times L \qquad (10.5)$$

In the case of sudden mining of the adjacent excavations, the complete induced load P_z is rapidly applied to the pillar. Figure 10.3 shows the deformation of the pillar up to the stage of maximum axial compression, and the corresponding load–deflection diagram. It is observed that the pillar load P_z is constant throughout the pillar axial deformation. The maximum axial compression is given by

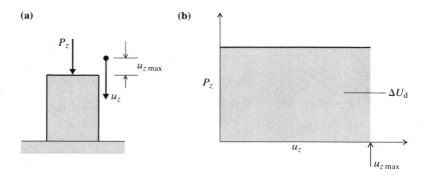

Figure 10.3 (a) Dynamic loading of a pillar, and (b) the corresponding load–deflection diagram.

$$u_{z\,\text{max}} = \varepsilon_{zz}\,L = (\sigma_{zz\,\text{max}}\,/E)\,L$$

and the strain energy increase is given by

$$\Delta U_\text{d} = \text{work done by the loading system}$$

i.e.

$$(\sigma_{zz\,\text{max}}\,/2E)\,AL = P_z u_{z\,\text{max}} \tag{10.6}$$

or

$$\sigma_{zz\,\text{max}} = 2P_z/A = 2\sigma_{zz}$$

Also

$$u_{z\,\text{max}} = 2u_{z\,\text{st}} \tag{10.7}$$

where σ_{zz} and u_z are the peak static values of axial stress and compression.

It is useful to examine the partitioning of energy in the pillar as it passes through the static equilibrium state on rebound from maximum axial compression. The problem geometry is illustrated in Figure 10.4. Since energy is conserved in the system, the total strain energy, ΔU_d, is distributed, at passage through the static equilibrium state, among the other forms of energy according to

$$\Delta U_\text{d} = (\text{increase in gravitational P.E.}) + \text{K.E.} + \Delta U_\text{st}$$

i.e. from equations 10.6 and 10.7

$$2P_z u_{z\,\text{st}} = P_z u_{z\,\text{st}} + \text{K.E.} + \tfrac{1}{2}\,P_z u_{z\,\text{st}}$$

or

$$\text{K.E.} = \tfrac{1}{2}\,P_z u_{z\,\text{st}}$$

The kinetic energy in the pillar (which is seen to be equal in magnitude to the released energy defined by equation 10.2) represents excess energy U_e in the system, compared with the case of static loading. In practice, this excess energy would be lost to the pillar during vibration, by transfer to the adjacent country rock and by dissipation through internal frictional damping. The question that arises is – what is the source of the excess energy, manifested as kinetic energy in the system, after rapid excavation of adjacent rock? The answer is related to the control which is exercised at a surface when it is created slowly. Controlled excavation implies a gradual reduction in the support forces which operated

254

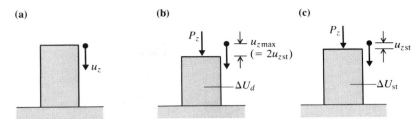

Figure 10.4 Problem geometry for examining energy partitioning in a pillar under dynamic load.

originally on the surface of interest. The gradual force reduction involves application of a decreasing restraining force as the surface displaces to its final equilibrium position. In this process, the excavated surface does work against the restraining force during the mining-induced displacement. In the case of unrestrained displacement of the excavated surface, as in sudden mining of the rooms in the example above, the excess energy manifested in the system represents the work which would have been done by the mined boundary against the support force, had the surface displacement been restrained during mining. This energy is retained instantaneously, and then transmitted to the surrounding rock.

This discussion suggests there are two factors to be considered in relation to energy changes associated with creating excavations. First, increase of static strain energy occurs in areas of stress concentration, equivalent (for an elastic rock mass) to the energy released during pseudo-static displacement of excavated surfaces. Second, sudden excavation of surfaces causes an energy imbalance in the system, and results in transient stresses different from the equilibrium static stresses. For an elastic system, the excess energy is equal in magnitude to the released energy.

10.2 Mining consequences of energy changes

Mining activity takes place in a medium subject to general triaxial stress. For purposes of illustration, attention will be restricted to two-dimensional problems involving biaxial states of stress. The analyses may be simply extended to accommodate three-dimensional problem geometries in multiaxial stress fields. The discussion is an elaboration of ideas proposed by Cook (1967b), Salamon (1974), Bray (1977) and Blight (1984).

The basic ideas introduced by the pillar loading problem can be extended to include energy changes at the boundaries of arbitrarily shaped excavations. Referring to Figure 10.5a, prior to excavation a surface S is subject, at any point, to tractions t_x, t_y. In the case of gradual excavation of the material within S, the surface tractions are gradually reduced to zero, and areas of induced stress are generated around the excavation, as indicated in Figure 10.5b. The strain energy density is inferred, from equation 10.5, to be proportional to the sum of second powers of the stress components. The zone of induced stress, therefore, is also the zone of increased strain energy density. Integration of the induced strain energy function over the zone of induced stress yields the static energy increase, ΔW_s, which is stored around the excavation.

In predicting the *in situ* performance of an excavation, it would be expected that the local response of the rock mass would depend on both the volume of

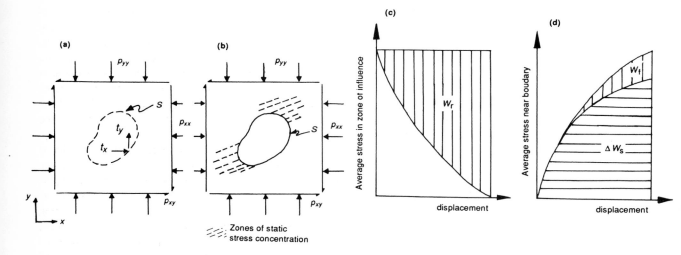

Zones of static
stress concentration

Figure 10.5 (a) Pre-mining and (b) post-mining, static states in a medium subject biaxial stress; (c), (d) balance in static energy release and storage (after Blight, 1984).

rock subject to induced stress, and the magnitude and distribution of the stress components in the affected volume. Both these notions are incorporated in the static strain energy increase, ΔW_s. For the elastic analysis described previously, the increase in static strain energy was equivalent to the energy W_r released by excavation. However, local rock fracture which frequently occurs around excavations consumes some of the released energy. These conditions are illustrated in Figure 10.5c, for the general case of an opening mined in a medium subject to a triaxial state of stress. If no fracture occurs, $\Delta W_s = W_r$. If fracture occurs, the rock fracture energy W_f reduces the stored energy, such that $W_r = \Delta W_s + W_f$. Ultimately, in the case of extensive rock fracture, all the released energy may be consumed in rock disintegration. For this reason, the released energy W_r can be considered as an index of the potential for local degradation of rock integrity, either in a stable way, by yield, or unstably, by bursting. Methods of calculation of W_r are therefore a matter of some interest.

In the case of sudden creation of an excavation, the pre-mining tractions on the surface S, illustrated in Figure 10.5a, are suddenly reduced to zero. The work which would have been done by the country rock, exterior to S, against gradually reducing support forces within S, appears as excess energy W_e at the excavation surface. It is subsequently released or propagated into the surrounding medium. In this process, the excavation surface executes oscillatory motion about the final equilibrium position, and dynamic stresses are associated with the transient displacement field. The magnitude of the excess energy can be readily understood to be reflected in the intensity and local extent of the dynamic stresses. These dynamic stresses can be expected to achieve their greatest magnitudes in the immediate periphery of the excavation, since the excess energy is momentarily concentrated in this domain. The excavation boundary, in fact, acts as a source for a stress wave which propagates through the rock medium.

The mining consequence of the excess energy can be perceived from the localised superposition of the associated dynamic stresses on the equilibrium static stresses. Even when the local static stress concentration may not be sufficient to cause failure in the rock mass, superposition of the dynamic stresses related to the excess energy may be sufficient to induce adverse mechanical response in the medium. Three simple modes of adverse response may be identified immediately: the combined dynamic and static stresses may exceed the strength of the rock mass; reduction of the normal stress on a plane of weakness reduces the shear resistance of the surface, and may lead to slip; tensile stresses may be induced, causing local relaxation in the rock structure. All of these response modes may be expressed as deteriorating ground conditions in the periphery of the mine opening.

This discussion suggests that, in the design of an opening, attention should be paid to both static and dynamic loading of rock around the excavation and in the zone of influence. Although the static stress distribution around an opening is determined readily, the potential for extensive rock mass disintegration under static conditions is indicated conveniently by the released energy W_r. Further, although dynamic stresses are not readily computed, the excess energy W_e is readily determined from excavation-induced tractions and displacements, and can serve as a useful index of dynamic stresses. In practice, the excess energy and the released energy are closely related in magnitude. The conclusion is that released energy constitutes a basis for excavation design, since it is indicative of both static and dynamic stresses imposed by excavation. This principle seems to be particularly appropriate in mining, where static stresses frequently approach the *in situ* strength of the host rock mass and the extent of rock mass failure needs to be considered.

10.3 Energy transmission in rock

Impulsive changes in the state of loading in a rock mass are associated with events such as sudden crushing of pillars, sudden slip on planes of weakness, or sudden loading or unloading of the surface of a blast hole or an excavation. Such changes result in generation and transmission of body waves in the medium. As will be discussed later, energy transmission in rock is accompanied by energy absorption, related to both the microscopic and macroscopic structure of rock. However, a damped, elastic progressive wave represents a fair conceptual model of energy transmission in a rock mass. It is therefore useful to consider initially the mechanics of elastic wave propagation in a medium. This topic is considered in detail by Kolsky (1963).

10.3.1 Longitudinal wave in a bar

Longitudinal wave propagation in a cylindrical bar is the simplest (one-dimensional) case of elastic energy transmission. Transient motion in a suspended bar may be initiated by an impulse applied at one end. A wave travels along the bar, as illustrated in Figure 10.6a, resulting in a transient longitudinal displacement, $u_x(t)$, at any point. To establish the nature of the transient motion and the associated transient state of stress, it is necessary to take account of the inertial effects associated with induced particle motion. An element of the bar of mass dM, shown in Figure 10.6b, is subject to a longitudinal acceleration $\ddot{u}_x(t)$. To take account of the impulsive displacement of the element, it is necessary to introduce

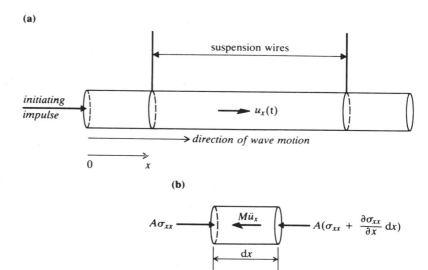

Figure 10.6 Problem definition and elementary body for analysis of a longitudinal wave in a bar.

an inertial (d'Alembert) force opposing the sense of motion, given in magnitude by $d\ddot{u}_x$. If this force is introduced, the forces on the element may be treated as an equilibrating system, i.e.

$$\sigma_{xx} A - dM\, \ddot{u}_x - [\sigma_{xx} + (\partial\, \sigma_{xx}/\partial_x)\, dx]\, A = 0 \qquad (10.8)$$

where A is the cross-sectional area of the bar.
Since $dM = \rho A dx$, where ρ is the material density, and

$$\partial\sigma_{xx}/\partial_x = (\partial/\partial x)\, E\varepsilon_{xx} = - E(\partial^2 u_x/\partial x^2)$$

equation 10.8 becomes

$$\partial^2 u_x/\partial t^2 = (E/\rho)\, \partial^2 u_x/\partial x^2 \qquad (10.9)$$

Equation 10.9 is the differential equation for particle motion in the bar, or the bar wave equation. The general solution of the equation is of the form

$$u_x = f_1(x - C_B t) + f_2\, (x + C_B t) \qquad (10.10)$$

where f_1 and f_2 are functions whose form is determined by the initial conditions, i.e. the manner of initiation of the wave. It is readily demonstrated, by differentiation, that the expression for u_x satisfies equation 10.9, provided C_B is defined by the expression

$$C_B = (E/\rho)^{\frac{1}{2}} \qquad (10.11)$$

C_B is called the bar velocity, and represents the velocity of propagation of a perturbation along the bar.

258

In equation 10.10, the term whose argument is $(x - C_B t)$ represents a wave propagating in the positive direction of the co-ordinate axis, i.e. a forward progressive wave. The term with argument $(x + C_B t)$ represents a wave propagating in the negative co-ordinate direction, i.e. a backward progressive wave. Each of the functions f_1 and f_2 is individually a solution to the wave equation, and since the constitutive behaviour of the system is linear, any linear combination of f_1 and f_2 also satisfies the governing equation.

During the propagation of the elastic wave, represented by equation 10.10, along a bar, each particle executes transient motion about its equilibrium position. The transient velocity, V, of a particle is associated with a transient state of stress, σ_{xx}, which is superimposed on any static stresses existing in the bar. For uniaxial longitudinal stress and using Hooke's Law, dynamic stresses and strains are related by

$$\sigma_{xx} = E\varepsilon_{xx} = -E\partial u_x/\partial x$$

or, from equation 10.10

$$\sigma_{xx} = -E\left[f_1'(x - C_B t) + f_2'(x + C_B t)\right] \tag{10.12}$$

Transient particle velocity is defined by

$$u_x = V = \partial u_x/\partial t$$

or, from equation 10.10

$$V = (-C_B) f_1'(x - C_B t) + C_B f_2'(x + C_B t) \tag{10.13}$$

Considering the forward progressive wave, the relevant components of equations 10.12 and 10.13, together with equation 10.11, yield

$$V = C_B \sigma_{xx}/E = \sigma_{xx}/\rho C_B$$

or

$$\sigma_{xx} = \rho C_B V \tag{10.14}$$

Thus the dynamic longitudinal stress induced at a point by passage of a wave is directly proportional to the transient particle velocity at the point. In equation 10.14, the quantity ρC_B is called the characteristic impedance of the medium. For the backward wave, it is readily shown that

$$\sigma_{xx} = -\rho C_B V \tag{10.15}$$

A case of some practical interest involves a forward wave propagating in a composite bar, as indicated in Figure 10.7. The bar consists of two components, with different material properties, denoted by subscripts 1 and 2. The interface between the component bars is welded. The forward, incident wave in medium 1 impinges on the interface, and is partly transmitted, generating a forward wave in medium 2, and partly reflected, generating a backward wave in medium 1. The

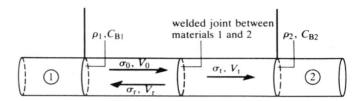

Figure 10.7 Geometry describing longitudinal wave transmission and reflection in a two-component bar.

quantities to be determined are the relative intensities of longitudinal stress in the forward and reflected waves.

Suppose the longitudinal stress and particle velocity in the forward incident wave are defined by magnitudes σ_0, V_0, and the corresponding magnitudes in the transmitted and reflected waves are given by σ_t, V_t and σ_r, V_r. The ratio of the characteristic impedances of the two media is defined by

$$n = \rho_2\, C_{B2}/\rho_1\, C_{B1} = \rho_2 C_2 / \rho_1 C_1 \tag{10.16}$$

The conditions to be satisfied at the interface between the bar components are continuity of longitudinal stress and displacement (and therefore particle velocity). These continuity conditions are expressed by the equations

$$\sigma_0 + \sigma_r = \sigma_t \tag{10.17}$$

$$V_0 + V_r = V_t \tag{10.18}$$

Introducing equations 10.14 and 10.15 to relate stresses and velocities for incident, reflected and transmitted waves, equation 10.18 becomes

$$\sigma_0/\rho_1 C_1 - \sigma_r/\rho_1 C_1 = \sigma_t/\rho_2 C_2$$

and introducing equation 10.17

$$\sigma_0/\rho_1 C_1 - \sigma_r/\rho_1 C_1 = (\sigma_0 + \sigma_r)/\rho_2 C_2$$

On rearrangement, this yields

$$\sigma_r = [(n-1)/(n+1)]\sigma_0 \tag{10.19}$$

It is then readily shown that

$$\sigma_t = [2n/(n+1)]\sigma_0 \tag{10.20}$$

Similarly, the relation between particle velocities may be shown to be

$$V_r = -[(n-1)/(n+1)] \tag{10.21}$$

$$V_t = [2/(n+1)] \tag{10.22}$$

It is useful to explore, briefly, some of the consequences of these expressions governing wave transmission and reflection. Suppose σ_0 is compressive. For the case $n > 1$, the reflected wave is characterised by a compressive stress. For

$n < 1$, the reflected wave induces a tensile stress. Thus an important general point to note is that internal reflections of a compressive wave in a medium may give rise to tensile stresses. Preceding chapters have described the low tensile strength of rock masses, and their inability to sustain tensile stress.

A case of particular interest occurs for a bar with a free end, i.e. a composite bar in which $\rho_2 = C_2 = 0$. Then $n = 0$, and equations 10.19 and 10.21 yield

$$\sigma_r = -\sigma_0 \tag{10.23}$$

$$V_r = V_0 \tag{10.24}$$

That is, a compressive pulse is reflected completely as a tensile pulse, while the sense of particle motion in the reflected pulse is in the original (forward) direction of pulse propagation. The generation of a tensile stress at a free face by reflection of a compressive pulse provides a plausible mechanism for development of slabs or spalls at a surface during rock blasting. The issue has been discussed in detail by Hino (1956), among others.

10.3.2 Plane waves in a three-dimensional medium

In the following discussion, a wave is assumed to be propagating in the x co-ordinate direction in a three-dimensional, elastic isotropic continuum. Passage of the wave induces transient displacements $u_x(t)$, $u_y(t)$, $u_z(t)$ at any point in the medium as indicated in Figure 10.8. The essential notion in the concept of a plane wave is that, at any instant in time, displacements at all points in a particular yz plane are identical, i.e. (u_x, u_y, u_z) are independent of (y, z). Alternatively, the definition of a plane wave may be expressed in the form

$$u_x = u_x(x), \quad u_y = u_y(x), \quad u_z = u_z(x) \tag{10.25}$$

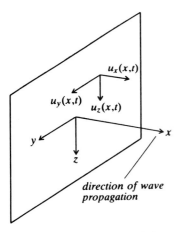

Figure 10.8 Specification of plane waves propagating in the x co-ordinate direction.

The derivation of the differential equations of motion for the components of a plane wave proceeds in a manner analogous to that for the longitudinal bar wave. From the general strain–displacement relations given in equations 2.35 and 2.36, the transient strains associated with a plane wave are obtained from equation 10.25 as

$$\varepsilon_{xx} = -\partial u_x / \partial x, \quad \varepsilon_{yy} = \varepsilon_{zz} = 0 \tag{10.26}$$

$$\gamma_{xy} = -\partial u_y / \partial x, \quad \gamma_{yz} = 0, \quad \gamma_{zx} = -\partial u_z / \partial x \tag{10.27}$$

For the case of axisymmetric uniaxial normal strain defined by equation 10.26, the equations of isotropic elasticity yield

$$\sigma_{yy} = \sigma_{zz} = \nu/(1 - \nu)\sigma_{xx}$$

Substitution of these expressions in the equation defining the x component of normal strain, i.e.

$$\varepsilon_{xx} = 1/E \, [\sigma_{xx} - \nu(\sigma_{yy} + \sigma_{zz})]$$

261

yields, after some manipulation

$$\varepsilon_{xx} = [(0.5 - \nu)/(1 - \nu)]\, \sigma_{xx}/G \tag{10.28}$$

For the shear strain components, Hooke's Law gives

$$\gamma_{xy} = 1/G\, \sigma_{xy}, \quad \gamma_{zx} = 1/G\, \sigma_{zx} \tag{10.29}$$

Equations 10.28 and 10.29, on rearrangement and introduction of the strain–displacement relations (equations 10.26 and 27), reduce to

$$\sigma_{xx} = -[(1 - \nu)/(0.5 - \nu)]G\, \partial u_x/\partial x \tag{10.30}$$

$$\sigma_{xy} = -G\, \partial u_y/\partial x, \quad \sigma_{zx} = -G\, \partial u_z/\partial x \tag{10.31}$$

In formulating governing equations for wave propagation, the requirement is to account for the inertial force associated with passage of the wave. Consider the small element of the body shown in Figure 10.9. If the net x-component of force on the body is X per unit volume, introduction of the d'Alembert force $\rho\ddot{u}_x$, in the sense opposing the net force, produces the pseudo-equilibrium condition described by the equation

$$X + \rho\ddot{u}_x = 0$$

or

$$X = -\rho\ddot{u}_x \tag{10.32}$$

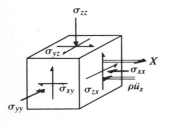

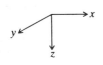

Figure 10.9 Force and stress components acting on an elementary free body subject to transient motion in the x co-ordinate direction.

Similar expressions can be established for the other co-ordinate directions.

For the geomechanics convention for sense of positive stresses being used here, the differential equations of equilibrium (equations 2.21), when combined with equation 10.32 (and similar equations for the other co-ordinate directions) become

$$\partial\sigma_{xx}/\partial x + \partial\sigma_{xy}/\partial y + \partial\sigma_{zx}/\partial z = X = -\rho\, \partial^2 u_x/\partial t^2$$

$$\partial\sigma_{xy}/\partial x + \partial\sigma_{yy}/\partial y + \partial\sigma_{yz}/\partial z = Y = -\rho\, \partial^2 u_y/\partial t^2 \tag{10.33}$$

$$\partial\sigma_{zx}/\partial x + \partial\sigma_{yz}/\partial y + \partial\sigma_{zz}/\partial z = Z = -\rho\, \partial^2 u_z/\partial t^2$$

The definition of the plane wave, and equations 10.30 and 10.31, reduce equations 10.33 to

$$\partial^2 u_x/\partial t^2 = [(1 - \nu)/(0.5 - \nu)]\,(G/\rho)\,(\partial^2 u_x/\partial x^2) = C_p^2\, \partial^2 u_x/\partial x^2 \tag{10.34}$$

$$\partial^2 u_y/\partial t^2 = (G/\rho)\, \partial^2 u_y/\partial t^2 = C_s^2\, \partial^2 u_y/\partial x^2$$

$$\tag{10.35}$$

$$\partial^2 u_z/\partial t^2 = (G/\rho)\, \partial^2 u_z/\partial t^2 = C_s^2\, \partial^2 u_z/\partial x^2$$

where

$$C_p = \{[(1 - \nu)/(0.5 - \nu)] \, (G/\rho)\}^{\frac{1}{2}}$$

$$C_s = (G/\rho)^{\frac{1}{2}}$$

Equations 10.34 and 10.35 are the required differential equations describing transient particle motion during passage of a plane wave. The constants C_p and C_s appearing in the equations are wave propagation velocities. In a manner analogous to the bar problem, the general solutions of the wave equations are found to be

$$u_x = f_1(x - C_p t) + F_1(x + C_p t) \qquad (10.36)$$

$$u_y = f_2(x - C_s t) + F_2(x + C_s t)$$

$$u_z = f_3(x - C_s t) + F_3(x + C_s t) \qquad (10.37)$$

As for the stress wave in a bar, each expression in equations 10.36 and 10.37 with argument $(x - Ct)$ corresponds to a forward progressive wave, and each with argument $(x + Ct)$ to a backward progressive wave.

Equation 10.36 describes particle motion which is parallel to the direction of propagation of the wave. Wave propagation occurs at a velocity C_p, given by the expression in equation 10.34. The waves are called P waves, or primary or longitudinal waves. Equations 10.37 describe particle motion which is transverse to the direction of wave propagation. Wave propagation occurs at a velocity C_s, as defined in equations 10.35. The waves are called S waves, or secondary or shear waves.

The expressions for P- and S-wave velocities indicate that, for $\nu = 0.25$, $C_p/C_s = 1.73$. Natural sources of wave motion normally generate both P and S waves. The higher velocity of the P wave means that it is received at some observation point remote from a wave source earlier than an S wave, allowing field identification of P- and S-wave velocities. It therefore offers the possibility of ready determination of the *in situ* dynamic elastic properties of rock masses.

Transmission of P and S waves in a non-homogeneous medium is subject to internal reflection, in the manner determined for the bar wave. For the case of normal incidence on an interface between domains with different elastic properties, an incident P wave generates transmitted and reflected P waves. For these waves, stresses and particle velocities can be calculated from expressions similar to equations 10.19–10.22, except that the ratio of characteristic impedances n_p is given by

$$n_p = \rho_2 C_{p2}/\rho_1 C_{p1}$$

Similar considerations apply to an S wave, except that the relevant ratio of characteristic impedances is now

$$n_s = \rho_2 C_{s2}/\rho_1 C_{p1}$$

Oblique incidence of P and S waves at an interface between dissimilar materials results in more complicated interaction than for normal incidence. Considering an incident P wave, transmitted and reflected P waves are generated in the usual way.

263

In addition, transmitted and reflected S waves (called PS waves) are produced, i.e. the interface acts as an apparent source for S waves. Similar considerations apply to an incident S wave, which gives rise to SP waves, in addition to the usual transmitted and reflected waves. As might be expected from the complexity of the wave motion induced at the interface, there is no simple expression for calculating the intensities of any of the transmitted and reflected waves.

10.3.3 Spherical and cylindrical waves

The mechanics of plain waves provides a useful basis for understanding wave propagation in a three-dimensional body, such as a rock mass. Several important sources of wave motion in rock are either cylindrical or concentrated, i.e. having a low length/diameter ratio. Due to their symmetry, spherical and cylindrical sources can be expected to produce spherically and cylindrically divergent P waves.

In the description of the plane P wave, the differential equation of motion (equation 10.34) can be recast in the form

$$\partial^2 u_x/\partial x^2 = C_p^2 \, \partial^2 u_x/\partial t^2 \tag{10.38}$$

for which the general solution is of the type given in equation 10.36. The wave equation, written in its most general form, may be expressed as

$$\partial^2\phi/\partial x^2 + \partial^2\phi/\partial y^2 + \partial^2\phi/\partial z^2 = (1/C^2)\partial^2\phi/\partial t^2 \tag{10.39}$$

where the nature of the variable ϕ is determined by the fundamental particle motion associated with passage of the wave. For a P wave, the parameter is the volumetric strain, or dilatation, Δ.

In spherical polar co-ordinates, the governing equation for a spherically divergent P wave is shown by Kolsky (1963) to take the form

$$\partial^2(ru_r)/\partial r^2 - 2\, ru_r/r^2 = (1/C_p^2)\, \partial^2\,(ru_r)/\partial t^2 \tag{10.40}$$

where r is the spherical co-ordinate radius. The general solution of equation 10.40 can be readily verified to be of the form

$$u_r = (1/r)f'\,(r - C_p t) - (1/r^2)f\,(r - C_p t) \tag{10.41}$$

where f is some arbitrary function, and f' its first derivative with respect to the argument $(r - C_p t)$. The form of equation 10.41 indicates that remote from the wave source (when $1/r \gg 1/r^2$), the term in f' predominates, and the spherical wave solution is approximated by

$$u_r = (1/r)f'(r - C_p t) \tag{10.42}$$

As discussed for the bar wave, for an elastic medium the wave function is invariant with respect to the local co-ordinate of the propagating wave, in this case $(r - C_p t)$. Equation 10.42 therefore implies that for the spherically divergent wave

264

$$u_r \propto 1/r \tag{10.43}$$

For a long cylindrical source, the wave equation takes the form

$$\partial^2(r^{\frac{1}{2}} u_r)/\partial r^2 - \frac{3}{4}(r^{\frac{1}{2}} u_r/r^2) = (1/C_p^2)\, \partial^2(r^{\frac{1}{2}} u_r)\,/\partial t^2 \tag{10.44}$$

There appears to be no completely general solution to this equation. For the case where r is large, equation 10.44 is approximated by

$$\partial^2(r^{\frac{1}{2}}u_r)/\partial r^2 = (1/C_p^2)\, \partial^2\,(r^{\frac{1}{2}} u_r)/\partial t^2$$

which, by comparison with the one-dimensional wave equation 10.38 and its solution, equation 10.36, has the solution

$$r^{\frac{1}{2}} u_r = f(r - C_p t) \tag{10.45}$$

This implies that, for the cylindrically divergent wave

$$u_r \propto 1/r^{\frac{1}{2}} \tag{10.46}$$

10.4 Spherical cavity in a hydrostatic stress field

The purpose in this section is to examine the relative magnitudes of the static and dynamic stresses associated with creating an excavation, to illustrate methods of evaluating released and excess energy, and to correlate the dynamic stresses with excess energy magnitude. The reason for choosing a spherical opening for study is that the problem is analytically tractable. A two-dimensional problem, such as a cylindrical excavation, cannot be treated productively due to the lack of a closed form solution for the response of a step increase in pressure to the internal surface of the hole. The problem of the spherical opening has been considered by Hopkins (1960), Timoshenko and Goodier (1970), and Bray (1979), on whose work the following discussion is based.

Figure 10.10 shows a diametral section of a spherical opening, of radius a, in a medium subject to a hydrostatic far-field stress, of magnitude p. Relative to spherical polar (r, θ, ϕ) co-ordinate axes, total stresses after excavation and excavation-induced displacement are given, according to Poulos and Davis (1974), by

$$\sigma_{rr} = p\,[1 - (a^3/r^3)]$$

$$\sigma_{\theta\theta} = \sigma_{\phi\phi} = p\,[1 + (a^3/2r^3)] \tag{10.47}$$

$$\sigma_{r\theta} = \sigma_{r\phi} = \sigma_{\theta\phi} = 0$$

$$u_r = -pa^3/4Gr^2 \tag{10.48}$$

$$u_\theta = u_\phi = 0$$

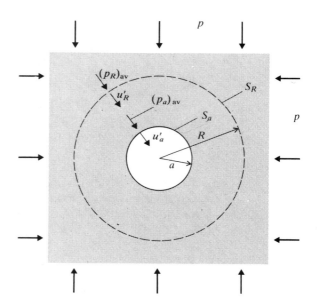

Figure 10.10 Diametral section through a sphere in a medium subject to hydrostatic stress.

Determination of the energy changes associated with the excavation of the opening requires estimation of work done by surface forces acting through induced displacements. Consider first the case of gradual excavation. At the excavation surface S_a ($r = a$), the radial pressure drops from p to zero, while the radial displacement is given by

$$u_r = -pa/4G$$

i.e. the induced displacement is directed radially inward.

The work W_1 done by the surface S_a against the support forces within S_a during their gradual reduction is the product of the average radial force and displacement; i.e.

$$W_1 = 4\pi a^2 \times p/2 \times pa/4G = \pi p^2 a^3/2G \qquad (10.49)$$

In the far field, excavation-induced displacement is diminishingly small, but the area affected is large, so the work done cannot be disregarded. Consider a spherical surface S_R in the medium, of radius R, and concentric with the opening. In the process of excavating the opening, the radial stress on the spherical surface drops from p to $p[1 - (a^3/R^3)]$, and the corresponding radial displacement u_R is $-pa^3/4GR^2$. The work W_2 done on the surface S_R by the exterior rock during the induced displacement is given by the product of the average radial force and displacement; i.e.

$$W_2 = 4\pi R^2 \times 1/2 \{p + p[1 - (a^3/R^3)]\} \ (pa^3/4GR^2) = (\pi p^2 a^3/G)[1 - (a^3/2R^3)] \qquad (10.50)$$

For the case of a remote surface, i.e. as $R \rightarrow \infty$, equation 10.50 becomes

$$W_2 = \pi p^2 a^3/G \qquad (10.51)$$

The increase in the static strain energy, ΔW_s, around the excavation is given by the difference between the work done on the rock medium at the remote surfaces, and the work done at the excavation surface by the medium against the support forces; i.e.

$$\Delta W_s = W_2 - W_1 = \pi p^2 a^3 / 2G \tag{10.52}$$

The energy released by excavating the opening is given by the complementary work done at the boundaries during their pseudo-static displacement; i.e.

$$W_r = W_1 = \pi p^2 a^3 / 2G \tag{10.53}$$

In the case of sudden generation of the spherical opening, the surface S_a will do no work during its radial displacement. Excess energy, W_e, present in the excavation peripheral rock, is expressed as a spherical stress wave which propagates away from the opening, so that the final conditions in the medium are identical to those for the case of gradual excavation. Thus, the increase in static strain energy ΔW_s is identical to that given by equation 10.52. Also, the work W_2 done at the far-field surface S_R is independent of the method of excavating the cavity. Therefore the excess energy W_e associated with the spherical stress wave is obtained from equations 10.51 and 10.52, i.e.

$$W_e = (\pi p^2 a^3 / G) - (\pi p^2 a^3 / 2G) = \pi p^2 a^3 / 2G \tag{10.54}$$

Equation 10.54 defining the excess energy, is identical to equation 10.53 defining the released energy. W_e is therefore confirmed to be the energy imbalance which arises from the impulsive unloading of a rock internal surface to form a traction free excavation surface.

In order to relate the excess energy to the magnitudes of the dynamic stresses induced by sudden excavation, it is necessary to consider details of wave propagation in an elastic medium. Corresponding to the differential equations of equilibrium for a static problem are the differential equations of motion for a dynamic problem. For the spherically symmetric problem, the equation of motion (equation 10.40) may be written as

$$\partial^2 u_r / \partial r^2 + (2/r)\, \partial u_r / \partial r - 2u_r / r^2 = (1/C_p^2)\, \partial^2 u_r / \partial t^2 \tag{10.55}$$

where C_p is the longitudinal wave velocity.

For the diverging wave, i.e. propagating radially outwards, the general solution to equation 10.55 is of the form

$$u_r = (1/r) f'(r - C_p t) - (1/r^2) f(r - C_p t) \tag{10.56}$$

where the nature of the function f is chosen to satisfy the initial and boundary conditions for a particular problem. Sharpe (1942) established a solution for equation 10.56 for the case of a varying pressure $p(t)$ applied to the surface of a spherical cavity, of radius a. For an exponentially decaying applied internal pressure, given by

$$p(t) = p_0 \exp(-\alpha t)$$

Sharpe found that the function f satisfying equation 10.56 is given by

$$f = \exp(-\alpha_0 T)(A \cos \omega_0 T + B \sin \omega_0 T) - A \exp(-\alpha T) \qquad (10.57)$$

where

$$\alpha_0 = (C_p/a)[(1 - 2\nu)/(1 - \nu)]$$

$$\omega_0 = (C_p/a)[(1 - 2\nu)^{\frac{1}{2}}/(1 - \nu)]$$

$$T = t - [(r - a)/C_p]$$

$$A = P_0 a/\{\rho[\omega_0^2 + (\alpha_0 - \alpha)^2]\} \qquad (10.58)$$

$$B = A(\alpha_0 - \alpha)/\omega_0 \qquad (10.59)$$

For the case of a spherical opening suddenly developed in a medium subject to hydrostatic stress p, the induced displacement field can be determined from Sharpe's solution by putting $P_0 = -p$ and $\alpha = 0$. This satisfies the required boundary condition that the excavation boundary be traction free, for $t > 0$. Simplifying the expressions for the case $\alpha = 0$, equations 10.58 and 10.59 yield

$$A = -a^3 p/4G$$

$$B = -(1 - 2\nu)^{\frac{1}{2}} a^3 p/4G$$

The expression for f (equation 10.57) then becomes

$$f = (-a^3 p/4G) \exp(-\alpha_0 T)[\cos \omega_0 T + (1 - 2\nu)^{\frac{1}{2}} \sin \omega_0 T] + (a^3 p/4G) \qquad (10.60)$$

The various partial derivatives of the function f required to establish displacement and strain components around the spherical opening are given by

$$f' = -(1 - 2\nu)^{\frac{1}{2}}(a^2 p/2G) \exp(-\alpha_0 T) \sin \omega_0 T \qquad (10.61)$$

$$f'' = [(1 - 2\nu)/(1 - \nu)](ap/2G) \exp(-\alpha_0 T)[\cos \omega_0 T - (1 - 2\nu)^{\frac{1}{2}} \sin \omega_0 T] \qquad (10.62)$$

Since the radial displacement is given by

$$u_r = (1/r)f' - (1/r^2)f$$

the strain components are given, for the spherically symmetric problem, by

$$\varepsilon_{rr} = -\partial u_r/\partial r = (2/r^2)f' - (1/r)f'' - (2/r^3)f \qquad (10.63)$$

$$\varepsilon_{\theta\theta} = -u_r/r = (1/r^3)f - (1/r^2)f' \qquad (10.64)$$

The total stresses in the medium are obtained directly from the induced strains, through application of the stress–strain relations and superposition of the field stresses; i.e.

$$\sigma_{rr} = \lambda\Delta + 2G\varepsilon_{rr} + p \tag{10.65}$$

$$\sigma_{\theta\theta} = \lambda\Delta + 2G\varepsilon_{\theta\theta} + p \tag{10.66}$$

where

$$\Delta = \varepsilon_{rr} + 2\varepsilon_{\theta\theta}$$

Introduction of the expressions for ε_{rr} and $\varepsilon_{\theta\theta}$ in equations 10.65 and 10.66, and subsequent substitution of the expressions for the function f and its derivatives (equations 10.60, 10.61 and 10.62), produce, after some simplification

$$\sigma_{rr} = pe^{-\alpha_0 T}\left[\left(\frac{a^3}{r^3} - \frac{a}{r}\right)\cos\omega_0 T + \left(\frac{a}{r} + \frac{a^3}{r^3} - \frac{2a^2}{r^2}\right)(1 - 2v)^{\frac{1}{2}}\sin\omega_0 T\right]$$

$$+ p\left(1 - \frac{a^3}{r^3}\right) \tag{10.67}$$

$$\sigma_{\theta\theta} = pe^{-\alpha_0 T}\left[-\left(\frac{v}{(1-v)}\frac{a}{r} + \frac{a^3}{2r^3}\right)\cos\omega_0 T + \left(\frac{v}{(1-v)}\frac{a}{r} - \frac{a^3}{2r^3} + \frac{a^2}{r^2}\right)\right.$$

$$\left. {}^*(1 - 2v)^{\frac{1}{2}}\sin\omega_0 T\right] + p\left(1 + \frac{a^3}{2r^3}\right) \tag{10.68}$$

Inspection of equations 10.67 and 10.68 shows that, at a relatively long elapsed time after excavation of the opening, the exponential terms vanish, and the static elastic solution is recovered. Equation 10.67 also indicates that, for $r = a$, σ_{rr} is identically zero, demonstrating that the boundary condition at the surface of the spherical opening is satisfied by the solution. It is also noted that, in each of equations 10.67 and 10.68, the first term on the right-hand side corresponds to the dynamic stress, and the second term to the static stress.

Insight into the magnitudes of the dynamic stresses and their temporal and local variations can be obtained directly from equations 10.67 and 10.68. The parameter T, which is the local reference time for a point in the medium, is defined by

$$T = t - [(r - a)/C_p] = (a/C_p)[(C_p t/a) - (r/a) + 1]$$

Therefore the parameters $\alpha_0 T$ and $\omega_0 T$ in equation 10.68 become

$$\alpha_0 T = [(1 - 2v)/(1 - v)][(C_p/a) - (r/a) + 1]$$

$$\omega_0 T = [(\dot{1} - 2v)^{\frac{1}{2}}/(1 - v)][(C_p t/a) - (r/a) + 1]$$

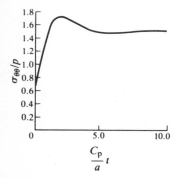

Figure 10.11 Temporal variation of boundary stress around a sphere suddenly excavated in a hydrostatic stress field.

Figure 10.12 Distribution of radial and circumferential stress after elapsed times of (a) a/C_p and (b) $2a/C_p$, around a sphere suddenly excavated in a hydrostatic stress field.

The case $a = 1$ m, $\nu = 0.25$, has been used to determine the temporal variation of the circumferential boundary stress, and the radial variation of the radial and circumferential stresses, at various elapsed times after the instantaneous generation of the spherical cavity. The temporal variation of the boundary stress ratio, shown graphically in Figure 10.11, indicates that $\sigma_{\theta\theta}/p$ decreases from its ambient value of unity immediately after creating the opening. The boundary stress ratio then increases rapidly to a maximum value of 1.72, at a scaled elapsed time which corresponds to the maximum radially inward displacement of the cavity surface. The boundary stress ratio subsequently relaxes, in a manner resembling an over-damped elastic vibration, to achieve the static value of 1.50. The transient over-stress at the boundary, which is about 15% of the final static value, is not insignificant. It is also observed that transient effects at the excavation boundary are effectively completed at a scaled time of about 8, corresponding, for $C_p = 5000$ m s^{-1}, to a real elapsed time of about 1.6 m s.

The radial variations of the radial and circumferential stress ratios, shown in Figure 10.12, confirm that the excavation process initiates a stress wave at the cavity surface. This radiates through the medium at the longitudinal wave velocity, before subsequent achievement of the static radial and circumferential stress distributions around the opening. This general view, that the excess energy mobilised locally by the sudden reduction of the surface forces, must be propagated to the far field to establish local equilibrium, is entirely compatible with earlier considerations of mining-induced energy changes.

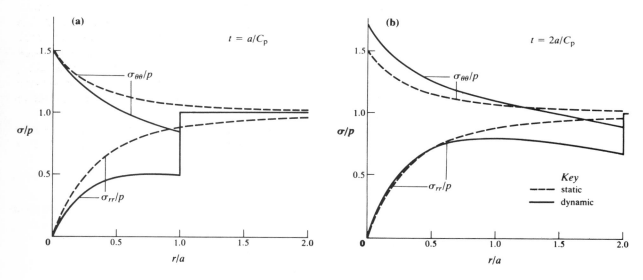

10.5 General determination of released and excess energy

In later discussion, it is shown that empirical relations can be established between released energy and the occurrence of crushing and instability around excavations. The preceding discussion indicated the relation between transient under-stressing and overstressing of the medium surrounding a suddenly developed excavation, and the excess energy mobilised at the excavation surface. Further,

270

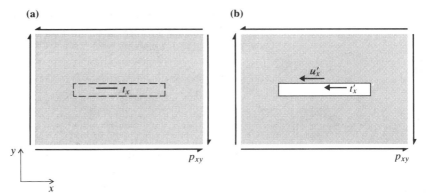

Figure 10.13 Effect of excavation of a narrow slot in a medium subject to pure shear stress.

consideration of the sudden development of a spherical cavity showed that excess elastic strain energy is propagated radially as a stress wave, at the P-wave velocity. For arbitrarily shaped excavations, the nature of the energy propagation must be more complex. This can be readily inferred from the case of development of a narrow excavation in a medium subject to pure shear stress, as illustrated in Figure 10.13. It can be readily appreciated that sudden creation of the slot will cause transverse displacement of the long surfaces of the excavation. The excess energy in this instance would be generated exclusively by transverse tractions and displacements, and it is suggested that energy propagation would occur via transverse, or S waves.

Calculation of the excess energy and energy released by excavation, for arbitrarily shaped openings, requires the use of a suitable computational method. The boundary element method is ideal for this purpose, since the solution procedure is formulated in terms of tractions and displacements induced at excavation surfaces by the mining process. Its other advantage is that no arbitrary surfaces, such as occur with finite element and finite difference methods, are introduced in the solution domain. The calculated energy changes are thus truly appropriate for an infinite or a semi-infinite body.

The nature of the problem to be solved is illustrated in Figure 10.14. Figure 10.14a illustrates a body of rock, subject to field stresses p_{xx}, p_{yy}, p_{xy}, in which it

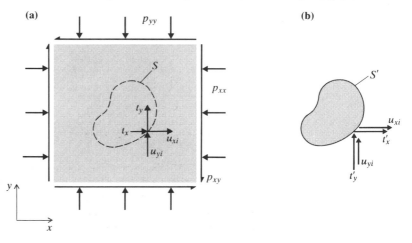

Figure 10.14 Problem geometry for determination of released energy from excavation-induced tractions and displacements.

271

is proposed to excavate an opening whose surface is S. Prior to excavation, any point on the surface S is subject to tractions t_x, t_y. For any corresponding point on the complementary surface S' which lies immediately within S and includes the material to be excavated, as shown in Figure 10.14b, the pre-mining tractions are t'_x, t'_y, and these are related to t_x, t_y by the equations

$$t_x + t'_x = 0, \ \ t_y + t'_y = 0 \tag{10.69}$$

Excavation of the material within S' is mechanically equivalent to applying a set of tractions t_{xi}, t_{yi} on S, and simultaneously inducing a set of displacements u_{xi}, u_{yi}. The magnitudes of the applied tractions are such as to make the surface S traction-free after excavation; i.e.

$$t_{xi} + t_x = 0, \ \ t_{yi} + t_y = 0 \tag{10.70}$$

When the excavation is created gradually, the surface S does work on the material within S', as the tractions on S' are gradually reduced to zero. The work, W_i, done by S against the tractions applied by S' is calculated from the average force and the displacement through which it acts; i.e.

$$W_i = \frac{1}{2} \int_s (t'_x u_{xi} + t'_y u_{yi}) \, \mathrm{d}S \tag{10.71}$$

From equations 10.69 and 10.70 it is seen that

$$t'_x = t_{xi}, \ \ t'_y = t_{yi}$$

so that equation 10.71 becomes

$$W_i = \frac{1}{2} \int_s (t_x u_{xi} + t_y u_{yi}) \, \mathrm{d}S$$

When the excavation is created suddenly, the surface S does no work against the forces applied to it by the interior material, and the work potential W_i is expressed as excess energy at the excavation periphery. Therefore, for an excavation of arbitrary shape, the excess energy and the released energy are obtained directly from the excavation-induced tractions and displacements, i.e.

$$W_e = W_r = \frac{1}{2} \int_s (t_x u_{xi} + t_y u_{yi}) \, \mathrm{d}S \tag{10.72}$$

Also, since the excess energy is mobilised at the excavation surface, this acts as the source for P and S waves, which radiate through the rock mass.

In mining small excavations, such as drives and crosscuts, the practice is to generate the complete excavation cross section rapidly, in an incremental longi-

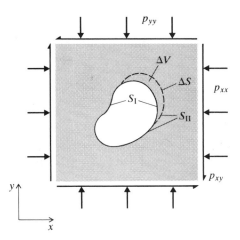

Figure 10.15 Problem geometry for determination of released energy for an incremental increase in a mined void.

tudinal extension of the opening. For the ore production excavations used in extracting an orebody, it is unusual for a complete stope to be mined instantaneously. The interest then is in the energy release rate for increments of extraction of the stope. Referring to Figure 10.15, the surface of the volume increment of excavation acts as a source for energy release. The area rate of energy release, dW_r/dS, becomes a more appropriate measure of the intensity of energy release. If the orebody is geometrically regular, e.g. of uniform thickness, the volume rate of energy release, dW_r/dV, is an index of the specific energy available for local crushing of rock around the excavation boundary. The value of this index is that it has the same dimensions as strain energy density, and therefore the same dimensions as stress.

The computational determination of W_r and its derivates is a simple matter using the boundary element method of analysis. It is a trivial exercise to integrate, numerically, the products of induced tractions and displacements over the surface of an excavation. If this is repeated for the successive stages of excavation, the released energy ΔW_r for an incremental increase in a mined void is obtained simply from the difference of the successive total amounts of released energy. The incremental area ΔS or volume ΔV of excavation is provided by the successive stages of the problem geometry, so that the derivates dW_r/dS or dW_r/dV are obtained directly.

10.6 Mine stability and rockbursts

In considering mine global stability, the concern is comprehensive control of rock mass displacement throughout the mine near-field domain. Assurance of mine global stability must be based on the principles of stability of equilibrium well known in basic engineering mechanics. They are discussed in detail in texts by Croll and Walker (1973) and Thompson and Hunt (1973). Essentially, the requirement is to make sure that any small change in the equilibrium state of loading in a structure cannot provoke a sudden release of energy or large change in the geometry of the structure. In a mine structure, small perturbations might

273

Figure 10.16 Probing the state of equilibrium of a body under load.

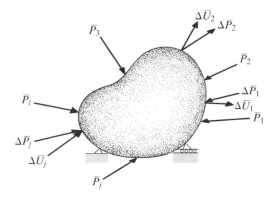

be caused by a small increase in the mined volume, transient displacements caused by blasting, or an episodic local failure. Increasing depth of mining, resulting in increased states of stress relative to rock strength, or the need for increased extraction ratios from near-surface orebodies, both promote the possibility of mine global instability. Under these circumstances, analytical techniques to identify the potential for mine instability and design concepts which will prevent the development of instability become important components of mining rock mechanics practice.

A general procedure for determining the state of equilibrium in a system is described by Schofield and Wroth (1968). The concepts are indicated schematically in Figure 10.16, where a body is in equilibrium under a set of applied forces P_i. Suppose a set of small, probing loads, ΔP_j, is applied at various parts of the structure, resulting in a set of displacements, ΔU_j. The work done by the small probing forces acting through the incremental displacements is given by

$$\ddot{W} = \tfrac{1}{2} \Sigma \, \Delta P_j \, \Delta U_j \qquad (10.73)$$

In this expression, $\ddot{W}$ represents the second order variation of the total potential energy of the system. The following states of equilibrium are identified by the algebraic value of $\ddot{W}$:

$$\text{(a) } \ddot{W} > 0 \text{ stable equilibrium}$$

$$\text{(b) } \ddot{W} = 0 \text{ neutral equilibrium} \qquad (10.74)$$

$$\text{(c) } \ddot{W} < 0 \text{ unstable equilibrium}$$

Using these definitions and a suitable analytical or computational model of a mine structure, it is possible, in theory at least, to assess the stability of an equilibrium state, by notional probing to determine the algebraic value of $\ddot{W}$.

Unstable equilibrium in a rock mass leads to unstable deformation, seismic events and seismic emissions from the source of the instability. Where a seismic event results in damage to rock around mine excavations it is conventionally called a rockburst. It is generally recognised (Gibowicz, 1988) that there are two

274

modes of rock mass deformation leading to instability and mine seismicity. One mode of instability involves crushing of the rock mass, and typically occurs in pillars or close to excavation boundaries. The second mode involves slip on natural or mining-induced planes of weakness, and usually occurs on the scale of a mine panel or district rather than on the excavation or pillar scale for the first mode.

(a)

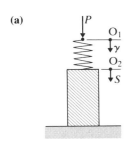

10.7 Instability due to pillar crushing

(b)

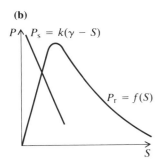

(c)

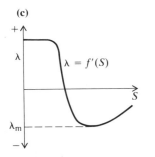

Figure 10.17 (a) Schematic representation of the loading of a rock specimen in a testing machine; (b) load–displacement characteristics of spring and specimen; (c) specimen stiffness throughout the complete deformation range (from Salamon, 1970).

Conditions for the crushing mode of instability in a mine structure arise in the post-peak range of the stress–strain behaviour of the rock mass. This aspect of rock deformation under load has been discussed in section 4.3.7. Cook (1965) recognised that rockbursts represent a problem of unstable equilibrium in a mine structure. He subsequently discussed the significance, for mine stability, of the post-peak behaviour of a body of rock in compression (Cook, 1967b). In the discussion in section 4.3.7, the term 'strain-softening' was used to denote the decreasing resistance of a specimen to load, at increasing axial deformation. It appears that much of the macroscopic softening that is observed in compression tests on frictional materials can be accounted for by geometric effects. These are associated with the distinct zones of rigid and plastic behaviour which exist in the cracked rock in the post-peak state (Drescher and Vardoulakis, 1982). Notwithstanding the gross simplification involved in the strain-softening model of rock deformation, it is useful in examination of the mechanics of unstable deformation in rock masses.

   The simplest problem of rock stability to consider is loading of a rock specimen in a conventional testing machine, as was discussed in an introductory way in section 4.3.7. The problem is represented schematically in Figure 10.17a, and has been discussed in detail by Salamon (1970). Figure 10.17b illustrates the load–displacement performance characteristics of the testing machine (represented as a spring) and the specimen. Adopting the convention that compressive forces are positive, the load (P) – convergence (S) characteristics of the rock specimen and the spring may be expressed by

$$P_r = f(S) \tag{10.75}$$

and

$$P_s = k(\gamma - S) \tag{10.76}$$

where the subscripts r and s refer to the rock and spring respectively, and γ is the displacement of the point O_1 on the spring. The specification of spring performance in equation 10.76 implies that spring stiffness k is positive by definition.

   For equilibrium at some stage of loading of the specimen through the spring, the net force at the rock–spring interface (i.e. $P_r - P_s$) must be zero. Suppose the equilibrium is probed by applying a small external force at the point O_2 in Figure 10.17a, causing an incremental convergence ΔS. From equations 10.75 and 10.76, the incremental changes in the forces in the rock and the spring are given by

$$\Delta P_r = f'(S)\,\Delta S = \lambda \Delta S$$

where λ is the slope of the specimen force–displacement characteristic, defined in Figure 10.17c, and

$$\Delta P_s = - k \Delta S$$

Thus the net probing force causing an incremental displacement ΔS is given by

$$\Delta P = \Delta P_r - \Delta P_s = (k + \lambda) \, \Delta S \tag{10.77}$$

Equation 10.77 relates applied external force to the associated convergence in the system, so that $(k + \lambda)$ can be interpreted as the effective stiffness of the spring–specimen system. The criterion for stability, defined by inequality (a) in expression 10.74, is that the virtual work term $\ddot{W}$, given by

$$\ddot{W} = \frac{1}{2} \Delta P \, \Delta S = \frac{1}{2} (k + \lambda) \, \Delta S^2 \tag{10.78}$$

be greater than zero. Thus equation 10.78 indicates that stable equilibrium of the spring–specimen system is assured if

$$k + \lambda > 0 \tag{10.79}$$

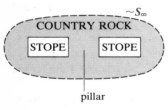

Figure 10.18 A mine pillar treated as a deformable element in a soft loading system, represented by the country rock.

A similar procedure may be followed in assessing the global stability of a mine structure, as has been described by Brady (1981). Figure 10.18 represents a simple stoping block, in which two stopes have been mined to generate a single central pillar. The mine domain exists within an infinite or semi-infinite body of rock, whose remote surface is described by S_∞. Suppose a set of probing loads $[\Delta r]$ is applied at various points in the pillar and mine near field, inducing a set of displacements $[\Delta u]$ at these points. The global stability criterion expressed by the inequality (a) in equation 10.74 then becomes

$$[\Delta u]^T [\Delta r] > 0 \tag{10.80}$$

In general, incremental displacements $[\Delta u]$ at discrete points in a rock mass may be related to applied external forces $[\Delta r]$ by an expression of the form

$$[K_g] [\Delta u] = [\Delta r] \tag{10.81}$$

where $[K_g]$ is the global (or tangent) stiffness matrix for the system. The stability criterion given by equation 10.80 is then expressed by

$$[\Delta u]^T [K_g] [\Delta u] > 0 \tag{10.82}$$

The term on the left of the inequality in expression 10.82 is a quadratic form in the vector $[\Delta u]$, and represents the virtual work term $\ddot{W}$ defined by equation 10.73. From the theory of quadratic forms (Jennings, 1977), the requirement that W be positive is equivalent to the condition that the global stiffness matrix be positive definite. This is assured if all principal minors of $[\mathbf{K_g}]$ are positive. The value of any principal minor of a matrix is obtained by omitting any number of corresponding rows and columns from the matrix, and evaluating the resultant determinant. In particular, each element of the principal diagonal of the global stiffness matrix is, individually, a principal minor of the matrix. Thus, a specific requirement for stability is that all elements of the principal diagonal of $[\mathbf{K_g}]$ be positive.

Techniques for practical assessment of mine global stability have been proposed by Starfield and Fairhurst (1968) and Salamon (1970). Both techniques relate to the geometrically simple case of stope-and-pillar layouts in a stratiform orebody, and can be shown readily to be particular forms of the criterion expressed by equation 10.82, involving the positive definiteness of $[\mathbf{K_g}]$. Starfield and Fairhurst proposed that the inequality 10.79 be used to establish the stability of individual pillars in a stratiform orebody, and therefore to assess the stability of the complete mine structure. Mine pillars are loaded by mining-induced displacement of the country rock, which are resisted by the pillar rock. The country rock therefore represents the spring in the loading system illustrated in Figure 10.17a, and the pillar represents the specimen. For a mining layout consisting of several stopes and pillars, the stiffness of pillar i, λ_i, replaces λ in inequality 10.79, while the corresponding local stiffness, $k_{\ell i}$, replaces k. Mine global stability is then assured if, for all pillars i

$$k_{\ell i} + \lambda_i > 0 \qquad (10.83)$$

It is to be noted, from Figure 10.17c, that in the elastic range of pillar performance, λ_i is positive, and for elastic performance of the abutting country rock, $k_{\ell i}$ is positive by definition. Pillar instability is liable to occur when λ_i is negative, in the post-peak range, and $|\lambda_i| > k_{\ell i}$.

In an alternative formulation of a procedure for mine stability analysis, Salamon (1970) represented the deformation characteristics of the country rock enclosing a set of pillars by a matrix $[\mathbf{K}]$ of stiffness coefficients. The performance of pillars was represented by a matrix $[\boldsymbol{\pi}]$, in which the leading diagonal consists of individual pillar stiffnesses λ_i, and all other elements are zero. Following the procedure used to develop equation 10.82, Salamon showed that the mine structure is stable if $[\mathbf{K} + \boldsymbol{\pi}]$ is positive definite. Brady and Brown (1981) showed that, for a stratiform orebody, this condition closely approximates that given by the inequality 10.83, when this is applied for all pillars.

In assessing the stability of a mine structure by repetitive application of the pillar stability criterion (inequality 10.83), the required information consists of the post-peak stiffnesses of the pillars, and the mine local stiffnesses at the various pillar positions. It must be emphasised that the idea that the post-peak deformation of a pillar can be described by a characteristic stiffness, λ', is a gross simplification introduced for the sake of analytical convenience. The idea is retained for the present discussion, because it presents a practical method of making a first estimate of pillar and mine stability, for geometrically regular mine structures.

The assumption is made in this treatment that the post-peak pillar stiffness, λ', is determined by the pillar elastic stiffness, λ, and the pillar geometry. It is

Figure 10.19 Stress–strain curves for specimens of Tennessee Marble with various length/diameter (*L/D*) ratios (after Starfield and Wawersik, 1972).

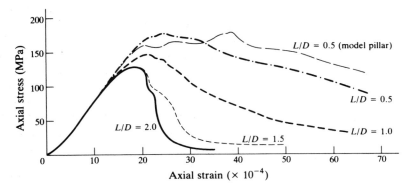

found in laboratory tests on intact rock specimens, such as those for which the results are presented in Figure 10.19, that the ratio λ/λ' decreases (becomes more negative) as the diameter/length ratio increases, corresponding to a change from steep to flat post-peak behaviour. Thus, the procedure in determining pillar stability is to calculate mine local stiffness k_ℓ, and pillar elastic stiffness λ, estimate pillar post-failure stiffness λ', and hence determine the stability index $k_\ell + \lambda'$.

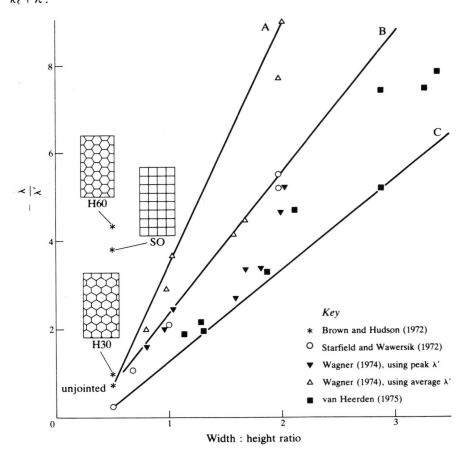

Figure 10.20 Elastic/post-peak stiffness ratios, for rock specimens and model pillars (after Brady and Brown, 1981).

278

Brady and Brown (1981) used a direct formulation of the boundary element method to determine mine local stiffness and pillar stiffness in the elastic range. Ratios of elastic/post-peak stiffness ratios for pillars of various width/height ratios were estimated from published data on laboratory specimens and from large-scale field tests. The results are presented in Figure 10.20. In order to establish the conditions under which the criterion for stability may be satisfied in practice, a series of stoping layouts was designed to achieve 75% extraction from a uniform stratiform orebody of 8 m thickness. Values of the pillar stability index, $k_\ell = \lambda'$, were estimated for the central pillar in stoping panels consisting of six stopes and five pillars. The pillar width/height ratio varied from 0.5 to 2.0, and stope spans from 12 m to 48 m, to provide the required extraction ratio. The results of computation of the pillar stability index are plotted in Figure 10.21. The plot suggests that pillar failure at any pillar width/height ratio may result in instability. This condition arises because mine local stiffness increases relatively slowly with decrease in stope span, and the country rock is therefore always readily deformable in comparison with the pillar. The inference from Figure 10.21 is that for massive orebody rock with the same elastic properties as the country rock, any pillar failures may result in instability, whatever the extraction ratio or pillar dimensions.

Another conclusion concerns the conditions under which stable pillar failure might be expected. For massive orebody and country rocks, it was suggested that unstable performance might not be possible, at a pillar width/height ratio of 0.5, if the orebody rock had a Young's modulus one-third that of the country rock. (The selected pillar width/height ratio of 0.5 was introduced, since it represents a general lower practical limit of pillar relative dimensions for effective performance in open stoping.)

The preceding conclusions apply in cases in which pillar failure occurs by the generation of new fractures in massive pillar rock. It is clear, however, that the post-peak deformation behaviour of rock and rock-like media is modified signi-

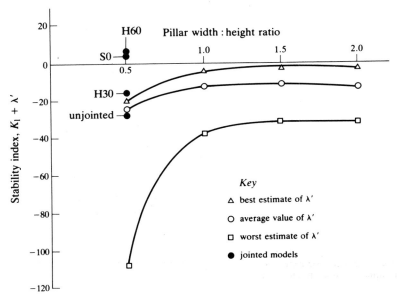

Figure 10.21 Variation of · pillar stability index with pillar width/ height ratio (after Brady and Brown, 1981).

ficantly by the presence of natural discontinuities. This is demonstrated in Figure 10.20, where the stiffness ratios for various jointed specimens are seen to be quite different from that for an unjointed specimen. As demonstrated in Figure 10.21, various joint patterns lead to a positive value of a pillar stability index, for a pillar width/height ratio of 0.5. It can be readily inferred, from the shape of the accompanying plots in Figure 10.21, that stable post-peak behaviour is assured at higher pillar width/height ratios. The results as presented imply that the natural discontinuities in a rock mass have a dominant effect on the post-peak deformation properties of the medium, and may control the potential for mine global instability. In general, joint sets and other features oriented to favour slip during the process of development of new fractures in a pillar can be expected to lead to stable yield of the pillar.

Analysis of mine stability for geometrically irregular mine structures is not amenable to simplification in the way described for the structures developed in stratiform orebodies. It is possible that a general computational method for global stability analysis may be formulated by incorporation of the localisation theories of Rudnicki and Rice (1975) and Vardoulakis (1979) in some linked computational scheme.

10.8 Thin tabular excavations

Interest in thin, tabular excavations arises since they are common and industrially important sources of ore. They are generated when coal seams or reef ore deposits are mined by longwall methods. Energy release has been studied extensively in relation to the mining of South African gold reefs, where, at the mining depths worked, static stresses are sufficient to cause extensive rock mass fracture around production excavations. Many of the original ideas associated with energy release evolved from studies of problems in deep mining in South Africa. For example, Hodgson and Joughin (1967) produced data on the relation between ground control problems in and adjacent to working areas in stopes and the rate of energy release. Some of these notions of the mining significance of energy release appear to have developed from macroscopic application of the principles of Griffith crack theory.

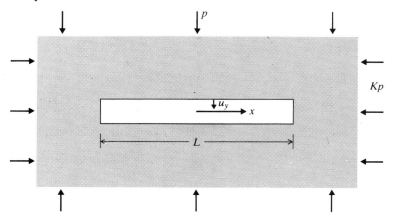

Figure 10.22 Representation of a narrow mine opening as a narrow slot.

The conventional treatment of a thin tabular excavation such as that by Cook (1967a), considers it as a parallel-sided slit, as shown in Figure 10.22. Sneddon (1946) showed that the mining-induced displacements of points on the upper and lower surfaces of the excavation are given by

$$u_y = \pm [(1 - \nu)/G] \, p \, [(L^2/4) - x^2]^{\frac{1}{2}} \tag{10.84}$$

where the negative sign corresponds to the upper excavation surface, and L is the stope span.

Using the methods described previously, it is readily shown that the released energy and the excess energy are given by

$$W_r = W_e = [(1 - \nu)/8G] \, \pi \, L^2 p^2 \tag{10.85}$$

Bray (1979) showed that the increase in static strain energy, ΔW_s, is also given by equation 10.85.

The expressions for W_r, W_e and ΔW_s (equation 10.85) apply up to the stage where the excavation remains open, i.e. until convergence between the footwall and hangingwall sides of the stope produces contact. This occurs when the convergence at midspan is equal to the mined stope height H. From equation 10.84, the critical span L_0 at which contact occurs is given by

$$H = [2 \, (1 - \nu)/G] \, pL_0/2$$

or

$$L_0 = GH/(1 - \nu)p \tag{10.86}$$

At this stage, the released energy W_r and stored strain energy ΔW_s are given by

$$W_r = \Delta W_s = (\pi/8) \, pL_0 H$$

As was noted earlier, the total released energy and strain energy increase are of limited practical significance, since a complete stope is not generated instantaneously. Mining interest is, instead, in the energy changes for incremental extension of the stope. Thus, while a stope remains open

$$dW_r/dL = [(1 - \nu)/4G] \, \pi L p^2 \tag{10.87}$$

For stoping spans greater than the critical span L_0, Bray (1979) showed that W_r approaches asymptotically to the expression

$$W_r = LHp \tag{10.88}$$

while ΔW_s approaches the maximum value

$$\Delta W_{s\,max} = \pi \, HL_0 P/4 \tag{10.89}$$

281

Therefore the incremental rates of energy storage and release are given by

$$dW_r/dL = Hp$$

$$d\Delta W_s/dL = 0$$

(10.90)

The nature of equations 10.90 indicates the key mechanical principle involved in longwall mining. If it were possible to mine a narrow orebody as a partially closed, advancing single slot, mining would occur under steady-state conditions. There would be no increase in stored strain energy and a constant rate of energy release. The process is equivalent to translation of a locally active domain (the stope face and its immediate environs) through the rock mass, as shown in Figure 10.23. There is no increase in stored strain energy since previously destressed rock is recompressed, by the advance of mining, to a state which would eventually approach, theoretically, its pre-mining state.

Figure 10.23 Schematic representation of the advance of the active zone in a longwall stope, after wall convergence.

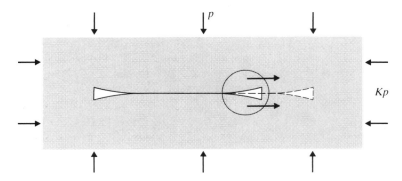

Unfortunately, the provision of sufficient work spaces to sustain the typical production rates required from a highly capitalised mine, requires that a tabular orebody cannot be mined as a single, advancing slot. When longwall stopes advance towards one another, high rates of energy release are generated by interaction between the respective zones of influence of the excavations. Energy release rates for these types of mining layouts, which also usually involve slightly more complex dispositions of stope panels in the plane of the orebody, are best determined computationally. The face element method described by Salamon (1964), which is a version of the boundary element method, has been used extensively to estimate energy changes for various mining geometries.

Cook (1978) published a comprehensive correlation between calculated rates of energy release and the observed response of rock to mining activity, for a number of deep, South African mining operations. The information is summarised in Figure 10.24. The data indicate a marked deterioration of ground conditions around work places in longwall stopes as the volume rate of energy release, dW_r/dV, increases. The inference is that the energy release rate may be used as a basis for evaluation of different mining layouts and extraction sequences, and as a guide to the type of local support required for ground control in working places.

In studies similar to those of gold reef extraction, Crouch and Fairhurst (1973) investigated the origin of coal mine bumps. They concluded that bumps could be

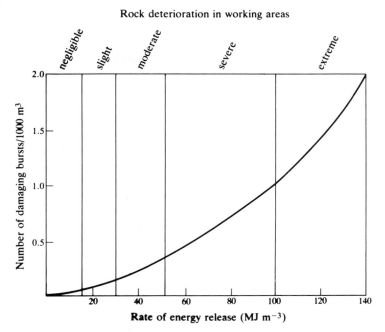

Rock deterioration in working areas

negligible   slight   moderate   severe   extreme

Figure 10.24 Relation between frequency of rockbursts, local ground conditions, and energy release rate in longwall mining of gold reefs (after Cook, 1978).

related to energy release during pillar yielding. It was suggested also that a boundary element method of analysis, similar in principle to the face element method, could be used to assess the relative merits of different extraction sequences.

10.9 Instability due to fault slip

The mechanism of mine instability considered previously results from the constitutive behaviour of the rock material, and may involve shearing, splitting or crushing of the intact rock. In hard rock mines, in addition to unstable material rupture, mine instability and seismicity may arise from unstable slip on planes of weakness such as faults or low-strength contacts between dykes and the country rock. For example, Stiller *et al.* (1983) record the similarity between many mine seismic events and natural earthquakes in terms of the seismic signatures associated with the various events. Rorke and Roering (1984) report first motion studies which suggest a source mechanism involving shear motion. A dominant role for unstable fault slip as the source of rockbursts has been proposed by Spottiswoode (1984), and is supported by interpretation of field observations of rock mass deformation attending rockbursts reported by Ortlepp (1978). Confirming the observations by Ortlepp, Gay and Ortlepp (1979) described in detail the character of faults induced by mining on which clear indications of recent shear displacement were expressed. The relation between rockbursts involving a crushing mode of rock mass deformation and those involving fault slip has been discussed by Ryder (1987).

The mechanics of unstable slip on a plane of weakness such as a fault has been considered by Rice (1983). The interaction between two blocks subject to relative

283

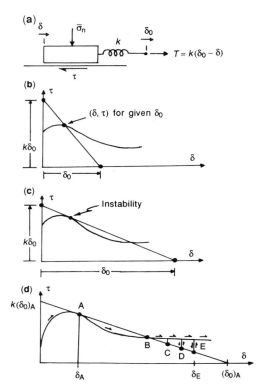

Figure 10.25 Conditions for stable and unstable slip in a single-degree-of-freedom spring–slider fault model (after Rice, 1983).

shear displacement at their contact surface is shown in Figure 10.25. The spring of stiffness, k, in Figure 10.25a represents the stiffness of the surrounding rock mass, and the stress–displacement curve for the slider models the non-linear constitutive relation for the fault surface. In Figure 10.25b, the spring stiffness is greater than the slope of the post-peak segment of the load–displacement curve for the fault. This permits stable loading and displacement of the fault in this range. Figure 10.25c represents loading through a softer spring. In this case, the notional equilibrium position is unstable, and dynamic instability is indicated.

To determine the final equilibrium position in the spring–slider system after unstable slip, it is necessary to consider the energy changes associated with the unstable motion. The area between the load–displacement curves for the spring and joint represents the energy released in the form of kinetic energy. In Figure 10.25d, if the final state of equilibrium is taken as point E, the energy released in block motion from A to B must be dissipated in various forms of damping. If the damping is due to frictional dissipation in the slider, the final state E is achieved by unloading along the slider's steep force–displacement curve.

This model of rock mass instability requires that the potential slip surface exhibit peak-residual behaviour, as discussed in section 4.7. Thus, in an analysis for prospective rock mass instability, .joint deformation involving displacement weakening, described by schemes such as the Barton–Bandis formulation or the continuous yielding model, must be taken into account. However, for faults which are at a residual state of shear strength, the displacement-weakening model is not tenable, and alternative concepts of unstable deformation must be considered.

The velocity dependence of the coefficient of friction for sliding surfaces has been known for many years (Wells, 1929). The proposal that a coefficient of dynamic friction for a fault less than the static coefficient was the cause of earthquake instabilities was made by Brace and Byerlee (1966). It was proposed that the static shear strength of a fault surface is defined by

$$\tau_s = \mu_s \, \sigma_n \qquad (10.91)$$

where μ_s is the coefficient of static friction.

The dynamic resistance to slip, τ_d, is taken to be described by

$$\tau_d = \mu_d \, \sigma_n \qquad (10.92)$$

where μ_d is the coefficient of dynamic friction.

Evaluation of equations 10.91 and 10.92 indicates a 'stress drop', given by $(\tau_s - \tau_d)$, in the transition from static to dynamic conditions on a fault subject to frictional sliding. Stress drops of 5–10% of the static shear strength have been observed in the laboratory. Applications of these notions of variable fault shear strength in rockburst mechanics have been discussed by Ryder (1987) and are considered in section 15.2.

An alternative treatment of dynamic instability, due to Dieterich (1978, 1979), Rice (1983), and Ruina (1983), among others, has been based on explicit relations between sliding velocity and fault shear strength. The analysis involves empirically derived expressions which describe the temporal evolution of shear resistance on a fault surface when it is subject to a step change in shear velocity. However, successful application of the various relations has yet to be demonstrated in practical rockburst analysis.

10.10 Seismic event parameters

Seismic events in mines are commonly characterised by parameters originally developed in earthquake seismometry. Source parameters such as energy output, seismic moment, source radius and radiated energy may be related to damage sustained by mine excavations.

For rockburst events involving fault slip, energy output may be estimated from the seismic moment M_0 (Aki and Richards, 1980), which is defined by

$$M_0 = G u_s A \qquad (10.93)$$

where G is the rock mass shear modulus, u_s is the average slip and A is the area over which slip occurs.

According to Hanks and Kanamori (1979), Richter magnitude M of a seismic event is related to the seismic moment by the expression

$$M = \frac{2}{3} \log M_0 - 10.7 \qquad (10.94)$$

Equations 10.93 and 10.94 suggest that the damage potential of a mine seismic event can be predicted if the magnitude and extent of slip can be determined by some analytical procedure.

In practice, seismic moment is determined experimentally from the wave displacement spectrum for a seismic event, obtained from the spectral analysis of body waves radiated from the source. As described by McGarr (1984), the velocity seismogram for a particular source and the displacement spectrum computed from it, shown in Figure 10.26, permit definition of two parameters: the corner frequency, f_0, and the low frequency plateau, $\Omega(0)$. Seismic moment M_0 is then estimated from the expression due to Hanks and Wyss (1972)

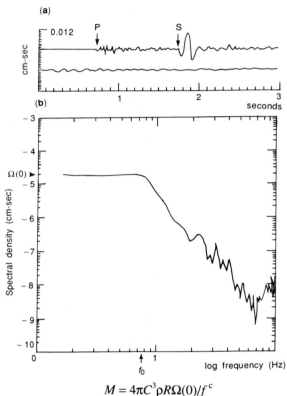

Figure 10.26 (a) Velocity seismogram of a seismic event; (b) S wave displacement spectrum for the event showing low-frequency plateau Ω_0 and corner frequency f_0 (after McGarr, 1984).

$$M = 4\pi C^3 \rho R \Omega(0)/f^c \qquad (10.95)$$

where C is either the longitudinal wave velocity C_p or the shear wave velocity C_s, ρ is the mass density of the rock, R is the distance to source hypocentre, $\Omega(0)$ is the level of the low frequency and f^c is a factor, dependent on wave type, which accounts for the radiation pattern of the seismic waves. Empirically determined values for f^c are, for P waves, $f^P = 0.39$, and for S waves, $f^s = 0.57$.

A commonly used model of seismic source mechanics is that due to Brune (1970, 1971). This assumes a homogeneous stress drop over a circular fault surface of radius r_0. The source radius is obtained from the corner frequency, f_0, using the expression

$$r_0 = 2.34 \, C_s/2\pi f_0 \qquad (10.96)$$

286

Although the source radius determined from this expression represents only an estimate of source geometry, it appears from field observations of fault traces that it yields reasonable results.

The seismic energy radiated from a seismic source may be obtained from the time history of the ground velocity at a point, using the expression due to Perret (1972), given by

$$E_\alpha = 4\pi\rho C_\alpha R^2 \int v_\alpha^2 \, dt \tag{10.97}$$

where $\alpha = p$ or s, denoting P or S waves respectively, and v_α represents ground velocity. The total radiated energy is then given by

$$E_s = E_p + E_s \tag{10.98}$$

In practice it appears that a relatively small proportion of the seismic energy is radiated in P waves, and it is usually neglected. Further, radiated energy has been related to local magnitude M_L by the empirical expression, due to Spottiswoode and McGarr (1975), by

$$M_L = \frac{2}{3} \log E_s - 3.2 \tag{10.99}$$

where E_s is in joules.

Seismic efficiency is the ratio of the total energy radiated as seismic energy, compared with the released energy, defined previously, associated with creation of mine excavations. For mining and seismic events at a depth of about 3 km, it was found (McGarr, 1976) that cumulative seismic energy was less than 1% of the total energy released by excavation. This suggests that, for the case of rockbursts associated with fault slip, most of the released energy appears to be dissipated by rock comminution during generation of faults and subsequent episodes of shear motion along them.

11 Rock support and reinforcement

11.1 Terminology

The term support is widely used to describe the procedures and materials used to improve the stability and maintain the load-carrying capability of rock near the boundaries of underground excavations. As will be shown in this chapter, the primary objective of support practice is to mobilise and conserve the inherent strength of the rock mass so that it becomes self-supporting. The procedures and materials used in this case may be more correctly described as **reinforcement**. The term **support** can then be reserved for use in those cases in which the rock mass is truly supported by structural elements which carry, in whole or part, the weights of individual rock blocks isolated by discontinuities or of zones of loosened rock.

It was once the custom to describe support as being temporary or permanent. **Temporary support** was that support or reinforcement installed to ensure safe working conditions during mining. For centuries, such support consisted of some form of timbering. If the excavation was required to remain open for an extended period of time, **permanent support** was installed subsequently. Quite often, the temporary support was partly or wholly removed to enable the permanent support to be installed. As will be demonstrated in section 11.2, this practice negates the advantage that can be obtained by applying the principles of rock–support interaction mechanics and so should be avoided.

Modern practice is to describe the support or reinforcement of permanent excavations as being primary or secondary. **Primary** support or reinforcement is applied during or immediately after excavation, to ensure safe working conditions during subsequent excavation, and to initiate the process of mobilising and conserving rock mass strength by controlling boundary displacements. The primary support or reinforcement will form part, and may form the whole, of the total support or reinforcement required. Any additional support or reinforcement applied at a later stage is termed **secondary**.

It was once common practice to regard stopes as temporary excavations having different support requirements from the more permanent mine installations such as major access ways, haulages, crusher chambers, workshops, pumping stations and shafts. Indeed, this distinction may still be made, particularly in the mining of narrow orebodies where the support techniques used in the vicinity of the face may be quite different from those used for permanent mine installations. However, many large-scale metalliferous mines now use mechanised stoping methods in which individual stopes may be very large and may have operational lives measured in years rather than weeks or months. In these cases, the support and reinforcement techniques used may have much in common with those used for permanent mine installations and in civil engineering construction.

Support or reinforcement may also be classified as being either active or passive. **Active support** imposes a predetermined load to the rock surface at the time of installation. It can take the form of tensioned rock bolts or cables, hydraulic props, expandable segmented concrete linings or powered supports for longwall faces. Active support is usually required when it is necessary to support the gravity loads imposed by individual rock blocks or by a loosened zone of rock. **Passive support** or reinforcement is not installed with an applied loading, but rather, develops its loads as the rock mass deforms. Passive support may be provided by steel arches, timbered sets or composite packs, or by untensioned grouted rock bolts, reinforcing bars or cables. Untensioned, grouted rock bolts, reinforcing bars and cables are often described as **dowels**.

The term **strata control** is used to describe the support and reinforcement techniques used in coal mining. The term is a good one because it evokes a concept of the control or limitation of strata displacements rather than one of support. Nevertheless, support in the strict sense is a major function of some strata control measures, most notably of hydraulic props used immediately behind the face in longwall mining.

Because this book is concerned with all types of modern underground mining, the terms support and reinforcement will be used in preference to strata control. In the present chapter, emphasis will be placed on the principles and major techniques used in good support and reinforcement practice for mining excavations having an extended life and for large underground excavations generally. Techniques used in particular types of mining, including the use of fill and longwall strata control measures, will be discussed in subsequent chapters.

11.2 Support and reinforcement principles

Consider the example illustrated in Figure 11.1 in which a heading is being advanced by conventional drill and blast methods. The pre-mining state of stress is assumed to be hydrostatic and of magnitude p_0. Blocked steel sets are installed after each drill and blast cycle. The following discussion concerns the development of radial displacement and radial support 'pressure' at a point on the excavation periphery at section X–X as the heading progressively advances to and beyond X–X. In this discussion, the term support will be used throughout although the process involved may be one of support and reinforcement or reinforcement alone. Following customary usage, the equivalent normal stress applied to the excavation periphery by the support system, will be termed the support pressure.

In step 1, the heading has not yet reached X–X and the rock mass on the periphery of the proposed profile is in equilibrium with an internal support pressure, p_i, acting equal and opposite to p_0.

In step 2, the face has been advanced beyond X–X and the support pressure, p_i, previously provided by the rock inside the excavation periphery, has been reduced to zero. However, the apparently unsupported section of the heading between the face and the last steel set installed, is constrained to some extent by the proximity of the face. Figure 11.2 shows the development with distance from the face of radial displacement at the periphery of a circular tunnel in an elastic

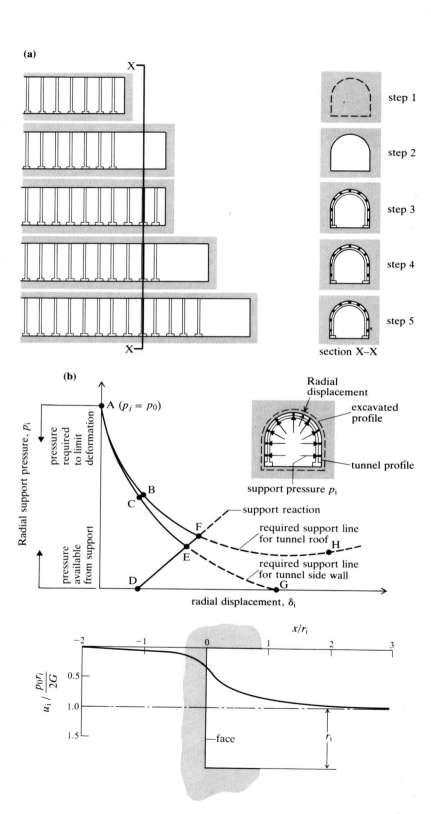

Figure 11.1 (a) Hypothetical example of a tunnel being advanced by full-face drill and blast methods with blocked steel sets being installed after each mucking cycle; (b) the radial support pressure–displacement curves for the rock mass and the support system (after Daemen, 1977).

Figure 11.2 Distribution near the face, of the radial elastic displacement, u_i, of the circular boundary of a tunnel of radius, r_i, in a hydrostatic stress field, p_0, normalised with respect to the plane strain displacement, $p_0 r_i / 2G$.

material subject to a hydrostatic *in situ* stress field. In this case, the zone of influence of the face may be defined as 2.25 radii, at which distance from the face, the radial displacement is within approximately 5% of the comparable plane strain value.

The graph in Figure 11.1 shows a plot of the radial support pressure, p_i, required at a point to limit the radial boundary displacement, δ_i, to the value given by the abscissa. If the restraint provided by the face at step 2 were not available, internal support pressures given by the ordinates of points B and C would be required to limit the displacements to their actual values. Different curves are shown for the side walls and for the roof. Extra support pressure is required to limit the displacement of the roof to a particular value because of the extra load imposed by the action of gravity on loosened rock in the roof.

By step 3, the heading has been mucked out and steel sets have been installed close to the face. At this stage, the sets carry no load because no deformation of the rock has occurred since their installation. This assumes that the rock mass does not exhibit time-dependent stress–strain behaviour. On the graph in Figure 11.1, the radial displacements of points in the roof and in the side wall, are still those given by points B and C.

In step 4, the heading is advanced about one and a half tunnel diameters beyond X–X by a further cycle of drilling and blasting. The restraint offered by the proximity of the face is now negligible, and there is further radial displacement of the rock surface at X–X as indicated by the curves CEG and BFH in Figure 11.1. This induces load in the steel sets which are assumed to show linear radial stress–displacement behaviour. Thus the supports typically load along a path such as DEF, known as the **support reaction** or **available support** line. The curve representing the behaviour of the rock mass is known as the **ground characteristic** or **required support** line. Equilibrium between the rock and the steel sets is reached at point E for the side wall and point F for the roof. It is important to note that most of the redistributed stress arising from creation of the excavation is carried by the rock and not by the steel sets.

If steel sets had not been installed after the last two stages of heading advance, the radial displacements at X–X would have increased along the dashed curves EG and FH. In the case of the side walls, equilibrium would have been reached at point G. However, the support pressure required to limit displacement of the roof may drop to a minimum and then increase again as rock becomes loosened and has to be held up. In this illustrative example, the roof would collapse if no support were provided.

The rational design of support and reinforcement systems must take into account the interaction between the support or reinforcing elements and the rock mass, described qualitatively for this simple example. It is clear from this analysis that control of rock displacements is the major rôle of support and reinforcement systems. As Figure 11.1 shows, enough displacement must be allowed to enable the rock mass strength to be mobilised sufficiently to restrict required support loads to practicable levels. However, excessive displacement, which would lead to a loosening of the rock mass and a reduction in its load-carrying capacity, must not be permitted to occur.

The stiffness and the time of installation of the support element have an important influence on this displacement control. Figure 11.3 shows a rock–

291

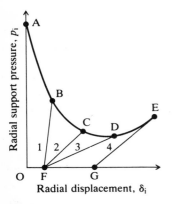

Figure 11.3 Illustration of the influence of support stiffness and of the timing of its installation on support performance.

support interaction diagram for a problem similar to that illustrated in Figure 11.1. The ground characteristic or required support line is given by ABCDE. The earliest practicable time at which support can be installed is after radial displacement of an amount OF has occurred.

Support 1 is installed at F and reaches equilibrium with the rock mass at point B. This support is too stiff for the purpose and attracts an excessive share of the redistributed load. As a consequence, the support elements may fail causing catastrophic failure of the rock surrounding the excavation.

Support 2, having a lower stiffness, is installed at F and reaches equilibrium with the rock mass at C. Provided the corresponding displacement of the periphery of the excavation is acceptable operationally, this system provides a good solution. The rock mass carries a major portion of the redistributed load, and the support elements are not stressed excessively. Note that if, as in the temporary/permanent support concept, this support were to be removed after equilibrium had been reached, uncontrolled displacement and collapse of the rock mass would almost certainly occur.

Support 3, having a much lower stiffness than support 2, is also installed at F but reaches equilibrium with the rock mass at D where the rock mass has started to loosen. Although this may provide an acceptable temporary solution, the situation is a dangerous one because any extra load imposed, for example by a redistribution of stress associated with nearby mining, will have to be carried by the support elements. In general, support 3 is too flexible for this particular application.

Support 4, of the same type and stiffness as support 2, is not installed until a radial displacement of the rock mass of OG has occurred. In this case, the support is installed too late, excessive convergence of the excavation will occur, and the support elements will probably become overstressed before equilibrium is reached.

Figure 11.4 Non-linear support reaction curves observed for some support types.

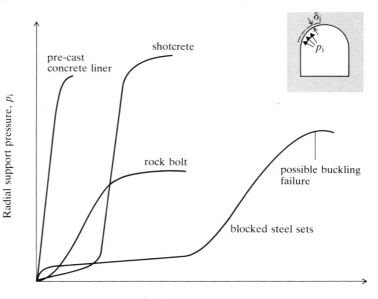

In Figures 11.1 and 11.3, constant support stiffnesses are assumed. In practice, the stiffnesses of support and reinforcing elements are usually non-linear. Figure 11.4 illustrates some of the effects that may arise. There is often initial non-linear behaviour because of poor or incomplete contact between the rock and the support system. Concrete and shotcrete may creep as they cure, as may grouted rock bolts and dowels. The support systems with the poorest stiffness characteristics are those using intermittent blocked steel or timber sets. Even if well installed, timber blocking provides a very flexible element in the system. Steel sets also suffer from the disadvantage that they often fail by sideways buckling.

From these considerations of rock–support interaction mechanics, it is possible to develop a set of principles to guide support and reinforcement practice. These principles are not meant to apply to the case of providing support for the self-weight of an individual block of rock, but to the more general case in which yield of the rock mass surrounding the excavation is expected to occur.

(a)   Install the reinforcement close to the face soon after excavation. (In some cases, it is possible, and advisable, to install the reinforcement before excavation. This case of pre-placed reinforcement will be discussed in section 11.4.)

(b)   There should be good contact between the rock mass and the reinforcement.

(c)   The deformability of the reinforcement should be such that it can conform to the displacements of the excavation surface.

(d)   Ideally, the reinforcing system should help prevent deterioration in the mechanical properties of the rock mass with time due to weathering.

(e)   Repeated removal and replacement of reinforcing elements should be avoided.

(f)   The reinforcing system should be easily adaptable to changing rock mass conditions and excavation cross section.

(g)   The reinforcing system should provide minimum obstruction of the working face.

(h)   The rock mass surrounding the excavation should be disturbed as little as possible during the excavation process.

11.3 Rock–support interaction analysis

In section 7.6, a solution was given for the radius of the yield zone and the stresses within the yield zone formed around a circular excavation in massive, elastic rock subjected to an initial hydrostatic stress field. Extensions of analyses of this type to include more realistic rock mass behaviour and to include the calculation of displacements at the excavation periphery, can be used to obtain numerical solutions to rock–support interaction problems.

In the axisymmetric problem considered in section 7.6 and illustrated in Figure 7.18, let the rock mass have a Coulomb yield criterion in which peak strength coincides with yield and the stress–strain behaviour is as shown in Figure 11.5. Note that dilatancy accompanies post-peak deformation of the rock mass. As before, the limiting states of stress in the elastic and fractured rock are given by

$$\sigma_1 = b\sigma_3 + C_0 \qquad (7.8)$$

and

$$\sigma_1 = d\sigma_3 \qquad (7.9)$$

The principal stresses within the fractured zone are

$$\sigma_3 = \sigma_{rr} = p_i \left(\frac{r}{a} \right)^{d-1}$$

and $\qquad (7.11)$

$$\sigma_1 = \sigma_{\theta\theta} = dp_i \left(\frac{r}{a} \right)^{d-1}$$

and the radius of the fractured zone is

$$r_e = a \left[\frac{2p - C_0}{(1+b)p_i} \right]^{1/(d-1)} \qquad (7.15)$$

The radial stress transmitted across the elastic-fractured zone interface at radius $r = r_e$ is

$$p_i = \frac{2p - C_0}{1 + b} \qquad (7.14)$$

In the elastic zone, the radial displacement produced by a reduction of the radial stress from p to p_i is

$$u_r = - \frac{(p - p_i)\, r_e^2}{2Gr}$$

At $r = r_e$

$$u_r = - \frac{(p - p_i)\, r_e}{2G}$$

If no fractured zone is formed, the radial displacement at the periphery of the excavation ($r_e = a$) is

$$u_i = - \frac{(p - p_i)a}{2G} \qquad (11.1)$$

Note that the radial displacement, u, is positive outwards from the centre of the excavation.

Within the fractured zone, for infinitesimal strain and with compressive strains positive, considerations of the compatibility of displacements give

$$\varepsilon_1 = \varepsilon_{\theta\theta} = - \frac{u}{r} \qquad (11.2)$$

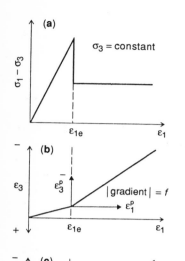

Figure 11.5 Idealised elastic-brittle stress–strain model (after Brown *et al.*, 1983).

294

and

$$\varepsilon_3 = \varepsilon_{rr} = -\frac{du}{dr} \qquad (11.3)$$

Assume that within the fractured zone

$$\varepsilon_3 = \varepsilon_{3e} - f(\varepsilon_1 - \varepsilon_{1e}) \qquad (11.4)$$

where ε_{1e}, ε_{3e} are the strains at the elastic-plastic boundary and f is an experimentally determined constant as defined in Figure 11.5. Substitution of

$$\varepsilon_{1e} = -\varepsilon_{3e} = \frac{(p - p_1)}{2G} \qquad (11.5)$$

into equation 11.4 and rearranging gives

$$\varepsilon_3 = -f\varepsilon_1 - (1 - f)\frac{(p - p_1)'}{2G} \qquad (11.6)$$

From equations 11.2, 11.3 and 11.6

$$\frac{du}{dr} = -f\frac{u}{r} + (1 - f)\frac{(p - p_1)}{2G}$$

The solution to this differential equation is

$$u = Cr^{-f} + \frac{(1 - f)(p - p_1)}{2G(1 + f)}$$

where C is a constant of integration which may be evaluated by substituting the value of ε_1 at $r = r_e$ given by equation 11.5. This leads to the solution

$$\frac{u}{r} = -\frac{(p - p_1)}{G(1 + f)}\left[\frac{(f - 1)}{2} + \left(\frac{r_e}{r}\right)^{1 + f}\right] \qquad (11.7)$$

Equation 11.7 can be used to plot a relation between radial displacement, generally represented by $\delta_i = -u_i$, and support pressure, p_i, at the excavation periphery where $r = a$. The differences between the displacements experienced by the rock in the roof, sidewalls and floor can be estimated by assuming that, in the roof, the resultant support pressure is the applied pressure, p_i, less a pressure that is equivalent to the weight of the rock in the fractured zone, $\gamma(r_e - a)$. In the sidewall, the support pressure is p, and in the floor, gravity acts on the fractured zone to increase the resultant support pressure to $p_i + \gamma(r_e - a)$.

Consider as an example, a circular tunnel of radius $a = 3$ m excavated in a rock mass subjected to a hydrostatic stress field of $p = 10$ MPa. The properties of the rock mass are $\gamma = 25$ kN m^{-3}, $G = 600$ MPa, $f = 2.0$, $\phi = 45°$, $\phi^f = 30°$ and $C = 2.414$ MPa, which give the parameter values $b = 5.828$, $C_0 = 11.657$ and $d = 3.0$. An internal radial support pressure of $p_i = 0.2$ MPa is applied.

295

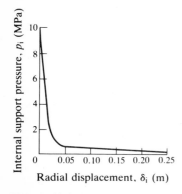

Figure 11.6 Calculated required support line for the sidewalls in a sample problem.

From equation 7.16, the radius of the fractured zone is calculated as

$$r_e = a \left[\frac{2p - C_0}{(1 + b)p_i} \right]^{1/(d-1)} = 7.415 \, \text{m}$$

and the radial pressure at the interface between the elastic and fractured zones is given by equation 7.14 as $p_1 = 1.222$ MPa. The radial displacement at the tunnel periphery is then given by equation 11.7 as

$$\delta_i = - u_i = 0.228 \, \text{m}$$

To determine the ground characteristic or required support curve, substitute successive values of p_i in equation 7.15 to obtain a series of values of r_e which are then substituted into equation 11.7 to obtain the corresponding values of $\delta_i = - u_i$. The results so obtained are tabulated in Table 11.1 and plotted in Figure 11.6. The critical support pressure below which a fractured zone will develop is found by putting $r_e = a$ in equation 7.15 which gives $p_{i\,cr} = 1.222$ MPa. In order to restrict radial displacements to values of δ_i, calculated for sidewall support pressures of p_i, roof and floor pressures of $p_i + \gamma (r_e - a)$ and $p_i - \gamma (r_e - a)$ will be required.

Figure 11.7 Idealised elastic-strain softening stress–strain model (after Brown et al., 1983).

Table 11.1 Required support line calculations for sample problem.

p_i(MPa)	10	4	2	1.222	1.0	0.5	0.2	0.1
r_e(m)	—	—	—	—	3.316	4.690	7.415	10.487
δ_i(m)	0	0.015	0.020	0.022	0.027	0.063	0.228	0.632
$\gamma(r_e - a)$(MPa)	0	0	0	0	0.008	0.042	0.110	0.187
$p_{roof} = p_i + \gamma(r_e - a)$ (MPa)	10	4	2	1.222	1.008	0.542	0.310	0.287
$p_{floor} = p_i - \gamma(r_e - a)$ (Mpa)	10	4	2	1.222	0.992	0.458	0.090	(−0.087)

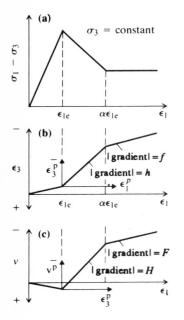

The analysis presented so far is a very simple one which applies to an axisymmetric problem and uses a simplified model of rock mass behaviour. Although several other solutions have been developed for different models of the stress–strain behaviour of the rock mass, they all use the basic concepts outlined here. Brown et al. (1983) have reviewed these various solutions, and have presented solutions for the cases in which the complete stress–strain behaviour of the rock mass may be idealised as shown in Figures 11.5 and 11.7, and where the Hoek–Brown non-linear strength criterion (equation 4.24) applies for peak and residual strengths. For the simpler model of Figure 11.5, a closed-form solution to the axisymmetric problem can be obtained. For the more complex model of Figure 11.7, an iterative finite difference solution must be used. The steps required to obtain solutions in both cases are set out in Appendix C. For non-axisymmetric problems, comparable closed-form or iterative solutions cannot be obtained, and a numerical method of computation, such as the finite element method, must be used.

The complete solution of a rock–support interaction problem requires determination of the support reaction or available support line in addition to the ground characteristic or required support line considered so far. Using methods introduced by Daemen (1975), Hoek and Brown (1980) have presented methods of calculating support reaction lines for concrete or shotcrete linings, blocked steel sets and ungrouted rock bolts or cables. Details of these calculations are also given in Appendix C.

Figure 11.8 shows the results of a set of calculations carried out for a sample problem using the material model of Figure 11.5. A 5.33 m radius access tunnel is driven in a fair quality gneiss at a depth of 120 m where the *in situ* state of stress is hydrostatic with $p = 3.3$ MPa. The properties of the rock mass are $\sigma_c = 69$ MPa, $m = 0.5$, $s = 0.0001$, $E = 1.38$ GPa, $\nu = 0.2$, $f = 4.2$, $m_r = 0.1$, $s_r = 0$ and $\gamma_r = 20$ kN m^{-3}. In this problem, the self-weight of the fractured rock around the tunnel has an important influence on radial displacements, as shown in Figure 11.8.

The support reaction or available support line for 8 I 23 steel sets spaced at 1.5 m centres with good blocking was calculated using the following input data: $W = 0.1059$ m, $X = 0.2023$ m, $A_s = 0.0043$ m^2, $I_s = 2.67 \times 10^{-5}$ m^4, $E_s = 207$ GPa, $\sigma_{ys} = 245$ MPa, $S = 1.5$ m, $\theta = 11.25°$, $t_B = 0.25$ m, $E_B = 10.0$ GPa and $\delta_{i0} = 0.075$ m. The available support provided by these steel sets is shown by line 1 in Figure 11.8 which indicates that the maximum available support pressure of about 0.16 MPa 'is quite adequate to stabilise the tunnel. However, because the set spacing of 1.5 m is quite large compared with the likely block size in the fractured rock, it will be necessary to provide a means of preventing unravelling of the rock between the sets.

The importance of correct blocking of steel sets can be demonstrated by changing the block spacing and block stiffness. Line 2 in Figure 11.8 shows the

Figure 11.8 Rock support–interaction analysis for a 5.33 m radius tunnel in fair quality gneiss at a depth of 120 m (after Hoek and Brown, 1980).

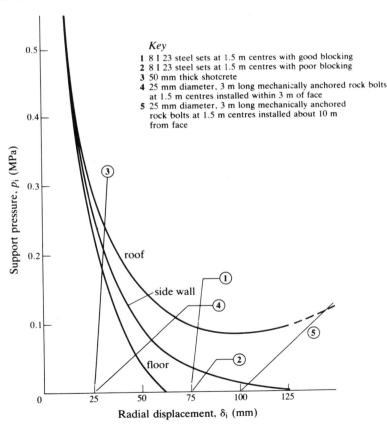

Key

1 8 I 23 steel sets at 1.5 m centres with good blocking
2 8 I 23 steel sets at 1.5 m centres with poor blocking
3 50 mm thick shotcrete
4 25 mm diameter, 3 m long mechanically anchored rock bolts at 1.5 m centres installed within 3 m of face
5 25 mm diameter, 3 m long mechanically anchored rock bolts at 1.5 m centres installed about 10 m from face

available support line calculated with $\theta = 20°$ and $E_B = 500$ MPa. The support capacity has now dropped below a critical level, and is not adequate to stabilise the tunnel roof.

Since it has already been recognised that some other support in addition to steel sets will be required, the use of shotcrete suggests itself. Line 3 in Figure 11.8 is the available support curve for a 50 mm thick shotcrete layer calculated using the following data: $E_c = 20.7$ GPa, $\nu_c = 0.25$, $t_c = 0.050$ m, $\sigma_{cc} = 34.5$ MPa. Because shotcrete may be placed close to the face soon after excavation, δ_{i0} was taken as 25 mm. It is clear that this shotcrete layer has adequate strength and stiffness to stabilise tunnel displacements. Indeed, it may well be too stiff and develop unacceptably high compressive stresses within the shotcrete ring. Brittle fracture of a shotcrete lining such as this should be avoided. Wire mesh reinforcement could increase the tensile and shear strengths of the shotcrete.

Pattern rock bolting is another possible means of providing primary support for this tunnel. Line 4 in Figure 11.8 is the available support curve calculated for a rockbolt system using the following parameters: $l = 3.0$ m, $d = 0.025$ m, $E_b = 207$ GPa, $dQ = 0.143$ m MN^{-1}, $T_{bf} = 0.285$ MN, $s_c = 1.5$ m, $s_\ell = 1.5$ m and $\delta_{i0} = 0.025$ m. It appears that this pattern of rock bolting provides a satisfactory solution. The strength of the rock mass is highly mobilised, and the rock bolts are not excessively loaded except in the roof where an adequate load factor may not exist. It would be preferable, therefore, to increase the density of bolting in the roof and to decrease that in the side walls and the floor. It will also be necessary to use mesh or a thin layer of shotcrete, to prevent unravelling of blocks of rock from between the rock bolts.

Line 5 illustrates the disastrous effect of delaying the installation of the rock bolts until excessive deformation of the rock mass has occurred. In this case, equilibrium of the rock in the roof and the support system cannot be reached and roof collapse will occur.

11.4 Pre-reinforcement

In some circumstances, it is difficult to provide adequate support or reinforcement to the rock mass sufficiently quickly after the excavation has been made. If suitable access is available, it is often practicable to pre-reinforce the rock mass in advance of excavation. In other cases, extra reinforcement may be provided as part of the normal cycle, in anticipation of higher stresses being imposed on the rock at a later stage in the life of the mine.

In mining applications, pre-reinforcement is generally provided by grouted rods or cables that are not pre-tensioned and so may be described as being passive rather than active. Such pre-reinforcement is effective because it allows the rock mass to deform in a controlled manner and mobilise its strength, but limits the amount of dilation and subsequent loosening that can occur. The effectiveness of this form of reinforcement is critically dependent on the bonding obtained between the reinforcing element and the grout, and between the grout and the rock.

Perhaps the major use made of pre-reinforcement in underground mining is in cut-and-fill stoping. Although limited use of pre-reinforcement had been made in other parts of the world, the major development of this technique took place

298

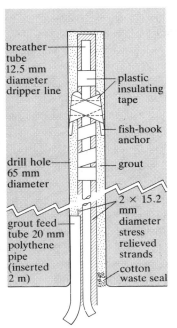

Figure 11.9 Arrangement of a cable dowel for use in an up-hole (after Hunt and Askew, 1977)

in a number of Australian metalliferous mines in the 1970s (Fuller, 1981). Cables used to reinforce stope crowns have been made from high tensile strength wire, mine rope, reinforcing bar and reinforcing strand. Figure 11.9 shows the components of a cable dowel based on two 15.2 mm diameter reinforcing strands, developed for use in approximately 20 m long up-holes at the Zinc Corporation and New Broken Hill Consolidated mines.

The use of cables to pre-reinforce the crowns of cut-and-fill stopes is illustrated in Figure 11.10. At a given stage of mining (Figure 11.10a), cables are installed to reinforce the rock mass over three or four lifts of mining. The cables are installed on approximately 2 m square grids; this spacing may be reduced or increased depending on the rock mass quality. Cables are installed normal to the rock surface when they are used for general pre-reinforcement. If shear on a particular discontinuity is to be resisted, the cables should be installed at an angle of 20°–40° to the discontinuity.

In the early mining application of fully grouted cable dowels or tendons, the full reinforcement potential of the dowel system was not realised in many cases. This was caused by failure of the grout–tendon bond and therefore ineffective load transfer between the deforming rock mass and the tendon. This problem has been solved by attaching anchors at various points along the full length of the tendon (Matthews *et al.*, 1983). The effect of the anchor is to change the load transfer mechanism from one of shear resistance at the grout–tendon interface, to one of axial compression of the grout immediately adjacent to the anchor. It has been found that, for these tendons with supplementary anchors, the surface condition of the tendon is not important. A significant improvement in the overall performance of these reinforcing systems has been observed.

In cut-and-fill stoping where pre-placed reinforcement may be up to 20 m long, accurate location of anchors at intervals suitable for assured control of the advancing stope crown is not possible. Thompson *et al.* (1987) describe a cable

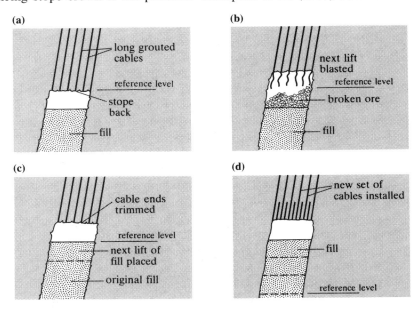

Figure 11.10 Use of cable dowel pre-reinforcement in cut-and-fill mining.

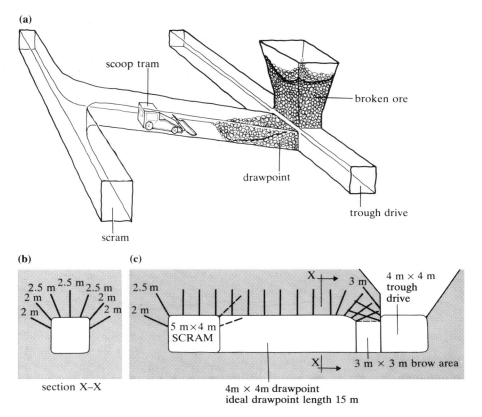

Figure 11.11 Use of grouted reinforcing bars to pre-support a drawpoint in a large mechanised mine. The brow area, shown shaded, is blasted last, after reinforcement has been installed from the drawpoint and from the trough drive (after Hoek and Brown, 1980).

with a high load capacity distributed over its complete length, which avoids this problem. The so-called 'birdcage' cable resembles the clusters of pre-stressing tendons described originally by Mathews and Meek (1975). The non-uniform cable cross section and the increased steel–grout interface area provide a strong and stiff reinforcement. An embedment length of less than 1 m is sufficient to provide an anchor capacity sufficient for cable rupture in pull tests, compared with greater than 1.5 m for standard cables.

Hoek and Brown (1980) describe a second type of pre-reinforcement in which grouted reinforcing bars are used to pre-reinforce drawpoints. For drawpoints that may be heavily loaded and subject to wear, their continued stability is vitally important in many underground mining operations. In particular, failure of the brow of the excavation can result in complete loss of control of the stope draw operation. Figure 11.11 shows a suitable method of pre-reinforcing the brow area with grouted reinforcing bars installed from the drawpoint and from the trough drive before the brow area is blasted.

11.5 Support and reinforcement design

Frequently, support and reinforcement design is based on precedent practice or on observations made, and experience gained, in trial excavations, or in the early stages of mining in a particular area. In that context, some simple design calcu-

300

lations may be performed for either general reinforcement of the rock mass or provision of support for a block or loosened zone in the excavation boundary. Alternatively, as discussed later, more comprehensive design calculations may employ computational models of rock–support interaction, and take account of deformation and strength properties of the reinforcement system and the post-peak properties of the rock mass.

11.5.1 General reinforcement

Rock-support interaction calculations. These may be carried out using the methods discussed in section 11.3 and the calculation procedures set out in Appendix C. Although idealisations of the problem have to be made, and some factors and techniques cannot be specifically allowed for in the calculations, use of this approach permits the designer to develop a clear understanding of the relative merits of candidate reinforcement systems in a particular application. In most cases, it will be necessary to carry out a series of calculations for a number of trial designs before an appropriate design can be selected for a field trial.

Empirical design rules. These rules, based on precedent practice, are available for determining reinforcement or support requirements for 'permanent' excavations. Cording *et al.* (1971) reviewed support provided in a large number of underground power stations throughout the world and found that the equivalent uniform support pressures in the roof and side walls could be calculated as

$$p_i = nB\gamma \tag{11.8a}$$

and

$$p_i = mH\gamma \tag{11.8b}$$

respectively, where $0.10 < n < 0.25$ for the roof, $0.05 < m < 0.12$ for the side walls, $B =$ the span of the excavation, $H =$ the height of the excavation, and $\gamma =$ the unit weight of the rock.

For example, if $B = 20$ m and $\gamma = 0.025$ MN m^{-3}, equation 11.8a gives the equivalent support pressure for the roof as being in the range 0.05 to 0.125 MPa. While not in itself providing a complete design method, this approach does provide a simple means by which the inexperienced designer can check his design against precedent practice.

The best known set of empirical design rules are those for pattern rock bolting developed by Lang (1961) during the construction of the Snowy Mountains hydroelectric scheme in Australia. These rules are still widely used for determining bolt lengths and spacings for 'permanent' excavations. Lang gives the minimum bolt length, L, as the greatest of

(a) twice the bolt spacing, s;
(b) three times the width of critical and potentially unstable rock blocks defined by the average discontinuity spacing, b;
(c) $0.5B$ for spans of $B < 6$ m, $0.25B$ for spans of $B = 18 - 30$ m.

For excavations higher than 18 m, the lengths of sidewall bolts should be at least one fifth the wall height.

301

Figure 11.12 Typical working sketch used during preliminary layout of a rockbolting pattern for an excavation in jointed rock (after Hoek and Brown, 1980).

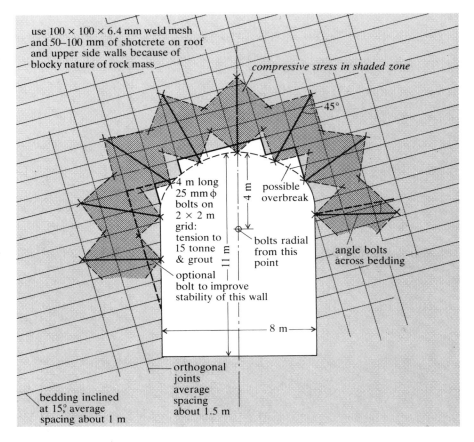

use 100 × 100 × 6.4 mm weld mesh and 50–100 mm of shotcrete on roof and upper side walls because of blocky nature of rock mass

compressive stress in shaded zone

45°

4 m long 25 mm ϕ bolts on 2 × 2 m grid: tension to 15 tonne & grout

4 m

possible overbreak

bolts radial from this point

angle bolts across bedding

optional bolt to improve stability of this wall

11 m

8 m

orthogonal joints average spacing about 1.5 m

bedding inclined at 15°, average spacing about 1 m

use spherical washer to angle bolts – angle not to exceed 15°

The maximum bolt spacing, s, is given as the least of $0.5L$ and $1.5b$. When weld mesh or chain mesh is used, a bolt spacing of more than 2 m makes attachment of the mesh difficult, if not impossible.

Figure 11.12 shows a preliminary layout of a rockbolting pattern for a horse-shoe-shaped excavation in jointed rock, prepared using Lang's rules. This figure also illustrates the basis on which Lang's rules were developed, namely the establishment of a self-supporting compressed ring or arch around the excavation. If a highly compressible feature such as a fault or clay seam crosses the compression ring, it is possible that the required compression will not develop and that the reinforcement will be inadequate.

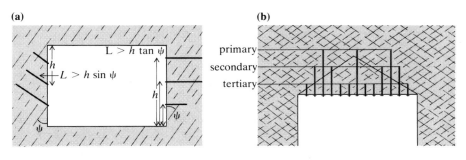

(a)

$L > h \tan \psi$

$L > h \sin \psi$

h

h

ψ

ψ

(b)

primary
secondary
tertiary

Figure 11.13 (a) Determination of bolt length in sidewalls; (b) primary, secondary and tertiary bolting patterns in a large-span excavation (after Farmer and Shelton, 1980).

302

Table 11.2 Rockbolt parameter design rules for rock masses with $\leqslant 2$ and $\geqslant 3$ discontinuity sets with clean, tight interfaces (after Farmer and Shelton, 1980).

Excavation span (m)	Number of discontinuity sets	Bolt design	Comments
<15	$\leqslant 2$ inclined at 0–45° to horizontal	$L = 0.3B$ $s = 0.5L$ (depending on thickness and strength of strata). Install bolts perpendicular to lamination where possible with wire mesh to prevent flaking	The purpose of bolting is to create a load-carrying beam over span. Fully bonded bolts create greater discontinuity shear stiffness. Tensioned bolts should be used in weak rock; subhorizontal tensioned bolts where vertical discontinuities occur
	$\leqslant 2$ inclined at 45–90° to horizontal	For side bolts: $L > h \sin \psi$ (installed perpendicular to discontinuity) $L > h \tan \psi$ (installed horizontally). See Figure 11.13a for h and ψ; L = bolt length; s = bolt spacing; B = excavation span	Roof bolting as above. Side bolts designed to prevent sliding along planar discontinuities. Spacing should be such that anchorage capacity is greater than sliding or toppling weight. Bolts should be tensioned sufficiently to prevent sliding
	$\geqslant 3$ with clean tight interfaces	$L = 2s$ $s = 3 - 4 \times$ block dimension. Install bolts perpendicular to excavation periphery with wire mesh to prevent flaking	Bolts should be installed quickly after excavation to prevent loosening and retain tangential stresses. Prestresses should be applied to create zone of radial confinement. Sidewall bolting where toe of wedge daylights in side wall
>15	$\leqslant 2$	$L_1 = 0.3 B_1$ primary bolting $s_1 = 0.5 L_1$ $L_2 = 0.3 s_1$ secondary bolting $s_2 = 0.5 L_2$ install wire mesh to prevent spalling	Primary bolting conforms to smaller excavation design. Secondary (and tertiary) bolting supplements primary design (Figure 11.13b)
	$\geqslant 3$ with clean tight interfaces	$L_1 = 0.3 B_1$ primary bolting $s_1 = 0.5 L_1$ $s_2 = 3 - 4 \times$ block size; secondary bolting $L_2 = 2 s_2$	Primary bolting should have sufficient capacity to restrain major blocks. Decisions on block size for secondary bolting should be left to the section engineer

Using Lang's rules and the experience of others as a basis, Farmer and Shelton (1980) compiled the rockbolt design guide-lines set out in Table 11.2 for excavations in rock masses having clean, tight discontinuity interfaces.

Rock mass classification schemes. Schemes such as those due to Barton *et al.* (1974) and Bieniawski (1973, 1976) were developed as methods of estimating support requirements for underground excavations, using precedent practice. Barton *et al.* proposed 38 categories of support based on their tunnelling quality index, Q, and the excavation support ratio, ESR, which varies with the use of the excavation and the extent to which some degree of instability is acceptable. These proposals have been discussed in detail by Hoek and Brown (1980) who point out the dangers involved in blindly adopting their provisions, particularly where the nature of the excavation and the properties of the rock mass differ from those in the case histories that were used in developing the recommendations.

Table 11.3 Guide for support of mining excavations based on modified geomechanics classification (after Laubscher and Taylor, 1976, with supplementary notes by Hoek and Brown, 1980).

Adjusted ratings	Original geomechanics ratings									
	90–100	80–90	70–80	60–70	50–60	40–50	30–40	20–30	10–20	0–10
70–100										
50–60		a	a	a	a					
40–50			b	b	b	b				
30–40				c, d	c, d	c, d, e	d, e			
20–30					g	f, g	f, g, j	f, h, j		
10–20						i	i	h, i, j	h, j	
0–10							k	k	l	l

a Generally no support but locally joint intersections might require bolting.

b Patterned grouted bolts at 1 m collar spacing.

c Patterned grouted bolts at 0.75 m collar spacing.

d Patterned grouted bolts at 1 m collar spacing and shotcrete 100 mm thick.

e Patterned grouted bolts at 1 m collar spacing and massive concrete 300 mm thick and only used if stress changes not excessive.

f Patterned grouted bolts at 0.75 m collar spacing and shotcrete 100 mm thick.

g Patterned grouted bolts at 0.75 m collar spacing with mesh reinforced shotcrete 100 mm thick.

h Massive concrete 450 mm thick with patterned grouted bolts at 1 m spacing if stress changes are not excessive.

i Grouted bolts at 0.75 m collar spacing if reinforcing potential is present, and 100 mm reinforced shotcrete, and then yielding steel arches as a repair technique if stress changes are excessive.

j Stabilise with rope cover support and massive concrete 450 mm thick if stress changes not excessive.

k Stabilise with rope cover support followed by shotcrete to and including face if necessary, and then closely spaced yielding arches as a repair technique where stress changes are excessive.

l Avoid development in this ground otherwise use support systems j or k.

Supplementary notes

1 The original geomechanics classification as well as the adjusted ratings must be taken into account in assessing the support requirements.

2 Bolts serve little purpose in highly jointed ground and should not be used as the sole support where the joint spacing rating is less than 6.

3 The recommendations contained in the table are applicable to mining operations with stress levels less than 30 MPa.

4 Large chambers should only be excavated in rock with adjusted total classification ratings of 50 or better.

Table 11.3 shows a guide to support of mining excavations developed by Laubscher and Taylor (1976) using the modified form of Bieniawski's geomechanics classification discussed in section 3.7.3 and listed in Table 3.8.

11.5.2 Support of individual blocks or zones

Design to suspend a roof beam in laminated rock. As illustrated in Figure 11.4, rockbolts may be used to suspend a potentially unstable roof beam in laminated rock. The anchorage must be located outside the potentially unstable zone. If it is assumed that the weight of the rock in the unstable zone is supported entirely by the force developed in the rockbolts then

$$T = \gamma D s^2$$

or   (11.9)

$$s = \left(\frac{T}{\gamma D} \right)^{\frac{1}{2}}$$

where T = working load per rock bolt, γ = unit weight of the rock, D = height of the unstable zone, and s = rockbolt spacing in both the longitudinal and transverse directions.

If, for example, $T = 10$ tonne $= 100$ kN, $\gamma = 25$ kN m^{-3} and $D = 4$ m, equation 11.9 gives $s = 1.0$ m.

In this application, care must be taken to ensure that the bolt anchors have an adequate factor of safety against failure under the working load, T. This design method is conservative in that it does not allow for the shear or flexural strength of the strata above the abutments.

Lang and Bischoff (1982) have extended this elementary analysis to incorporate the shear strength developed by the rock mass on the vertical boundaries of the rock unit reinforced by a single rockbolt. The rock is assumed to be de-

Figure 11.14 Rockbolt design to support the weight of a roof beam in laminated rock.

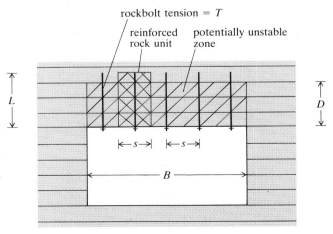

stressed to a depth, D, as in Figure 11.14, but variable vertical stresses, σ_v, and horizontal stresses, $k\,\sigma_v$, are assumed to be induced within the de-stressed zone. Typically, k may be taken as 0.5. The shear strength developed at any point on the perimeter of the reinforced rock unit is given by $c + \mu k \sigma_v$, where c is the cohesion and $\mu = \tan\phi$ is the coefficient of friction for the rock mass. Lang and Bischoff's analysis leads to the result

$$\frac{T}{A\,R} = \frac{\alpha}{\mu k}\left(1 - \frac{c}{\gamma R}\right)\left[\frac{1 - \exp\left(-\mu k D/R\right)}{1 - \exp\left(-\mu k L/R\right)}\right] \qquad (11.10)$$

where T = rockbolt tension, A = area of roof carrying one bolt ($= s^2$ for a $s \times s$ bolt spacing), R = shear radius of the reinforced rock unit, $= A/P$, where P is the shear perimeter ($= 4s$ for a $s \times s$ bolt spacing), α = a factor depending on the time of installation of the rockbolts ($\alpha = 0.5$ for active support, and $\alpha = 1.0$ for passive reinforcement), and L = bolt length which will often be less than D, the height of the de-stressed zone of rock.

Lang and Bischoff suggest that, for preliminary analyses, the cohesion, c, should be taken as zero. Design charts based on equation 11.10 show that, particularly for low values of ϕ, the required bolt tension, T, increases significantly as L/s decreases below about two, but that no significant reduction in T is produced when L/s is increased above two. This result provides some corroboration of Lang's empirical rule that the bolt length should be at least twice the spacing. For a given set of data, equation 11.10 will give a lower required bolt tension than that given by equation 11.9. Clearly, Lang and Bischoff's theory applies more directly to the case of the development of a zone of reinforced, self-supporting rock, than to the simpler case of the support of the total gravity load produced by a loosened volume of rock or by a roof beam in laminated rock.

Design to support a triangular or tetrahedral block. In Chapter 9, the identification of potential failure modes of triangular and tetrahedral blocks was discussed, and analyses were proposed for the cases of symmetric and asymmetric triangular roof prisms. These analyses take account of induced elastic stresses and discontinuity deformability, as well as allowing for the self-weight of the block and for support forces. The complete analysis of a non-regular tetrahedral wedge is more complex. An otherwise complete solution for the tetrahedral wedge which does not allow for induced elastic stresses is given by Hoek and Brown (1980).

The analyses presented in Chapter 9 may be incorporated into the design procedure. Consider the two-dimensional problem illustrated in Figure 11.15 to which the analysis for an asymmetric triangular prism may be applied. If it is assumed that the normal stiffnesses of both discontinuities are much greater than the shear stiffnesses, equation 9.18 may be used. Substituting $H_0 = 20$ MN (corresponding to a boundary stress of 5 MPa), $\alpha_1 = 40°$, $\alpha_2 = 20°$, $\phi_1 = \phi_2 = 40°$ in equation 9.18 gives the vertical force required to produce limiting equilibrium of the prism as $p_\ell = 3.64$ MN per metre thickness. Since the weight of the prism is $W = 0.26$ MN per metre thickness, it is concluded that the prism will remain stable under the influence of the induced elastic stresses.

306

Figure 11.15 Example of a triangular roof prism.

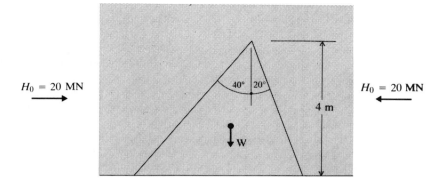

$H_0 = 20$ MN $\quad\rightarrow$

40° 20°

4 m

$H_0 = 20$ MN $\quad\leftarrow$

W

If the wedge is permitted to displace vertically so that joint relaxation occurs, the limiting vertical force is given by equation 9.19, with values of H being determined from equation 9.11. In the present case, the post-relaxation limiting vertical force is $P_\ell = 0.18$ MN per metre thickness. This is less than the value of W and so, without reinforcement, the block will be unstable. The reinforcement force, R, required to maintain a given value of factor of safety against prism failure, F, is given by $R = W - P_\ell / F$. If $F = 1.5$, then $R = 0.14$ MN per metre thickness. This force could be provided by grouted dowels made from steel rope or reinforcing bar.

If the stabilising influence of the induced horizontal stresses were to be completely removed, it would be necessary to provide support for the total weight of the prism. For a factor of safety of 1.5, the required equivalent uniform roof support pressure would be 0.08 MPa, a value readily attainable using pattern rock bolting.

Figure 11.16 shows a case in which a two-dimensional wedge is free to slide on a discontinuity AB. If stresses induced around the excavation periphery are ignored, tensioned rock bolt or cable support may be designed by considering limiting equilibrium for sliding on AB. If Coulomb's shear strength law applies for AB, the factor of safety against sliding is

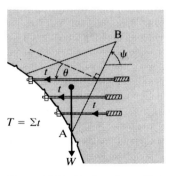

$T = \Sigma t$

Figure 11.16 Design of a rockbolt or cable system to prevent sliding of a triangular prism (after Hoek and Brown, 1980).

$$F = \frac{cA + (W \cos \psi + T \cos \theta)\tan \phi}{W \sin \psi - T \sin \theta}$$

where W = weight of the block, A = area of the sliding surface, T = total force in the bolts or cables, ψ = dip of the sliding surface, θ = angle between the plunge of the bolt or cable and the normal to the sliding surface, c, ϕ = cohesion and angle of friction on the sliding surface.

Thus the total force required to maintain a given factor of safety is

$$T = \frac{W(F \sin \psi - \cos \psi \tan \phi) - cA}{\cos \theta \tan \phi + F \sin \theta}$$

A factor of safety of 1.5 to 2.0 is generally used in such cases. The value of T required to maintain a given value of F will be minimised if $\theta = F \cot \phi$.

307

A comprehensive analysis of rock reinforcement must be based on loads mobilised in reinforcement elements by their deformation and by relative displacement between host rock and components of the reinforcement. For local reinforcement, represented by a reinforcing bar or bolt fully encapsulated in a strong, stiff resin or grout, a relatively large axial resistance to extension can be developed over a relatively short length of the shank of the bolt, and a high resistance to shear can be developed by an element penetrating a slipping joint. Spatially extensive reinforcement, represented by cement-grouted cables or tendons, offers little resistance to joint shear, and development of full axial load may require deformation of the grout over a substantial length of the reinforcing element. The following analysis of these reinforcement systems is based on that due to St John and Van Dillen (1983) and Lorig (1985).

Local reinforcement. Analysis of local reinforcement is conducted in terms of the loads mobilised in the reinforcement element by slip and separation at a joint and the deformation of an 'active length' of the element, as shown in Figure 11.17. This reflects experimental observations by Pells (1974), Bjurstrom (1979), and Haas (1981) that, in discontinuous rock, reinforcement deformation is concentrated near an active joint. The conceptual model of the local operation of the active length is shown in Figure 11.18a, where local load and deformation response is simulated by two springs, one parallel to the local axis of the element and one perpendicular to it. When shear occurs at the joint, as shown in Figure 11.18b, the axial spring remains parallel to the new orientation of the active length, and the shear spring is taken to remain perpendicular to the original axial orientation. Displacements normal to the joint are accompanied by analogous changes in the spring orientations.

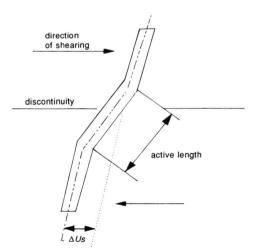

Figure 11.17 Local reinforcement action through an active length of bolt.

The loads mobilised in the element by local deformation are related to the displacements through the axial and shear stiffnesses of the bolt, K_a and K_s respectively. These can be estimated from the expressions (Gerdeen *et al.*, 1977)

$$K_a = \pi k d_1 \tag{11.11}$$

308

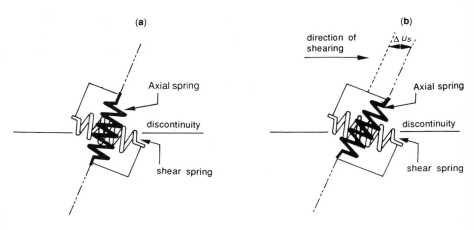

(a)

Axial spring

discontinuity

shear spring

(b)

direction of
shearing

Δu_s

Axial spring

discontinuity

shear spring

Figure 11.18 Models for axial and shear action of reinforcement at a slipping joint.

$$K_s = E_b I \beta^3 \tag{11.12}$$

where

$$k = \left[\frac{1}{2} G_g E_b / (d_2 / d_1 - 1) \right]^{\frac{1}{2}}$$

G_g = shear modulus of grout, E_b = Young's modulus of bolt, d_2, d_1 = hole and bolt diameter, respectively

$$\beta = K / (4 E_b I)^{\frac{1}{2}}$$

$$K = 2 E_g / (d_2 / d_1 - 1)$$

and I = second moment of area of the reinforcing element, and E_g = Young's modulus of grout.

The mobilised axial and shear forces are both assumed to approach limiting values asymptotically. A continuous-yielding model for axial performance (with an analogous expression for shear) is described by the expression

$$\Delta F_a = K_a \left| \Delta u_a \right| f(F_a) \tag{11.13}$$

where ΔF_a is an incremental change in axial force due to an incremental axial displacement Δu_a, and $f(F_a)$ is a function defining the load path by which F_a approaches its ultimate value, P_{ult}^a.

The expression for $f(F_a)$ is conveniently defined by

$$f(F_a) = \left| P_{ult}^a - F_a \right| (P_{ult}^a - F_a) / (P_{ult}^a)^2 \tag{11.14}$$

The values for the ultimate axial and shear loads that can be sustained ideally should be determined from appropriate laboratory tests on the rock–grout–shank

309

system, since it depends on factors such as grout properties, roughness of the grout–rock interface, adhesion between the grout and the shank, and the thickness of the grout annulus. If measured values for the ultimate loads are not available, approximate values can be estimated from the expressions

$$P_{\text{ult}}^{\text{s}} = 0.67 d_1^2 (\sigma_{\text{b}}\,\sigma_{\text{c}})^{\frac{1}{2}} \tag{11.15}$$

$$P_{\text{ult}}^{\text{a}} = \tau_{\text{peak}}\,\pi d_2 L \tag{11.16}$$

where σ_{b} = yield strength of bolt, σ_{c} = uniaxial compressive strength of the rock, τ_{peak} shear strength of grout or grout–rock interface, and L = bond length. In equation 11.16, it is assumed that shear occurs at the hole boundary. If shear occurs at the bolt–grout interface, the appropriate diameter in equation 11.16 is d_1.

Considering equations 11.11–11.16, it is observed that any increment of relative displacement at a joint can be used to determine incremental and then total forces parallel and transverse to the axis of the reinforcement element. From the known orientation of the element relative to the joint, these forces can be transformed into components acting normal and transverse to the joint. In this form, they can be introduced into a suitable finite difference code, such as the distinct element scheme described in section 6.7, which simulates the behaviour of a jointed rock mass.

Spatially-extensive reinforcement. In a rock mass subject to fracture and yield, the spatially-extensive model of reinforcement is more appropriate than the local deformation model. Because local resistance to shear in the reinforcement is not significant in this case, a one-dimensional constitutive model is adequate for describing its axial performance. The finite difference representation of reinforcement shown in Figure 11.19 involves division of the complete cable or tendon into separate segments and assignment of equivalent masses to the nodes. Axial extension of a segment is represented by a spring of stiffness equivalent to the axial stiffness, limited by a plastic yield condition.

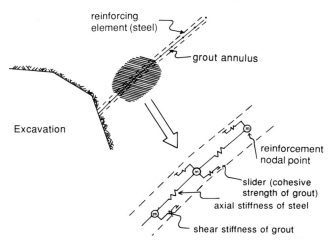

Figure 11.19 Model of spatially extensive reinforcement in rock subject to diffuse deformation.

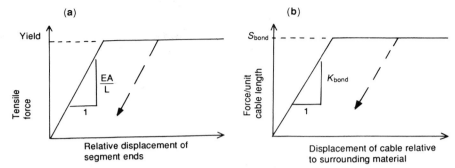

Figure 11.20 (a) Axial performance of a reinforcement segment; (b) shear performance of grout annulus between tendon and borehole surfaces.

Interaction between the tendon and the rock is modelled by a spring–slider unit, with the stiffness of the spring representing the elastic deformability of the grout, and with the limiting shear resistance of the slider representing the ultimate shear load capacity of either the grout annulus itself, the tendon–grout contact, or the grout–rock contact.

The elastic–plastic performance of a tendon segment in axial extension is shown in Figure 11.20a. For the elastic response, the bar stiffness is related to the properties and dimensions of the tendon

$$K_a = E_b A/L \tag{11.17}$$

where A and L are the cross-sectional area and length of the tendon segment.

The yield load is related directly to the yield strength of the tendon and the cross-sectional area. If, after yield, the segment is subjected to a phase of unloading, the unloading stiffness is taken to be equal to the loading stiffness.

The elastic–perfectly plastic performance of the tendon segment represents the simplest possible constitutive model. With minimal increase in complexity, a kinematic hardening model or a continuous weakening model analogous to that described by equations 11.13 and 11.14 could be introduced.

The elastic–perfectly plastic performance of the grout annulus is represented in Figure 11.20b. As a result of relative shear displacement u_a between the tendon surface and the borehole surface, the shear force f_a mobilised per unit length of cable is related to the stiffness K_{bond}, i.e.

$$f_a = K_{bond} u_a \tag{11.18}$$

Usually, K_{bond} would be measured directly in laboratory pull-out tests. Alternatively, it may be calculated from the expression

$$K_{bond} = 2\pi G_g / [\ell \text{ń} (1 + 2t/d_1)] \tag{11.19}$$

where t is the thickness of the grout annulus.

The ultimate load capacity of the grout is defined by equation 11.16, with length L of unity, and τ_{peak} given by

$$\tau_{peak} = \tau_1 Q_b \tag{11.20}$$

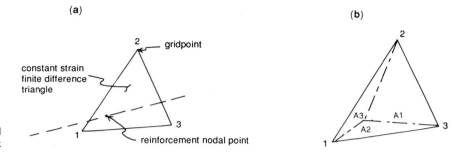

Figure 11.21 (a) Basis of natural co-ordinates for interpolating rock displacements.

where τ_1 is approximately one-half of the smaller of the uniaxial strengths of the grout and the rock, and Q_b is a factor defining the quality of the bond between the grout and the rock. (For perfect bond quality, $Q_b = 1$.)

Calculation of the loads generated in reinforcement requires determination of the relative displacement between the rock and a node of a reinforcement segment. Consider the constant strain triangular zone shown in Figure 11.21a, with components of displacement u_{xi}, u_{yi}, for example, at the corners i ($i = 1, 3$). The natural co-ordinates η_i of a reinforcement node p lying within the triangle are given by the relative areas of the triangular areas defined in Figure 11.21b. Thus

$$\eta_i = A_i/A \qquad i = 1, 3 \tag{11.21}$$

where A is the area of the triangle with corners 1, 2, 3.

The displacements u_x^p and u_y^p at node p are interpolated linearly from the displacements at the corners, using the natural co-ordinates as weight factors, i.e.

$$u_x^p = \eta_i u_{xi}, \; u_y^p = \eta_i u_{yi} \quad i = 1, 3 \tag{11.22}$$

where summation is implied on repeated subscripts.

In the finite difference analysis, incremental forms of equation 11.22 are used in successive computation cycles to calculate incremental nodal displacements, from which the new configuration of an element can be determined. The axial component of relative displacement at a node can then be calculated from the absolute displacement of a node and the absolute displacement of the adjacent rock. The axial force is obtained from equation 11.18 and the active length adjacent to the node, taking account of the limiting condition defined by equation 11.20. The force F_x^p, F_y^p mobilised at the grout–rock interface at a node is distributed to the zone corners using the natural co-ordinates of the node as the weight factor; i.e.

$$F_{xi}^p = \eta_i F_x^p \qquad\qquad \text{etc.} \qquad (11.23)$$

where F_{xi}^p are forces assigned to the zone corner.

This formulation of reinforcement mechanics may be readily incorporated in a dynamic relaxation, finite difference method of analysis of a deformable medium, such as the code called FLAC described by Cundall and Board (1988). The solution of a simple problem involving long, grouted, untensioned cable bolts illustrates application of the method of analysis. The problem involves a circular

312

hole of 1 m radius, excavated in a medium subject to a hydrostatic stress field, of magnitude 10 MPa. For the elasto–plastic rock mass, the shear and bulk elastic moduli were 4 GPa and 6.7 GPa respectively, and Mohr–Coulomb plasticity was defined by a cohesion of 0.5 MPa, angle of friction of 30°, and dilation angle of 15°. Reinforcement consisted of a series of radially oriented steel cables, of 15 mm diameter, grouted into 50 mm diameter holes. The steel had a Young's modulus of 200 GPa, and a yield load of 1 GN. Values assigned to K_{bond} and S_{bond} were 45 GN m^{-2} and 94 kN m^{-1}. These properties correspond to a grout with a Young's modulus of 21.5 GPa and a peak bond strength (τ_{peak}) of 2 MPa.

The problem was analysed as a quarter plane. The near-field problem geometry is illustrated in Figure 11.22, where the extent of the failure zone that develops both in the absence and presence of the reinforcement is also indicated. The

Figure 11.22 Problem geometry and yield zones about a circular excavation (a) without and (b) with reinforcement (after Brady and Lorig, 1988).

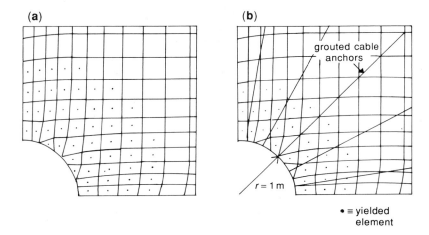

● ≡ yielded element

distributions of stress and displacement around the excavation are shown in Figure 11.23 for the cases where the excavation near-field rock is both unreinforced and reinforced. Examination of the distributions of radial displacement u_r and radial and tangential stresses σ_r and σ_t indicates the function of the radial reinforcement. It substantially reduces the radial displacement u_r, and generates a higher magnitude of σ_r in the fractured zone, resulting in a higher gradient in

Figure 11.23 Distributions of (a) stress and (b) displacement around a circular excavation for unreinforced and reinforced near-field rock (after Brady and Lorig, 1988).

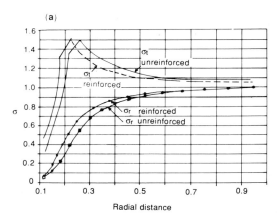
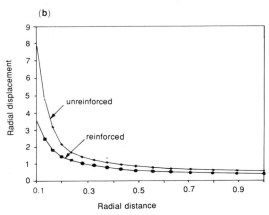

313

the σ_t distribution. The effect is to shift the plastic–elastic transition closer to the excavation boundary. Thus, both closure of the excavation and the depth of the zone of yielded rock are reduced.

The density of reinforcement used in this demonstration problem is greater than would be applied in mining practice. However, it confirms the mode of action of reinforcement and the prospect for application of computational methods in reinforcement design.

11.6 Materials and techniques

The emphasis in this chapter has been on the principles of the support and reinforcement of rock masses and on design analyses. If support or reinforcement is to be fully effective it is also necessary that suitable materials be used for the particular application, and that these materials be installed using satisfactory techniques. For full details of these materials and techniques the reader should consult books such as those by Proctor and White (1977), Schach *et al.* (1979) and Hoek and Brown (1980), and the proceedings of specialty conferences such

Figure 11.24 Rockbolt with a slot and wedge anchor (after Hoek and Brown, 1980).

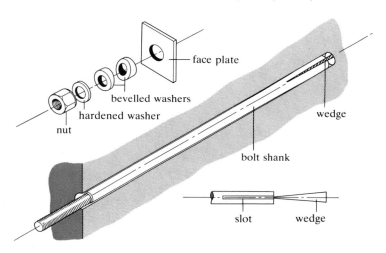

Background

Probably the earliest type of mechanically anchored rockbolt. Very simple and inexpensive to manufacture and widely used throughout the world. The end of the bolt shank is slotted as illustrated and the wedge is driven home by pushing the assembly against the end of the drill hole. The wedge expands the end of the bolt shank and anchors it in the rock.

Also illustrated are two bevelled washers which are used to accommodate an inclined rock face. The hardened washer is used when the bolt is tensioned by applying a measured torque to the nut.

Advantages:

Simple and inexpensive. In hard rock it provides an excellent anchorage and permits immediate tensioning of the bolt.

Disadvantages:

Due to the small contact area between the expanded anchor and the rock, local crushing of the rock with consequent slip of the anchor can occur when the intact rock strength is less than about 10 MPa.

Applications:

Because of the unreliability of the anchor in poor-quality rock, the slotted bolt and wedge has given way to the more versatile expansion shell anchor. Relatively rarely used today.

314

as that on the use of shotcrete for underground structural support published by the American Society of Civil Engineers (Anon, 1974). In this section, a brief account only will be given of the essential features of rockbolts, dowels, shotcrete, wire mesh, props and steel sets. A discussion of the various strata control measures used in longwall coal mining will be given in Chapter 15.

11.6.1 Rockbolts and dowels

A single **tensioned rockbolt** consists of an anchorage, a shank, a face plate, a tightening nut and, in some cases, a deformable bearing plate. The major features of a number of typical rockbolt systems are illustrated and discussed in Figures 11.24–11.27

Figure 11.25 Rockbolt with an expansion shell anchor, tensioned and grouted (after Hoek and Brown, 1980).

grout inlet and return tubes attached to bolt shank

grout return tube in slot along length of bolt shank

grout return through centre hole in bolt shank

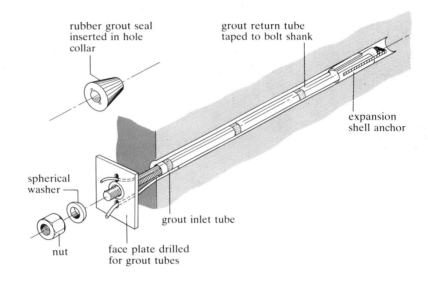

Background

Expansion shell mechanical anchors were developed to provide more reliable anchorage in a wider range of rock conditions than that for which a slotted bolt and wedge system can be used. This drawing illustrates a number of components which can be used in different combinations. The expansion shell anchor is one of a large number of different types, all of which operate in basically the same way. A wedge, attached to the bolt shank, is pulled into a conical anchor shell forcing it to expand against the drillhole walls.

The rubber grout seal is used to centre the bolt in the hole and to seal the collar of the hole against grout leakage. An alternative system is to use a quick-setting plaster to seal the hole collar. Different grout tube arrangements are illustrated. In all cases the grout is injected into the collar end of the hole (except in down-holes) and the return pipe is extended for the length of the hole. Grout injection is stopped when the air has been displaced and when grout flows from the return tube.

Advantages:

Bolt can be tensioned immediately after installation and grouted at a later stage when short-term movements have ceased. Very reliable anchorage in good rock and high bolt loads can be achieved.

Disadvantages:

Relatively expensive. Correct installation requires skilled workmen and close supervision. Grout tubes are frequently damaged during installation and check by pumping clean water before grouting is essential.

Applications:

Very widely used for permanent support applications in civil engineering. Mechanically anchored bolts without grout are widely used in mining.

315

Early rockbolt anchorages were of the mechanical slot-and-wedge and expansion shell types. It is often difficult to form and maintain mechanical anchorages in very hard and in soft rocks. Mechanical anchors are also susceptible to blast-induced damage. Anchors formed from Portland cement grout or resin are generally more reliable and permanent.

Rockbolt shanks are made of high tensile strength or mild steel, and are in the order of 15–30 mm in diameter. The shank may be of plain circular cross section (Figure 11.24), deformed over the whole or part of its length (Figure 11.26) or threaded (Figure 11.27). For permanent installations, the shank of the bolt may be grouted after the anchor has been formed and the rock bolt tensioned. Tensions in the order of 5–20 tonne are usually applied, the magnitude depending on the type of steel used, the bolt diameter and the nature of the anchorage.

Dowels, which are not pre-tensioned, may be made of similar materials to rock bolts and are grouted along their full lengths. An example of a dowel made from reinforcing bar is shown in Figure 11.28. In some applications, the body of the

Figure 11.26 Rockbolt with a deformed section used to form a grouted anchorage (after Hoek and Brown, 1980).

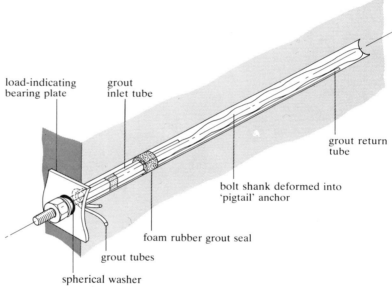

load-indicating bearing plate

grout inlet tube

grout return tube

bolt shank deformed into 'pigtail' anchor

foam rubber grout seal

grout tubes

spherical washer

Background: Drawing is a composite of various systems used in rock bolting, particularly in the Australian mining industry. Grouted anchors have the advantage that they can be used in very poor quality rock masses. One system for grout injection is illustrated. An alternative system is to inject a dry sand/cement mixture through one pipe and a measured water quantity through a second pipe, withdrawing both pipes as the anchor is formed.

The load-indicating bearing plate illustrated is one of several designs which give visual load indication by progressive deformation with load.

Advantages: Inexpensive system with good anchorage characteristics in a wide range of rock conditions. Load-bearing plate gives good visual indication of bolt load and adds 'spring' to bolt for certain applications.

Disadvantages: Care required to form good anchor. Bolt cannot be tensioned until grout has set. Stiffness of bolt and bearing plate may be too low for some applications.

Applications: Principally used in the mining industry where relatively short-term support requirements do not require complete grouting of bolt shank for corrosion protection. Ungrouted bolt length acts as a spring in cases where large stress changes are anticipated during the life of the bolt.

dowel may be made of other materials such as wood or fibreglass, or may take the form of a cable as discussed in section 11.4 and illustrated in Figure 11.9. Recent developments in dowel systems include the proprietary friction anchor or 'split-set' system developed by Scott (1977) (Figure 11.29), and the 'Swellex' system developed by the Atlas Copco Company (Figure 11.30). These rock mass reinforcement systems are being used increasingly in underground mining operations throughout the world.

Grouting serves two important purposes in rock bolt and dowel installations. First, it bonds the bolt shank to the rock making it an integral part of the rock mass and improving the interlocking of the elements of the rock mass. This mechanism is essential to the satisfactory functioning of dowels. Second, grouting provides protection against corrosion. Rockbolts intended for long-term application should be grouted.

Figure 11.27 Resin grouted rockbolt made from a threaded bar (after Hoek and Brown, 1980).

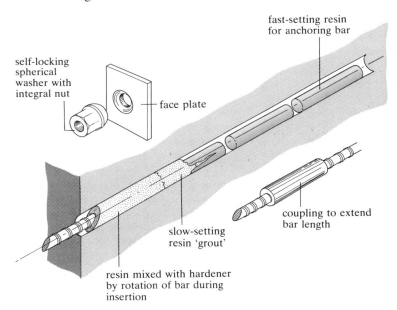

self-locking spherical washer with integral nut

face plate

fast-setting resin for anchoring bar

coupling to extend bar length

slow-setting resin 'grout'

resin mixed with hardener by rotation of bar during insertion

Background:

The most sophisticated rockbolt system currently in use, combines most of the advantages of other bolt systems. Resin and a catalyst are contained in plastic 'sausages', the catalyst being separated in a glass or plastic container in the resin. These capsules are pushed into the hole with a loading stick and the bar is then inserted. Rotation of the bar during insertion breaks the plastic containers and mixes the resin and catalyst.

In the application illustrated, a fast-setting resin capsule is inserted first and forms a strong anchor which permits tensioning of the bolt a few minutes after mixing. Slow-setting resin then 'grouts' the remainder of the bar.

The bar illustrated has a very coarse rolled thread which gives good bonding and allows the length of the installation to be adjusted very easily.

Advantages:

Very convenient and simple to use. Very high-strength anchors can be formed in rock of poor quality and, by choosing appropriate setting times, a 'one-shot' installation produces a fully grouted tensioned rockbolt system.

Disadvantages:

Resins are expensive and many suffer from a limited shelf-life, particularly in hot climates.

Applications:

Increasingly used in critical applications in which cost is less important than speed and reliability.

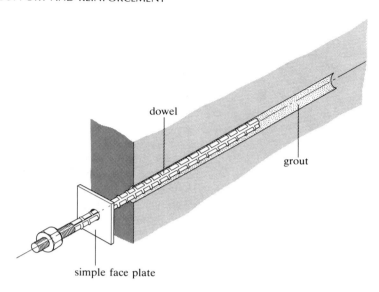

dowel

grout

simple face plate

Figure 11.28 Grouted dowel made from reinforcing bar (after Hoek and Brown, 1980).

Background:

Developed as an inexpensive alternative to the Scandinavian 'Perfobolt' system where use of untensioned dowels is appropriate. A thick grout is pumped into the drillhole by means of a simple hand pump or a monopump. The dowel is pushed into the grout as shown in the illustration. For up-holes, the dowel is sometimes held in place by a small wooden or steel wedge inserted into the collar of the hole. A face plate and nut can be added if required although, for very light support, a plain dowel is sometimes used.

Advantages:

Simple and inexpensive.

Disadvantages:

Cannot be tensioned and hence must be installed before significant deformation of the rock mass has taken place.

Applications:

Widely used in the mining industry for light support duties and in civil engineering for mesh fixing and for supporting ventilation tubing, pipework and similar services.

The essential features of the grouting technique are outlined in Figure 11.25. The hole should be flushed with clean water before grouting to clean the hole and to check the effectiveness of the collar seal. The mix design should be such that the grout is readily pumpable but does not shrink or bleed. A water : cement ratio of 0.35 by weight gives satisfactory results. If a filler such as flyash is used, the water : cement ratio may be higher. Additives may be used to improve fluidity or reduce shrinkage.

11.6.2 Shotcrete

Shotcrete is pneumatically applied concrete used to provide passive support to the rock surface. Gunite, which pre-dates shotcrete in its use in underground construction, is pneumatically applied mortar. Because it lacks the larger aggregate sizes of up to 25 mm typically used in shotcrete, gunite is not able to develop the same resistance to deformation and load-carrying capacity as shotcrete. For many years, shotcrete has been used with outstanding success in civil engineering underground construction in a wide variety of ground types. It is so successful because it satisfies most of the requirements for the provision of satisfactory primary support or reinforcement discussed in section 11.2. In recent years, shotcrete has found increasing use in underground mining practice, mainly for

318

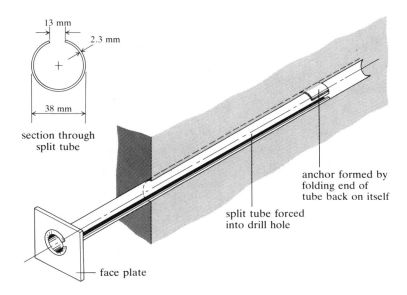

13 mm

2.3 mm

38 mm

section through
split tube

anchor formed by
folding end of
tube back on itself

split tube forced
into drill hole

face plate

Figure 11.29 Friction anchor or split-set dowel (after Hoek and Brown, 1980).

Background: Developed by Scott (1977) in conjunction with the Ingersoll-Rand Company in the USA. This device has gained considerable popularity in the mining industry.

The $1\frac{1}{2}$ in (38 mm) diameter split tube is forced into a $1\frac{3}{8}$ in (35 mm) diameter drillhole. The spring action of the compressed tube applies a radial force against the rock and generates a frictional resistance to sliding of the rock on the steel. This frictional resistance increases as the outer surface of the tube rusts.

Advantages: Simple and quick to install and claimed to be cheaper than a grouted dowel of similar capacity.

Disadvantages: Cannot be tensioned and hence is activated by movement in the rock in the same way as a grouted dowel. Its support action is similar to that of an untensioned dowel and hence it must be installed very close to a face. The drillhole diameter is critical and most failures during installation occur because the hole is either too small or too large.

In some applications, rusting has occurred very rapidly and has proved to be a problem where long-term support is required. The device cannot be grouted.

Applications: Increasingly used for relatively light support duties in the mining industry, particularly where short-term support is required.

permanent mine installations, but sometimes for stope support. Almgren (1981), for example, describes the use of shotcrete in cut-and-fill stopes at the Garpenberg mine in central Sweden. In this case, the batching plant is located on surface with the dry mix (see below) being transported to the point of application by pipeline over vertical and horizontal distances of up to 300 and 500 m, respectively.

Shotcrete is prepared using either the dry-mix or the wet-mix process. In the dry-mix process, dry or slightly dampened cement, sand and aggregate are mixed at the batching plant, and then entrained in compressed air and transported to the discharge nozzle. Water is added through a ring of holes at the nozzle. Accurate water control is essential to avoid excessive dust when too little water is used or an over-wet mix when too much water is added. In the wet-mix process, the required amount of water is added at the batching plant, and the wet mix is pumped to the nozzle where the compressed air is introduced. A comparison of

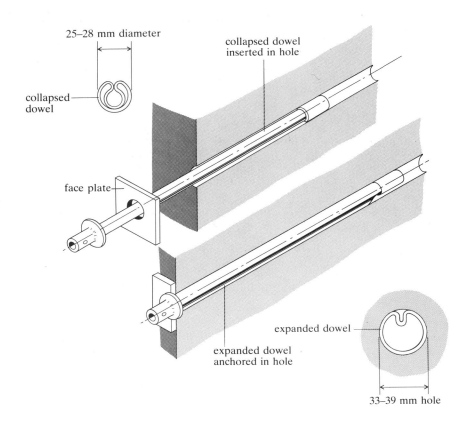

Figure 11.30 Swellex dowel (after Hoek, 1982).

Background: Developed by the Atlas Copco Company. A thin-walled tube folded into a collapsed shape of between 25 and 28 mm in diameter is inserted into a 33–39 mm diameter borehole and expanded by injection of water at a pressure of about 20 MPa delivered by a small portable pump. Expansion of the dowel results in an overall length reduction which pulls the face plate tight against the rock and induces a small tension in the dowel.

Advantages: Simple and inexpensive. The anchoring force is high. A small initial axial force is applied.

Disadvantages: The strength of the system is limited by the strength of the thin-walled tube. As with other dowel systems, it must be installed close to the face. Cannot be grouted, but rusting is inhibited by the presence of a sealed volume of water inside the expanded dowel and an external protective coating. Long-term durability not yet demonstrated.

Applications: Increasingly used for medium-term reinforcement in the mining industry.

the dry- and wet-mix processes is given in Table 11.4. The dry-mix method is more widely used, mainly because the equipment required is lighter and less expensive, and because the dry material can be conveyed over longer distances, an important advantage in mining applications. However, wet-mix methods do have important advantages for underground mining applications in terms of reduced dust levels, lower skill requirements and the need for less equipment at the application site.

320

Table 11.4 Comparison of dry- and wet-mix shotcreting processes (after Anon, 1982).

Factor	Dry mix	Wet mix
machinery	Lower total expenditure. Conveyance by compressed air/low efficiency. Maintenance relatively simple and infrequent	Less machinery at work site. Diesel or electric power more efficient than compressed air. Less wear rate in pump, hoses and nozzles. 60% less air consumed
mixing	At work site. Pre-mixed, dry ingredients can be used but cannot be left open in humid or wet environments. Performance impaired by wet sand	Accurate mixing remote from work site. Multiple handling avoided. Can use bulk ready mix. Wet sand acceptable
output	Rarely exceeds $5 \, m^3 \, h^{-1}$ in place. Can be conveyed over longer distances than wet mixes, maximum 400 m horizontally and 100 m in elevation. 50–100 m normal	Higher than similar dry mix machines; $2–10 \, m^3 \, h^{-1}$ with hand-held nozzle. Up to $18 \, m^3 \, h^{-1}$ when boom mounted
rebound	Can be 15–40% from vertical walls, 20–50% from overhead. Forms rebound pockets. Loss of aggregate makes compliance with mix specification difficult, and excess cement is usually added	Low rebound with correct mix, can be less than 10%. Little loss of aggregate. Rebound pockets do not occur
impact velocity	Higher – better adhesion; easier to use overhead	Generally adequate for tunnel/mine work
addition of reagents	Powders in mixer or gun. Liquids at nozzle	Generally as liquid
dust	Suppression additives, pre-wetting with 5–15% moisture or 'semi-wet' methods can reduce dust formed	Very little dust formed: can be 1/6 of that from dry-mix spraying. Better visibility. No danger of lamination by dust
versatility	Can be used for sand blasting, guniting, refractory materials, repairs, resurfacing	Can be used as concrete pump for pouring in place

Shotcrete mix design is a difficult and complex process involving a certain amount of trial and error. The mix design must satisfy the following criteria (Hoek and Brown, 1980):

(a) Shootability – the mix must be able to be placed overhead with minimum rebound.

(b) Early strength – the mix must be strong enough to provide support to the ground at ages of a few hours.

(c) Long-term strength – the mix must achieve a specified 28 day strength with the dosage of accelerator needed to achieve the required shootability and early strength.

(d) Durability – adequate long-term resistance to the environment must be achieved.

(e) Economy – low-cost materials must be used, and there must be minimum losses due to rebound.

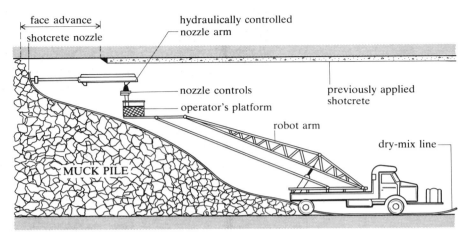

Figure 11.31 Application of dry-mix shotcrete close to face.

A typical mix contains the following percentages of dry components by weight:

cement	15–20%
coarse aggregate	30–40%
fine aggregate or sand	40–50%
accelerator	2–5%

The water : cement ratio for dry-mix shotcrete lies in the range 0.3–0.5 and is adjusted by the operator to suit local conditions. For wet-mix shotcrete, the water : cement ratio is between 0.4 and 0.6.

Figure 11.31 illustrates the application of shotcrete using a dry mix and remote control of the nozzle. The efficacy of the shotcreting process depends to a large extent on the skill of the operator. The nozzle should be kept as nearly perpendicular to the rock surface as possible and at a constant distance of about 1 m from it. A permanent shotcrete lining is usually between 100 mm and 500 mm thick, the larger thicknesses being placed in a number of layers. The addition of 20–50 mm long and 0.25–0.8 mm diameter steel fibres has been found to improve the toughness, shock resistance, durability, and shear and flexural strengths of shotcrete, and to reduce the formation of shrinkage cracks. Fibre-reinforced shotcrete will accept larger deformations before cracking occurs than will unreinforced shotcrete; after cracking has occurred, the reinforced shotcrete maintains its integrity and some load-carrying capability. However, fibre-reinforced shotcrete is more expensive and more difficult to apply than unreinforced shotcrete.

11.6.3 Wire mesh

Chain-link or welded steel mesh is used to restrain small pieces of rock between bolts or dowels, and to reinforce shotcrete. For the latter application, welded mesh is preferred to chain-link mesh because of the difficulty of applying shotcrete satisfactorily through the smaller openings in chain-link mesh. For underground use, weld mesh typically has 4.2 mm diameter wires spaced at 100 mm centres.

322

Figure 11.32 Reinforced haulage at a depth of 1540 m following a seismic event of magnitude 4.0. Severely damaged rock is well contained by mesh and rope lacing (after Ortlepp, 1983).

In underground metalliferous mining, rock blocks or fragments of fractured rock are often held in place by a pattern of hoist rope lacing installed between rockbolts or anchor points. Rope lacing may be used to stiffen mesh in those cases in which the mesh is unable to provide adequate restraint to loosened rock. Ortlepp *et al.* (1972–3) and Ortlepp (1983) give a number of examples of the use of mesh and lacing in conjunction with rockbolts and grouted cables and steel rods to stabilise tunnels in the deep-level gold mines of South Africa. Figure 11.32 shows the appearance of an intensively reinforced haulage at a depth of 1540 m following a seismic event of magnitude 4.0 which had its source on a fault intersecting the haulage near the location of the photograph. The haulage was reinforced with 2.5 m long grouted steel rope tendons and 7.5 m long prestressed rock anchors which provided an overall support capacity of 320 kNm^3. The 3.2 mm diameter by 65 mm square galvanised mesh was backed by 16 mm diameter scraper rope. Across the intersection with the fault, the severely damaged rock was well contained by the mesh and lacing even though several of the prestressed anchors had failed.

11.6.4 Steel and timber supports

The simplest form of support is the **post and cap**, or prop and lid. It consists of a single upright with a plate above. The post may be made of timber or steel, or may be a hydraulic prop. Timber posts, sometimes called **sticks**, are widely used to assist in supporting the mined-out area behind the stope face in the mining of

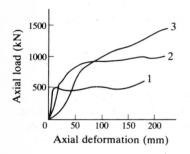

Figure 11.33 Axial load–deformation curves for 1, a 150 mm diameter pipe stick; 2, a 150 mm diameter pipe prop; 3, a 600 mm × 600 mm sandwich pack (after Steinhobel and Klokow, 1978–9).

Figure 11.34 Toussaint–Heintzmann yielding arch: (a) cross section; (b) clamp joint; (c) alternative joint; (d) arch configuration before and after yielding; (e) idealised load–radial displacement response.

narrow, reef-like metalliferous ore bodies. These sticks are usually 100–200 mm in diameter and are designed to support, or help support, the dead weight of the first 1–2 m of rock in the hangingwall. A typical 200 mm diameter stick can have a short-term load-carrying capacity of 60 tonne (Korf, 1978–9), but for long-term use and low rates of load application, lower design capacities should be used.

Supports that are stiffer and have higher load-carrying capacities than timber posts or sticks of the same diameters, are provided by **pipe sticks** in which a stick is press fitted into a steel pipe, typically 150 mm in diameter with a 4 mm thick wall (Steinhobel and Klokow, 1978–9). Figure 11.33 shows typical axial load–displacement curves for a 150 mm diameter pipe stick, a 150 mm diameter pipe prop, and a 600 mm × 600 mm timber sandwich pack.

The major disadvantages of sticks are that they cannot be installed with a pre-load to provide truly active support and that they must undergo considerable deformation to develop their full load-carrying capacity. Sometimes, it can be difficult to set sticks adequately, with the result that local falls, leading to move widespread loosening of the hanging wall, can occur.

Hydraulic props are set with a pre-load to provide active support and suffer from none of the disadvantages of sticks, although they are obviously very much more expensive. The load-carrying capacities of individual props may vary from as little as 5 tonne for a very light prop to over 100 tonne for a 0.3 m diameter prop. In the deep-level gold mines of South Africa, rapid-yielding hydraulic props are widely used to provide concentrated active support of the hanging wall close to the face. Their rapid-yielding capability allows the energy released by rockbursts to be absorbed rapidly and safely, thereby minimising the damage caused.

Steel arches or **steel sets** are used where high load-carrying capacity elements are required to support tunnels or roadways. A wide range of rolled steel sections are available for this application. Where the rock is well jointed, or becomes fractured after the excavation is made, the spaces between the sets may be filled

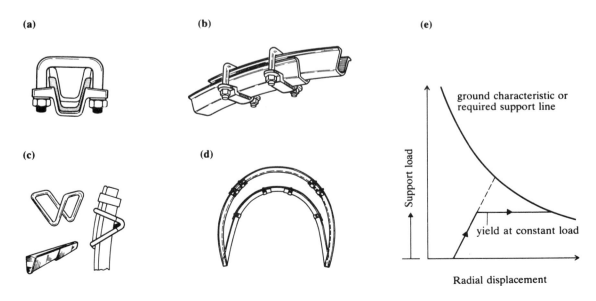

324

with steel mesh, steel or timber lagging, or steel plates. Proctor and White (1977) provide the most detailed account available of the materials and techniques used in providing steel support.

Steel sets provide support rather than reinforcement. They cannot be pre-loaded against the rock face and, as pointed out in section 11.3, their efficacy largely depends on the quality of the blocking provided to transmit loads from the rock to the steel set. Steel arches are widely used to support roadways in coal mines where they are often required to sustain quite large deformations. These deformations may be accommodated by using yielding arches containing elements designed to slip at predetermined loads (Figure 11.34), or by permitting the splayed legs of the arches to punch into the floor. Where more rigid supports are required as, for example, in circular transportation tunnels, circular steel sets are used.

12 Mining methods and method selection

12.1 Mining excavations

Recovery of mineral from subsurface rock involves the development of physical access to the mineralized zone, liberation of the ore from the enclosing host rock and transport of this material to the mine surface. Excavations of various shapes, sizes, orientations and functions are required to support the series of operations which comprise the complete mining process. A schematic layout of an underground mine is shown in Figure 12.1. Three types of excavations employed in the process are recognised in the figure. These are the ore sources, or stopes, the stope access and service openings, or stope development, and the permanent access and service openings. Irrespective of the method used to mine an orebody, similarities exist between the functions and the required geomechanical performances of the different types of non-production excavations.

A stope is the site of ore production in an orebody. The set of stopes generated during ore extraction usually constitutes the largest excavations formed during the exploitation of the deposit. The stoping operation, i.e. ore mobilisation from its *in situ* setting and its subsequent extraction from the mine void, is the core of the mine production process. Clearly, control of rock performance within the orebody, and in the rock mass adjacent to the orebody, is critical in assuring the efficient geomechanical and economic performance of the individual stopes, and of the mine as a whole.

The size of stopes means that their zone of influence is large relative to virtually all other mine excavations. Stope design therefore exercises a dominant rôle in the location, design and operational performance of other excavations which sustain mining activity. The principles of stope layout and design are integrated with the set of engineering concepts and physical operations which together compose the mining method for an orebody. This chapter considers in an introductory way the relation between the geomechanical properties of an orebody (and its host rock mass) and the appropriate method of mining it.

The second type of mine excavation identified in Figure 12.1 is represented by stope access and service openings. These consist of in-stope development such as drill headings and slot raises, horizontal and vertical openings for personnel access to stope work places, and ore production and transport openings such as drawpoints, tramming drives and ore passes. These excavations are developed within the orebody rock, or within the orebody peripheral rock. Their operational life approximates that of adjacent stoping activity, and in some cases, such as drill headings, the excavations are consumed in the stoping process.

326

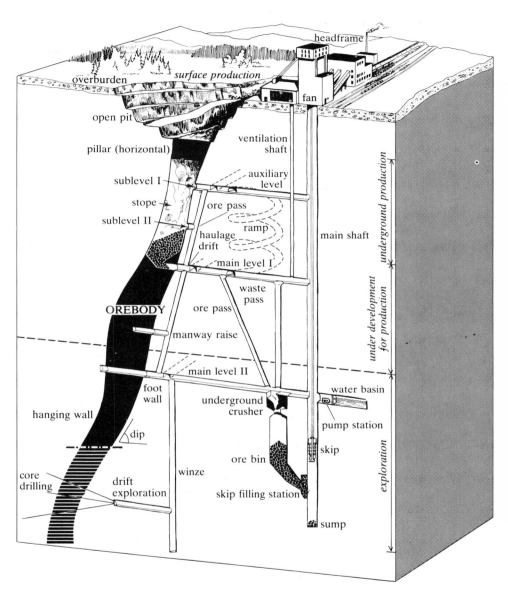

Figure 12.1 Schematic layout of an idealised underground mine (after Hamrin, 1982).

The location of stope development in the zone of geomechanical influence of the stope, and, typically, the immediate proximity of a stope and its related service openings, may impose severe and adverse local conditions in the rock medium. The procedures discussed in preceding chapters may be applied to the design of these openings, provided account is taken of the local stress field generated by stoping activity, or of rock mass disturbance caused by stoping-induced relaxation in the medium. In all cases, design of a stoping layout requires detailed attention to the issues of position, shape and, possibly, support and reinforcement of stope development openings, to assure their function is maintained while adjacent extraction proceeds. These excavation design issues are

327

obviously related closely to the geomechanical principles on which the mining method is based.

Permanent access and service openings represent the third class of mining excavations illustrated in Figure 12.1. This class consists of openings which must meet rigorous performance specifications over a time span approaching or exceeding the duration of mining activity for the complete orebody. For example, service and ore hoisting shafts must be capable of supporting high-speed operation of cages and skips continuously. Ventilation shafts and airways must conduct air to and from stope blocks and service areas. Haulage drives must permit high-speed operation of ore trains and personnel transport vehicles. In these cases, the excavations are designed and equipped to tolerances comparable with those in other areas of engineering practice. The practical mining requirement is to ensure that the designed performance of the permanent openings can be maintained throughout the mine life. In rock mechanics terms, this requirement is expressed as a necessity to locate the relevant excavations (and associated structures) in areas where rock mass displacements, strains and tilts are always tolerable. The magnitudes of these mining-induced perturbations at any point in the rock medium surrounding and overlying an orebody are determined, in part, by the nature and magnitude of the displacements induced by mining in the immediate vicinity of the orebody. They can be estimated using one of the analytical or computational methods described in Chapter 6.

In previous chapters, attention has been devoted to the design of mine excavations in different types of rock masses. The preceding discussion indicates that the location and design of all elements of a mine structure are related to the strategy adopted for excavating the orebody. This implies that the formulation of the complete layout for a mine must evolve from consideration of the geomechanical consequences of the selected method for recovering the ore from the orebody. The need is apparent for an elaboration of the principles and scope for application of the various mining methods, and this is now presented. It is intended as a prelude to discussion of the procedures applied in the design of elements of the mining layouts generated by the various mining methods, which are the concern of some subsequent chapters.

12.2 Rock mass response to stoping activity

The dimensions of orebodies of industrial significance typically exceed hundreds of metres in at least two dimensions. During excavation of an orebody, the spans of the individual stope excavations may be of the same order of magnitude as the orebody dimensions. It is convenient to describe the performance of the host rock mass during mining activity in terms of the displacements of orebody peripheral rock induced by mining, expressed relative to the minimum dimension d_m of excavations created in the orebody. It is also useful to consider the rock mass around an orebody in terms of near-field and far-field domains. In a manner analogous to definition of the zone of influence of an excavation, the near field of an orebody may be taken as the rock contained within the surface distance $3d_m$ from the orebody boundaries.

328

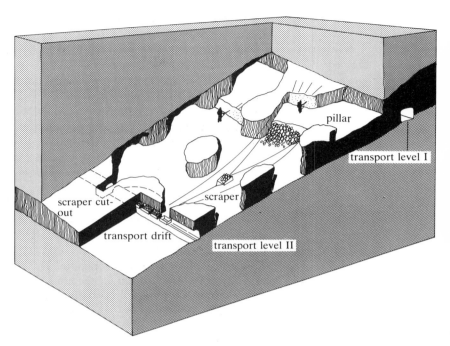

Figure 12.2 Elements of a supported method of mining (after Hamrin, 1982).

Different mining methods are designed to produce different types and magnitudes of displacements, in the near-field and far-field domains of an orebody. For example, the mining method illustrated schematically in Figure 12.2 is designed to restrict rock displacements in both the near field and the far field of the orebody to elastic orders of magnitudes. Following the usual notions of engineering mechanics, prevention of displacements is accompanied by increase in the state of stress in and around the support units preserved to control near-field rock deformation. The result is to increase the average state of stress in the orebody near field. The orebody peripheral rock is fully supported, by pillar remnants in the orebody, against large-scale displacements during stoping activity. Such a fully supported mining method represents one conceptual extreme of the range of geomechanical strategies which may be pursued in the extraction of an orebody.

A fundamentally different geomechanical strategy is implemented in the mining method illustrated in Figure 12.3. In this case, mining is initiated by generating rigid-body displacements of rock above an excavation in the orebody. In this region, the initial displacements are of the same order of magnitude as the vertical dimension of the excavation. As extraction proceeds, the rigid-body displacement field propagates through the orebody, to the near and far fields. The success of this operation relies on spatially continuous and progressive displacement of near-field and far-field rock during ore extraction. In this caving method of mining, rock performance is the geomechanical antithesis of that generated in the supported method described earlier.

An alternative method of describing rock mass behaviour in the mining methods represented by Figures 12.2 and 12.3 is in terms of the constitutive behaviour of the host medium for mining. By restricting rock mass displacements, in both

329

the near field and far field, to elastic orders of magnitude, the supported method of working shown in Figure 12.2 is intended to maintain pseudo-continuous behaviour of the host rock medium. The caving method illustrated in Figure 12.3, by inducing rigid-body displacements in units of the rock mass, exploits the discontinuous behaviour of a rock medium when confining stresses are relaxed. The supported methods of working can succeed only if compressive stresses, capable of maintaining the continuum properties of a rock mass, can be sustained by the near-field rock. Caving methods can proceed where low states of stress in the near field can induce discontinuous behaviour of both the orebody and overlying country rock, by progressive displacement in the medium. Thus under ideal mining conditions, supported stoping methods would impose fully continuous rock mass behaviour, while caving methods would induce fully discontinuous behaviour.

The geomechanical differences between supported and caving methods of mining may be described adequately by the different stress and displacement

Figure 12.3 Elements of a caving method of mining (after Dravo Corporation, 1974).

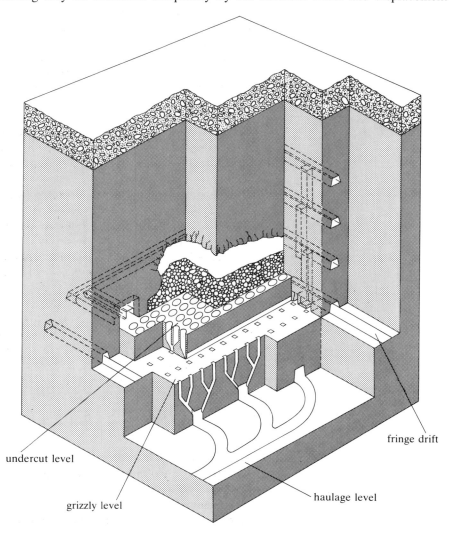

undercut level

grizzly level

fringe drift

haulage level

fields induced in the orebody near-field and far-field domains. Added insight into the distinction between the two general mining strategies may be obtained by considering the energy concentration and redistribution accompanying mining.

In supported methods, mining increases the elastic strain energy stored in stress concentrations in the support elements and the near-field rock. The mining objective is to ensure that sudden release of the strain energy cannot occur. Such a sudden release of energy might involve sudden rupture of support elements, rapid closure of stopes, or rapid generation of penetrative fractures in the orebody peripheral rock. These events present the possibility of catastrophic changes in stope geometry, damage to adjacent mine openings, and immediate and persistent hazard to mine personnel.

For caving methods, the mining objective is the prevention of strain energy accumulation, and the continuous dissipation of pre-mining energy derived from the prevailing gravitational, tectonic and residual stress fields. Prior to caving, rock in, around and above an orebody possesses both elastic strain energy and gravitational potential energy. Mining-induced relaxation of the stress field, and vertical displacement of orebody and country rock, reduce the total potential energy of the rock mass. The objective is to ensure that the rate of energy consumption in the caving mass, represented by slip, crushing and grinding of rock fragments, is proportional to the rate of extraction of ore from the active mining zone. If this is achieved, the development of unstable structures in the caving medium, such as arches, bridges and voids, is precluded. Volumetrically uniform dissipation of energy in the caving mass is important in developing uniform comminution of product ore. The associated uniform displacement field prevents impulsive loading of installations and rock elements underlying the caving mass.

The contrasting mining strategies of full support and free displacement involve conceptual and geomechanical extremes in the induced response of the host rock mass to mining. In practice, a mining programme may be based on different geomechanical concepts at different stages of orebody extraction. For example, the extraction of an orebody may exploit completely natural support in the initial stoping phase, using orebody remnants as pillar elements. In the early stages of pillar recovery, various types of artificial support may be emplaced in the mined voids, with the objective of controlling local and regional rock mass displacements. In the final stages of pillar recovery, pillar wrecking and ore extraction may be accompanied by caving of the adjacent country rock. It is clear that the transition from one geomechanical basis to another can have important consequences for the integrity and performance of permanent openings and other components of a mine structure. This indicates that the key elements of a complete mining strategy for an orebody should be established before any significant and irrevocable commitments are made in the pre-production development of an orebody.

12.3 Orebody properties influencing mining method

A mining method consists of a sequence of production unit operations, which are executed repetitively in and around the production blocks into which an orebody is divided. The operations of ore mobilisation, extraction and transport are common to all mining methods, while other operations may be specific to a particular method. Differences between mining methods involve different techniques of performing the unit operations. The different operating techniques employed in the various methods are the result of the different geometric, geomechanical and geologic properties of the orebody and the host rock medium. Other more general engineering and social questions may also be involved. In the following discussion, only the former issues, i.e. readily definable physicomechanical orebody properties, are considered.

12.3.1 Geometric configuration of orebody

This property defines the relative dimensions and shape of an orebody. It is related to the deposit's geological origin. Orebodies described as seam, placer or stratiform (stratabound) deposits are of sedimentary origin and always extensive in two dimensions. Veins, lenses and lodes are also generally extensive in two dimensionally, and usually formed by hydrothermal emplacement or metamorphic processes. In massive deposits, the shape of the orebody is more regular, with no geologically imposed major and minor dimensions. Porphyry copper orebodies typify this category.

Both the orebody configuration and its related geological origin influence rock mass response to mining, most obviously by direct geometric effects. Other effects, such as depositionally associated rock structure, local alteration of country rock, and the nature of orebody–country rock contacts, may impose particular modes of rock mass behaviour.

12.3.2 Disposition and orientation

These issues are concerned with the purely geometric properties of an orebody, such as its depth below ground surface, its dip and its conformation. Conformation describes orebody shape and continuity, determined by the deposit's post-emplacement history, such as episodes of faulting and folding. For example, methods suitable for mining in a heavily faulted environment may require a capacity for flexibility and selectivity in stoping, to accommodate sharp changes in the spatial distribution of ore.

12.3.3 Size

Both the absolute and relative dimensions of an orebody are important in determining an appropriate stoping method. A large, geometrically regular deposit may be suitable for mining using a mechanised, mass-mining method, such as block caving. A small deposit of the same ore type may require selective mining and precise ground control to establish a profitable operation. In addition to its direct significance, there is also an interrelation between orebody size and the other geometric properties of configuration and disposition, in their effect on mining method.

12.3.4 Geomechanical setting

Specific geomechanical issues determining an appropriate mining method for a deposit have been defined, in part, in preceding chapters discussing the properties of rock materials and rock masses. The response of a rock mass to a particular mining method reflects the mechanical and structural geological constitution of the orebody rock and the surrounding country rock. Rock material properties include strength, deformation characteristics (such as elastic, plastic and creep properties) and weathering characteristics. Rock mass properties are defined by the existence, and geometric and mechanical properties, of joint sets, faults, shear zones and other penetrative discontinuities. The pre-mining state of stress in the host rock is also a significant parameter.

In addition to the conventional geomechanical variables, a number of other rock material properties may influence the mining performance of a rock mass. Adverse chemical properties of an ore may preclude caving methods of mining, which generally require chemical inertness. For example, a tendency to re-cement, by some chemical action, can reduce ore mobility and promote bridging in a caving mass. Similarly, since air permeates a caving medium, a sulphide ore subject to rapid oxidation may create difficult ventilation conditions in working areas, in addition to being subject itself to a degradation in mechanical properties.

Other more subtle ore properties to be noted are the abrasive and comminutive properties of the material. These determine the drillability of the rock for stoping purposes, and its particle size degradation during caving, due to autogeneous grinding processes. A high potential for self-comminution, with the generation of excessive fines, may influence the design of the height of draw in a caving operation and the layout and design of transport and handling facilities in a stoping operation.

In some cases, a particular structural geological feature or rock mass property may impose a critical mode of response to mining, and therefore have a singular influence on the appropriate mining method. For example, major continuous faults, transgressing an orebody and expressed on the ground surface, may dictate the application of a specific method, layout and mining sequence. Similar considerations apply to the existence of aquifers in the zone of potential influence of mining, or shattered zones and major fractures which may provide hydraulic connections to water sources. The local tectonic setting, particularly the level of natural or induced seismic activity, is important. In this case, those methods of working which rely at any stage on a large, unfilled void would be untenable, due to the possibility of local instability around open stopes induced by a seismic event. A particular consequential risk under these conditions is air blast, which may be generated by falling stope wall rock.

12.3.5 Orebody value and spatial distribution of value

The monetary value of an orebody, and the variation of mineral grade through the volume of the orebody, determine both mining strategy and operating practice. The critical parameters are average grade, given various cut-off grades, and grade distribution. The former parameter defines the size and monetary value of the deposit as the market price for the product mineral changes. It also indicates the degree of flexibility required in the selected method of mining the orebody, since it is necessary that marginal ore be capable of exclusion from the production operation, in response to changing market conditions. The significance of dilution of the ore stream, arising, for example, from local failure of stope wall rock and its incorporation in the extracted ore, is related to the value per unit weight of ore. In particular, some mining methods are prone to dilution, and marginal ore may become uneconomic if mined by these methods.

Grade distribution in an orebody may be uniform, uniformly varying (where a spatial trend in grade is observed), or irregular (characterised by high local concentrations of minerals, in lenses, veins or nuggets). The concern here is with the applicability of mass mining methods, such as caving or sublevel stoping, or the need for complete and highly selective recovery of high-grade domains within a mineralised zone. Where grade varies in some regular way in an orebody, the obvious requirement is to devise a mining strategy which assures recovery of higher-grade domains, and yet allows flexible exploitation of the lower-grade domains.

12.3.6 Engineering environment

A mining operation must be designed to be compatible with the mine exterior domain and to maintain acceptable conditions in the interior domain. Mine interaction with the external environment involves effects on local groundwater flow patterns, changes in the chemical composition of groundwater, and possible changes in surface topography through subsidence. Different mining methods interact differently with the external environment, due to the disparate displacement fields induced in the far-field rock. In general, caving methods of mining have a more pronounced impact on the mine external environment, through subsidence effects, than supported methods. In the latter case, it is frequently possible to cause no visible disturbance or rupture of the ground surface, and to mitigate the surface waste disposal problem by emplacement of mined waste in stope voids. In fact, stope backfill generated from mill tailings is an essential component in many mining operations.

Specific mining methods and operating strategies are required to accommodate the factors which influence the mine internal environment. Mine gases such as methane, hydrogen sulphide, sulphur dioxide, carbon dioxide or radon may occur naturally in a rock mass, or be generated from the rock mass during mining activity. Pre-mining rock temperatures are related to both rock thermal properties such as thermal conductivity, and regional geophysical conditions. The thermal condition of mine air is subject to local climatic influences. Supported and caving methods require different layouts for ventilation circuits, and present diverse opportunities for gas generation and liberation in a ventilation air stream. In all cases, the requirement is to understand the interaction between the rock domain, active at any stage of mining, and the thermodynamic process of ventilation which operates in that domain and sustains the operation. As a general rule, supported and caving methods may impose grossly dissimilar loads on a ventilation air stream, due to the different opportunities offered for gas liberation, losses from the ventilation stream, and heat pick-up during air circulation.

12.4 Underground mining methods

A comprehensive discussion of the general features of various underground mining methods is beyond the scope of this text. Such a general discussion is provided in the text by Hamrin (1982). Attention is confined here to the relations between working method, the rock mass conditions essential to sustain the

334

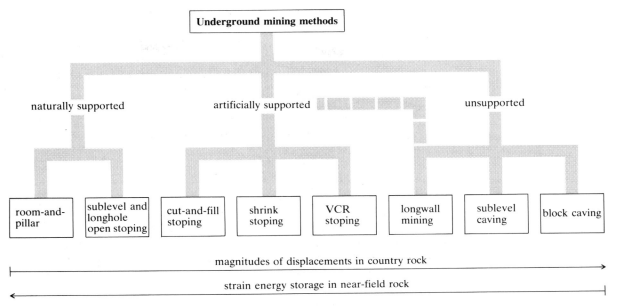

Figure 12.4 A hierarchy of underground mining methods and associated rock mass response to mining.

method, and the key orebody properties defining the scope for application of the method.

The mining methods commonly employed in industrial practice are defined in Figure 12.4. Other mining methods, mostly of historical or local significance, such as top slicing or cascade stoping, could be readily incorporated in this categorisation. The gradation of rock performance, ranging from complete support to induced failure and granular flow, and in spatial energy change from near-field storage to far-field dissipation, is consistent with the notions discussed earlier.

12.4.1 Room-and-pillar mining (Figure 12.5)

Ore is produced from rooms or entries, each of which serves the multiple rôles of ore source, access opening, transport drift and airway. Pillars are generated as ore remnants between entries, to control both the local performance of immediate roof rock and the global response of the host rock medium. It is preferable to arrange pillars in a regular grid array, to simplify planning, design and operation. Since personnel operate continuously under exposed roof spans, close observation of the performance of roof spans and pillars is required. Immediate roof rock may be unsupported, or supported or reinforced artificially, using methods described elsewhere. The pillars may be permanently unmined. Alternatively, pillar ore may be recovered in the orderly retreat from a mine panel or district, inducing collapse of the immediate roof of the mined void and caving of the superincumbent strata.

Room-and-pillar mining is applied in flat-lying stratiform or lenticular orebodies, although variations of the method can accommodate an orebody dip up to about 30°. The orebody must be relatively shallow, to prevent commitment of excessive ore in pillars. Mechanised mining operations require a fairly uniform

335

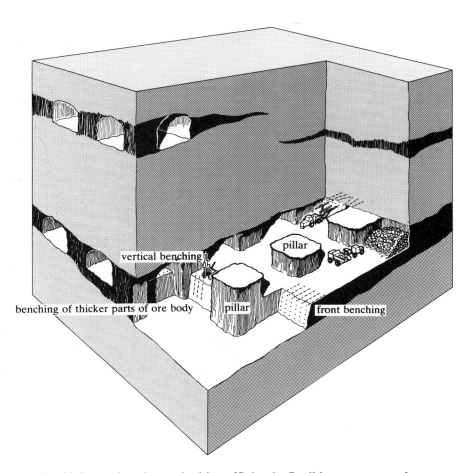

Figure 12.5 Schematic layout for room-and-pillar mining (after Hamrin, 1982).

orebody thickness, but the method is sufficiently flexible to accommodate some local variations in thickness of the mineralised zone. It is one of the two methods suitable for recovery of thin, flat-lying deposits. Orebody heights greater than about 6 m are generally worked by multiple passes. A top slice is mined conventionally, and the underlying ore is then mined by an underhand method, such as downhole benching.

The geomechanical setting suitable for implementation of room-and-pillar mining consists of a strong, competent orebody and near-field rock medium, with a low frequency of cross jointing in the immediate roof rock.

Close control of product ore grade is possible in room-and-pillar mining, since the method admits highly selective extraction of pockets of ore. Variability of grade distribution can be accepted, with low-grade ore being left as irregularly distributed pillars. Barren rock produced during mining can be readily stowed in mined voids.

12.4.2 Sublevel open stoping (Figure 12.6)

Ore is produced from a stope block in which extensive development has been undertaken prior to stoping activity. Stope pre-production development consists of an extraction level, access raises and drifts, drill drifts, slot raise and stope

336

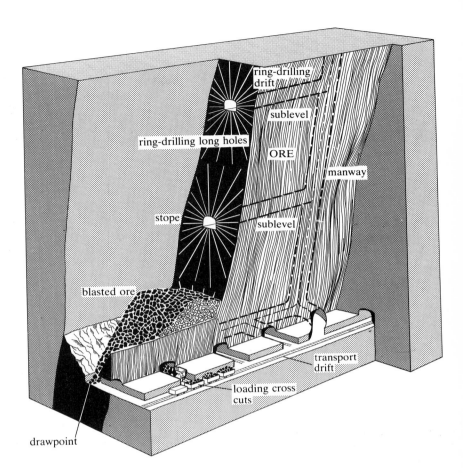

Figure 12.6 Schematic layout for sublevel open stoping with ring-drilled blast holes (after Hamrin, 1982).

return airway. An expansion slot is developed by enlarging the slot raise, using parallel hole blasting, to the width of the stope. Ore is fragmented in the stope using ring-drilled or long parallel blast holes, exploiting the expansion provided by the stope slot. Broken ore reports to the drawpoints for extraction. Stope faces and side walls remain unsupported during ore extraction, while local and near-field support for the country rock is developed as pillars are generated by stoping.

Open stoping is applied in massive or steeply dipping stratiform orebodies. For an inclined orebody, resulting in inclined stope walls, the inclination of the stope foot wall must exceed the angle of repose of the broken rock by some suitable margin. This is required to promote free flow of fragmented rock to the extraction horizon. Since open stoping requires unsupported, free-standing stope boundary surfaces, the strength of orebody and country rock must be sufficient to provide stable walls, faces and crown for the excavation. The orebody boundaries must be fairly regular, since selective mining is precluded by the requirement for regular stope outlines, which are associated with the use of long blast holes. Blast hole penetration of stope walls, due to drilling inaccuracy, leads to dilution. Dilution from this source is, relatively, a more significant problem in narrow orebodies. The resulting minimum orebody width for open stoping is about 6 m.

Pillar recovery is common practice in open stoping. Backfill of various qualities may be placed in the primary stope voids, and pillar mining performed by exploiting the local ground control potential of the adjacent fill. Alternatively, pillars may be blasted into adjacent stope voids, with the possibility of extensive collapse of the local country rock. Successful ore recovery would then require draw of fragmented ore from beneath less mobile, barren country rock.

12.4.3 Cut-and-fill stoping (Figure 12.7)

In the most common form of cut-and-fill stoping, mining proceeds up-dip in an inclined orebody. (An alternative, less common, method involves down-dip advance of mining.) The progress of mining is linked to a closely controlled cycle, involving the sequential execution of the following activities:

(a) drilling and blasting, in which a slice of rock, typically about 3 m thick, is stripped from the crown of the stope;
(b) scaling and support, consisting of the removal of loose rock from the stope crown and walls, and the emplacement of lightweight support;
(c) ore loading and transport, with ore moved mechanically in the stope to an ore pass and then by gravity to a lower transport horizon;
(d) backfilling, when a layer of granular material, of depth equal to the thickness of the ore slice removed from the stope crown, is placed on the stope floor.

An important aspect of cut-and-fill stoping is that miners work continuously in the stope, and execute all production activities under the stope crown which is advanced by mining. The success of the method therefore involves achieving

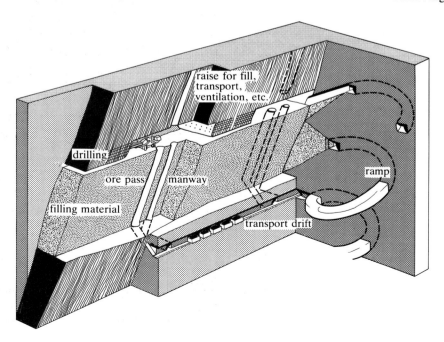

Figure 12.7 Layout for overhand cut-and-fill mining (after Hamrin, 1982).

reliable control of the performance of rock in the immediate periphery of the work area, as indicated in Figure 12.8. This is realised by controlled blasting in the stope crown, application of a variety of local rock support and reinforcement techniques, and more general ground control around the stope derived from the use of backfill.

Cut-and-fill stoping is applied in veins, inclined tabular orebodies and massive deposits. In the last case, the orebody is divided into a set of stope blocks, separated by vertical pillars, geometrically suitable for application of the method. It is suitable for orebodies or stope geometries with dips in the range 35°–90°, and applicable to both shallow and deep orebodies. Orebody or stope spans may range from 4 m to 40 m, although 10–12 m is regarded as a reasonable upper limit. The use of a stope-scale support system (the backfill) renders cut-and-fill stoping suitable for low rock mass strength conditions in the country rock, but better geomechanical conditions are required in the orebody.

Cut-and-fill stoping is a relatively labour-intensive method, requiring that the *in situ* value of the orebody be high. The ore grade must be sufficiently high to accommodate some dilution of the ore stream, which can occur when backfill is included with ore during loading in the stope. On the other hand, the method

Figure 12.8 Layout for shrink stoping (after Hamrin, 1982).

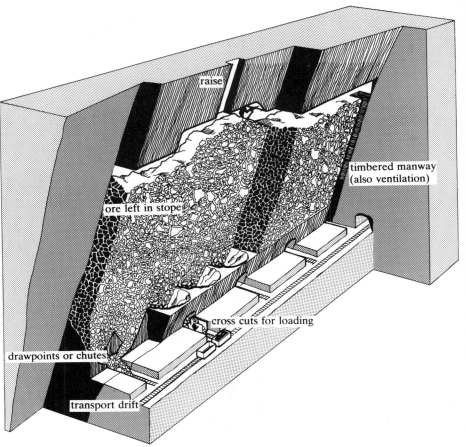

provides both flexibility and selectivity in mining. This permits close control of production grades, since barren lenses may be left unmined, or fragmented but not extracted from the stope. It is also possible to follow irregular orebody boundaries during mining, due to the high degree of selectivity associated with drilling and blasting operations.

Significant environmental benefits of cut-and-fill stoping are related to the use of backfill. In the mine internal environment, close control of rock mass displacements provides a ventilation circuit not subject to losses in fractured near-field rock. Similarly, maintenance of rock mass integrity means that the permeability and hydrogeology of the mine far field may be relatively unaffected by mining. The advantages of the method, for the mine external environment, include limited possibility of mining-induced surface subsidence. Reduction in surface storage of mined wastes follows from possible replacement of a high proportion of these materials in the stoping excavations. De-slimed mill tailings are particularly suitable for backfilling, since the material may be readily transported hydraulically to working areas. This eliminates the need for extra mine development for transfer of backfill underground.

The amount of stope development for cut-and-fill stoping is small, compared with open stoping. The reasons for this are that the production source is also the working site, and some access openings may be developed as stoping progresses. On the other hand, stope block pre-production development may be comparable with that required for open stoping. Cut-and-fill stoping can commence only after developing a transport level, ore passes, sill incline and sill drift, service and access raises or inclines, and return air raises. Openings with these functions can be identified readily in Figure 12.7.

12.4.4 Shrink stoping (Figure 12.8)

The technique of shrink stoping, as implied in Figure 12.8, involves vertical or subvertical advance of mining in a stope, with the broken ore used as both a working platform and temporary support for the stope walls. The method is directly comparable with cut-and-fill stoping, with broken ore temporarily fulfilling some of the functions of backfill.

Shrink stoping operations follow the cyclic pattern of drilling and blasting, ore extraction, and scaling and supporting. Ore is broken in the stope by stripping the crown of the excavation, with miners working directly under the stope crown. To maintain adequate work space in the stope, broken ore must be drawn from the base of the stope after each blast. The amount drawn is related to the swell, or increase in void ratio, which occurs when rock is blasted. Since the void ratio of blasted, displaced rock is about 50–55%, only 30–35% of the freshly broken ore can be drawn after any stope blast. This aspect of the method clearly has production scheduling disadvantages. Once the stope has been mined to its full design height (which may take many months or years), ore is drawn until either the stope is empty or until dilution due to stope-wall collapse becomes excessive.

The orebody type, orientation, geomechanical properties and setting suitable for shrink stoping are virtually the same as described for cut-and-fill stoping. However, in the case of shrink stoping, there are additional restraints on the physicochemical properties of the ore to be satisfied. The orebody rock must be

340

completely inert, with no tendency for oxidation, hydrolysis, dissolution or development of cementitious materials. It must also be strong and resistant to crushing and degradation during draw. All these properties are necessary to ensure that once the ore is mobilised by blasting, it will remain mobile and amenable to free granular flow during its residence time in the stope. Chemical and physical degradation of ore, recementing and binding of ore are all promoted by mine casual water (introduced by drilling, for example), which may percolate through the broken ore mass.

Pre-production development for shrink stoping resembles that for cut-and-fill stoping, except that no ore-pass development is required in the orebody foot wall. Instead, an extraction system must be developed at the base of the stope. This may consist of a slusher drift, with associated finger raises, or drawpoints suitable for mucking with mobile front-end loaders.

12.4.5 Vertical crater retreat (VCR) stoping (Figure 12.9)

VCR stoping is a variation on the shrink stoping procedure, made possible by fairly recent developments in both large-diameter blasthole drilling technology and explosive formulation. Details of the method are provided by Mitchell (1981). It is a non-entry stoping method, the use of long, subvertical blast holes eliminating the need for entry of operating personnel into the developing mine void. Upward advance of the stope crown is effected by a series of cratering blasts using short, concentrated charges in the large-diameter blast holes. After each episode of blasting, sufficient ore is drawn from the stope to provide a suitable expansion void for the subsequent blast.

VCR stoping is applicable in many cases where conventional shrink stoping is feasible, although narrow orebody widths (less than about 3 m) may not be

Figure 12.9 Elevation and section of a VCR stope using long, parallel blast holes (after Hamrin, 1982).

(a) Longitudinal section

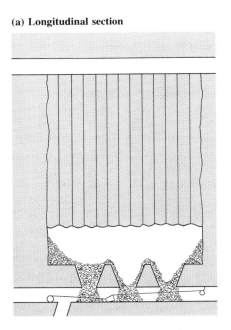

(b) Cross section

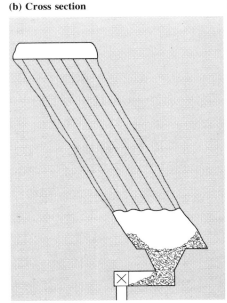

341

tractable. The method is also particularly suitable for mining configurations in which sublevel development is difficult or impossible. These geometric conditions arise frequently in pillar recovery operations in massive orebodies.

12.4.6 Longwall mining

Longwall mining is the preferred method of mining a flat-lying stratiform orebody when a high area extraction ratio is required and a pillar mining method is precluded. The method is applicable to both metalliferous mining in a hard-rock environment and coal mining in soft rock. In both mining situations, the method preserves continuous behaviour of the far-field rock. Different modes of response are induced in the stope near-field rock. For both cases, longwall mining requires an orebody dip of less than 20°, with a reasonably uniform distribution of grade over the plane of the orebody. A high degree of continuity of the orebody is necessary. A particular structural geological requirement is that the throw of any faults transgressing a mining block must be less than the thickness of the orebody.

In hard-rock mining, the stoping method seeks to maintain pseudo-continuous behaviour of the near-field rock mass, although significant fracturing may be induced in the stope peripheral rock. Thus a basic requirement is that the orebody footwall and hangingwall rocks be strong and structurally competent. The schematic representation of the method shown in Figure 12.10 indicates the main elements of the operation. Mining advances along strike by blasting rock from the face of a stope panel. Ore is drawn by scraper down dip into a transport gully,

Figure 12.10 Key elements of longwall mining in hard rock (after Hamrin, 1982).

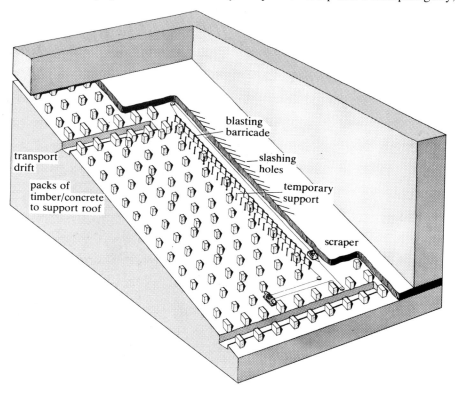

excavated in the foot wall, through which it may be scraped to an ore pass. Temporary support, perhaps provided by yielding props, is emplaced near the mining face, while resilient supports, such as timber and concrete brick packs, are constructed in the void behind the face. Both support measures are designed to prevent the development, near the working area, of discontinuous behaviour in the stope peripheral rock.

The principle of longwall mining in hard rock has been discussed in section 10.8. Considering an isolated longwall stope as a single slot, deflection and closure of the foot wall and hanging wall occur as mining advances. After contact is established between the stope wall rocks, the state of stress around the working area of a single stope is invariant during further stope advance. Stable ground conditions in the working area can then be maintained by emplacement of comparatively light but resilient support in the active mining domain.

Longwall mining in coal seams follows a somewhat different principle from that employed in hard-rock mining. The operations and equipment applied in the method are illustrated in Figure 12.11. The coal is won, and the stope face advanced, by mechanically ploughing or shearing the coal seam by translation of the cutting device parallel to the stope face. This operation simultaneously loads the broken coal on to an armoured conveyor, by which it is transported to the roadway lying parallel to the direction of face advance. All face operations take place within a working domain protected by a set of hydraulic roof supports. Since the load capacity of the powered supports is always small compared with the overburden load, the local performance of roof rock adjacent to mining activity needs to be examined closely.

Figure 12.11 Longwall mining in coal and soft rock (after Hamrin, 1982).

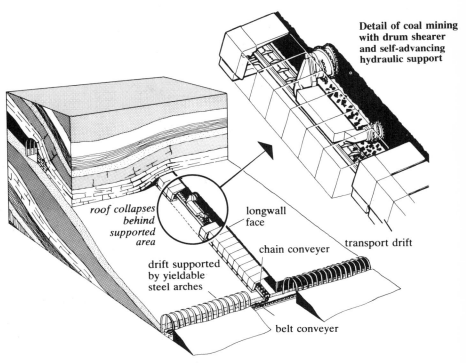

Detail of coal mining with drum shearer and self-advancing hydraulic support

roof collapses behind supported area

longwall face

transport drift

chain conveyer

drift supported by yieldable steel arches

belt conveyer

The geomechanical phenomenology of longwall coal mining, illustrated in Figure 12.11, reflects quite different modes of induced behaviour of the immediate roof rock of the seam and of the main roof stratum. The undermined immediate roof rock for the seam detaches from the hangingwall rock, separates into constituent blocks behind the line of supports, and falls (or caves) into the mined void. The caving process is accompanied by an increase in the void ratio of the displaced rock, from effectively zero, to about 50%. The caved immediate roof occupying the mined void acts as a natural bed of backfill against which the main roof deflects. Therefore the major roles of the near-field rock are to cave and swell to fill the mined void, to restrain the displacement of the main roof, and to maintain the mechanical integrity of the main roof. As the stope face advances, the resultant deflection of the main roof causes consolidation of the caved waste rock. It is the mobilised consolidation pressure in the waste which can maintain pseudo-continuous behaviour of the main roof stratum after relatively large strains have occurred.

Longwall coal mining can be undertaken in geomechanical conditions different from those favouring room-and-pillar mining. As implied by the preceding discussion, the preferred condition is that the immediate roof rock for the coal seam consists of relatively weak shales, siltstones or similar lithologies, with sufficient jointing to promote easy caving. Quite different properties are required of the main roof, which must be sufficiently competent to bridge the span between the mine face and the consolidating bed of caved roof rock. The seam floor rock must have sufficient bearing capacity to support the loads applied by the roof support system at the face line.

A characteristic of longwall mining, common to both coal and metalliferous mining, is that support loads are mobilised in the rock mass by large-scale displacement of pseudo-continuous country rock. Thus although gravitational potential energy is dissipated by restrained displacement of the country rock, the method also results in the accumulation of strain energy in the near-field rock. The method clearly falls between the extremes of fully supported and complete caving methods of mining.

12.4.7 Sublevel caving (Figure 12.12)

This mining method exploits a true caving technique, in that the aim of mining activity is to induce free displacement of the country rock overlying an orebody. Operations in the orebody are undertaken in headings developed at comparatively small vertical intervals. Ore is fragmented using blast holes drilled upwards in fans from these headings. Since the ore is blasted against the caved waste, explosive consumption in blasting is high. This is due to the necessity to consolidate the caved waste as blasting occurs in order to generate an expansion (swell) volume for the fragmenting ore. Ore is extracted selectively, with a front-end loader operating in the drill heading, from the local concentration of fragmented orebody rock contained within the caved waste. As broken ore is extracted at the drawpoint, fragmented ore and enclosing caved waste displace to fill the temporary void. The success of draw, and of the method itself, is determined by the relative mobilities of caved waste and fragmented ore.

The main conceptual components of sublevel caving are indicated in Figure 12.12. Mining progresses downwards in an orebody, with each sublevel

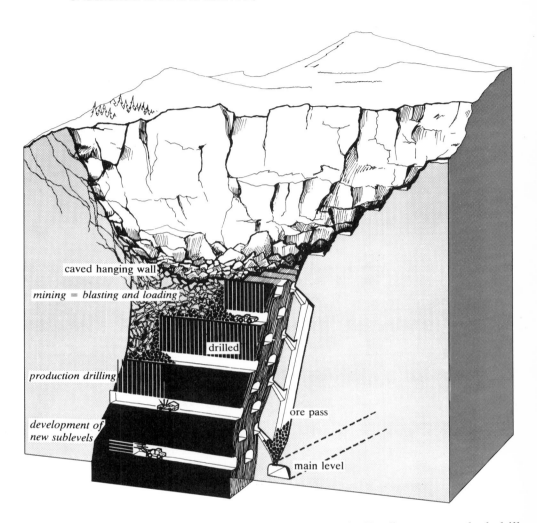

Figure 12.12 Mining layout for sublevel caving (after Hamrin, 1982).

being progressively eliminated as mining proceeds. Headings serve as both drill drifts and transport openings, in which ore is trammed to an ore pass located outside the orebody boundary. Since the gravitational flow of the granular medium formed by the broken ore and caved waste controls the ultimate yield from the orebody, generation of the caving mass and the disposition of the drill headings are the important fundamental and practical aspects of the mining method.

Current industrial application of sublevel caving is limited. It is suitable only for steeply dipping orebodies, with reasonably competent orebody rock enclosed by incompetent overlying and wall rocks. The ore must be of sufficient grade to accept dilution, perhaps exceeding 20%, arising from entrainment of barren country rock in the ore stream. The method produces significant disturbance of the ground surface, imposing some possible limitations on its applicability, from considerations of local topography and hydrology. A general, pragmatic observation is that sublevel caving is declining in industrial popularity, due to the low ore recovery (rarely greater than 65%) and high costs of production. These arise from the relatively high development requirement per tonne produced, and the specific intensity of drilling and blasting required to generate mobile, granular ore within a caving medium. Close control of draw is required to prevent excessive dilution of the ore stream. Finally, geomechanics problems may arise in production headings due to concentration of the field stresses in the effective lower abutment for mining.

12.4.8 *Block caving* (Figure 1.4)

The preceding discussion of sublevel caving indicated that the mining process involved transformation of the *in situ* ore into a mechanically mobile state by drilling and blasting, and subsequent recovery of the ore from a small domain embedded in the caving country rock. In block caving, mobilisation of the ore into a caving medium is achieved without recourse to drilling and blasting of the ore mass. Instead, the disintegration of the ore (and the country rock) takes advantage of the natural pattern of fractures in the medium, the stress distribution around the boundary of the cave domain, the limited strength of the medium, and the capacity of the gravitational field to displace unstable blocks from the cave boundary. The method is therefore distinguished from all others discussed until now, in that primary fragmentation of the ore is accomplished by natural mechanical processes. The elimination of drilling and blasting obviously has positive advantages in terms of orebody development requirements and other direct costs of production.

The geomechanical methodology of block caving entails the initiation and propagation of a caving boundary through both the orebody and the overlying rock mass. The general notions are illustrated in Figure 1.4. At a particular elevation in the orebody, an extraction layout is developed beneath a block or panel of ore which has plan and vertical dimensions suitable for caving. An undercut horizon is developed above the extraction level. When the temporary pillar remnants in the undercut excavation are removed, failure and progressive collapse of the undercut crown occurs. The ore mass swells during failure and displacement, to fill the void. Removal of fragmented ore on the extraction horizon induces flow in the caved material, and loss of support from the crown of the caved excavation. The rock forming the cave boundary is itself then subject to failure and displacement. Vertical progress of the cave boundary is therefore directly related to the extraction of fragmented ore from the caved domain and to the swell of ore in the disintegration and caving process. During vertical flow of rock in the caved domain, reduction of the fragment size occurs, in a process comparable to autogenous grinding.

Block caving is a mass mining method, capable of high, sustained production rates at relatively low cost per tonne. It is applicable only to large orebodies in which the vertical dimension exceeds about 100 m. The method is non-selective, except that high recovery of ore immediately above the undercut horizon is virtually certain. In general, a fairly uniform distribution of values throughout the orebody is required to assure realisation of the maximum ore potential of the deposit.

It has been observed that initial and induced geomechanical conditions in an orebody determine the success of block caving. Productive caving in an orebody is prevented if the advancing cave boundary can achieve spontaneously a mechanically stable configuration, such as an arched crown, or if caved fragment sizes are too large to be drawn through the raises and drawpoints of the extraction system. Details of the mechanics of caving will be discussed in Chapter 15. At this stage it is noted that factors to be considered in evaluating the caving potential of an orebody include the pre-mining state of stress, the frequency of joints and other fractures in the rock medium, the mechanical properties of these features, and the mechanical properties of the rock material. It also appears that the orientations of the natural fractures are important. Kendorski (1978) suggested that initiation and propagation of caving require a well-developed, low-dip joint set (dip less than $30°$). The most favourable rock structural condition for caving is represented by a rock mass containing at least two prominent subvertical joint sets, plus the subhorizontal set.

12.5 Mining method selection

The mining principles and methods which have been described have evolved to meet the geomechanical and operational problems posed in the recovery of ore deposits characterised by a broad set of geological and geometric parameters. A common industrial requirement is to establish the mining method most appropriate for an orebody, or segment of an orebody, and to adapt it to the specific conditions applying in the prospective mining domain. In addition to orebody characteristics which influence method selection, the various mining methods have, themselves, particular operational characteristics which directly affect their scope for application. These operational characteristics include mining scale, production rate, selectivity, personal ingress requirements and extraction flexibility. The final choice of mining method will reflect both the engineering properties of the orebody and its setting, and the engineering attributes of the various methods. For example, a non-selective method such as block caving would not be applied in a deposit where selective recovery of mineralised lenses is required, even if the deposit were otherwise suitable for caving.

It sometimes appears that method selection for a particular mining prospect can present acute technical difficulty. With the exception discussed below, this is not usually the case. In fact, the choice of potential methods of working a deposit is quickly circumscribed, as candidate methods are disqualified on the basis of specific properties of the orebody and its surroundings. It follows that the development of various selection schemes, based on determination of a 'score' which purports to reflect the gross mining characteristics of an orebody, is unnecessary. Such an approach implies that, for an orebody, any mining method is a candidate method. This is clearly at variance with the philosophy and historical development of mining engineering. Mining methods were developed to accommodate and exploit particular mining conditions. A more appropriate procedure to be developed for method selection might involve the formal application of the eliminative logic invoked in computer-based expert systems.

One case in which method selection may present difficulty is that of large orebodies, which are typically prospects for recovery by the two main mass mining methods, block caving and open stoping. In general, block caving would be the preferred method, due to the low labour requirements, low cost per tonne and other favourable engineering aspects. The basic prerequisites to be satisfied are that caving can be initiated in the orebody, and that it will propagate steadily through the orebody as ore is drawn from the extraction level. Prediction of the caving potential of an orebody is not a simple matter, as instanced by complete failure of an ore block to cave in some cases, and the implementation of 'assisted caving' schemes in others. The practical solution to this problem, in the current state of understanding of caving mechanics, may be found in the application of a modified geomechanics classification of the host rock mass for mining.

The most extensive observations of the relation between the geomechanical state of a rock mass and its potential for caving or stoping, have been reported by Laubscher (1977). The classification scheme evolved by Laubscher represents the results of correlation of method performance and rock mass conditions in and around asbestos and gold orebodies in Zimbabwe. Laubscher's scheme is a natural extension of other geomechanics classification techniques, and has been discussed in Chapter 3. The application of the scheme in the selection of a mass mining method, and other aspects of mine planning and design, has been described by Laubscher (1981).

As described in Chapter 3, the Laubscher classification generates a quantitative assessment or index of rock mass properties (a number in the range 0–100), which is used to establish a descriptive class number (class 1 to class 5). Each class therefore represents an index range of 20. Class 1 rock masses are represented by *in situ* conditions of high material strength, low joint frequency, high joint shear strength and low fissure water pressure. The opposite conditions are expressed in class 5 rock masses. From the brief outline of caving mechanics given earlier, it is clear that rock masses in the higher-numbered classes, composed of more fractured and friable media, are more amenable to caving. Laubscher's observations posit a direct relation between class number and performance factors such as the propensity of an ore mass to sustain caving (i.e. cavability), fragment size of the product ore, the need for secondary breaking in drawpoints (which is inversely related to natural fragmentation), and undercut dimension requirements to initiate caving. The last parameter is expressed as an equivalent hydraulic radius, i.e. a ratio of undercut area to undercut perimeter, to take account of the geometry of that excavation. The results of these observations are given in Table 12.1.

An interpretation of the data in Table 12.1 is that block caving qualifies as the preferred mining method for geomechanical classes of 3 or higher. For classes 1 and 2, a method such as open stoping would be preferred. Such conclusions are, of course, subject to modification in the light of specific conditions expressed in a particular mine environment. For example, in the case of marginal caving propositions, in the class range 2–3, account may be taken of the orientation of the joint sets and their influence on caving. The existence of subhorizontal joint sets, to conform to the caving requirement noted by Kendorski (1978), may be a critical factor in deciding in favour of a caving operation in an orebody.

The information in Table 12.1 on undercut dimensions is particularly useful in preliminary assessment of a prospective caving layout. For example, for a caving

Table 12.1 Caving performance of various geomechanical classes of rock masses (after Laubscher, 1981)

geomechanical class	1	2	3	4	5
cavability	not a candidate	poor	fair	good	very good
fragment size	—	large	medium	small	very small
secondary blasting	—	high	medium	small	very small
undercut dimensions (m) (equivalent hydraulic radius)	—	30	30–20	20–8	8

panel, with an undercut excavation square in plan, and a rock mass of class 4, the suggested average equivalent radius of 14 m yields an undercut side length of 56 m. Such simple calculations of undercut design need to be complemented by more detailed analysis involving the specific conditions in the rock mass, such as *in situ* state of stress and rock mass strength. However, they provide a basis derived from engineering experience which can allow some initial decisions on the applicability of caving in the geomechanical conditions prevailing in an orebody.

13 Naturally supported mining methods

13.1 Components of a supported mine structure

A mining method based on natural support seeks to control rock mass displacements throughout the zone of influence of mining, while mining proceeds. This implies maintenance of the local stability of rock around individual excavations and more general control of displacements in the mine near-field domain. As a first approximation, stope local stability and near-field ground control might be considered as separate design issues, as indicated schematically in Figure 13.1. Stopes may be excavated to be locally self-supporting, if the principles described in Chapters 7–9 are applied in their design. Near-field ground control is achieved by the development of load-bearing elements, or pillars, between the production excavations. Effective performance of a pillar support system can be expected to be related to both the dimensions of the individual pillars and their geometric disposition in the orebody. These factors are related intuitively to the load capacity of pillars and the loads imposed on them by the interacting rock mass.

Generation of pillar support elements in an orebody results in either temporary or permanent sterilisation of a fully proven and developed mining reserve. An economic design of a support system implies that ore committed to pillar support be a minimum, while fulfilling the paramount requirement of assuring the global stability of the mine structure. Therefore, detailed understanding of the properties and performance of pillars and pillar systems is essential in mining practice, to achieve the maximum, safe economic potential of an orebody.

Room-and-pillar mining, and the several versions of longhole and open stoping, represent the main industrial implementations of the principle of natural support. In these various methods, clear differences exist between the ways in

Figure 13.1 Schematic illustration of problems of mine near-field stability and stope local stability, affected by different aspects of mine design.

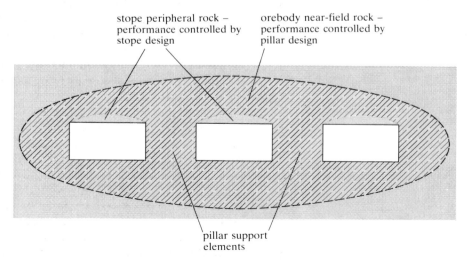

stope peripheral rock – performance controlled by stope design

orebody near-field rock – performance controlled by pillar design

pillar support elements

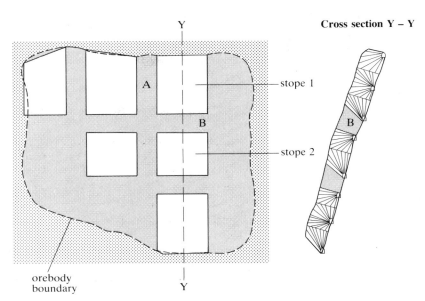

Cross section Y – Y

Figure 13.2 Pillar layout for extraction of an inclined orebody, showing biaxially confined transverse and longitudinal pillars, 'A' and 'B', respectively.

which pillars are generated and the various macroscopic states of loading and confinement applied to individual pillars. For example, pillars in flat-lying, stratiform orebodies are frequently isolated on four sides, providing a resistance to imposed displacement that is essentially uniaxial. Of course, interaction between the pillar ends and the country rock results in non-homogeneous, triaxial states of stress in the body of the pillar, even though it is uniaxially loaded by the abutting rock. A set of uniaxially loaded pillars is illustrated in Figure 12.5. An alternative situation is shown in Figure 13.2. In this case, the pillars generated by open stoping are biaxially loaded by the country rock.

The terminology used in denoting the support mode of a pillar reflects the principal direction of the resistance imposed by it to displacement of the country rock. Each of the pillars illustrated in Figure 12.5 is a vertical pillar. For a biaxially loaded or confined pillar, the direction corresponding to the smaller dimension of loading is used to denote its primary mode of support. The pillar labelled 'A' in Figure 13.2 is a horizontal, transverse pillar, while the pillar labelled 'B' is a horizontal, longitudinal pillar. Pillar 'B' could also be called the floor pillar for stope '1', or the crown pillar for stope '2'. If the longitudinal pillar persisted along the strike of the orebody for several stope and pillar blocks, it might be called a chain pillar.

In the mine structures illustrated in Figures 12.5 and 13.2, failure of pillars to sustain the imposed states of stress may result in extensive collapse of the adjacent near-field rock. If the volume of the unfilled mined void is high, the risk is that collapse may propagate through the pillar structure. In an orebody that is extensive in two dimensions, this possibility may be precluded by dividing the deposit into mine districts, or panels, separated by barrier pillars. A plan view of such a schematic layout is shown in Figure 13.3. The barrier pillars are designed to be virtually indestructible, so that each panel performs as an isolated mining domain. The maximum extent of any collapse is then restricted to that of any mining panel. Obviously, the principles applied in the design of panel pillars will

351

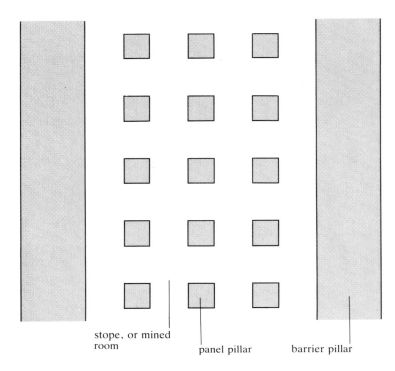

stope, or mined room

panel pillar

barrier pillar

Figure 13.3 Layout of barrier pillars and panel pillars in a laterally extensive orebody.

be different from those for barrier pillar design, due to their different functions. In the following discussion, attention is confined to the performance and design of panel pillars, since their rôle is that most frequently and generally exploited in stoping practice.

13.2 Field observations of pillar performance

The most convenient observations of pillar response to induced mining loads and displacements are made in room and pillar operations, since the method allows direct access to the pillar sites. Detailed measurements and observations of pillar response have been reported, for both non-metal and metalliferous mines, by a large number of authors, including Bunting (1911), Greenwald et al. (1939, 1941), Wagner (1974, 1980), Hardy and Agapito (1977), and Van Heerden (1975). Useful summaries of qualitative and quantitative information on pillar performance are given by Salamon and Munro (1967) and Coates (1981). The following discussion is concerned initially with pillars subject to uniaxial loading.

Stoping activity in an orebody causes stress redistribution, and an increase in pillar loading, illustrated conceptually in Figure 13.4. For states of stress in a pillar less than the *in situ* rock mass strength, the pillar remains intact and responds elastically to the increased state of stress. Mining interest is usually concentrated on the peak load-bearing capacity of a pillar. Subsequent interest may then focus on the post-peak, or ultimate load-displacement behaviour, of the pillar.

352

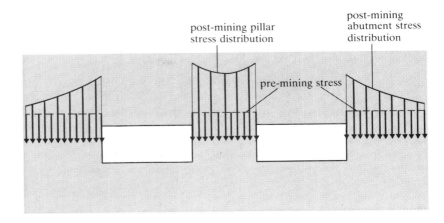

Figure 13.4 Redistribution of stress in the axial direction of a pillar accompanying stope development.

The structural response of a pillar to mining-induced load is determined by the absolute and relative dimensions of the pillar, the structural geology of the pillar rock mass, and the nature of surface constraints applied to the pillar by the country rock. Three main modes of pillar behaviour under stresses approaching the rock mass strength have been recognised, which may be reproduced qualitatively by laboratory tests on model pillars in a displacement controlled testing machine. These failure modes are illustrated in Figure 13.5.

In relatively massive rock, the dominant pillar rupture mode involves spalling from the pillar surfaces, as indicated in Figure 13.5a. Fretting or necking of the pillar occurs. In practice, the generation of small, surface spalls on a pillar may be the first signs of local overstressing in the mine structure. Pil-

Figure 13.5 Principal modes of deformation behaviour of mine pillars.

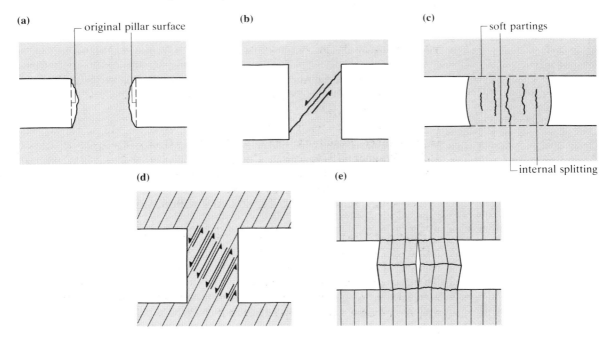

lars in massive rock with low height/width ratio almost invariably fail in this manner.

The effect of pillar relative dimensions on failure mode is illustrated in Figure 13.5b. For regularly jointed orebody rock, a high pillar height/width ratio may favour the formation of inclined shear fractures transecting the pillar. There are clearly kinematic factors promoting the development of penetrative, localised shear zones of this type. Their occurrence has been reproduced in model tests by Brown (1970), under the geometric conditions prescribed above.

The third major mode of pillar response is expressed in an orebody with highly deformable planes of weakness forming the interfaces between the pillar and the adjacent country rock. Yield of the soft layers generates transverse tractions over the pillar end surfaces and promotes internal axial splitting of the pillar. This may be observed physically as lateral bulging or barrelling of the pillar surfaces. Geomechanical conditions favouring this mode of response may occur in stratiform orebodies, where soft bedding plane partings define the foot wall and hanging wall for the orebody. The failure condition is illustrated in Figure 13.5c.

Other specific modes of pillar response may be related directly to the structural geology of the pillar. For example, a pillar with a set of natural transgressive fractures, as illustrated in Figure 13.5d, can be expected to yield if the angle of inclination of the fractures to the pillar principal plane (that perpendicular to the pillar axis) exceeds their effective angle of friction. The amount of slip on the fractures required for yield, and subsequent relaxation of the elastic state of stress in the pillar, need only be of elastic orders of magnitude. A pillar (or other rock remnant in a mine structure) with a well-developed foliation or schistosity parallel to the principal axis of loading will fail in a buckling mode, as illustrated in Figure 13.5e. This mechanism resembles the formation of kink bands.

Field observations of pillars subject to biaxial loading are usually difficult, due to the nature of the mine geometries in which they occur. Limited observations reported by Brady (1975, 1977) suggest that the general modes of response for biaxially confined pillars resemble those for a uniaxially confined state. A similar conclusion is implied in the work by Wagner (1974).

13.3 Tributary area analysis of pillar support

In the analysis and prediction of the behaviour of a set of pillars in a mine structure, the computational techniques described in Chapter 6 could be used for detailed determination of the state of stress throughout the rock mass. However, some instructive insights into the properties of a pillar system can be obtained from a much simpler analysis, based on elementary notions of static equilibrium. These are used to establish an average state of stress in the support elements, which can then be compared with some average strength of the rock mass.

Figure 13.6a shows a cross section through a flat-lying orebody, of uniform thickness, being mined using long rooms and rib pillars. Room spans and pillar spans are w_o and w_p respectively. For a sufficiently extensive set of rooms and

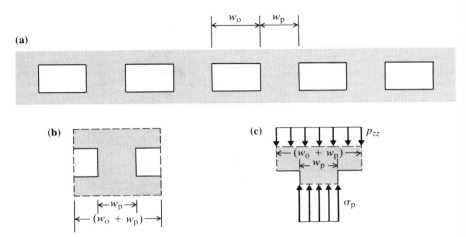

Figure 13.6 Bases of the tributary area method for estimating average axial pillar stress in an extensive mine structure, exploiting long rooms and rib pillars.

pillars, a representative segment of the mine structure is as shown in Figure 13.6b. Considering the requirement for equilibrium of any component of the structure under the internal forces, and unit thickness in the antiplane direction, the free body shown in Figure 13.6c yields the equation

$$\sigma_p w_p = p_{zz}(w_o + w_p)$$

or

$$\sigma_p = P_{zz}(w_o + w_p)/w_p \tag{13.1}$$

In this expression, σ_p is the average axial pillar stress, and p_{zz} is the vertical normal component of the pre-mining stress field. The width $(w_o + w_p)$ of the representative free body of the pillar structure is often described as the area which is tributary to the representative pillar. The term 'tributary area method' is therefore used to describe this procedure for estimating the average state of axial stress in the pillar. A quantity of practical interest in mining an orebody of uniform thickness is the area extraction ratio, r, defined by (area mined)/(total area of orebody). Considering the representative element of the orebody illustrated in Figure 13.6c, the area extraction ratio is also defined by

$$r = w_o/(w_o + w_p)$$

so that

$$1 - r = w_p/(w_o + w_p)$$

Insertion of this expression in equation 13.1 yields

$$\sigma_p = p_{zz}[1/(1 - r)] \tag{13.2}$$

The mining layout shown in plan in Figure 13.7, involving pillars of plan dimensions a and b, and rooms of span c, may be treated in an analogous way. The area tributary to a representative pillar is of plan dimensions $(a + c)$, $(b + c)$, so that satisfaction of the equation for static equilibrium in the vertical direction requires

355

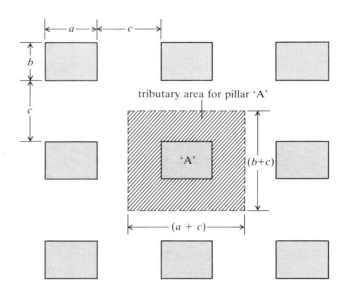

Figure 13.7 Geometry for tributary area analysis of pillars in uniaxial loading.

$$\sigma_p \, ab = p_{zz}(a + c)\,(b + c)$$

or

$$\sigma_p = p_{zz}(a + c)\,(b + c)/ab \qquad (13.3)$$

The area extraction ratio is defined by

$$r = [(a + c)\,(b + c) - ab]/(a + c)\,(b + c)$$

and some simple manipulation of equation 13.3 produces the expression

$$\sigma_p = p_{zz}[\,1/(1 - r)]$$

which is identical with equation 13.2.

For the case where square pillars, of plan dimensions $w_p \times w_p$, are separated by rooms of dimension w_o, equation 13.3 becomes

$$\sigma_p = p_{zz}[(w_o + w_p)/w_p]^2 \qquad (13.4)$$

Of course, average axial pillar stress is still related to area extraction ratio through equation 13.2.

Equations 13.1, 13.3, and 13.4 suggest that the average state of axial stress in a representative pillar of a prospective mining layout can be calculated directly from the stope and pillar dimensions and the pre-mining normal stress component acting parallel to the pillar axis. It is also observed that for any geometrically uniform mining layout, the average axial pillar stress is directly determined by the area extraction ratio. The relation between pillar stress level and area extraction ratio is illustrated graphically in Figure 13.8. The principal observation from the plot is the high incremental change in pillar stress level, for small change in extraction ratio, when operating at high extraction ratio. For example, a change

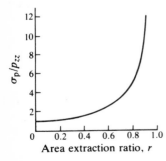

Figure 13.8 Variation of pillar stress concentration factor with area extraction ratio.

356

in r from 0.90 to 0.91 changes the pillar stress concentration factor from 10.00 to 11.11. This characteristic of the equation governing stress concentration in pillars clearly has significant design and operational implications. It explains why extraction ratios greater than about 0.75 are rare when natural pillar support is used exclusively in a supported method of working. Below this value of r, incremental changes in σ_p / p_{zz} with change in r are small. For values of r greater than 0.75, the opposite condition applies.

When calculating pillar axial stress using the tributary area method, it is appropriate to bear in mind the implicit limitations of the procedure. First, the average axial pillar stress is purely a convenient quantity for representing the state of loading of a pillar in a direction parallel to the principal direction of confinement. It is not simply or readily related to the state of stress in a pillar which could be determined by a complete analysis of stress. Second, the tributary area analysis restricts attention to the pre-mining normal stress component directed parallel to the principal axis of the pillar support system. The implicit assumption that the other components of the pre-mining stress field have no effect on pillar performance is not generally tenable. Finally the effect of the location of a pillar within an orebody or mine panel is ignored.

The tributary area method provides a simple method of determining the average state of axial stress in a pillar. Prediction of the *in situ* performance of a pillar requires a method of assessing the strength or peak resistance of the pillar to axial compression. Retrospective analysis of the *in situ* performance of pillars, using the tributary area method to estimate imposed pillar stresses, suggests that the strength of a pillar is related to both its volume and its geometric shape. The effect of volume on strength can be readily understood in terms of a distribution of cracks, natural fractures and other defects in the rock mass. Increasing pillar volume ensures that the defect population is included representatively in the pillar. The shape effect arises from three possible sources: confinement which develops in the body of a pillar due to constraint on its lateral dilation, imposed by the abutting country rock; redistribution of field stress components other than the component parallel to the pillar axis, into the pillar domain; change in pillar failure mode with change in aspect (i.e. width/height) ratio. The second of these factors is, in fact, an expression of an inherent deficiency of the tributary area method.

As noted by Hardy and Agapito (1977), the effects of pillar volume and geometric shape on strength S are usually expressed by an empirical power relation of the form

$$S = S_o v^a (w_p/h)^b = S_o v^a R^b \tag{13.5}$$

In this expression, S_o is a strength parameter representative of both the orebody rock mass and its geomechanical setting, v, w_p and h are pillar volume, width and height respectively, R is the pillar width/height ratio, and a and b reflect geostructural and geomechanical conditions in the orebody rock.

Examination of equation 13.5 might suggest that if strength tests were performed on a unit cube of orebody rock (i.e. 1 m^3, each side of length 1 m), the value of the representative strength parameter S_o could be measured directly. Such an interpretation is incorrect, since equation 13.5 is not dimensionally balanced. Acceptable sources of a value for S_o are retrospective analysis of a set of observed pillar failures, in the geomechanical setting of interest, or by carefully designed *in situ* loading tests on model pillars. The loading system described by Cook, N.G.W. *et al.* (1971), involving the insertion of a jack array in a slot at the midheight of a model pillar, appears to be most appropriate for these tests, since it preserves the natural boundary conditions on the pillar ends.

An alternative expression of size and shape effects on pillar strength is obtained by recasting equation 13.5 in the form

$$S = S_o \, h^\alpha \, w_p^\beta \qquad (13.6)$$

For pillars which are square in plan, the exponents α, β, a, b in equations 13.5 and 13.6 are linearly related, through the expressions

$$a = \frac{1}{3}(\alpha + \beta), \quad b = \frac{1}{3}(\beta - 2\alpha)$$

Salamon and Munro (1967) summarise some estimated values of the pillar strength exponents for square pillars, determined from various sources. The values are presented in Table 13.1.

Table 13.1 Exponents determining pillar strength from its volume and shape (equations 13.5 and 6) (from Salamon and Munro, 1967).

Source	α	β	a	b	Subject medium
Salamon and Munro (1967)	-0.66 ± 0.16	0.46	-0.067 ± 0.048	0.59 ± 0.14	South African coal; *in situ* failures
Greenwald *et al.* (1939)	-0.83	0.50	-0.111	0.72	Pittsburgh coal; model tests
Steart (1954); Holland and Gaddy (1957)	-1.00	0.50	-0.167	0.83	West Virginia coal; laboratory tests
Skinner (1959)	–	–	-0.079	–	hydrite; laboratory tests

Equation 13.6 suggests pillar strength is a simple function of pillar width and height. However, a study reviewed by Wagner (1980) indicated that the operating area (defined by the pillar dimensions perpendicular to the pillar axis) is important. Measurement of the load distribution in a pillar at various states of loading, as shown in Figure 13.9, showed that failure commenced at the pillar boundary and migrated towards the centre. At the stage where structural failure of the pillar had occured, the core of the pillar had not reached its full load-bearing potential. Further, it was proposed that the relative dimensions of the pillar operating area had a substantial influence on pillar strength. This led to definition of the effective width, w_p^e, of a pillar of irregular shape, given by

$$w_p^e = 4A_p/C \qquad (13.7)$$

where A_p is the pillar operating area and C is the pillar circumference.

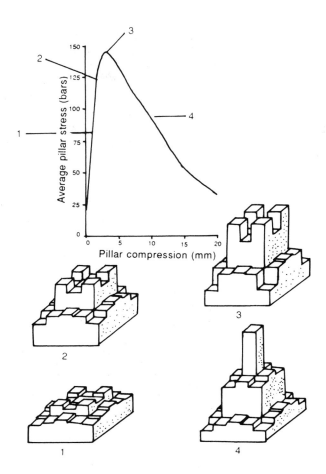

Figure 13.9 Distribution of vertical stress in a coal pillar at various stages of pillar failure (after Wagner, 1980).

In the application of this expression for pillar effective width, pillar strength may be estimated from equations 13.5 and 13.6, with w_p replaced by w_p^e. It is notable that equation 13.7 indicates that, for long rib pillars, with $l_p \gg w_p$, $w_p^e = 2w_p$. This is consistent with the field observation that rib pillars are significantly stronger than square pillars of the same width.

When equation 13.6 is applied to pillars with width-to-height ratio greater than about four or five, pillar strength is underestimated substantially. For these pillars with so-called squat aspect ratios, Wagner and Madden (1984) propose that equation 13.5 can be modified to incorporate terms which reflect more accurately the effect of aspect ratio on strength. The modified pillar strength expression has the form

$$S = S_0 v^a R_0^b \ \{(b/\varepsilon) \ [(R/R_0)^\varepsilon - 1] + 1 \ \} \ , \quad R > R_0 \tag{13.8}$$

In this expression, ε is a parameter with magnitude $\varepsilon > 1$ which describes the rate of strength increase when aspect ratio R is greater than a nominal aspect ratio R_0 at which equation 13.6 is no longer valid. Values suggested for R_0 and ε which lead to conservative estimates of squat pillar strength are 5 and 2.5 respectively.

359

13.4 Design of a stope-and-pillar layout

Design of a supported mining layout should seek to achieve the highest possible extraction ratio of mineral, while assuring locally stable stope spans and general control of near-field rock. In typical design practice, involving irregular stope-and-pillar geometry, it is usually preferable to apply one of the computational methods described in Chapter 6. These may be used to determine stress and displacement distributions associated with various extraction strategies, stope-and-pillar geometries, and stope mining sequences. However, it is useful to explore some general aspects of stope-and-pillar design, and mine layout, using the tributary area method. This is appropriate since there should be a convergence between the outputs of the independent methods of design analysis, for simple geometric conditions in a mine structure. Some broad geomechanical principles of mine layout may then be proposed from these generic studies.

When the tributary area method of stress analysis is used in the design of a mining layout in a flat-lying, stratiform orebody, five parameters are involved in the design analysis. The field stress component, p_{zz}, acting perpendicular to the plane of the rebody is determined by the geomechanical setting. The four variables to be established in the design process are the working or pillar height h, the room or stope span w_o, pillar width w_p, and the factor of safety, F, against pillar failure. Although the following discussion considers square pillars, of side length w_p, it applies equally to long, rib pillars.

As has been noted previously, the stope span which will ensure the local stability of the stope walls can be determined using the design procedures appropriate for isolated excavations. That is, stope span may be established independently of the other design variables.

The selection of an appropriate factor of safety against pillar failure is based upon engineering experience. In his retrospective analysis of the *in situ* performance of South African coal pillars, Salamon produced the data shown in Figure 13.10. The histograms illustrate the frequency distributions of pillar collapse and intact, elastic pillar performance as a function of factor of safety. In

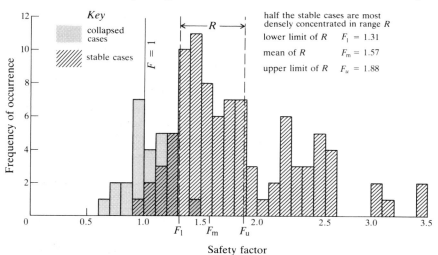

Figure 13.10 Histograms showing frequencies of intact pillar performance, and pillar failure, for South African coal mines (after Salamon and Munro, 1967).

particular, the distribution of intact pillar performance is concentrated in the range of F from 1.3 to 1.9. A reasonable design value of F in this case is suggested to be 1.6. In any other mining setting, a similar approach could be used to establish an appropriate factor of safety.

These observations indicate that the remaining parameters to be determined in the design process are the pillar dimensions, w_p, and the working height, h. At first sight, it may appear that a degree of arbitrariness is involved in the solution to the layout design problem. Consider the following example. A 2.5 m thick horizontal orebody is located at a depth of 80 m, with the rock cover having a unit weight of 25 kN m^{-3}. An initial mining layout is based on 6.0 m room spans and 5.0 m square pillars, with the full orebody thickness of 2.5 m being mined. The pillar strength is defined empirically by the formula

$$S = 7.18h^{-0.66}\, w_p^{0.46} \tag{13.9}$$

where S is in MPa, and h and w_p are in m.

The tributary area analysis of this prospective layout is as follows:

(a) pre-mining stress

$$P_{zz} = 80 * 25\ \text{KPa} = 2.0\ \text{MPa}$$

(b) average axial pillar stress

$$\sigma_p = 2.0 * [(6.0 + 5.0)/5.0]^2\ \text{MPa} = 9.68\ \text{MPa}$$

(c) pillar strength

$$S = 7.18 * 2.5^{-0.66} * 5.0^{0.46}\ \text{MPa} = 8.22\ \text{MPa}$$

(d) factor of safety

$$F = 8.22/9.68 = 0.85$$

The low factor of safety provided by this prospective layout indicates that redesign is necessary to achieve the required factor of 1.6. The options are (i) to reduce the room span, thereby reducing the pillar stress level, (ii) to increase the pillar width, or (iii) to reduce the pillar (and mining) height. Options (ii) and (iii) are intended to increase the pillar strength. The mining geometric parameters can be recalculated, exercising these options, to provide a pillar factor of safety of 1.6. For options (i) and (iii), solutions for the revised stope span and stope height are obtained explicitly. For option (ii), a non-linear equation in w_p is obtained, which may be solved by Newton–Raphson iteration (Fenner, 1974). The following results are obtained from these calculations:

$$\text{Option (i): } w_o = 3.0\,\text{m}, w_p = 5.0\,\text{m}, h = 2.5\,\text{m}$$

$$\text{Option (ii): } w_o = 6.0\,\text{m}, w_p = 7.75\,\text{m}, h = 2.5\,\text{m}$$

$$\text{Option (iii): } w_o = 6.0\,\text{m}, w_p = 5.0\,\text{m}, h = 0.96\,\text{m}$$

Each of the mining geometries defined by the dimensions stated above satisfies the pillar strength criterion. However, the question that remains is – which geometric layout provides the greatest recovery from the orebody? Clearly, the stoping geometry which assures the geomechanical security of the layout and also yields the greatest volumetric recovery of mineral from the deposit, represents the preferred mine excavation design. Option (iii) is immediately unacceptable on the basis of recovery, because it implies leaving mineral in the roof or floor of the orebody over the complete mining area, as illustrated in Figure 13.11c. The choice of pursuing options (i) or (ii), which are both admissible

Figure 13.11 Options in the design of an extraction layout in a 2.5 m thick orebody, to achieve a particular factor of safety for pillars.

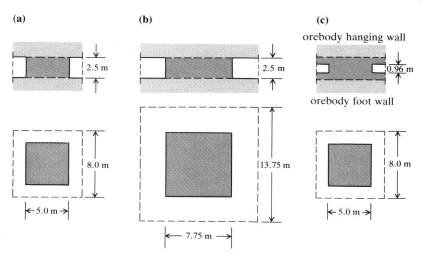

geomechanically, and illustrated in Figures 13.11a, b, can be made on the basis of the volume extraction ratio for the mining reserve. Of course, where pillars are to be recovered in a subsequent phase of the operation, and questions of primary recovery become less critical, the choice between the two options may involve other operational and planning issues. If pillars are to be only partly recovered, the effect of stope and pillar dimensions on volume extraction ratio needs to be considered carefully. This issue has been considered in detail by Salamon (1967), whose analysis is elaborated below. It is particularly apposite in relation to the yield from thick coal seams, or orebodies where rock mass strength is low relative to *in situ* stresses.

The various expressions for pillar strength and pillar axial stress, when taken together, indicate that pillar factor of safety, F, is a function of pillar dimensions, stope span, and pillar height (or orebody stoping width), i.e.

$$F = f(w_p, w_o, h)$$

The objective is to determine the mining dimensions w_p, w_o, h, such that in any single-phase mining operation, the mechanical integrity of the pillar support is assured, and volume extraction ratio is maximised. A graphical, heuristic procedure is used to demonstrate how this objective may be realised.

Figure 13.12 Representation of a partial extraction operation (a) in terms of an equivalent volume, total area extraction (b).

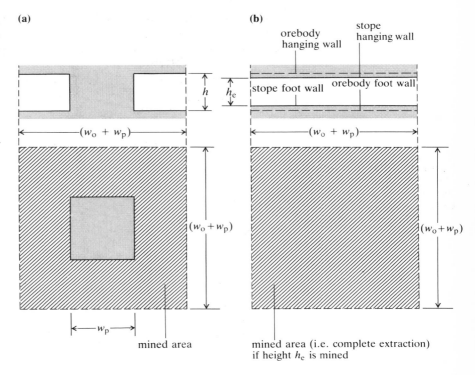

Considering a representative element of a mine structure, shown in Figure 13.12a, the volume V_e of coal extracted from a block of plan dimensions $w_o + w_p$, and working height h, is

$$V_e = h\left[(w_o + w_p)^2 - w_p^2\right] \qquad (13.10)$$

To obtain the same extraction volume V_e by complete extraction over the area of the element, i.e. over $(w_o + w_p)^2$, a stope height h_e, called the equivalent working height, could be mined. This situation is illustrated in Figure 13.12b. The equivalent working height is given by the expression

$$h_e (w_o + w_p)^2 = V_e = h\left[(w_o + w_p)^2 - w_p^2\right]$$

or

$$h_e = h\left[1 - (w_p/(w_o + w_p))^2\right] \qquad (13.11)$$

The yield of a naturally supported operation can be measured conveniently in terms of the equivalent working height, h_e, of an associated, fictitious operation in which extraction is complete over the area of a representative element of the deposit. Thus, any change in stoping geometry which increases the equivalent working height represents an increase in the yield performance of the operation. The effect of varying the mining geometry may be assessed by considering an arbitrarily thick orebody, selecting particular stope spans and working heights,

363

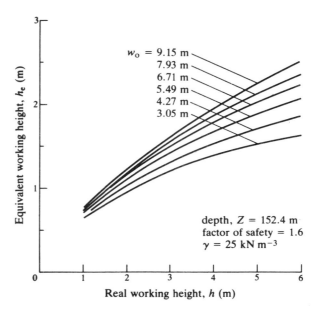

w_o = 9.15 m
7.93 m
6.71 m
5.49 m
4.27 m
3.05 m

depth, Z = 152.4 m
factor of safety = 1.6
γ = 25 kN m^{-3}

Equivalent working height, h_e (m)

Real working height, h (m)

and calculating pillar dimensions (as discussed in option (ii) in the preceding design exercise) to achieve a required factor of safety for the pillar-supported structure. Salamon carried out this type of exercise, for the field conditions of a mining depth of 152 m, and a required factor of safety of 1.6. The results of this exercise are presented in Figure 13.13, in which equivalent working height is shown as a function of actual working height, for selected stope spans. The observation from the plots is that independent increases in stope span w_o and real working height h both lead to increased equivalent working height, and therefore to increased yield from the orebody. The significant engineering inference is that recovery from an orebody may be maximised, while assuring the integrity of the support system, if the following conditions are met simultaneously:

(a) the maximum (i.e. complete) thickness of the orebody is mined;
(b) the maximum room span consistent with assuring local stability of wall rock is mined.

These conclusions may appear self-evident. The reality is that they are a direct result of the nature of the pillar strength formula. A different relation between pillar strength and pillar shape and dimensions could conceivably have led to different geometric requirements for maximising volumetric yield from an orebody.

Having shown how maximum mineral potential of an orebody can be achieved in a pillar-supported operation, it is useful to explore the way in which maximum yield varies with geomechanical setting. The volumetric extraction ratio R can be seen, from Figure 13.12b, to be given by the ratio of equivalent working height h_e and orebody thickness M; i.e. introducing equation 13.11

$$R = h_e/M = h/M \left[1 - (w_p/(w_o + w_p))^2\right] \qquad (13.12)$$

364

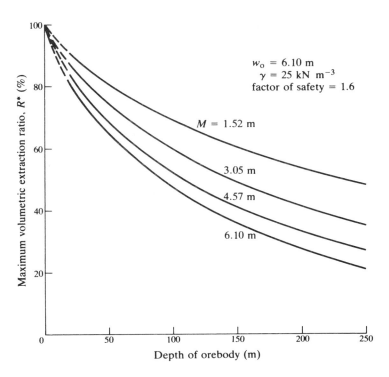

Figure 13.14 Maximum volumetric extraction ratio for various depths of orebody and orebody thickness (after Salamon, 1967).

Consider the hypothetical case of a set of orebodies, of thickness M ranging from 1.5 m to 6.0 m, at varying depths below ground surface. Suppose the maximum stable room span w_o is 6.0 m, the unit weight of the cover rock is 25 kN m^{-3}, and the pillar strength formula is defined by equation 13.8. In each orebody, both the full orebody thickness and the maximum stable room span can be used in the determination of pillar plan dimensions which will yield a factor of safety for pillars of 1.6. (The procedure described for option (ii) of the initial design exercise is employed to calculate w_p.) Since any such stoping geometry will provide the maximum mineral yield, the maximum volumetric extraction ratio, R^*, for the particular case of orebody depth and thickness can be calculated directly from equation 13.12. The results of a set of such computations are presented graphically in figure 13.14.

Two particular observations may be made from Figure 13.14. The obvious one is that, for any orebody thickness, the maximum safe extraction from a pillar-supported operation decreases significantly with increasing depth of the orebody below ground surface. Thus, if one were dealing with a gently dipping orebody, an increasing proportion of the ore reserve would be committed to pillar support as mining progressed down dip. The other observation concerns the low maximum extraction ratio possible with a thick seam or orebody using intact pillar support and a single phase mining strategy. For the 6 m thick seam, at a depth of 244 m below ground surface, the yield from single phase mining is less than 25% of the total mineral reserve.

Some general conclusions regarding supported methods of mining may be formulated from these exploratory studies based on the tributary area method of pillar design. The first is that, if no pillar recovery operations are to be conducted, a pillar layout must be based on the largest stable spans for stopes, to assure maximum recovery of the reserve. Secondly, fully supported methods using intact, elastic pillars, are limited economically to low stress settings, or orebodies with high rock mass strength. Finally, thick seams and orebodies consisting of relatively weak rock masses may be mined more appropriately and productively in successive phases which are themselves based on different design principles, rather than in a single phase of supported mining.

The usual problem in a pre-feasibility study, preliminary design or initial design of a supported mining layout is selection of an appropriate pillar strength formula and of relevant values for a characteristic strength parameter and the scaling exponents. A reasonable approach may be to employ equation 13.5 to estimate pillar strength. The quantity S_o may be replaced by the rock mass uniaxial strength C_o, defined in section 4.9. Values for the exponents a and b might reasonably be taken to be the average of those given in Table 13.1. Improved values for these parameters may then be established as mining progresses in the orebody, by observations of pillar response to mining, or by large-scale *in situ* tests. Judicious reduction in dimensions of selected pillars may be performed in these large-scale tests, to induce pillar failure.

13.5 Bearing capacity of roof and floor rocks

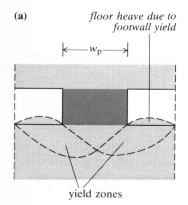

(a)

floor heave due to footwall yield

yield zones

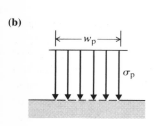

(b)

Figure 13.15 Model of yield of country rock under pillar load, and load geometry for estimation of bearing capacity.

The discussion of pillar design using the tributary area method assumed implicitly that a pillar's support capacity for the country rock was limited by the strength of the orebody rock. Where hangingwall and footwall rocks are weak relative to the orebody rock, a pillar support system may fail by punching of pillars into the orebody peripheral rock. The mode of failure is analogous to bearing capacity failure of a foundation and may be analysed in a similar way. This type of local response will be accompanied by heave of floor rock adjacent to the pillar lines, or extensive fretting and collapse of roof rock around a pillar.

The load applied by a pillar to footwall or hangingwall rock in a stratiform orebody may be compared directly with a distributed load applied on the surface of a half-space. Schematic and conceptual representations of this problem are provided in Figure 13.15. Useful methods of calculating the bearing capacity, q_b, of a cohesive, frictional material such as soft rock are given by Brinch Hansen (1970). Bearing capacity is expressed in terms of pressure or stress. For uniform strip loading on a half-space, bearing capacity is given by classical plastic analysis as

$$q_b = \tfrac{1}{2} \gamma w_p N_\gamma + c N_c \qquad (13.13)$$

where γ is the unit weight of the loaded medium, c is the cohesion and N_c and N_γ are bearing capacity factors.

The bearing capacity factors are defined, in turn, by

$$dN_c = (N_q - 1) \cot \phi$$

$$N_\gamma = 1.5\,(N_\gamma - 1)\,\tan\phi$$

where ϕ is the angle of friction of the loaded medium, and N_q is given by

$$N_q = e^{\pi\tan\phi}\,\tan^2\left[(\pi/4) + (\phi/2)\right]$$

Equation 13.13 describes the bearing capacity developed under a long rib pillar. For pillars of length l_p, the expression for bearing capacity is modified to reflect the changed pillar plan shape; i.e.

$$q_b = \tfrac{1}{2}\,\gamma\,w_p\,N_\gamma\,S_\gamma + c\cot\phi\,N_q S_q - c\cot\phi \qquad (13.14)$$

where S_γ and S_q are shape factors defined by

$$S_\gamma = 1.0 - 0.4\,(w_p/l_p)$$

$$S_q = 1.0 + \sin\phi\,(w_p/l_p)$$

The factor of safety against bearing capacity failure is given by

$$\text{F of S} = q_b/\sigma_p$$

i.e. it is assumed that the average axial pillar stress is equivalently applied as a uniformly distributed normal load to the adjacent country rock. The coarseness of this assumption justifies the practical choice of a factor of safety greater than 2.0.

13.6 The Elliot Lake room-and-pillar mines

The history of mining the uranium-bearing orebodies of the Elliot Lake district of western Ontario, Canada, is interesting because of the evolution of the mining layout and rock response as mining progressed down dip. Rock mechanics aspects of mine performance have been described by Hedley and Grant (1972) and Hedley *et al.* (1984). More than 30 years' observations of roof and pillar performance are recorded for the orebodies.

As described by Hedley *et al.* (1984), the conglomerate stratiform orebodies at Elliot Lake are set on the north and south limbs of a broad syncline. Figure 13.16 is a north–south cross section, looking east, showing the Quirke and Denison mines on the north limb. The orebodies are from 3 m to 8 m thick and dip south at about 15°–20°, persisting to a depth of 1050 m. They are separated from the basement rock by a quartzite bed, and overlain successively by beds of quartzite, argillite, a massive 250 m bed of quartzite, and conglomerate and limestone formations. Although the orebody rock is unbedded, the hangingwall contact is commonly a prominent bedding plane, with an argillaceous parting. Diabase dykes and numerous normal faults transgress the orebodies, and several thrust faults are prominent features. The rock material strengths of the orebody,

367

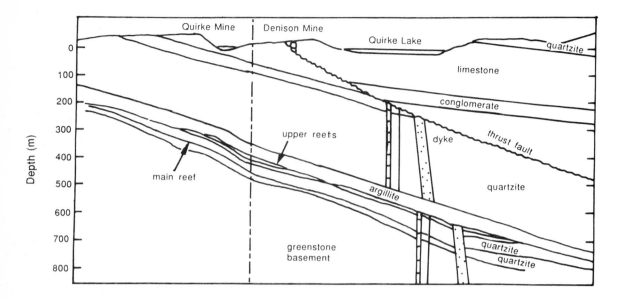

Figure 13.16 North–south vertical cross section (looking east) through the Quirke and Denison Mines (after Hedley *et al.*, 1984).

footwall and hangingwall rocks are generally greater than 200 MPa. The pre-mining state of stress is defined by a vertical principal stress, σ_v, equal to the depth stress, an east–west horizontal stress about σ_v, and a north–south component about equal to σ_v.

At the Quirke Mine, the mining method resembled that shown in Figure 12.2, with transport drifts developed along strike, at 47 m vertical intervals. This resulted in stopes with a down-dip dimension of about 76 m. Crown and sill pillars protected the rail haulages. Rib pillars, instead of the scattered, irregularly

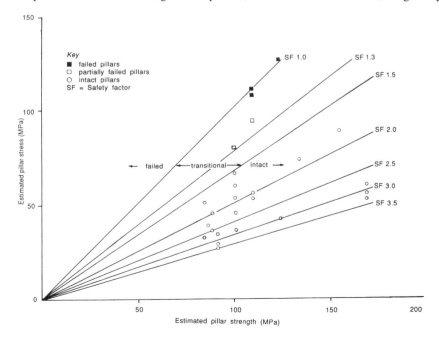

Figure 13.17 Relation between pillar factor of safety and pillar performance (after Hedley and Grant, 1972).

shaped panel pillars shown in Figure 12.2, separated the stopes which were mined up dip from the haulage level.

In the early development of the mining layout, stope spans (i.e. measured in the strike direction), were originally in the range 6 m–30 m, but operating experience of local structural control of roof performance, and local instability, resulted in adopting working spans in the range 15 m–20 m. Pillar design received considerably more attention, due to the relation between pillar dimensions and performance, mine stability and extraction ratio. For the long strike and rib pillars, some early observations of pillar crushing and fretting led to proposal of an expression for pillar strength given by

$$S = 133h^{-0.75} w_p^{0.5} \qquad (13.15)$$

where S is pillar strength in MPa, and w_p and h are pillar width and height in m. When pillar stresses were estimated from the normal component of the field stress acting perpendicular to the plane of the orebody, this expression resulted in the relation between pillar factor of safety and pillar performance shown in Figure 13.17. The conclusion from this is that a factor of safety of 1.5 was required to assure intact performance of pillars.

When the pillar strength expression given in equation 13.15 and a safety factor of 1.5 are used to calculate pillar widths for 3 m and 6 m thick orebodies, the pillar widths corresponding to various depths below surface are as shown in Figure 13.18a. For a mine layout based on stopes extending 76 m down-dip, and rib pillars 23 m apart on strike, the effect on volume extraction is shown in Figure 13.18b. The progressive reduction in extraction ratio required to maintain the required factor of safety of 1.5 would have adverse consequences for mine profitability.

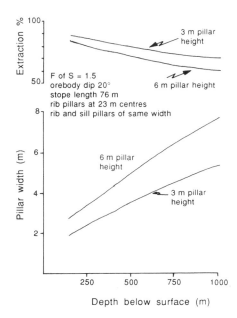

Figure 13.18 Effect of mining depth on pillar width and volume extraction ratio (after Hedley and Grant, 1972).

369

Until 1981, the record of mining at the Quirke Mine was one of satisfactory pillar performance and ground control, with extraction ratio in the range 70–85%, depending on orebody thickness and panel depth below ground surface. At that time, pillars in a flat dipping section of the deposit located 450 m below ground surface, mined about four years previously by a trackless method, began to disintegrate. The trackless mined area is shown in Figure 13.19. The mining sequence in the area involved the extraction of seven level stopes up-dip of the trackless area, the eight level stopes, the trackless area, and then the nine level stopes. Mining of the nine level stopes was in progress when pillar disintegration commenced in the trackless area. Pillar disintegration progressed from this area, involving initially stable pillar failure, and subsequently unstable failure and pillar bursting in the seven level sill pillars.

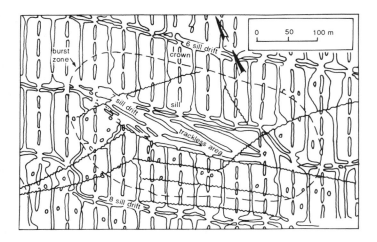

Figure 13.19 Stope-and-pillar layout around the trackless area of the Quirke Mine (after Hedley et al., 1984).

In the analysis of conditions in the collapse area, Hedley et al. found that the dimensions of pillars around the burst-prone and trackless area (3 m wide and from 4.3 m to 6.1 m high) resulted in pillar strengths ranging from 78 to 60 MPa. Other sill and crown pillars in the area were 4.7 m and 3.0 m wide respectively, and 3.0 m high. The respective pillar strengths were 123 MPa and 100 MPa. For an average extraction ratio of 80%, the average pillar stress was 58.5 MPa, providing factors of safety in the range 1.34 to 1.02 in the area of initial pillar collapse, and 2.11 and 1.72 respectively for the sill and crown pillars. Clearly, the safety factor for pillars in the trackless area was far below the value identified in the earlier studies as required to maintain pillar integrity. The evolution of pillar stress with mining sequence shown in Figure 13.20 was determined from a displacement discontinuity analysis. From this, it was proposed that the marginal change in pillar stress associated with progressive down-dip extraction of the nine level stopes was sufficient to initiate pillar failure in the trackless area.

Perhaps the most important principle illustrated by this study is the need to carefully consider orebody thickness, and therefore pillar height, in design of a room-and-pillar layout. Although the area extraction ratio appears marginally greater in the trackless area, the most significant effect on pillar factor of safety was the local increase in orebody thickness and pillar height, causing a marked

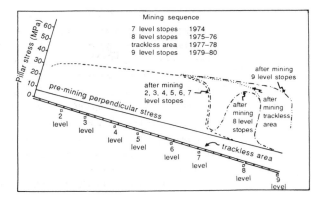

Figure 13.20 Evolution of pillar axial stress during mining sequence around trackless area (after Hedley *et al.*, 1984).

local reduction in pillar strength. Thus, when orebody thickness is variable, stope and pillar design based on maintaining a constant area extraction ratio will not provide assurance of intact pillar performance.

13.7 Stope-and-pillar design in irregular orebodies

The various assumptions and simplifications made in the formulation of the tributary area method for design of pillar mining layouts have already been noted. In spite of these implicit limitations, the method provides a useful technique for a non-rigorous design of pillar support in geometrically regular mineral deposits. Coal seams, stratiform metalliferous orebodies and lenticular deposits are suitable subjects for its application. In practice, many metalliferous orebodies are irregular in shape, and several may occur in close proximity to one another. Metalliferous mining is also characterised by routine recovery of pillar ore. In coal mining, it is frequently necessary to co-ordinate the extraction of several contiguous seams. In all these cases, a more comprehensive and versatile design methodology is required to plan the stope-and-pillar layout and the extraction method itself.

The discussion in Chapter 1 indicated that in the more discerning treatments of mine planning and design, a mine is regarded as a rock structure which evolves geometrically with the progressive extraction of the orebody. The formal design of such a structure requires that account be taken of all geomechanical properties of the mine setting relevant to the prospective performance of stopes and pillars. Application of one of the computational methods described in Chapter 6, in conjunction with proven or inferred rock mass properties, allows prediction of orebody and near-field response to various operationally and temporally feasible mining strategies. Usually, stope excavation sequence also becomes an important component of these studies.

Implementation of this more advanced design practice, necessary for irregular or large orebodies, is based on an observational principle. It involves direct knowledge, interpolation or estimation of geomechanical conditions in the orebody and near-field rock mass. Thus, all the site characterisation information specified for design of an isolated excavation, described in previous chapters, is required for design of the developing mine structure. This includes the local pre-mining triaxial stress field, the *in situ* strength and deformation properties of the lithologic units in the mine domain, and the location, attitude, and mechanical properties of major penetrative structural features transgressing the zone of influence of mining. It is also implied that this information is updated and revised as the rock mass properties are expressed during mining activity, and as mine site properties are revealed by mine development.

Design of a stope-and-pillar layout commences by locating pillars in segments of the orebody inferred to be free of adversely oriented structural features. For example, a pillar which is intended to provide intact, elastic resistance to country rock displacements needs to be located in rock free of features which may allow slip to develop in its interior. Local slip on transgressive features in a pillar may present more pervasive problems. Although theoretically a plastic response mode, penetrative slip may be associated with collapse of large rock units from the pillar surface into adjacent stopes and general degradation of the pillar rock mass.

In close-spaced, parallel orebodies, an additional requirement in siting pillars is alignment of pillar axes. Pillars whose axes are offset may result in a stress distribution which imposes excessive shear stress on planes of weakness oriented parallel to the orebodies. These notions are illustrated in Figure 13.21.

Generation of a suitable extraction scheme for an orebody is an iterative process. Having established suitable locations for the major support elements of a mine structure, a preliminary design for a stope-and-pillar layout can be proposed. Such a layout must be compatible with existing development, yield stope

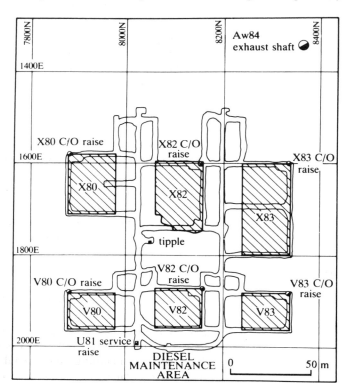

Figure 13.21 Stope-and-pillar layout to maintain favourable states of stress in rock remnants (after Goddard, 1981).

sizes which satisfy mine production requirements, and involve stope-and-pillar dimensions which conform with established notions for maintaining local stability. An analysis of stress distribution in this prospective structure may then be conducted, using a computational technique such as the boundary element method and assuming linear elastic behaviour of the rock mass. In these preliminary design analyses, the assumption of more complex constitutive behaviour for the medium is rarely justified, since the initial problem is to recognise obvious defects or limitations of the design. In addition, the size of the numerical problem may prevent the use of a procedure such as the finite element method, even though this may appear more appropriate conceptually.

Having determined the stress distribution in the prospective structure, it is a simple matter to map regions of tensile stress, or regions in which compressive stresses exceed the rock mass strength. Refinements to the design can then be made until unacceptable states of stress are eliminated from the structure, or some compromise is achieved between geomechanical and other engineering requirements of the mine operation. In the latter case, it may be necessary to accept unfavourable states of stress in particular parts of the structure, in order to create a mine geometry which will sustain independently established working methods and production levels. In either case, some final layout of the mine structure of pillars and stopes may be proposed, for which it is necessary to produce detailed designs of access and service openings, and stope development.

Whatever the expected final state of stress in a mine structure, it is necessary to develop a sequence in which stopes are to be mined, and, perhaps, pillars are to be recovered, which satisfies both geomechanical and general mine engineering constraints. In the case where the integrity of the final mine structure is in doubt, the question of extraction sequence becomes acute. It is therefore reasonable to postulate some general principles which can be applied in the formulation of a logical order for stope-and-pillar extraction.

The obvious mining objective in the development of a stope block extraction sequence is assurance of the recovery of the highest grade blocks in the orebody. This is not to say that a strategy of 'high grading' should be adopted. Instead, the intention is to mine the higher grade blocks completely, in a manner which provides both local and global maximisation of mineral yield.

Geomechanics issues need to be considered in conjunction with the preceding mining objective, in establishing an acceptable mining sequence. The general extraction strategy should involve the early mining of stope blocks with little intrinsic integrity or support potential, and the orderly retreat of mining towards more stable ground. This implies that it is possible to recognise, on the basis of structural geological conditions in the orebody and of an evolving stress distribution in the components of the mine structure, a preferred initial point of attack, a logical development of the stope-and-pillar layout, and an unambiguous order for pillar extraction. In the case of a single orebody, it is frequently possible to achieve such engineering rationality. For multiple orebodies, or a large orebody mined by many stopes and pillars, the complexity of the mine structure and the possible extraction schemes preclude rigorous definition of the optimum scheme in this way. In these cases, it is necessary to propose several extraction sequences which satisfy scheduling and general mine engineering requirements, and also involve the maintenance of an orderly direction of retreat for stoping. The geomechanically preferred scheme may then be established by prediction of rock mass response for each scheme, by analyses of stress in each evolving structure.

Implicit in the development of an extraction sequence for an orebody is the protection of the major installations, external to the deposit, which sustain operations within it. These include ventilation shafts and airways, workshops, haulages, and drives and passes for placement of backfill. Although some of these installations are inevitably damaged or destroyed as mining proceeds, the objective is to maintain their integrity for as long as is demanded by extraction operations. This is accomplished, in the geomechanical assessment of any proposed extraction scheme, by estimation of potential rock mass response around these excavations. The analyses of stress conducted in the sequencing studies may be used for this purpose. Other modes of rock mass response, including rigid body instability of large wedges identified in the zone of influence of mining, or slip and separation on penetrative planes of weakness, must also be considered in a global analysis of the mine structure and its near field.

A final objective in the development of an orebody extraction sequence is to limit the amount of work, by mine personnel, in areas subject to high stress or potential instability. The extraction scheme should seek to avoid the generation of narrow remnants, by such activity as the gradual reduction in the dimensions of a pillar. As an alternative, a large-scale blast involving the fragmentation of pillar ore in a single episode is usually preferable.

13.8 Open stope-and-pillar design at Mount Charlotte

The Mount Charlotte Mine is located on the Golden Mile at Kalgoorlie, Western Australia. Its value as a case study is that it illustrates some of the problems of stope-and-pillar design in a relatively complicated geometric and structural geological setting, and the need to consider appropriate modes of rock mass response in stope and excavation design and extraction sequencing. The following account of rock mass performance is based on that by Lee *et al.* (1990).

The orebody considered in the study by Lee *et al.* was the Charlotte orebody. As shown in Figure 13.22a, the long axis of the orebody strikes north. It is bounded on the north–east by the Flanagan fault, and on the south–west by the Charlotte fault. A fracture system related to the Beta fault trends north–south and dips west at about 45°. The rock materials are strong and stiff, and the rock mass is infrequently jointed. Estimated friction angles for the Charlotte and Flanagan faults were about 20°, and about 25° for the Beta fault.

From beneath the original open-cut mine of the surface outcrop, the orebody had been mined progressively downwards, by open stoping and subsequent pillar blasting. After rib and crown pillars in a block were fragmented in single mass blasts, the broken ore was drawn from beneath coarse granular fill which was introduced at the surface and which rilled into the active extraction zone. The stoping block considered in the study was the G block, located between the 19 and 24 mine levels, which were about 630 m and 750 m below ground surface. The mining sequence for the G block shown in Figure 13.22b indicates large stopes (G1, G2 etc.) and rib pillars beneath a continuous crown pillar. The stopes were 40–80 m wide, 35 m long and up to 100 m high. Rib pillars were 32 m long,

(a)

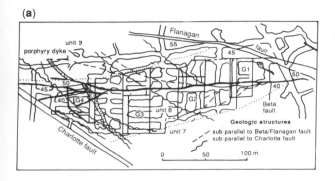

(b)

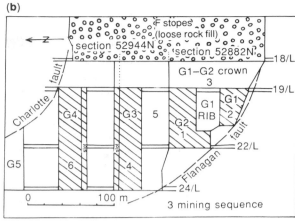

(c)

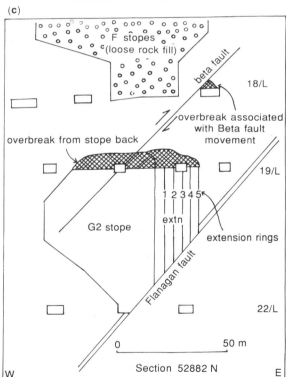

Figure 13.22 (a) Plan view of 19 level of the Mount Charlotte Mine, showing dominant geological structure and stope outlines; (b) longitudinal section through G block, showing stopes, pillars and planned extraction sequence; (c) rock performance during eastern extension of G2 stope (after Lee *et al.*, 1990).

while the crown pillar was 32 m thick. The dimensions of stable stope spans were established from prior experience, while pillar dimensions were determined by analysis intended to limit pillar crushing or spalling at pillar lines.

During the extraction of the eastern extension of the G2 stope, shown in Figure 13.22c, a fault slip rockburst resulted in substantial spalling from the underside of the crown pillar. Although such events had occurred previously in the mine, it was recognised that bursts involving the Beta fault, as this event did, had important implications for extraction of the taller and wider stopes to the north

375

which were transgressed by the same fault. The analysis reported by Lee *et al.* was intended to provide a proven model of rock mass performance in the block, from which the consequences for future stoping could be evaluated.

The analysis of rock performance in the block was conducted in two stages. Preliminary plane strain studies were conducted with the boundary element code

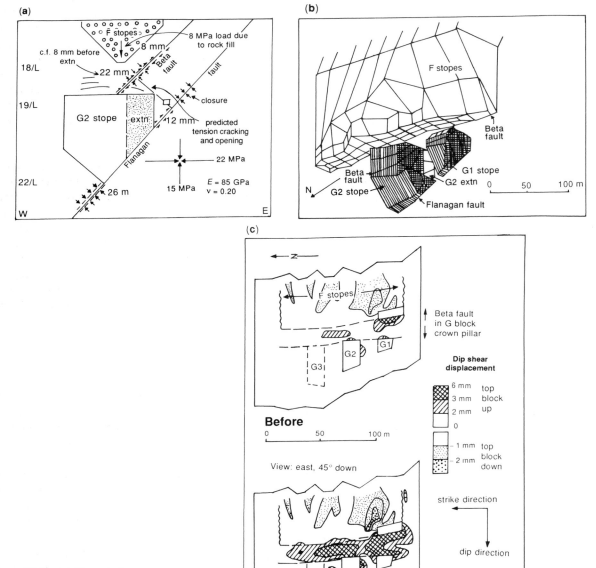

Figure 13.23 (a) Plane strain analysis of G2 stope cross section; (b) problem geometry for three-dimensional analysis; (c) contours of dip slip on the Beta fault before and after extension of G2 stope (after Lee *et al.*, 1990).

BITEMJ (Crotty, 1983), which can simulate slip and separation on faults. These analyses indicated sense and magnitude of slip on the Beta fault consistent with *in situ* measurements, as indicated in Figure 13.23a. However, because the plane analysis was incapable of examining the development of slip along the strike of the Beta fault as stoping progressed, a more comprehensive three-dimensional analysis of stress and displacement was conducted. The problem geometry, as defined for input to the linked boundary element–finite element scheme BEFE (Beer, 1984), is shown in Figure 13.23b. The results of the analyses are illustrated in Figures 13.23c and d, for the state of fault slip before and after mining the eastern extension of the G2 stope. These are contour plots of the magnitudes of dip shear displacement on the Beta fault, mapped on the plane of the fault. It is observed that a substantial increase in the zone and magnitude of slip is indicated over the part of the fault transgressing the crown pillar, attending the mining of the eastern extension of the G2 stope. Such correspondence of observed and calculated rock mass response provided a basis for prediction of rock performance during mining of stopes further north in the block.

Several principles are illustrated by the Mount Charlotte study. First, effective stope-and-pillar design in irregular orebodies or complex structural settings usually requires computational methods for design analysis. Second, two-dimensional analysis has an important rôle in assessing and characterising rock mass performance, before more complicated and expensive three-dimensional methods are employed. Third, field observations combined with appropriate analysis provide a basis for confident prediction of host rock performance as stoping proceeds through a selected extraction sequence. If analysis suggests a particular stope and layout or stoping sequence may induce intolerable rock mass response, a verified mine model similar to that developed for the Charlotte orebody permits easy assessment of alternative layouts and sequences.

13.9 Yielding pillars

The discussion in section 13.4 indicated that when the magnitudes of the pre-mining stresses increased relative to the *in situ* strength of the orebody rock, an excessive proportion of an ore reserve is committed to pillar support. The solution to this problem varies with the type of deposit. For a metalliferous orebody where reserves are limited and the post-peak behaviour of pillars is uncertain, ore in pillars which were initially designed to perform in an intact, elastic mode may be recovered by the extensive use of backfill. This procedure is described more fully in Chapter 14. Where reserves are more extensive, such as coal seams or other stratiform deposits, pillars may be designed to operate in a plastic mode, i.e. at a factor of safety less than unity.

Panel pillars in a yielding mine structure are designed to provide locally resilient support for rock in the periphery of mine excavations. Thus, room and stope spans can be designed on the same principles as applied in a conventional design. The potential for pillar yield can be assessed directly from the rock mass properties which control stable post-peak behaviour. These have been described in section 10.7. The design of the panel pillars must take account of the global stability of the structure, using the techniques discussed previously. Finally, since the ultimate load-bearing capacity of a yielding structure resides in the barrier pillars, these must be designed to be of virtually infinite strength. This implies that the barrier pillars each have a sufficient width/height ratio to create a central core of confined rock capable of sustaining the load shed by the yielding panel pillars. The design of such a pillar is best accomplished using a finite element or finite difference code, so that local yield in the barrier pillar can be incorporated explicitly in the design analysis.

Problems

1 A horizontal stratiform orebody at a depth of 150 m below ground surface is planned for extraction using 6.0 m room spans and pillars 7.0 m square in plan. The full stratigraphic thickness of 3 m is to be mined. The unit weight of the overburden rock is 22.5 kN m^{-3}. Analysis of pillar failures in the orebody indicates that pillar strength is defined by

$$S = 10.44 h^{-0.7} w_p^{0.5}$$

where S is in MPa, and pillar height h and width w_p are in m.

Determine the factor of safety against compressive failure of pillars in the planned layout. If the factor of safety is inadequate, propose a mining layout which will achieve a maximum volume extraction ratio, for a selected factor of safety of 1.6. State the assumption made in this calculation.

2 A flat-lying coal seam 3 m thick and 75 m below ground surface has been mined with 5.0 m rooms and 7.0 m square pillars, over the lower 2.2 m of the seam. Determine the factor of safety of the pillars, and assess the feasibility of stripping an extra 0.6 m of coal from the roof. The strength of the square pillars, of width w_p and height h, is given by

$$S = 7.5^{-0.66} w_p^{0.46}$$

where S is in MPa, and h and w_p are in m.

The unit weight of the overburden rock is 25 kN m^{-3}.

3 The orebody described in Problem 1 above is underlain by a clay shale, for which $c = 1.2$ MPa, $\phi = 28°$, and $\gamma = 22$ kN m^{-3}. If the mining layout is based on 6.0 m room spans and pillars 10.0 m square in plan, determine the factor of safety against bearing capacity failure of the floor rock.

4 A pillar with a width/height ratio of 1.2 is to be subjected to stress levels exceeding the peak rock mass strength. For the elastic range, it is calculated that the pillar stiffness, λ, is 20 GN m^{-1}. The ratio λ'/λ varies with the width/height ratio of the pillar, where λ' is the pillar stiffness in the post-peak regime. For various width/height ratios, *in situ* tests produced the following results:

378

Width/height ratio	1.0	1.33	1.85	2.14
λ'/λ	-0.60	-0.43	-0.29	-0.23

Analysis of convergence at the pillar position for distributed normal loads, P, of various magnitudes applied at the pillar position, yielded the following results:

P(MN)	0.0	125.0	220.0	314.0
Convergence(m $\times 10^{-3}$)	32.0	22.0	14.0	6.0

Assess the stability of the pillar.

14 Artificially supported mining methods

14.1 Techniques of artificial support

An analysis was presented in Chapter 13 of the yield potential of an orebody when mined with a naturally supported method. It showed that under certain circumstances, the maximum extraction possible from the deposit may be unacceptably low. This conclusion applies particularly when the compressive strength of the rock mass is low relative to the local *in situ* state of stress. The discussion in section 13.4 implied that a limited capacity of either orebody rock or its peripheral rock to maintain reasonable unsupported spans also results in limited yield from a deposit. Thus, problems of low yield from naturally supported mine structures can be ascribed directly to geomechanical limitations, either in maintaining the local stability of stope wall rock, or in controlling displacements in the near-field domain.

Artificial support in a mine structure is intended to control both local, stope wall behaviour and also mine near-field displacements. Two main ground control measures are used. Potentially unstable rock near the boundary of mine excavations may be reinforced with penetrative elements such as cable-bolts, grouted tendons, or rock anchors. The principles of this method were introduced in section 11.4. This type of ground control generates a locally sound stope boundary within which normal production activity can proceed. The second, and most widely used, artificial support medium is backfill, which is placed in stope voids in the mine structure. In this case, a particular stoping geometry and sequence needs to be established to allow ore extraction to proceed and the granular fill medium to fulfil its support potential.

Three mechanisms can be proposed to demonstrate the support potential of mine backfill. They are illustrated in Figure 14.1. By imposing a kinematic constraint on the displacement of key pieces in a stope boundary, backfill can prevent the spatially progressive disintegration of the near-field rock mass, in low stress settings. Second, both pseudo-continuous and rigid body displacements of stope wall rock, induced by adjacent mining, mobilise the passive resistance of the fill. The support pressure imposed at the fill–rock interface can allow high local stress gradients to be generated in the stope periphery. This mechanism is discussed in section 7.6, where it is shown that a small surface load can have a significant effect on the extent of the yield zone in a frictional medium. Finally, if the fill mass is properly confined, it may act as a global support element in the mine structure. That is, mining-induced displacements at the fill–rock interface induce deformations in the body of the fill mass, and these are reflected as reductions in the state of stress throughout the mine near-field domain. These three mechanisms represent fill performance as superficial, local, and global support components in the mine structure. The mode of support performance in any instance can be assumed to be related to both the mode of rock mass deformation and backfill properties.

380

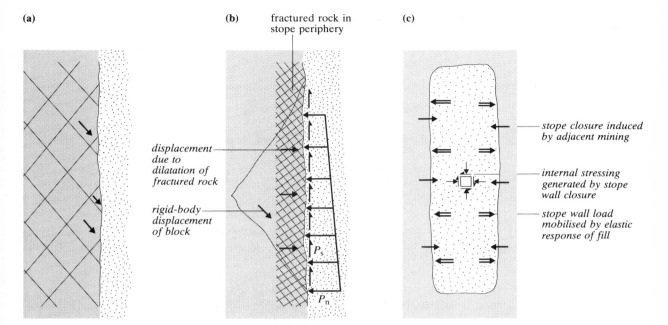

(a) **(b)** fractured rock in stope periphery **(c)**

displacement due to dilatation of fractured rock

rigid-body displacement of block

P_s

P_n

stope closure induced by adjacent mining

internal stressing generated by stope wall closure

stope wall load mobilised by elastic response of fill

Figure 14.1 Modes of support of mine backfill: (a) kinematic constraint on surface blocks in destressed rock; (b) support forces mobilised locally in fractured and jointed rock; (c) global support due to compression of the fill mass by wall closure.

Two mining methods may exploit techniques for reinforcement of stope peripheral rock. The crowns of underhand cut-and-fill stopes are frequently subject to mining-induced fractures, or unfavourable structural geology with associated local instability. The technique of ground control under these conditions is discussed in detail later. The walls of large open stopes are also candidates for local reinforcement, when required by the structural geology of the peripheral rock. There are now many examples of the use of long cable reinforcement of the boundaries of large stopes.

Backfill can be used as a support medium in mining practice in two ways. In conventional cut-and-fill stoping (by overhand or underhand methods), fill is introduced periodically, during the progressive extension of the stope. The operational effectiveness of the fill is related to its capacity to produce a stable working surface soon after its emplacement in the stope. Where backfill is used in an open stoping operation, fill placement in a particular stope is delayed until production from it is complete. Successful performance of the fill mass requires that during pillar recovery, free-standing walls of fill, capable of withstanding static and transient loads associated with adjacent mining activity, can be sustained by the medium. In both cases, the function and duty of the fill mass can be prescribed quantitatively. It is necessary to design the backfill, as one can design any other geomechanical component of the mine structure, to satisfy the prescribed duty rôle.

It has already been observed that mine backfill is frequently a granular cohesionless medium. The height of a fill mass in a stope can exceed several hundred metres. It is also well known that the shear strength of a granular medium is determined directly by the pore-water pressure according to the effective stress law. Therefore, great care must be exercised in fill design and in mining practice to ensure that significant pore pressure cannot develop in a body of backfill. The particular problem is the potential for catastrophic flow of fill under high hydros-

381

tatic head, should high pore pressure lead to complete loss of shear resistance and subsequent liquefaction of the medium. The practical requirement is to ensure that, in the strict soil mechanics sense, any *in situ* loading of backfill occurs under drained conditions.

14.2 Backfill properties and placement

Techniques for preparation of suitable mine backfills, their placement in mined voids, and for mining adjacent to fill masses, have been the subject of extensive investigation for about 25 years. Comprehensive reviews are provided in the proceedings of various symposia (Aust. IMM (Anon), 1973; Can. IMM (Anon), 1978; Stephansson and Jones, 1981) and by Thomas *et al.* (1979). The key elements of the subject are described here.

Materials used as mine backfill or components of a fill mass are of four types: deslimed mill tailings; natural sands; aggregates, development mullock and similar coarse, cohesionless media; cementing agents. A survey of Canadian mining practice (Thomas *et al.*, 1979) indicated that about 76% of all mine fill is derived from mill tailings, and that 84% of all fill is transported and distributed in stopes as an hydraulic suspension. The properties of this so-called hydraulic fill, more appropriately called sandfill, therefore deserve special attention.

14.2.1 Sandfill
Sandfill is prepared from concentrator tailings by hydrocyclone treatment to remove the slimes, or clay-size fraction. Typical products of such a classification process, for various mine fill preparation plants, are given in Table 14.1.

Table 14.1 Size analyses of sandfills prepared from mill tailings (from Thomas *et al.* 1979).

Size fracation (μm)	Weight %					
	Mine 1	Mine 2	Mine 3	Mine 7	Mine 8	Mine 9
+ 3340	0.0	0.00	0.0	0.00	0.07	0.00
2360–3340	0.0	0.00	0.0	0.00	1.09	0.00
1180–2360	0.0	0.00	0.0	0.00	9.01	0.00
850–1180	0.0	0.00	0.0	0.00	8.16	0.13
600–850	0.0	0.00	0.0	0.00	11.00	1.39
425–600	0.7	0.05	0.0	0.20	8.28	6.02
300–425	2.5	0.21	0.0	0.59	6.44	12.86
212–300	8.3	1.62	0.8	3.16	6.37	19.73
150–212	15.7	2.33	3.2	8.69	7.08	18.88
106–150	17.1	11.08	17.2	22.50	5.71	18.33
75–106	20.9	11.67	22.8	17.30	4.79	10.04
53–75	11.0	20.52	13.2	17.53	3.57	4.90
40–53	7.5	9.32	14.9	6.78	1.17	4.04
30–40	7.1	14.25	14.9	15.93	3.36	1.09
20–30	2.4	15.72	8.0	5.14	4.04	0.69
15–20	1.4	7.91	2.2	1.11	4.39	0.43
10–15	0.8	1.32	0.6	0.30	2.88	0.23
–10	4.8	4.01	2.2	0.79	12.60	1.25

It is seen that, in most cases, the highest proportion of the fill product lies in the 40–150 µm range of particle sizes. In soil mechanics terms, this corresponds to a fairly narrow grading of coarse silt and fine sand. On average, the proportion of − 10 µm material is less than 4%, so that, by inference, the clay-size fraction is very low. Natural sands which have been used or considered for use as mine backfill are generally coarser than artificial sands. Particle sizes are concentrated in the range 150–600 µm, corresponding to a medium sand.

The permeability of a fill mass determines the drainage condition it experiences internally under imposed surface load. Investigations by McNay and Corson (1975) indicate that successful sandfills have a permeability coefficient in the range $7 \times 10^{-8} - 7.8 \times 10^{-5} \, \text{m s}^{-1}$. These values correspond to soil gradings of medium silt to coarse sand, which are consistent with the classifications defined by the particle size data.

Sandfill is a cohesionless material, with a purely frictional resistance to deformation. The apparent angle of friction is dependent on the angularity of the particles and the packing density of the medium. Hydraulic placement of a sandfill results in a loose fill structure, with a void ratio of about 0.70. This corresponds to an *in situ* dry unit weight, γ_d, of about $15.7 \, \text{kN m}^{-3}$, or a dry mass density of placed fill of about $1.6 \, \text{t m}^{-3}$. In this state, the peak angle of friction of many artificial sandfills is about 37°. In practice, sandfill at low water content also displays an apparent cohesion, due to suction developed in the pores of the dilatant medium when subjected to a change in boundary load or confinement. This may allow free-standing vertical walls of sandfill, of limited height (perhaps 3–4 m) to be maintained temporarily under some mining conditions.

14.2.2 Cemented sandfill

The lack of true cohesion restricts the scope for mining application of sandfill. This is overcome in practice by the addition of various cementing agents to the sand mass. The obvious choice is Portland cement which, although an expensive commodity, can provide a significant cohesive component of strength at a relatively low proportional addition to the medium. The results given in Table 14.2 indicate the cohesion attained in Portland cement–sandfill mixtures after curing times of 7 and 28 days. The relatively low uniaxial compressive strength determined from these figures (e.g. 5.75 MPa for a 16% Portland cement mix at 28 days) is partly the result of the excess water used in preparing and transporting

Table 14.2 · Some typical strength parameters for cemented sandfill (from Thomas *et al.*, 1979).

Cement content (wt %)	Curing time (days)	Specimens tested	Cohesion c (MPa)	Friction angle, ϕ (deg)
4	7	22	0.13	30
	28	23	0.15	
8	7	24	0.24	33
	28	24	0.31	
16	7	24	1.02	36
	28	24	1.46	
0 (fines added)	205	11	0.03	32
4	207	12	0.21	37

a cemented sandfill mix. The water content of such a mix is always far in excess of that required for hydration of the Portland cement.

The expense of Portland cement as a fill additive has led to its total or partial replacement by other cementing agents. Thomas and Cowling (1978) reported on the pozzolanic properties of such materials as quenched and finely ground copper reverberatory furnace slags. Other materials such as fly ash and iron blast furnace slags are also known to be pozzolanic, and suitable for incorporation in a fill mass to augment the cohesion conferred by Portland cement. In fact, quenched slags, ground to a fineness of $300 \, m^2 \, kg^{-1}$ or greater, may present real advantages as pozzolans. Their slow reactivity may serve to heal any damage caused in the fill mass due to disturbance of the rapid curing Portland cement component of the mix. However, it should be noted that the addition of any fine material to sandfill reduces the permeability of the medium. The design of a fill mix should take due account of the consequences of this reduction.

Sandfill, or sandfill with cementitious additions, is transported underground as a suspension at about 70% solids (i.e. weight of solids/total weight of mix). Flow velocities exceeding about $2 \, m \, s^{-1}$ are required to maintain homogeneous dispersion of the fill components in the slurry. The slurry is discharged into the stope at selected points, chosen to achieve some specified distribution of fill in the mined void. As described in detail by Barrett (1973), segregation occurs after discharge, with the coarser particles settling close to the discharge point, and finer particles transported in the low velocity surficial flow of the transport medium. This leads to cement-lean and cement-rich zones in any horizontal plane through a fill mass. A further degree of heterogeneity of the fill mass arises from the different local settling rates of coarse and fine particles. The low settling rate for the finely ground cementitious additives to sandfill results in the development of a sedimentary structure in the mass, with the top of any fill bed having a high cement content, while the cement content of its base is low.

De-watering is an essential phase of fill placement. Water is removed from newly placed fill by two mechanisms. Following settlement and consolidation of fill solids soon after placement, excess water collects on the fill surface. The provision of vertical drainage facilities in the mass, such as perforated pipes or timbered raises, allows decantation of surface water and its rapid removal at the stope base. Alternatively, surface water may flow through the porous fill bed, and be collected at the stope base in horizontal drains for discharge from the stope. Percolation of the excess water through the mass is clearly dependent on adequate permeability in the vertical direction in the backfill medium.

14.2.3 Rockfill

When backfilling large stopes, the demand for fill material can exceed the available supply, which is limited by the mine production rate and mill capacity. The simultaneous placement in a mine void of aggregate or similar dry rockfill and cemented sandfill can reduce the unit cost of filling the void with cohesive fill. This is achieved by reducing the total amount of cement addition, and extending the capacity of a backfill preparation plant to meet mine demand. The composite fill is placed by discharging cemented sandfill slurry and rockfill into the stope

384

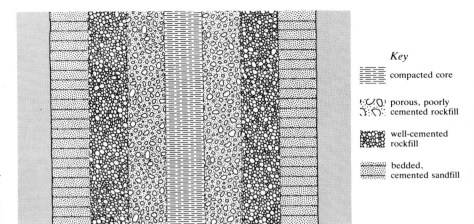

Figure 14.2 Simplified view of structure of a composite rockfill-cemented sandfill mass, due to heterogeneous distribution of fill components (after Gonano, 1977).

Key

━━━ compacted core

porous, poorly cemented rockfill

well-cemented rockfill

bedded, cemented sandfill

simultaneously. The variety of mechanical processes accompanying rockfill placement, including high-velocity impact and compaction, bouncing and rilling, lead to a non-homogeneous fill mass. The structure of a composite of rockfill and cemented sandfill (with a rockfill/sandfill ratio at placement between two and three) has been described by Gonano (1975), and is illustrated schematically in Figure 14.2. The various zones, of different constitution and texture, and various degrees of cemented infilling of rockfill interstices, have different mechanical properties. For example, the zone with a porous, open structure is poorly cemented, with low cohesion. However, such a zone ensures that the interior of the fill mass can drain adequately during fill placement. The development of a highly heterogeneous fill mass, whose structure is controlled by placement conditions, indicates that, in practice, careful attention must be paid to location of discharge points into the stope void. For example, the generation of a poorly cemented zone near the surface of a pillar, which is to be recovered subsequently, would represent a failure in the fill design procedure.

The varieties of compositions and structural domains in cemented rockfill make laboratory determination of their representative properties difficult. Gonano (1975, 1977) described the procedures used for large-scale, *in situ* determination of a well-cemented zone in a composite cemented sandfill/rockfill mass at the Mount Isa Mine, Australia, and comparable tests on a cemented sandfill. The results from these tests are given in Table 14.3. It is observed from these data that the main effect of the rockfill inclusion in the sandfill medium is a significant increase in the cohesion of the mass. Clearly, there could be significant mining advantages if the design of the fill placement system could produce, preferentially, a composite fill of the type tested at locations requiring a high-strength fill mass.

Table 14.3 *In situ* properties of composite backfills (from Gonano, 1977).

Fill type	c' (MPa)	ϕ' (deg)	E (MPa)
8% cemented sandfill (CSF)	0.22	35	285
composite of 8% CSF and rockfill	0.60	35.4	280

385

14.3 Design of mine backfill

Design of backfill for stoping operations must consider two sets of requirements. First, the *in situ* fill mass must meet various criteria for self-support, such as specified stable free-standing, vertical or horizontal (undercut) spans, to permit excavation of adjacent rock. Second, its role as a functional structural element in the mine must be considered. It is possible to design backfill to perform particular functions, such as provision of support for stope boundaries or lateral confinement of pillars to improve their post-peak deformation characteristics.

Design of backfill to assure stable performance under the operating conditions imposed by extraction of adjacent rock involves an analysis of stress and displacement in the fill mass, taking account of fill properties, geometry of the fill and rock structure, and spans of fill exposed by rock excavation. The study by Barrett *et al.* (1978) illustrates the procedures. They showed that, for cemented fill masses, it may be necessary to simulate the stope filling sequence as well as the extraction sequence of the adjacent rock to identify potential failure zones in the fill.

With regard to the structural role of backfill, the report by Blight (1984) describes the interaction of both soft and stiff backfills with mine pillars. It indicates the substantial benefits which can be derived when pillar deformation occurs against the resistance provided by adjacent confined backfill. The way in which both pillar strength and post-peak behaviour are modified by the passive resistance generated in the backfill is illustrated in Figure 14.3a. Although relatively little lateral stress was generated in the soft fill, there was sufficient to maintain a post-peak strength in the model pillar of 85% of the maximum strength. For the stiff fill, there was a threefold increase in peak strength of the pillar attended by the mobilisation of lateral resistance to pillar deformation. To achieve these benefits in practice, it is essential to tight-fill the void, and to place the fill prior to inelastic lateral deformation of pillars, which mobilises the passive resistance of the fill. Further, although stiff backfill appears very attractive, economical practices which permit its routine application have yet to be developed.

Benefits of backfill for ground control in deep underground reef mines are reported by Jager *et al.* (1987). They indicate how fill may be used in place of

Figure 14.3 (a) Effect of soft and stiff fills on model pillar strength and deformation properties (after Blight, 1984); (b) effect of soft fills on pillar post-peak behaviour (after Swan and Board, 1989)

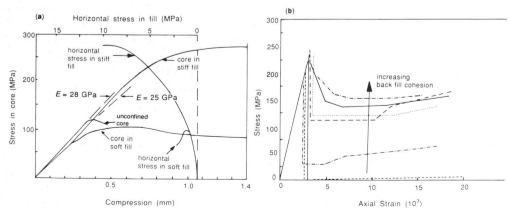

regional stabilising pillars without increasing the energy release rate, and for local support near stope faces and accesses. In these cases, substantial benefit is obtained from soft fill. These matters are discussed in more detail in section 15.2.

Although the structural value of backfill in deep-level mines is readily understood, it is less clear if similar benefit may be achieved in relatively shallow operations where rock displacements may be small. The investigations of fill interaction with pillars reported by Swan and Board (1989) are therefore of considerable interest for pillar and fill design in conventional stope-and-pillar mining methods. They examined the performance of pillars of brittle rock supported by soft fills under conditions of low confinement. Laboratory tests were conducted in a soft testing machine to model the loading conditions experienced by a pillar *in situ*, and numerical simulations were performed for pillars confined by soft fill and subject to strain-softening deformation behaviour.

The results of a series of axial compression tests on model pillars confined by weakly cemented backfill are shown in Figure 14.3b. Three different types of pillar performance are observed: (a) pillars with no residual strength; (b) pillars showing brittle failure but substantial residual strength; and (c) pillars with modified post-peak stiffness and substantial residual strength. In these tests, in no case was peak strength improved by pillar confinement by fill during loading. However, the observed maintenance of a substantial proportion of pillar peak strength in the post-peak range, and modification of the post-peak deformation characteristic to favour stable yield, can contribute to mitigation of pillar burst problems and to mine-scale ground control.

Achievement of pillar performance modes (b) and (c) noted above was found to require two conditions: (1) the fill must be cohesive (i.e. cemented) and (2) no significant void must exist between the fill and the loading platen. Satisfaction of these requirements *in situ* is not straightforward. Original stope design and fill placement practices can have a considerable effect on the distribution of cementing agent in the fill mass, as well as on the effectiveness of achieving backfill contact with the stope crown.

Understanding the mechanics of pillar interaction with soft fill can provide a basis for deriving maximum benefit from cement additions to sandfill. Pillar deformation mechanics under confinement by soft, cohesive backfill are illus-

Figure 14.4 (a) Model of a strike pillar in an inclined orebody; (b) calculated pillar performance for various types of confinement by backfill (after Swan and Board, 1989).

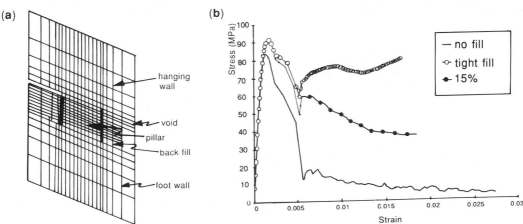

(a)

(b)

387

trated by analysis of a strike pillar in an inclined tabular orebody, shown in Figure 14.4a. Elastic-strain softening constitutive behaviour was assumed for the rock mass, using the procedure described by Cundall and Board (1988). The backfill was simulated as an elasto-plastic material with a Mohr–Coulomb yield criterion and a non-associated flow rule. Two initial porosity conditions for the fill were considered, one corresponding to a need for 15% consolidation prior to development of resistance to hangingwall displacement, the second to full resistance starting from initial deflection of the wall rock.

The load–deformation characteristics for a pillar, calculated for the various conditions of fill confinement of the pillar, are shown in Figure 14.4b. It is observed that the key features of the experimental performance of pillars subject to fill confinement are expressed in the simulations – i.e. limited effect of the soft fill on pillar peak strength, considerable effect on the residual strength, and discernible effect on post-peak stiffness. In a large-scale room and pillar operation in a mine in the Elliot Lake district, it appears that these effects of fill on pillar performance are observed in mine practice (Townsend, 1988). Such correspondence between laboratory experiment, solid mechanics analysis and field observation of pillar performance provides some confidence in a capacity to design backfill to achieve an intended mode of pillar or country rock response.

14.4 Cut-and-fill stoping

Cut-and-fill stoping is performed mainly in relatively narrow, steeply dipping orebodies, where the stope peripheral rock cannot sustain stable, free-standing spans suitable for open stoping. Whether stoping is overhand (up-dip advance) or underhand (down-dip advance), the method involves the incremental development of a slot-shaped excavation, as shown in Figure 14.5. Fill is emplaced in the mined excavation after each increment of sub-vertical stope advance. Because miners work beneath rock surfaces (in overhand stoping) or cemented sandfill

Figure 14.5 Schematic view of a cut-and-fill stope illustrating the development, by mining, of a slot-shaped excavation.

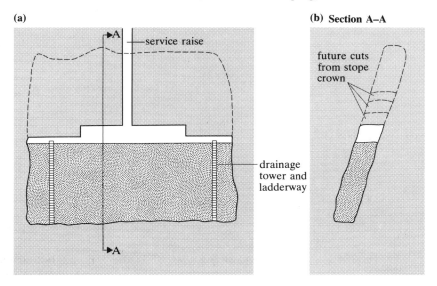

(a)

A

service raise

drainage tower and ladderway

A

(b) Section A–A

future cuts from stope crown

masses (in underhand stoping) which are regenerated with each mining cycle, the success of the method is crucially dependent on effective ground control. The geomechanics of cut-and-fill stoping have therefore been the subject of extensive study. Integration of geomechanics practice in cut-and-fill mining is illustrated in proceedings of a related symposium (Stephansson and Jones, 1981).

Some useful insights into the geomechanics of cut-and-fill stoping can be gained from an elastic analysis of the state of stress around a stope. Of particular interest is the evolution of stresses in the crown and sidewalls of the excavation, during its vertical extension. Following the ideas established in section 7.6, that any conceivable, realistic support pressure at an excavation surface can have only a negligible effect on the elastic stress distribution in a rock mass, it is possible in this exploratory analysis to neglect the presence of the backfill in the mined and filled zone.

The stress distribution can be readily determined around a series of excavations with the geometry illustrated in Figure 14.6, using a plane strain analysis. The stope is taken to have a crown which is semi-circular in cross section. As stoping progresses vertically in the plane of the orebody, the state of stress in the peripheral rock can be related directly to change in relative dimensions (i.e. height/width ratio) of the excavation. Points of particular value in indicating the stope boundary state of stress are point A, in the centre of the stope sidewall, and point B, in the centre of the crown of the excavation. From the discussion in section 7.4 concerning excavation shape and boundary stresses, engineering estimates of boundary stress concentration factors can be obtained from the equations

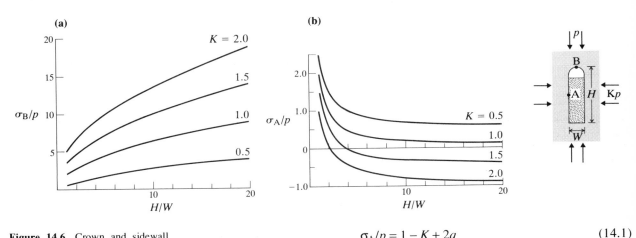

Figure 14.6 Crown and sidewall stresses developed around a cut-and-fill stope at various stope geometries, and for various field stress conditions.

$$\sigma_A/p = 1 - K + 2q \tag{14.1}$$

$$\sigma_B/p = K - 1 + K(2H/\rho_B)^{\frac{1}{2}}$$

and since $\rho_B = W/2$

$$\sigma_B/p = K - 1 + K[2H/(W/2)]^{\frac{1}{2}} = K - 1 + 2K(H/W)^{\frac{1}{2}} \tag{14.2}$$

where $q = W/H$, K is the ratio of horizontal and vertical field stresses, and σ_A, σ_B are boundary stresses at A, B respectively.

389

Equation 14.1 evaluates stope sidewall stress from the shape of the inscribed ellipse, while equation 14.2 evaluates stope crown stress by considering that some local curvature develops in the stope crown. This represents a lower bound estimate of crown stresses. The inscribed ellipse would predict a considerably higher state of stress in the stope crown, and would overestimate the real value.

The results of calculating crown and sidewall stresses, for a range of stope height/width ratios, are given in Figures 14.6a and b. Since the value of K existing naturally in a rock mass, and particularly for settings with sub-vertical mineral veins, is frequently greater than unity, the following conclusions can be drawn from these figures. First, low states of stress, which are frequently tensile, are generated in the sidewalls of the excavation. Since a jointed or fractured rock mass will de-stress and generally disintegrate in a notionally tensile field, it is clear from the calculated sidewall stresses why narrow orebodies are frequently candidates for a fill-based, supported method of mining. The obvious function of emplaced fill is to prevent spatially progressive disintegration, at the low local stresses, of the stope wall rock. This conclusion is supported by more extensive computational analyses of cut-and-fill stoping, including those by Pariseau *et al.* (1973) and Hustrulid and Moreno (1981).

The main conclusion from Figure 14.6 concerns the geomechanical setting in which active cut-and-fill mining occurs, i.e. around the stope crown. At the usual stope height/width ratio and field stress ratios at which mining proceeds, crown stress concentration factors exceeding 10 are to be expected. The implication of this can be appreciated from the following example. Suppose mining is proceeding in a medium where the vertical and horizontal field stresses are 14 MPa and 21 MPa respectively. (This might be about 500 m below ground surface.) Figure 14.6a indicates that at a stope height/width ratio of 10, the stope crown stress would be 140 MPa. Very few jointed rock masses could be expected to have an *in situ* uniaxial compressive strength, C_0, of this magnitude. Local fracturing is indicated in the stope crown. It is also clear that, since the crown stress concentration factor increases with increasing H/W ratio, the geomechanical state of the active mining domain (corresponding to an increase in size of the local fractured region) must deteriorate as stoping proceeds up-dip. This observation probably represents the most important rock mechanics issue in conventional cut-and-fill stoping. It should be noted that the development of a local compressive fracture domain in the crown of a cut-and-fill stope does not automatically imply that crown collapse will occur. However, fracture development certainly increases the possibility of local falls of rock, and this condition may be exacerbated by unfavourable structural geology in the orebody rock. Observations of induced fracture patterns in cut-and-fill stope crowns, and their rôle in crown stability, have been made by the authors at a number of mines.

The picture that emerges from this discussion is one of increasing need for crown support and reinforcement as mining progresses up-dip. At low stope height/width ratio, the stope crown may require little or no support. When the development of induced fractures is expressed as incidents of local instability in the stope crown, rockbolting may be used for securing loose or potentially unstable surface rock. The generation of penetrative fractures across and at depth in the stope crown, and their interaction with the rock structure, may create conditions under which rockbolting cannot assure crown stability. A number of

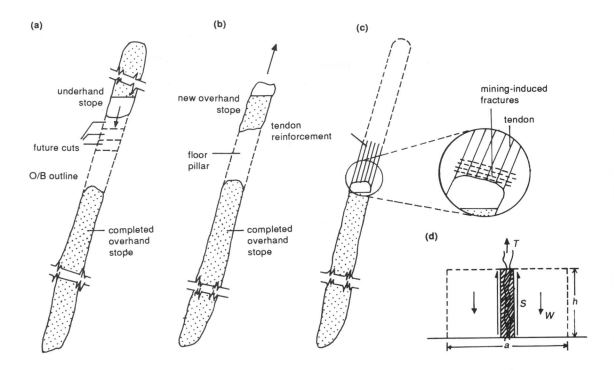

(a)

underhand stope

future cuts

O/B outline

completed overhand stope

(b)

new overhand stope

tendon reinforcement

floor pillar

completed overhand stope

(c)

mining-induced fractures

tendon

(d)

Figure 14.7 Mining strategies for responding to excessive degradation of conditions in the crown of a cut-and-fill stope: (a) abandonment of stope, and underhand stoping from a higher elevation; (b) development of a new overhand stope at a higher elevation, leaving floor pillar; (c) pre-reinforcement of stope crown; (d) free-body diagram for assessment of tendon support design.

mining options exist, which allow further exploitation of the orebody. For example, the overhand stope may be abandoned, and an underhand stope commenced at a higher elevation, as indicated in Figure 14.7a. Alternatively, a pillar may be left above the stope crown and overhand stoping resumed at the higher elevation, in the manner illustrated in Figure 14.7b. The resulting floor pillar might be recovered by some other method subsequently. Finally, a more practical alternative may be to reinforce the stope crown in such a way as to allow mining to proceed even though an extensive fractured zone exists above the active mining domain. This is illustrated in Figure 14.7c.

Large-scale reinforcement is used routinely in many overhand cut-and-fill operations. Reinforcement technology and field practices are discussed in Chapter 11. The function of the cable or tendon reinforcement system is to maintain the integrity of the fractured mass in the crown of the stope. The system acts in a passive mode, so it is necessary to consider the loads mobilised in the reinforcement by the displacements of the host rock. The method of analysis of reinforcement mechanics described in Chapter 11 may be used to design a stope crown reinforcement system. Further, a reasonable check on the design may be based on the ultimate requirement to suspend any potentially unstable rock in the crown of the excavation. This simple procedure may be illustrated by the example shown in Figure 14.7d. Suppose intrascope inspection of holes shows transverse cracking to a depth, h, of 1.5 m into the crown, that the unit weight of the rock mass is 30 kN m^{-3}, and that tendons each with a yield load capacity of 260 kN are to be used. The design of the tendon assembly is such that grout failure produces a shear surface of diameter, d, equal to 50 mm.

For a yield load of 260 kN, and a factor of safety of 1.5, the allowable load per tendon is 173.3 kN. If the tendons are emplaced at $a \times a$ m centres, the weight of rock W to be supported per tendon is

$$W = \gamma a^2 h = 30 \times 1.5 \times a^2 \text{ kN}$$

$$= 173.3 \text{ kN}$$

Hence

$$a = 1.96 \text{ m}$$

This ensures that the load capacity of the tendon can support potentially unstable rock. It is also necessary to demonstrate that the grout annulus passing through the potentially unstable block can support its dead-weight load. If the shear strength of the grout is 1.4 MPa, the maximum shear resistance S of the grout column is given by

$$S = \pi \, dh \times 1.4 \text{ MN} = \pi \times 0.05 \times 1.5 \times 1.4 \text{ MN}$$

$$= 329.9 \text{ kN}$$

The factor of safety against grout column shear failure is therefore 1.90.

The preceding discussion of the geomechanics of cut-and-fill stoping, and of associated ground control practice, took no account of the structural geology of the orebody. When mining in jointed rock, the design of the active mining zone should follow the rules established in Chapter 9, relating to a single excavation in a jointed medium. The particular requirement is that the stope boundary be mined to a shape conformable with the dominant structural features in the medium. Maintaining the natural shape for a stope, with the excavation boundary defined by joint surfaces, restricts the potential for generating unstable wedges in the crown and sidewalls of the active domain.

It was noted in Chapter 12 that shrink stoping can be regarded as a variant of cut-and-fill stoping. At any stage in the upward advance of mining, the broken remnant ore in a stope performs the same role as backfill in cut-and-fill stoping. The performances of crown and sidewalls of cut-and-fill and shrink stopes during mining are also directly comparable. The additional geomechanical aspect of shrink stoping is expressed during the final draw from the stope. Since the stope sidewalls are under low confining stress, or de-stressed, removal of the superficial support applied by the resident, fragmented ore allows local, rigid body displacements to develop in the stope wall rock. If the zone of de-stressing is extensive, or the rock mass highly fractured, draw from the stope can be accompanied by dilution of the ore by caved hangingwall rock.

14.5 Backfill applications in open stoping

Open stoping is a naturally supported mining method, in which control of rock mass displacements is achieved by the generation of ore remnants to form support elements in the orebody. As was observed in Chapter 13, any mining setting in which field stresses are high relative to rock mass strength requires the commitment of a high proportion of the proven mineral reserve to pillar support. In metalliferous mining, where reserves are always limited, the life of a mine may be linked directly to efficient and economical recovery of a high proportion of pillar ore. Because the location of pillars in an orebody is in some way related to the maximum stable stope spans that can be sustained by the orebody peripheral rock, it follows that pillar extraction may introduce the possibility of orebody wall rock or crown collapse. Under these conditions, the need is apparent for artificial support elements distributed in the mine structure during pillar mining, and operating on a scale comparable with that of the natural pillar system. The current position in technically advanced countries is that very little metalliferous mining, undertaken using pillar support, is not accompanied by subsequent stope filling and pillar mining. In general, the stope filling operation in this method of mining is not as closely integrated in mine production activity as it is in cut-and-fill stoping. However, in both cut-and-fill stoping and open stoping with delayed filling and pillar recovery, the support potential of the fill is exploited to achieve a high proportional extraction of the ore reserve.

Although the modern use of backfill is as a structural component in pillar recovery, its application in underground mining evolved from a need for achieving regional ground control above a mining area. According to Dickhout (1973), backfill was first used to control surface displacements above a mining domain in 1864. Much of its subsequent use until recently appears to have been in this rôle, in restricting the scope for sudden, large displacements in the orebody near-field and in controlling the mine internal environment. Of course, these functions are still served by any emplaced fill. Mining under sensitive surface features is frequently permitted by close integration of mining and backfilling. The control of a mine ventilation circuit is also simplified considerably when mined voids are backfilled.

The modern structural function of backfill is to facilitate mining of pillar ore without dilution by waste material or losing control of the orebody near-field rock. This implies that emplaced fill is capable of forming artificial pillars that will prevent the generation of unstable spans of the orebody peripheral rock. Since the Young's modulus of a backfill is always low compared with that of a rock mass suitable for open stoping, the operating mode of the fill is unlikely to involve the global mobilisation of support forces in the fill mass. Instead, satisfactory performance of fill probably involves a capacity to impose local restraint on the surface displacements of rock units in the orebody periphery. This is inferred since the fill is likely to become effective when the wall rock is in a state of incipient instability, and relatively small resisting forces mobilise significant frictional resistance within the mass of wall rock. This model of fill performance suggests that it is necessary to understand local deformation mechanics, in both fill and rock, at the low state of stress existing near the fill–rock interface, if fill design is to be based on logical engineering principles.

In seeking to recover pillar ore under the ground control imposed by backfill, the fill mass must satisfy a range of performance criteria. The primary one is the local resilience of the medium, implied in the previous discussion of fill support mechanics. Although relatively large strains may be imposed near the fill–rock interface, the local resistance and integrity of the fill must be maintained.

In many mining applications an essential property of the fill mass is a capacity to sustain the development of a large, unsupported fill surface. This is illustrated in Figure 14.8, in which a pillar is to be recovered between fill masses emplaced in the adjacent stope voids. Successful mining of the pillar ore might be achieved by the slot and mass blast method illustrated in Figure 14.8b. Extraction of the ore from the stope without dilution requires maintenance of the integrity of the complete fill mass. This in turn demands that the fill has sufficient strength to sustain gravity loads, and also any stresses imposed by displacement of the adjacent stope walls. Possible modes of fill failure leading to dilution of pillar ore include surface spalls and the development of deep-seated slip surfaces. The cohesive fills described in section 14.2 have been formulated specifically to control these modes of fill response.

When the slot and mass blast method, or an alternative vertical retreat method, is used for pillar mining, the result is a large bin filled with broken ore, which is to be drawn empty. Depending on the layout of the drawpoints at the stope base, plug flow of ore may occur past the fill surface. The requirement is that the fill demonstrate sufficient resistance to attrition to prevent both excessive dilution of ore and destruction of the fill pillar by progressive erosion.

The temporal rate of increase of strength may be an important factor in cemented backfill applications, particularly when developed reserves of ore are limited. Strength gain in the curing of cemented fill is determined by the kinetics of hydration and subsequent crystal growth in the reactions of the chemical species in cement and its silicate substitutes. It can be controlled in part by the chemical composition of the cementing mixture (i.e. Portland cement and finely ground, reactive silicates), but it is also related to the thermal and hydraulic setting in which mining occurs. The ability of cemented backfill to undergo autogenous healing, during placement of superincumbent layers of the fill mass, is also determined by the composition of the cementitious materials, and the physical conditions existing during curing.

Pillar recovery adjacent to backfill usually involves detonation of explosive charges in rock close to the fill–rock interface, and questions arise about the effect of the associated dynamic loading on the integrity of the fill mass. Using the terminology adopted in Chapter 10, the difference in the characteristic impedances of rock and backfill is such that very little dynamic strain energy can be transmitted into the fill medium. In fact, this effect should promote the clean separation, during blasting, of rock and fill at their interface. In the case where a blast hole is located close to the interface, i.e. within about 10–15 blasthole diameters, the risk arises of detonation product gases acting directly on the fill surface. There is then an obvious risk of superficial damage to the fill structure.

A key issue in the effective use of backfill for artificial support during pillar mining is complete integration of the initial, primary stoping phase with the subsequent phase, mining of pillar ore. This requires that the general mining strategy be established early in the life of the orebody. It then allows explicit decisions to be made on the proportion of ore to be extracted in the primary stoping phase. This is a question that has to be considered carefully, since, although primary open stoping usually proceeds under conditions of easy mining

(a)

Longitudinal vertical projection

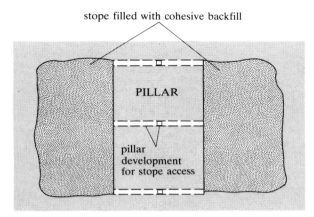

stope filled with cohesive backfill

PILLAR

pillar
development
for stope access

(b)

Longitudinal vertical projection

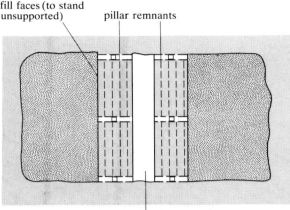

fill faces (to stand
unsupported)

pillar remnants

expansion slot for pillar mass blast

Plan view of orebody at midheight

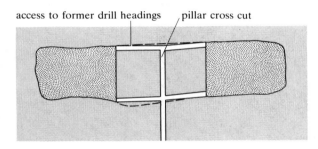

access to former drill headings   pillar cross cut

Plan view of orebody at midheight

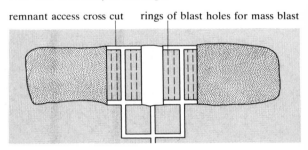

remnant access cross cut   rings of blast holes for mass blast

Figure 14.8 A method of pillar recovery in open stoping, using a cohesive backfill to control stope-wall behaviour.

with little dilution, primary stope voids require stabilised backfill for later mining of pillars. Excessive primary stoping can thereby encumber pillar mining with costs (related to the cement addition to the primary stope backfill) which would render large-scale pillar extraction unprofitable.

Some interesting case studies illustrate the improved ground control and increased recovery of the mineral reserve derived from the use of stabilised backfill and exploiting the principles discussed above. Koskela (1983) describes several open stope and filling operations in Finnish mines. In the Valisaari orebody of the Vihanti Mine, the typical extraction sequence may be inferred from Figure 14.9. A central stope is mined and filled with a backfill obtained from de-slimed concentrator tailings, and stabilised with a pozzolan of finely ground, granulated blast furnace slag activated with slaked lime. The fill mix consists of lime/slag in the proportion 1.5/100, added to sandfill in the slag/sandfill proportion of 1/11. The pulp density of the mix is controlled at 40% water per unit weight of mixture. When the backfill mix is introduced into the centre of the stope from hangingwall fill points, fine material containing a high proportion of slag reports to the stope walls. The centre of the fill mass is therefore weaker than backfill abutting the stope boundaries.

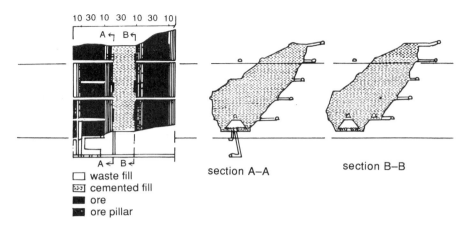

Figure 14.9 Sublevel open stoping and filling in the Valisaari orebody, Vihanti Mine (after Koskela, 1983).

Filling practice involves the construction of fill dams or bulkheads in the original stope accesses and installation of filter and drainage lines. Bulkheads are designed to withstand 0.5 MPa earth pressure on the lower sublevels, and 0.25 MPa in the upper sublevels. Coarse backfill is introduced at the bottom of the stope to provide base drainage, and filter lines are erected from the stope base to each sublevel. The filter lines are made from agricultural drainage pipe.

Backfill prepared and placed in this way has a dry unit weight (γ_d) of about 16.5 kN m^{-3}. One year after emplacement, the *in situ* strength of the mass is 1.05 MPa. The peak strength is achieved in 3 to 5 years. At that time, during adjacent stoping the fill mass may support vertical fill face exposures 100 m high and 60 m wide.

14.6 Reinforcement of open stope walls

Several well-executed investigations have demonstrated the performance and benefit of cable reinforcement of stope boundaries. These include test stopes at the Homestake Mine (Donovan *et al.*, 1984), and several Australian mines (Thompson *et al.*, 1987). The evaluation of several cable bolt reinforcement patterns in stopes at the Pyhasalmi Mine, Finland, reported by Lappalainen and Antikainen (1987) is illustrated in Figure 14.10. The materials and practices used in such large-scale reinforcement have been described in Chapter 11.

In order to assess the relative effectiveness of various stope wall reinforcement designs, the finite difference method of analysis of reinforcement mechanics described in Chapter 11 has been used by Brady and Lorig (1988) to analyse hangingwall reinforcement in an inclined open stope, as shown in Figure 14.11a. The stope resembles that mined in the field reinforcement trial described by Greenelsh (1985). In Figures 14.11c and 14.11d, the designs were based on a constant 150 m of tendon for each reinforcement pattern, so that any differences in performance may reflect the intrinsic effectiveness of the pattern. The design in Figure 14.11e required 200 m of reinforcement.

Assessment of the relative effectiveness of the various patterns is indicated by the degrees of control exercised at the hangingwall surface by the reinforcement. The plots of hangingwall deflection for the reinforcement conditions indicated in Figures 14.11b–e are provided in Figure 14.11f. For the unreinforced hanging-

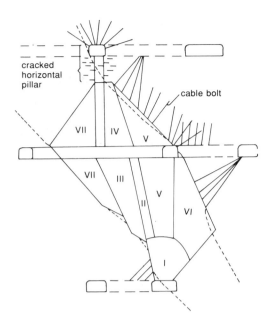

Figure 14.10 Some applications of cable bolts around open stopes (after Lappalainen and Antikainen, 1987).

wall, the analysis showed that the wall rock was de-stressed. The related displacement distribution indicates wall failure, and the uneven shape of the wall deflection curve reflects the five-stage extraction sequence used for the stope. The distributions of wall deflection for the radial reinforcement pattern and the longitudinal reinforcement pattern are similar. Both have achieved substantial control of hangingwall displacement, but instability is inferred near stope midheight. This is consistent with the observed field performance of the stope. The wall deflection plot for the uniform distribution of reinforcement (Figure 14.11e) indicates that this pattern is ineffective in controlling hangingwall displacements.

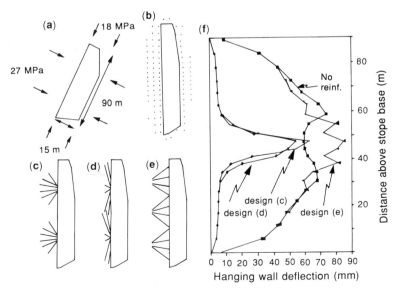

Figure 14.11 (a) Stope design for wall reinforcement study; (b) plastic domain around unreinforced stope; (c–e) three cases for assessment of reinforcement; (f) hangingwall deflections for various reinforcement patterns.

397

Three conclusions are proposed from these analyses. First, the density of reinforcement is quite low in these mining applications, compared with what would be applied around civil excavations subject to boundary instability. Second, the relatively uniform distribution of sparse reinforcement is ineffective in achieving stope wall control. Finally, where reinforcement is necessarily sparse on average, concentration into appropriately located zones may enhance its ground control potential. It is notable that the third conclusion was proposed independently by Lappalainen and Antikainen (1987) from their field observations of reinforcement performance at the Pyhasalmi Mine, providing a measure of confirmation of the reliability of the analysis.

15 Longwall and caving mining methods

15.1 Classification of longwall and caving mining methods

The essential geomechanical features of longwall and caving methods of mining were outlined in Chapter 12. Longwall and caving methods are distinguished from other mining methods by the fact that the near-field rock undergoes large displacements so that mined voids become self-filling. In caving methods, the far-field rock may also undergo large displacements. In increasing order of magnitude of country rock displacement, and in decreasing order of strain energy stored in the near-field rock, the basic mining methods to be considered in the present category are longwall mining in hard rock, longwall coal mining, sublevel caving and block caving.

In this chapter, the geomechanics issues involved in each of these methods of mining will be discussed in turn. If a relevant issue has been dealt with elsewhere, the discussion will not be repeated, but a cross reference will be given. One of the major geomechanics concerns in longwall and caving methods, mining-induced surface subsidence, will not be considered here, but will be the subject of Chapter 16.

15.2 Longwall mining in hard rock

As was noted in section 12.4.6, longwall methods are used to mine narrow, flat-dipping, metalliferous orebodies of large areal extent. The near-field rock is usually strong, and mining often takes place at considerable depth where *in situ* stresses are high. The deep-level mines of South Africa provide the classic example of these conditions. The key elements of longwall mining in hard rock are illustrated in Figure 12.10. A short account of the major variants of this method is given by Haycocks (1973). Some of these variants are described locally as breast stoping. In the present account, longwall mining will be used as a generic term to describe this group of mining methods.

The basic geomechanical objective of the mining and support systems used in this case is to preserve the pseudo-continuous behaviour of the near-field rock. This pseudo-continuous behaviour may be disturbed by two influences. First, natural or mining-induced discontinuities in the rock mass may isolate rock blocks that become free to fall from the hanging wall. This condition is exacerbated by the tendency for tensile boundary stresses to be induced in the near-field rock above the mined-out void. Stone (1978) and Heunis (1980) refer to structurally controlled instabilities in the deep-level gold mines and in the shallower platinum mines of South Africa, respectively. Second, rockburst phenomena can be associated with slip on discontinuities or with the stress-induced fracturing that is commonly observed to occur around longwall faces in high-stress settings.

A rockburst has been defined as the uncontrolled disruption of rock associated with a violent release of energy additional to that derived from falling rock fragments (Cook, N.G.W. *et al.*, 1966). Rockbursts are a sub-set of a broader range of seismic events, and are associated with conditions of unstable equilibrium as defined in sections 10.6–10.9. Mining gives rise to seismic events ranging in energies in the approximate range 10^5 to 10^9 J. Sudden, violent events which might cause considerable damage to workings will radiate not less than about 10^4 J (Salamon, 1983). Rockbursts may have damaging effects on the rock surrounding other mine openings as well as on the rock in the vicinity of a longwall face. The reinforcement system used to limit the effects of rockbursts in a haulage in a South African gold mine was described in section 11.6.3.

The theory of elasticity has been used with outstanding success to develop an understanding of the causes of rockbursts in longwall mining in hard rock, and to develop mining strategies which limit the incidence and effects of rockbursts. As shown in Figures 10.22 and 10.23, the longwall stope is represented as a narrow slot in a stressed elastic medium. The stresses and displacements induced by the creation of a new excavation, or by the extension of a longwall stope, may be calculated most conveniently using one of the forms of the boundary element method outlined in section 6.5. Closed-form solutions may be obtained for some simple problem configurations (Salamon, 1974).

As observed in Chapter 10, it is now generally recognised that there are two basic modes of rock mass instability leading to rockbursts. Fault-slip events resemble natural earthquakes, and usually occur on a mine panel or mine scale. Unstable crushing of a pillar or abutment (sometimes called a strain burst) occurs on a stope or excavation scale. There is clearly a potential for much greater release of energy in a fault-slip rockburst than in a pillar burst due to the larger volume of rock involved. However, in an operating stope, a local pillar or face burst may be as destructive as a large slip on an adjacent fault. Techniques are required to identify mining layouts which may be subject to each type of burst, and to develop preferred extraction layouts and sequences to restrict burst frequency.

The concept of the 'stress drop' on a fault subject to frictional sliding was introduced in Chapter 10. It is defined by $(\tau_s - \tau_d)$, the difference between the limiting static and dynamic shear strengths at the prevailing normal stress, in the transition from static to dynamic conditions on the fault. The average stress drop, τ_e, illustrated in Figure 15.1, has been suggested to be in the range 0.1–10 MPa (Spottiswoode and McGarr, 1975). Stress drops of 5–10% of the static shear strength of a fault have been observed in the laboratory.

Application of notions of stress drop in rockburst mechanics has been discussed by Ryder (1987). It was proposed that the Excess Shear Stress on a fault, defined by the stress drop $(\tau_e = \tau_s - \tau_d)$, may be used as an indicator of the potential for unstable slip on a fault, as it is the forcing function for the motion.

In an analysis of rock mass deformation associated with a major seismic event at a deep gold mine, Ryder (1987) calculated the state of stress on the affected fault using a boundary element method. Excess Shear Stress contours were mapped on to the plane of the fault, and compared with the shear displacements on the fault which attended the event. The plots of Excess Shear Stress and shear displacements shown in Figure 15.2 indicate that the region of maximum fault

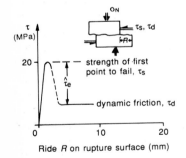

Figure 15.1 Shear stress drop in the transition from static to dynamic conditions on a fault (after Ryder, 1987).

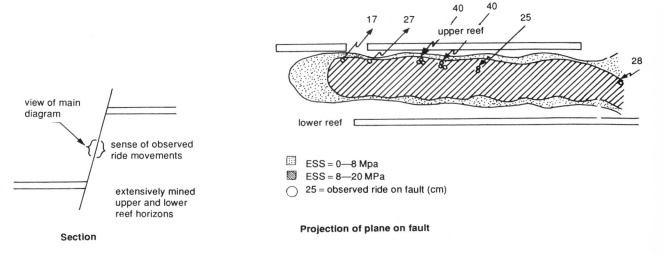

ESS = 0—8 Mpa
ESS = 8—20 MPa
25 = observed ride on fault (cm)

Projection of plane on fault

view of main diagram

sense of observed ride movements

extensively mined upper and lower reef horizons

Section

Figure 15.2 Retrospective analysis of a large seismic event in terms of excess shear stress (after Ryder, 1987).

ride is concentrated in the zone of relatively high Excess Shear Stress. The good correlation between Excess Shear Stress (ESS) and observed fault slip suggests the prospect of predicting conditions under which seismic events may occur from the ESS parameter. However, it may be noted that stress analysis for this purpose should take account of progressive displacement on planes of weakness liable to slip, and not be based on the assumption of elastic continuous rock mass deformation during advance of mining.

It was shown in section 10.5 that for the crushing mode of rock bursting, there is a well-developed correlation between rockburst damage and the spatial rate of energy release or energy released per unit reef area or volume mined. The essential principle used in this case in the design of mining layouts and in excavation scheduling in rockburst-prone areas follows from this observation. This is that the spatial rate of energy release associated with an increment of mining should be kept within predetermined limits. It was also shown in section 10.9 that if a narrow orebody could be mined as a partially closed, advancing single slot, mining would occur under steady-state conditions with a uniform rate of energy release. Generally, it is not practical to mine in this way, and more complex layouts with multiple longwall faces must be used. Under these circumstances, energy release rates can reach critical levels, particularly where the face approaches highly stressed dykes (Figure 3.4), faults or other excavations, or where an attempt is made to mine highly stressed remnants (Cook, N.G.W. *et al.*, 1966). The general strategy of layout planning now used to control rockburst incidence is to limit and to even out energy release rates in space and time.

An early attempt was made to limit the incidence and severity of rockbursts around longwall faces by application of the concept of **de-stressing** (Figure 15.3). Roux *et al.* (1957) report the results obtained when de-stressing was used routinely in 32 stopes at East Rand Proprietary Mines over a 19 month period. Holes 51 mm in diameter and 3 m long were drilled into the face on 1.5 m centres. The bottom half of each hole was charged and the remainder stemmed with sand tamping. As illustrated in Figure 15.3, fracturing the rock ahead of the face in

401

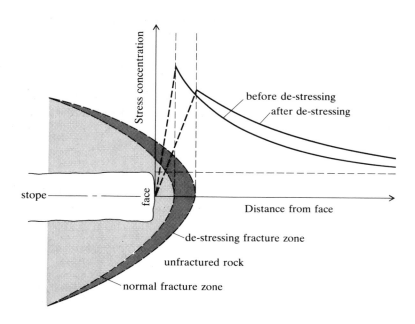

Figure 15.3 Stress concentration ahead of a de-stressed face (after Roux *et al.*, 1957).

this way, successfully reduced the stress level in the rock to be mined next. It was found that the incidence and severity of the rockbursts associated with advance of the face were reduced. Despite these early encouraging results, rock-bursting continued as mining progressed to deeper levels. The 3 m de-stressed zone did not provide a large enough buffer to prevent damage resulting from large bursts occurring close to the face. In order to be fully effective, a de-stressing programme would have to be implemented on a massive and costly scale. De-stressing or rock mass modulus reduction techniques have been evaluated for combating rockbursts in other mining applications, notably in North America (e.g. Cook and Bruce, 1983). However, the persistence of rockbursts in routine mining suggests both the mechanics and practices of de-stressing need further attention in research and field evaluation.

It is now recognised that energy release rate can be controlled most effectively by limiting the displacement of the excavation peripheral rock in the mined-out area. This control may be achieved in several ways.

(a) Provide active support for the hangingwall in the immediate vicinity of the face using hydraulic props, with pack or stick support in the void behind (Figure 12.10). As well as contributing to the limitation of the overall displacements, these forms of support, if sufficiently closely spaced, will prevent falls of rock blocks isolated by natural and mining-induced discontinuities. As noted in section 11.6.4, rapid yielding props can help minimise rockburst damage by absorbing released energy.

(b) Practise partial extraction by leaving regularly spaced pillars along the entire length of longwall stopes, generally oriented on strike. An elastic analysis by Salamon (1974) shows that if a large area of a flat-lying, narrow, tabular orebody is mined to a height, H, the difference in the quantity of energy released per unit length of stope by extraction of a single panel of span, L,

402

and that released by partial extraction in a series of panels of span, l, spaced on centres of S (Figure 15.2a), is

$$\delta W_r = pL \ \{ \ H + [\ (1 - v) \ Sp/\pi G] \ \ell n \ (\cos \alpha) \ \}$$ (15.1)

where p is the vertical *in situ* stress and $\alpha = \pi l/2S$. It was shown previously that, for a span greater than the critical, the quantity of energy released per unit of stope by total extraction is given by equation 10.88; i.e.

$$W_r = LHp$$

Dividing equation 15.1 by equation 10.88 and substituting the expression for the critical span given by equation 10.86; i.e

$$L_0 = GH/[(1 - v) \ p]$$

leads to the result

$$\delta W_r/W_r = 1 + (S/\pi L_0) \ \ell n \ (\cos \alpha)$$ (15.2)

Figure 15.4b shows a plot of $\delta W_r/W_r$ against extraction ratio, l/S, for varying values of S/L_0. Note that, even for quite high extraction ratios, major reductions in energy release rates are achieved by using partial extraction.

The use of stabilising pillars as they are known has been effective in reducing seismicity in a number of deep South African gold mines (e.g. Hagan, 1988). Details of one such example are given in section 18.3.2. The major disadvantage of the method is that in deep mines with potentially high stress conditions something like 15% of the ore reserves may be sterilised in stabilising pillars. For high extraction ratios, the pillars may be subject to particularly high stress concentrations at their edges and may even suffer bearing capacity failure.

Figure 15.4 (a) Mining plan, and (b) normalised reduction in energy change, for partial extraction of a flat-lying, narrow, tabular orebody (after Salamon, 1974).

(a)

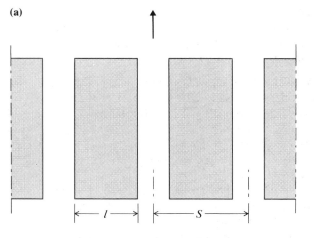

(b)

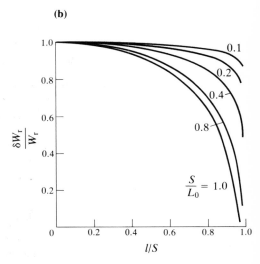

(c) Backfilling the mined void with tailings, sand or waste rock has great potential for limiting convergence and for providing both regional and local support. Backfill has two significant virtues. First, its presence reduces the permissible convergence, limiting both the ultimate convergence volume and the displacement induced at each increment of mining. Second, the rock mass must do work on the fill to deform it; in the process large amounts of energy can be absorbed (Salamon, 1983). Although the benefits likely to be associated with the use of backfill were recognised in the 1960s (Cook, N.G.W. *et al.*, 1966), the method was not adopted in practice because of apparently high costs and perceived operational problems. In particular, it was believed that it would be difficult to achieve the fill stiffnesses required if backfilling was to be effective in limiting convergence and reducing energy release rates sufficiently (Heunis, 1980). As the result of a major research and development programme undertaken in the 1980s, many of these problems have been overcome and a number of South African mines now have backfill systems in operation.

Computer simulations carried out by Jager *et al.* (1987) and others suggest that backfill alone can provide the regional support required at depths of less than 3 km. Below 3 km the major benefits of backfilling are to reduce the stresses acting on stabilising pillars or to enable the spans between pillars, and hence the extraction ratios, to be increased. Improvements recorded in those mines using backfilling include improved access and hangingwall conditions, reductions in stoping width and hence in dilution, reductions in rockburst damage to stope face areas when the backfill is kept close to the stope face, decreases in the numbers of accidents caused by rock falls and improved productivity. This initial experience has shown that backfill should be placed at as low a porosity and as close to the face as possible. Backfilling should be used in conjunction with a good temporary face support system and should be incorporated into the standard mining cycle (Jager *et al.*, 1987).

15.3 Longwall coal mining

15.3.1 Basic geomechanics considerations

The elements of the longwall method of coal mining are illustrated in Figure 12.11. As with longwall mining in hard rock, the primary objective of mining design is to achieve pseudo-continuous deformation of the main upper strata overlying the seam. The differences between the layouts and the mining and strata control techniques used in the two cases arise essentially from the appreciably lower strengths of coal measures rocks. In the United Kingdom, for example, the uniaxial compressive strengths of the roof, seam and floor rocks are typically in the range 20–40 MPa. The pre-mining stress field is difficult to measure in these weak, sedimentary rocks, but back analyses based on observed excavation performance suggest that, in the UK coal fields, the stress field is approximately hydrostatic with magnitude $p = \gamma h$, where γ is the weighted average unit weight of the superincumbent strata and h is the mining depth.

Because of the low strengths of the coal measures rocks, it is found that fractured or yielded zones develop around openings made at depths greater than 100–200 m. This influences the distribution of stress and displacement around the mining excavations, and imposes stringent requirements on support or strata control measures. These measures fall into five distinct categories, each with different objectives:

(a) **Face supports** are required to prevent the gravity fall of blocks of rock detached from an arched zone immediately behind the face and so maintain safe working conditions and offer protection to face equipment. The face support system may also aid in controlling the development of caving behind the face.

(b) **Support of the mined out void**, usually by the stowing of solid material, may be introduced to limit the total displacement of the superincumbent strata and the associated surface subsidence.

(c) **Roadways** which provide access to the face for men and materials and transportation routes for the mined coal, must be formed and supported in such a way that displacements are kept within operationally acceptable limits.

(d) **Rib pillars** may be left between adjacent longwall faces to protect an existing roadway from the excessive displacement that can be associated with the mining of an adjacent face, to isolate a particular panel where unfavourable geological structures or fire, water or gas hazards exist, or to control surface subsidence by practising partial extraction.

(e) **Rockbolts or dowels** may be used to reinforce shafts, roadways, the coal-face or advance headings. In the UK, the dowels used at the coal face are 1.2–2.4 m long wooden rods, grouted in place with resin and hardener pre-packed in capsules. Wood is used since it can be cut by the coal shearer as it passes along the face. Wooden dowels are also used for reinforcing roadway floors.

Following consideration of the redistribution of vertical stress associated with longwall extraction of coal, face support, roadway formation and support, and rib pillar design will be discussed in turn. The related question of surface subsidence resulting from longwall mining is considered in detail in section 16.5.

15.3.2 Distribution of vertical stress around a longwall face

Figure 15.5 shows a widely accepted conception of the distribution of vertical stress, σ_{zz}, around a single longwall coal face. The vertical stress is zero at the face and at the rib side. The stress rapidly increases with distance into the yield zone in the unmined coal, producing a peak stress at a distance into the coal which varies with rock properties. The peak stress so produced is in the order of four to five times the overburden stress, $p = \gamma h$, where h is the mining depth and γ is the weighted average unit weight of the superincumbent strata. With increasing distance into the unmined coal, the vertical stress reduces towards the overburden stress.

In the mined-out area, the effects of arching and of face and roadway support are such that the vertical stresses are negligibly small immediately behind the face and near the rib side. As shown in sections $Y - Y$ and $X_2 - X_2$ in Figure 15.5,

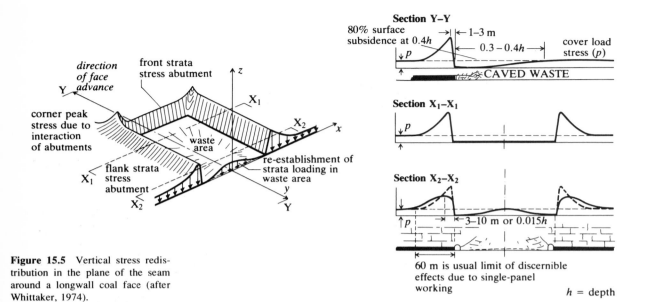

Figure 15.5 Vertical stress redistribution in the plane of the seam around a longwall coal face (after Whittaker, 1974).

the vertical stress increases with distance from the face and the rib side. Where full closure of the mined void occurs, the full overburden stress, p, is transmitted.

Finite element calculations made by Everling (1973) gave results which are not in accord with one feature of Figure 15.5. The high corner peak stress due to abutment interaction postulated by Whittaker (1974) was not reproduced. Rather, the maximum vertical stress calculated at the corner for a single face was lower than the stresses on the side and front abutments. A higher corner peak stress occurred only at the corner between a mined-out panel and an adjacent current longwall face. Park and Gall (1989) performed a more advanced three-dimensional finite element analysis of the stresses around a particular longwall face. Progressive yield of the coal was modelled using the Hoek–Brown rock mass strength criterion introduced in section 4.9.1. The analysis took into account changes in modulus and the redistribution of stresses occurring following yield. The results obtained were qualitively similar to the distribution shown in Figure 15.5 although the concentration of vertical stress at the corner was not as marked.

Wilson (1977, 1981) has developed an instructive analysis of the distribution of vertical stress around a longwall extraction. This analysis uses a stress balance method in which the total vertical force applied over a large plan area must remain equal to that caused by the overburden, even after part of the seam has been removed. It is assumed that, compared with the total vertical force to be redistributed, the vertical forces transmitted by the roadway supports are small, and may be ignored.

Figure 15.6 shows Wilson's approximations to the vertical stress distributions acting at cross sections such as $X_2 - X_2$ in Figure 15.5. The distribution of vertical stress postulated in the caved waste is based on an assumption, derived from field observations, that the stress reaches the overburden stress at a distance of $0.3h$ from the rib side (Fig. 15.6a). The distribution of stress within the yielded zone at the rib side is calculated using an elastic-plastic analysis of the type introduced

406

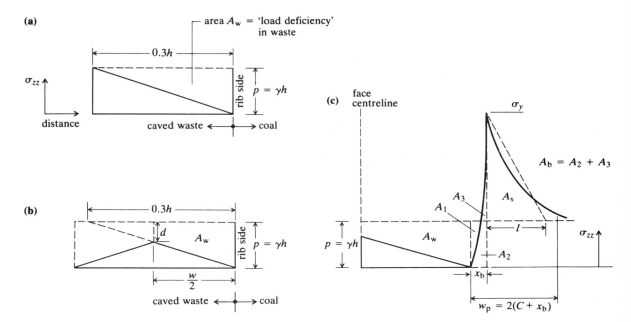

Figure 15.6 Idealised distribution of vertical stress, σ_{zz}, in the caved waste for (a) face length, $w > 9.6h$, and (b) $w < 0.6h$; (c) stress area balance across the rib side (after Wilson, 1981).

in section 7.6 and extended in section 11.3. As shown in Figure 15.6c, the resulting non-linear stress distribution is approximated by a triangular distribution. Table 15.1 sets out the values calculated by Wilson for the vertical stress, the peak abutment stress, the width of the ribside yield zone and the total vertical force carried by the yield zone for two sets of strata conditions. In Figure 15.6 and Table 15.1, the following symbols are used:

$\sigma_{zz} = $ vertical stress;
$\sigma_y = $ peak abutment or yield stress;
$p = \gamma h = $ vertical stress remote from the excavation;
$C_0 = $ *in situ* uniaxial compressive strength of the strata;
$b = $ constant in the principal stress form of the Coulomb shear strength equation,
$\quad \sigma_1 = C_0 + b\sigma_3$;
$m = $ height of the extraction;
$x = $ distance from the rib side;
$x_b = $ width of yield zone;
$p^* = $ support pressure, p_i, plus the unconfined compressive strength of the broken material at the rib side, taken as 0.1 MPa, i.e.
$\quad p^* = p_i + 0.1$ MPa

$$F = \frac{b-1}{\sqrt{b}}\left(1 + \frac{b-1}{\sqrt{b}} \tan^{-1}\sqrt{b}\right), \text{ where } \tan^{-1}\sqrt{b} \text{ is expressed in radians.}$$

The stress in the zone beyond the peak stress decays asymptotically towards the overburden stress, p. The stress decay curve is assumed to be of the form

$$(\sigma_{zz} - p) = (\sigma_y - p) \exp\left(\frac{x_b - x}{C}\right)$$

407

where C is a constant having the units of distance. By equating areas under the curves in Figure 15.6c so as to maintain a vertical stress balance, the value of C is calculated as

Table 15.1 Expressions used in calculating distributions of vertical stress at rib sides (after Wilson, 1981).

	Weak seam between strong roof and floor	Roof, seam and floor of similar strength
vertical stress	$\sigma_{zz} = bp^* \exp\left(\dfrac{xF}{m}\right)$	$\sigma_{zz} = bp^*\left(\dfrac{2x}{m} + 1\right)^{b-1}$
peak abutment or yield stress	$\sigma_y = C_0 + bp$	$\sigma_y = C_0 + bp$
width of yield zone	$x_b = \dfrac{m}{F}\ell n\left(\dfrac{p}{p^*}\right)$	$x_b = \dfrac{m}{2}\left[\left(\dfrac{p}{p^*}\right)^{1/(b-1)} - 1\right]$
vertical force carried by the yield zone	$A_b = \dfrac{m}{F}b(p - p^*)$	$A_b = \dfrac{m}{2}p^*\left[\left(\dfrac{p}{p^*}\right)^{b/(b-1)} - 1\right]$

$$C = \frac{A_w + px_b - A_b}{\sigma_y - p} \tag{15.3}$$

where A_w is the 'load deficiency' associated with each rib side as shown in Figures 15.6a and 15.6b.

For $w > 0.6h$,

$$A_w = 0.15\,\gamma h^2 \tag{15.4a}$$

and for $w < 0.6h$,

$$A_w = \frac{1}{2}\,w\gamma\left(h - \frac{w}{1.2}\right) \tag{15.4b}$$

where w is the width of the extracted area.

Figure 15.7 shows a comparison between the calculated and measured change of vertical stress ahead of a longwall retreating face at Bates Colliery, Northumberland, UK. The following parameters were used in the calculation: $\gamma = 0.025$ MN m^{-3}, $h = 229$ m, $m = 1.32$ m, $b = 4.0$, $C_0 = 5$ MPa and $p_i = 0$. The seam, roof and floor rocks were of similar strength and the face width exceeded 0.6 h. Substitution in the appropriate equations gives $p = 5.73$ MPa, $p^* = 0.1$ MPa, $\sigma_y = 27.9$ MPa, $x_b = 1.88$ m, $A_b = 14.5$ MNm^{-1}, $A_w = 196.7$ MNm^{-1} and $C = 8.70$ m. The stress and convergence measurements confirmed that there is a discernible stress rise up to 30 m from the face.

It must be emphasised that Wilson's method of calculating the distribution of vertical stress in the mining domain following extraction, is not mechanically rigorous, but is based on a number of assumptions and approximations. The assumed value of p^*, for example, can have an important influence on the results obtained from the calculation sequence. The value of Wilson's approach is that

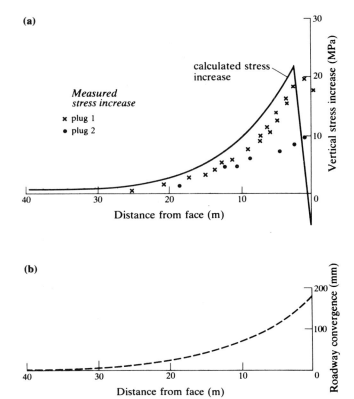

(a)

calculated stress increase

Measured stress increase

x plug 1
• plug 2

Vertical stress increase (MPa)

Distance from face (m)

(b)

Roadway convergence (mm)

Distance from face (m)

Figure 15.7 Measured and calculated vertical stress increase (a) and (b), measured roadway convergence ahead of 10 west face, Bates Colliery (after Wilson, 1981).

it provides a readily understood and applied method of estimating vertical stress distributions in a case in which a more rigorous elastic-plastic stress analysis requires a high degree of specialist knowledge. A more rigorous analysis may not produce results of greater practical utility than those given by Wilson's approach because of the difficulties associated with the determination of the required input parameters, notably the *in situ* stresses.

15.3.3 Face support

Active support of the newly exposed roof strata immediately behind the face is required to prevent the collapse of detached blocks of roof rock and to promote effective caving of the waste. Support systems used for this purpose consist of a series of chocks each containing four or six yielding, hydraulic legs. The upper part of each chock is formed by roof beams or a canopy (Figure 15.8a). Where the waste is friable or breaks into small pieces, or where significant horizontal components of waste displacement towards the face occur, shield supports are used. These supports include not only a roof canopy, but also a heavy rear shield connected to the base by linkages (Figure 15.8b). Peng (1978) gives details of the range of powered face supports available.

The essential geomechanics concern in the design and operation of face support systems, is the value of the support thrust to be provided against the roof, and the associated question of the yield loads of the hydraulic legs. An inadequate setting or yield load may permit excessive convergence to occur at the face, and

409

(a)

hydraulic flow lines — roof beam or canopy

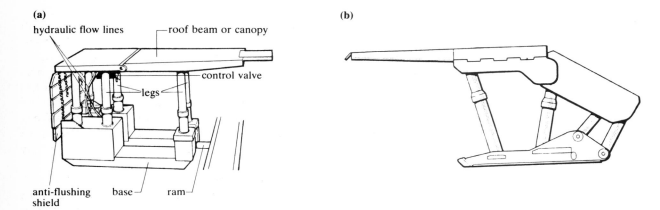

— control valve

— legs

anti-flushing shield — base — ram

(b)

Figure 15.8 (a) Basic components of a four-leg powered support; (b) a shield-type face support (after National Coal Board, 1979).

may permit uneven caving to develop. For example, strong sandstone roofs require high shear forces at the cave line to promote caving. Conversely, excessive setting loads may damage weak roof or floor rocks.

Figure 15.9 illustrates the basis for the determination of required support thrust developed by Ashwin *et al.* (1970). It is assumed that the face support system will be required to support the weight of a detached block of rock of height $2H$ where H is the mining height. Assuming a unit weight of roof rock of 0.02 MN m^{-3}, this gives the minimum support load required as $0.04\,H$ MN per m^2 of roof area. Application of a factor of safety of 2.0 gives the minimum 'setting load density' as $0.08\,H$ MN m^{-2}. Nominal setting load densities that are 1.33 times these values are recommended to compensate for losses in the hydraulic system supplying the setting thrust. Nominal yield load densities are 1.25 times the nominal setting load densities. The support load densities calculated by this method agree well with those found to give satisfactory performance in practice.

Figure 15.9 Assumed caving mechanism and loading on a face powered support (after Whittaker, 1974).

detached block of roof strata the dead weight of which is assumed to rest on the support —

volume increase of caved waste by 50% allows the upper beds to span the ground carried by the supports —

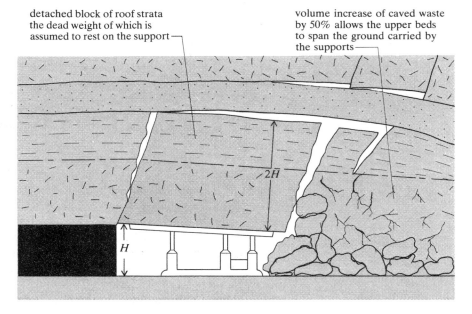

$2H$

H

410

15.3.4 Roadway formation and support

Attention will be focused here on the face or gate roads used to service a longwall face. Differences exist between the formation, service requirements and support systems used in the gate roads for advancing and for retreating longwall faces. Gate roads for advancing faces are situated in the waste area behind the face. After formation, they are subjected to stresses induced as the face advances and the surrounding waste caves. Gate roads serving retreat faces are driven ahead of the face in otherwise undisturbed strata with coal on either side of them. They reduce in length as the face retreats, and may be permitted to collapse behind the face. They are subject to the maximum displacement field travelling with the face.

Figure 15.10 illustrates some of the possible locations of roadways for longwall advancing faces. The roadways may be formed in front of the face as advance headings, at the face, or behind the face. Depending on the relative strengths of the roof and floor rocks and on the height of the seam, they may be placed at the immediate rib side, or may have a ribside pack, and may be enlarged into the roof or floor. Advancing longwall roadways may have to remain serviceable for several months. Normally, it is only the first 50 m or so behind the face that are likely to be subject to high loads and associated deformation. When the face has travelled further away than this, roadway closure and damage should be small, unless other excavations are made which influence the roadway.

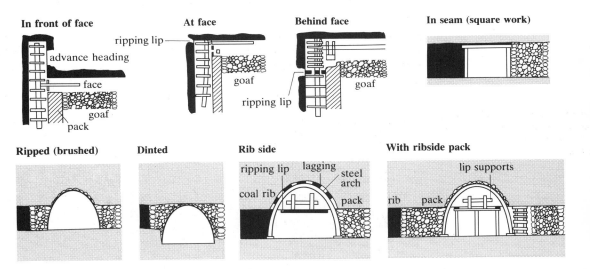

Figure 15.10 Roadways for a longwall advancing coal face (after Thomas, 1978).

The geomechanical objective of roadway location and the method of formation should be to limit the loads and deformations to which the roadway is subjected. Thomas (1978) suggests that, under a weak roof at depths in excess of 200–300 m, the ideal position for a roadway formed behind an advancing face is 4–6 m from the coal rib side (Fig. 15.11a). An arch-shaped roadway closely conforms to the trajectories of the redistributed stresses within the broken rock and is the preferred shape. However, if the roof rock is relatively strong, a rectangular roadway cross section may be used with the roadway located next to the rib side (Fig. 15.11b).

411

(a)

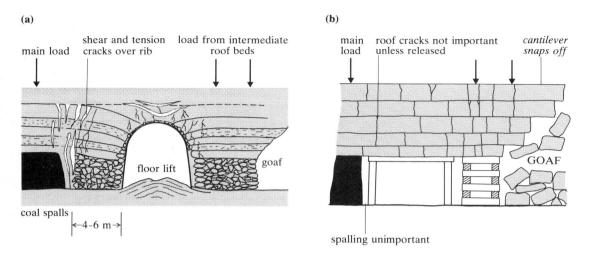

main load

shear and tension
cracks over rib

load from intermediate
roof beds

floor lift

goaf

coal spalls

←4–6 m→

(b)

main
load

roof cracks not important
unless released

*cantilever
snaps off*

GOAF

spalling unimportant

Figure 15.11 Preferred roadway position for (a) a weak roof and (b) a strong roof (after Thomas, 1978).

A feature of all of the roadway types illustrated so far, is the provision of packs to carry loads at the waste side of the roadway and, in some cases, between the rib side and the roadway. It is essential that these packs be of sufficiently high stiffness and strength to shield the roadway supports from excessive loads. Ideally, a pack should be built completely to the roof, be able to offer immediate resistance to closure and bed separation at the road head, and be able to undergo limited yield, without catastrophic failure, when subject to excess loads. Traditional hand-built packs consist of stone walls with dirt or rubble used to fill the void. A number of substitutes for these traditional pack materials have been used. These include timber chocks, softwood dirt-filled packs, hessian or paper bags filled with debris, cribs formed from aerated concrete blocks, and pumpable cementitious materials. An interesting innovation was the development by the UK National Coal Board's Mining Research and Development Establishment, of a high water content cementitious pack material known as 'Aquapak' (Nixon and Mills, 1981). The mix materials are pumped into a PVC brattice cloth bag hung inside a weld-mesh cage. The mix achieves an initial set within 10–20 min of placement, and is capable of withstanding an applied uniaxial stress of 0.3 MPa within 2 h.

In weak rocks at depths in excess of about 150 m, it is possible for a roadway to undergo as much as 1–1.5 m of vertical closure. Such closures may be accommodated by using yielding arches or permitting the arch legs to punch into the floor. In the UK, the floor rocks are usually seat earths which can yield under the induced stresses to produce floor heave (Figure 15.11a). In such cases, the floor has to be re-excavated to its original shape. The use of dowel reinforcement can limit floor heave. A more complete solution would be to use a circular roadway section with closed-ring support. Another major concern in maintaining roadway serviceability is asymmetric roadway deformation induced by high non-vertical thrusts.

The method and location or timing of the formation of a gate roadway have an important influence on its service performance. Three main systems of roadway formation are used in longwall advance mining.

(a) **Conventional rip**, in which the roadway is formed behind the face line. The roadway is fully formed and the steel arches and packs are installed to a line

412

corresponding to the rear of the existing face supports. The area ahead of the arches and packs is mined to the standard extraction height and supported with hydraulic props and link bars. This method has the disadvantage of poor pack formation.

(b) **Advance heading**, in which the roadway is formed ahead of the face line. The roadway is fully formed ahead of mining, but the packs on the goaf side of the arches cannot be placed until mining passes a given point. The arch legs on this side temporarily rest on steel girders supported by hydraulic props. This method suffers from the disadvantage that high rates of convergence occur under the influence of the front abutment vertical stress concentration.

(c) **Half-heading**. The initial heading is excavated to seam extraction height and to half the final roadway width. This reduces the area of exposed roof at the face end and enables packs to be formed up to the line of the rear of the face supports, before significant convergence has occurred. The complete roadway is then formed some distance behind the face, by excavating into the unmined coal. This results in less roadway convergence than in the other two cases because the roadway is formed largely under a canopy of intact rock, and because the packs are established and taking load before the gate is formed.

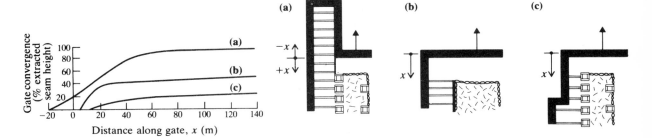

Figure 15.12 The main design features of advance heading (a), conventional rip (b), and half-heading (c) methods of gate roadway formation and their influence on gate convergence measured in a typical case (after Whittaker and Pye, 1977).

Figure 15.12 illustrates the main features of these three methods of gate roadway formation, and shows an example of how gate convergence in each case varies with distance along the gate.

15.3.5 Rib pillar design

Traditionally, rib pillar design has been based on precedent practice and empirical design formulae. For example, the sizes of pillars between wide areas of extraction in the UK, are often fixed using the 'one-tenth depth plus 15 yd' rule. Similar rules are used in other parts of the world (Peng, 1978). Wilson's stress balance approach, outlined in section 15.3.2, permits pillar design to be carried out on a more rational basis.

The width of a long protection pillar, or rib pillar, may be determined using the zone-of-influence concept. In the present case, this will require that the roadway on one side of the pillar will not be influenced by the stress concentration arising from the extraction on the other side of the pillar. In Figure 15.6c, the exponential vertical stress decay curve is replaced by a triangular distribution of base width, l. The vertical stress balance requires that

$$\tfrac{1}{2}l(\sigma_y - p) = C(\sigma_y - p)$$

where C is the constant evaluated from equation 15.3. Thus $l = 2C$, and the required width of the pillar is

$$w_p = 2(C + x_b) \tag{15.5}$$

a width x_b having been added at either extremity of the pillar so as to ensure that the roadway is placed in a low-stress area at the edge of the yielded zone. Wilson (1981) has shown that, in a number of cases, pillar sizes calculated using equation 15.5 agree well with those given by empirical rules developed on the basis of precedent practice.

Example.   In a particular case, the width of extraction of adjacent longwall faces is $w = 300$ m, the depth of mining is $h = 650$ m, the unit weight of the overburden is $\gamma = 0.023\,\mathrm{MN\,m^{-3}}$ and the height of the roadways is 3.66 m. The coal is assumed to have an *in situ* compressive strength of $C_0 = 5$ MPa, and the 'triaxial stress factor' is $b = 3.75$. The resistance of the roadway side is $p^* = 0.1$ MPa.

  Substitution in the formulae given in Table 15.1 for the case of roof, seam and floor of similar strength, gives $x_b = 9.5$ m, $\sigma_y = 61.1$ MPa and $A_b = 169\,\mathrm{MN\,m^{-1}}$. Note that in the calculations for this case, m is taken as the height of the roadways. Since $w < 0.6h$, A_w is found from equation 15.4b as $1380\,\mathrm{MN\,m^{-1}}$. Substitution in equation 15.3 gives $C = 29.3$ m. Thus the width of the pillar required to protect the roadway is

$$w_p = 2(C + x_b)$$
$$= 77.6\,\mathrm{m}, \;\; \text{say } 80\,\mathrm{m} \tag{15.5}$$

In this case, the 'one-tenth depth plus 15 yd' rule gives $w_p = 78.7$ m.

15.4 Sublevel caving

The essential features of sublevel caving, and the conditions best suited to its use, were outlined in section 12.4.7. The **longitudinal sublevel caving** method was developed for the mining of steeply dipping, narrow orebodies. In this case, the production headings are driven on strike as shown in Figure 15.13 For wider orebodies, a **transverse sublevel caving** method may be used with the production headings being driven across the orebody from foot wall to hanging wall as shown in the generalised mining layout in Figure 12.12. In either case, the major geomechanics issues involved are the gravity flow of blasted ore and of caved waste, the design of the mining layout to achieve maximum recovery and minimum dilution while ensuring stability of the production and service openings, and surface subsidence. The discontinuous surface subsidence resulting from the progressive hangingwall caving associated with sublevel caving is discussed in section 16.4.2. The gravity flow of broken ore and sublevel caving layout design are discussed in the next two sections.

414

Cross section (through A–A)

Vertical longitudinal projection

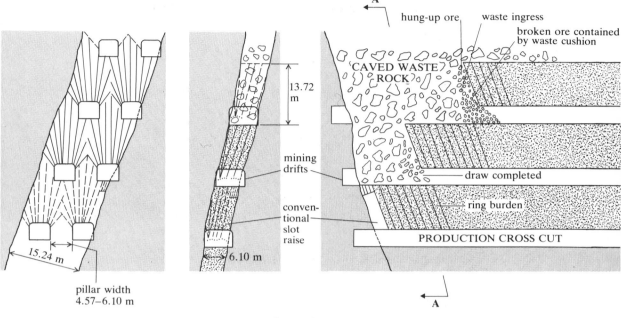

Figure 15.13 A typical longitudinal sublevel caving layout (after Sarin, 1981).

15.4.1 Gravity flow of caved ore

Despite its importance to the performance of a range of underground mining systems, the mechanics of the gravity flow of blasted and/or caved ore is not well understood. At a fundamental level, the subject has been studied using physical model experiments, by analogy with the flow of other granular materials in bins and bunkers (Kvapil, 1965; Jenike, 1966), by mathematical or numerical modelling using the theory of plasticity (Pariseau and Pfleider, 1968), by numerical modelling using probability theory (Jolley, 1968) and by full-scale field studies based on marker recovery (Janelid and Kvapil, 1966; Just, 1981).

In a detailed review of previous research on the subject, Yenge (1980) concluded that the flow of caved ore cannot be described satisfactorily by theories developed for the flow of other granular materials such as sand, powders and cereal grains, because the particle sizes, discharge rates and boundary conditions in the mining problem are not analogous to those applying in the other cases. The particle sizes of caved ores are generally larger in comparison with the dimensions of the 'container' than are those of the other particulate materials for which well-developed flow theories exist. The progressive breakdown of blocks of broken ore during gravity flow is a further feature of the mining problem which makes direct analogy with other granular flow problems difficult. For similar reasons, the use of the theory of plasticity cannot be expected to be totally successful, although recent advances in strain-softening and non-associative plasticity theory, and in the application of the concepts of critical state soil mechanics to rock fracture, offer a possible basis for future analytical development.

The most satisfactory concept of the gravity flow of ore in sublevel caving operations is that developed by Janelid and Kvapil (1966). Central to their

415

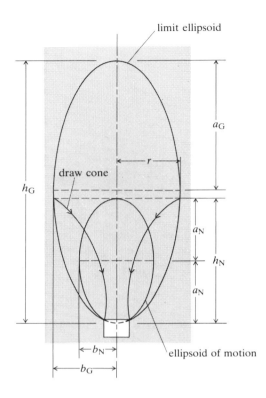

Figure 15.14 Janelid and Kvapil's flow ellipsoid concept.

approach is the concept of the flow ellipsoid illustrated in Figure 15.14. Broken material is contained within a bin or bunker. When the bottom outlet is opened, the material will begin to flow out under the influence of gravity. After a given time, all discharged material will have originated from within an approximately ellipsoidal zone known as the **ellipsoid of motion**. Material between the ellipsoid of motion and a corresponding **limit ellipsoid** will have loosened and displaced, but will not have reached the discharge point. The material outside the limit ellipsoid will remain stationary. As draw proceeds, an originally horizontal line drawn through the broken material in the bunker will deflect downwards in the form of an inverted 'cone'. The shape of this **draw cone** indicates how the largest displacements occur in a central flow channel.

Laboratory and field observations have shown that the ellipsoid of motion is not always a true ellipsoid. Its shape is a function of the distribution of particle sizes within the flowing mass and of the width of the discharge opening. The smaller the particle size, the more elongated is the ellipsoid of motion for the same discharge opening width. The upper portion of the ellipsoid of motion tends to be flattened or broadened with respect to a true ellipsoid, particularly for large draw heights and irregular particle sizes (Figure 15.15). Despite these observed complexities, Janelid and Kvapil's concept of the ellipsoid of motion provides a useful basis on which to develop some understanding of the mechanics of the gravity flow of broken ore in sublevel caving.

The shape of a given ellipsoid of motion can be described by its **eccentricity**

modified shape

ellipsoidal shape

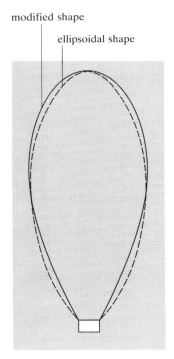

Figure 15.15 Modified draw shape for large draw heights (after Just, 1981).

$$\varepsilon = \frac{1}{a_N} (a_N^2 - b_N^2)^{1/2}$$

where a_N and b_N are the major and minor semi-axes of the ellipsoid, it being assumed that the horizontal cross section of the ellipsoid is circular. In practice, ε varies between 0.90 and 0.98 with values in the range 0.92 to 0.96 being found to apply most commonly. If E_N is the volume of material discharged from an ellipsoid of motion of known height h_N, then the corresponding value of the semi-minor axis of the ellipsoid can be calculated as

$$b_N = \left(\frac{E_N}{2.094 \, h_N} \right)^{1/2} \tag{15.6}$$

or as

$$b_N = \frac{h_N}{2} (1 - \varepsilon^2)^{1/2} \tag{15.7}$$

For a given ellipsoid of motion of volume E_N, there will be a corresponding limit ellipsoid of volume E_G, beyond which the material remains stationary. The material contained between the boundaries of the two ellipsoids will loosen and displace, but will not report to the discharge point. Janelid and Kvapil (1966) described this loosening by a factor

$$\beta = \frac{E_G}{E_G - E_N}$$

417

They found that β varies between 1.066 and 1.100. For most broken ores, β tends towards the lower end of this range which gives

$$E_G \simeq 15E_N \tag{15.8}$$

Assuming that the limit ellipsoid has the same eccentricity as the ellipsoid of motion, equations 15.6, 15.7 and 15.8 can be used to calculate its height as

$$h_G \simeq 2.5h_N \tag{15.9}$$

As material is progressively discharged, the size of the ellipsoid of motion, and of the corresponding limit ellipsoid, continues to grow. A dimension required in the design of sublevel caving layouts, is the radius of the limit ellipsoid at the height h_N (Figure 15.14)

$$r \simeq [h_N(h_G - h_N)(1 - \varepsilon^2)]^{1/2} \tag{15.10}$$

The analysis so far assumes that flow is symmetric about a vertical axis. In sublevel caving, the boundary conditions are often such that the ellipsoid of motion and the corresponding limit ellipsoid are not fully developed. Figure 15.16 shows a cross section of draw patterns observed in model studies of longitudinal sublevel caving for the Granduc Mine, Canada. In this case, the narrowness and dip of the orebody inhibit the development of fully ellipsoidal motion. In a vertical section through the longitudinal axis of a production heading in the general case, the ellipsoid of motion is truncated by the wall of the unblasted ore (Figure 15.17). In addition, the centre line of the ellipsoid is deviated away from the wall by an angle, η, which varies with the ring gradient, α, and with the angle of friction developed between the broken and unbroken ore. For an approximately vertical wall, η is typically about $5°$

Figure 15.16 Influence of orebody width and dip on the draw patterns observed in model studies of longitudinal sublevel caving. Numbers on the flow 'ellipsoids' are the number of 4 m³ buckets of ore removed from the draw point at a given time (after Sarin, 1981).

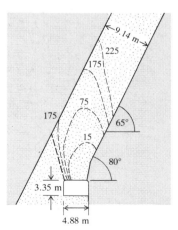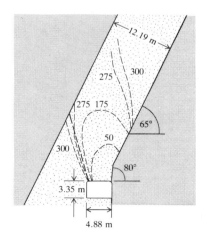

The flow pattern in the plane of the wall of the slice is shown in Figure 15.18. The shapes of the truncated ellipsoid of motion and limit ellipsoid illustrated in Figure 15.17 vary with the particle size of the broken ore, the height of the flow,

418

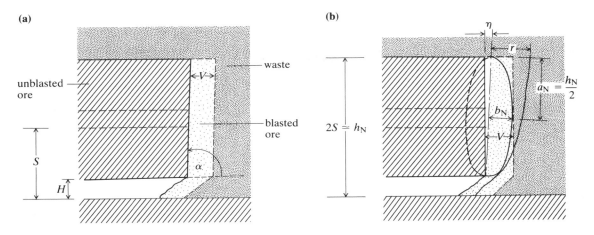

(a)

(b)

Figure 15.17 Vertical section through the longitudinal axis of an extraction drift, showing (a) ring burden, V, and slice gradient, α, and (b) truncated ellipsoid of motion.

the extraction width and the extraction velocity. The flow becomes more narrow, or more parallel, as the height of the flow zone increases, as the extraction width increases and as the extraction velocity increases. If it is assumed that the flow may be described by equations 15.6 to 15.10, and that $h_N \simeq 2S$ where S is the slice height, then the semi-width of flow, r, can be found from equation 15.10 as

$$r \simeq S \, [6 \, (1 - \varepsilon^2)]^{1/2} \qquad (15.11)$$

where ε is the unknown eccentricity, dependent on the height of flow and the particle size. Figure 15.19 is a chart developed by Janelid and Kvapil (1966) for making a preliminary estimate of the eccentricity for 'hard' ore. Using this as a

Figure 15.18 Flow pattern in the plane of the wall of a vertical slice.

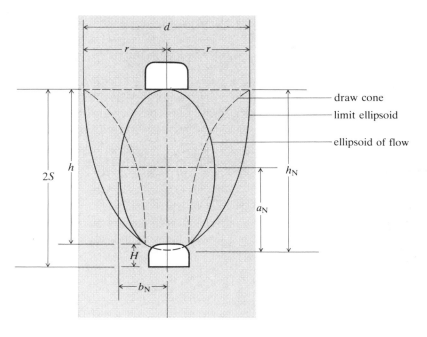

419

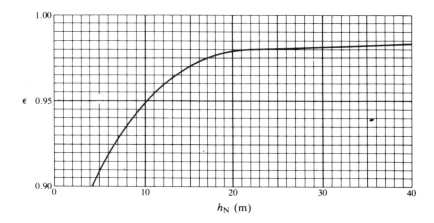

Figure 15.19 Chart giving an approximation to eccentricity, ε, as a function of the height of the ellipsoid of flow, h_N (after Janelid and Kvapil, 1966).

starting point, the semi-width of flow, r, and other geometric parameters, may be estimated. A more exact determination of the flow parameters for a particular case can only be obtained from large-scale tests.

15.4.2 Design of sublevel caving layouts

Following Janelid and Kvapil (1966), the parameters that must be determined in the design of a sublevel caving layout are described by the following symbols and nomenclature:

A = width of the slice (Figure 15.20)
H = height of the extraction drift (Figure 15.20)
B = width of the extraction drift (Figure 15.20)
P = width of the pillars between extraction drifts (Figure 15.20)
V = ring burden (Figure 15.17a)
K = factor dependent upon the particle size distribution in the broken ore
a_N, b_N = major and minor semi-axes of the ellipsoid of motion (Figure 15.17b)
c = width of the extraction point (Figure 15.21)
r = semi-width of the limit ellipsoid at the height h_N (Figure 15.18)
x = digging depth of the loader (Figure 15.22)
α = gradient of the slice (Figure 15.17a)
ϕ = angle of repose of the ore in the draw point (Figure 15.22)
ε = eccentricity of the ellipsoid of motion.

The interrelations between these design parameters are complex, so that universally applicable design equations are difficult to establish. However, the theory of gravity flow outlined in the previous section does lead to some relations that are of use in layout design.

Consideration of the progress of ore and waste flows indicates that a staggered arrangement of extraction drifts such as that shown in Figure 15.20 is generally preferable to one in which the drifts on successive sublevels are aligned vertically. With this arrangement, $h_N \simeq 2s$ as shown in Figure 15.20. The optimum burden for any slice height may be related to the minor semi-axis of the ellipsoid of flow as shown in Figure 15.17b. To reduce ore loss, it is necessary that

420

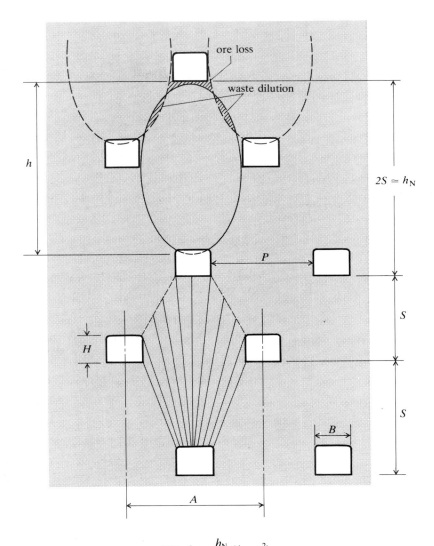

Figure 15.20 Geometry of a sublevel caving layout in the plane of the wall of a slice.

or

$$V \geqslant b_N = \frac{h_N}{2}(1-\varepsilon^2)$$

$$V \geqslant S(1-\varepsilon^2) \tag{15.12}$$

Values of V greatly in excess of those given by the right-hand side of inequality 15.12 will lead to high dilution rates. If V is fixed by operational considerations, inequality 15.12 can be used to establish the slice height, S.

To minimise dilution, the width of the slice, A, should be less than, or at most equal to, the width of the flow, i.e.

$$A \leqslant 2r$$

where r may be estimated from equation 15.11, so that

421

$$A \leqslant 4.9S \, (1 - \varepsilon^2)^{1/2} \tag{15.13}$$

If the flow ellipsoids for adjacent extraction drifts do not intersect, the volume of unrecovered ore will increase. If, on the other hand, the ellipsoids overlap, the dilution will increase (Figure 15.20). Just *et al.* (1973) developed equations for calculating the ore recovery and waste dilution percentages from the volumes of intersecting and overlapping draw ellipsoids. For the case shown in Figures 15.17 and 15.18 where the ellipsoid has a circular cross section, the volume of material drawn is

$$V_d = \frac{\pi}{3} \, a_N \, b_N^2 \left[\left(1 - \frac{h - a_N}{\sqrt{a_N}} \right)^2 \left(2 - \frac{h - a_N}{\sqrt{d}} \right) - \frac{1}{2} \left(2 - \frac{h}{a_N} \right)^2 \left(1 + \frac{h}{a_N} \right) \right]$$

and the corresponding volume of waste is

$$V_w = \frac{\pi}{3} \, a_N b_N^2 \left(\frac{1 + h - a_N - V/\sin \eta}{\sqrt{d}} \right)^2 \left(\frac{2 - h - a_N - V/\sin \eta}{\sqrt{d}} \right)$$

where

$$d = a_N^2 + \frac{b_N^2}{\tan^2 \eta}$$

From these and other equations, design charts may be prepared to assist in evaluating alternative layouts in a given case. Chatterjee *et al.* (1974, 1979) extended this analysis to permit digital simulations of complete sublevel caving systems to be carried out.

If the slice width, A, is determined from inequality 15.13 or from the results of a simulation or a field trial, the trial pillar width is given as

$$P = A - B \tag{15.14}$$

The width of the extraction drift, B, should be such as to promote a satisfactory shape of the gravity flow involving minimum dilution and inducing even runoff of lumpy ore with minimum hang ups and blockages. Janelid and Kvapil (1966) suggest that an approximate value of B can be determined from

$$B > 11.2 \sqrt{KD}$$

where D is the maximum particle size of the broken ore, and K varies from about 0.5 to 1.5 depending on the particle size distribution of the broken ore. The best results are obtained if ore is drawn from across the full width of the drift rather than from a narrower extraction point. This promotes more closely parallel flow of the ore and ensures that the ore–waste interface is flatter, resulting in less dilution (Figure 15.21). Parallel flow may be achieved if $B \simeq A$. In this case, equation 15.14 indicates that the pillar width, P, approaches zero, which is clearly untenable. Under conditions of closely parallel flow, the required relation becomes $A \simeq B \simeq P$. Increasing B to achieve parallel flow can introduce addi-

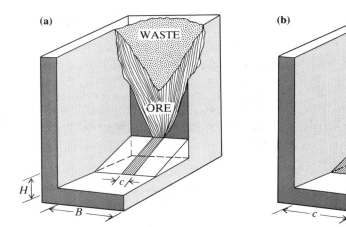

Figure 15.21 Flow patterns for draw from (a) a narrow extraction point, and (b) the full width of the extraction drift (after Janelid and Kvapil, 1966).

tional difficulties and costs in maintaining drift and brow stability. These factors have to be weighed against the disadvantages of the lower recovery and higher dilution that may result if narrower extraction drifts are used.

Using Rankine earth pressure theory, Janelid and Kvapil (1966) suggested that, as shown in Figure 15.22, the optimum digging depth of the loader is

$$x = H\left[\cot\phi - \tan\left(45° - \frac{\phi}{2}\right)\right] \qquad (15.15)$$

If, for example, $\phi = 35°$ and $H = 3.0$ m, equation 15.15 gives $x = 2.72$ m. The digging depths used in practice are usually less than those given by equation 15.15.

Finally, the slice or ring gradient, α, must be chosen so as to minimise the intermixing of ore and waste while maintaining operational practicability. The relative particle sizes of broken ore and caved waste have an important influence on the optimum value of α. A finer material will tend to migrate through an underlying coarser material under the influence of gravity. Thus if the ore has a

Figure 15.22 Theoretical determination of the optimum digging depth, x, of a loader extracting ore from an ore pile in an extraction drift (after Janelid and Kvapil, 1966).

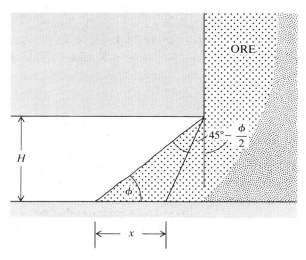

larger particle size than the waste, the ring should be drilled with $\alpha < 90°$ (Figure 15.17). Operationally, it is usually most convenient to drill the ring with α slightly less than 90°. This is common practice when the ore and waste have similar particle sizes. If the ore is finer than the waste, loss of ore by migration into the waste can be reduced if $\alpha > 90°$. However, values of α in excess of 90° can introduce inefficiencies in drilling and can exacerbate difficulties experienced with brow instability. Accordingly, such values are used only in exceptional circumstances. In practice, the ring gradient can vary between about 60° and 110°.

The methods described for designing sublevel caving layouts are based on the geometry of the gravity flow of broken ore. It must be recognised, however, that, particularly as mining depths increase, the 'honeycomb' effect produced by the relatively closely spaced production openings, can lead to stress-induced instability in the rock surrounding the drifts and in the pillars between the drifts. Yenge (1980) and Tucker (1981) describe the transverse fracturing of pillars in longitudinal sublevel caving operations at the Pea Ridge iron ore mine, USA, where high horizontal *in situ* stresses exist. Such occurrences may require reorientation of the mining direction, an increase in sublevel and/or drift spacing, or the reinforcement of production headings. Sarin (1981), for example, refers to the routine use of shotcrete, timber, rock bolts and grouted reinforcing bars to reinforce production headings at a number of sublevel caving operations in Canada.

The distribution of elastic stress induced in layouts such as that shown in Figure 15.20 may be readily determined using the numerical methods outlined in Chapter 6. Strength/stress checks carried out using an appropriate rock mass strength criterion, can then give an indication of the likely extent of the fracture zones that will develop around the headings and in the pillars. However, numerical analysis of the problem can be complicated by the relatively complex three-dimensional nature of the mining layout (Figure 15.13 and 15.16, for example) and by the fact that additional stresses may be imposed during the drawing of ore.

Figure 15.23 illustrates a further cause of stress concentration around the production levels in sublevel caving operations. As mining progresses downwards with the removal of ore and the attendant caving of the waste rock, the *in situ* stresses are relieved. Some vertical stress will be transmitted to the new mining horizon by the caved waste, but by far the greater effect will be the concentration of horizontal stresses in the lower abutment zone as shown in Figure 15.23. If the major principal stress is the horizontal stress transverse to the orebody, fewer stability problems are likely to be associated with transverse than with longitudinal sublevel caving layouts. Other than in the immediate vicinity of an advancing face, induced stresses will be less severe around the boundaries of production headings driven parallel to rather than normal to, the direction of the maximum stress. Furthermore, in the case of transverse sublevel caving, the high horizontal stresses in the production area will be partly relieved when the first slice is drawn and caving of the waste rock is initiated. Tucker (1981) reports that at Pea Ridge, the incidence of stress-induced instability was significantly reduced when longitudinal sublevel caving was replaced with a transverse method.

424

15.5 Block caving

The essential features of the block caving method of mining and the basic geomechanics issues involved, were discussed in section 12.4.8. The key issue to be addressed when considering the use of block caving is the **cavability** of the orebody and the waste rock. This is a function of the geomechanical properties of the rock mass and of the *in situ* and induced stresses. Mine design in block caving involves important geomechanics considerations in the choice of block dimensions, the choice of an extraction system, the determination of drawpoint size and spacing, the design of the ore pass and haulage system, and the scheduling of development and production operations. Finally, the results obtained from a block caving operation depend on the **fragmentation** and **draw control** achieved. The geomechanics of each of these major aspects of the block caving method will be discussed in the subsequent sections. The surface subsidence associated with block caving is considered in Chapter 16.

15.5.1 Cavability

As noted by Bucky (1956), it has long been recognised that 'the ability of a block to cave or fragment is a function of its strength in tension or shear and the value of the applied forces'. Subsequently, many attempts have been made to develop classification systems for use in cavability determinations. These approaches have used empirical assessments of rock mass strengths in relation to the existing stresses, and measures of rock structure such as *RQD*. In a detailed study of the subject, Mahtab and Dixon (1976) showed that the 'natural' features influencing cavability are the *in situ* stress field, the strength of the rock material, the geometry of the discontinuities in the rock mass, and the shear strengths of the discontinuities. Cavability can be be enhanced by a set of 'induced' features comprising the undercut span, boundary slots and boundary mass weakening.

Mahtab and Dixon found that the existence of closely spaced discontinuities having low resistance to in-plane shear stresses, and favourably oriented for slip, is an important determinant of cavability. Accumulated experience strongly supports the contention of Kendorski (1978), referred to in section 12.4.8, that the initiation and propagation of caving requires a well-developed, low-dip

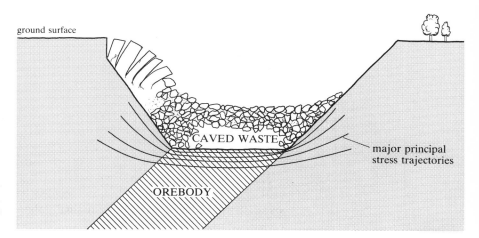

Figure 15.23 Schematic illustration of the transverse horizontal stress concentration produced in the mining area by a notch effect.

discontinuity set. The most favourable rock structure is one in which a low-dip discontinuity set is augmented by two steeply dipping sets which provide conditions suitable for the vertical displacement of blocks. Mahtab *et al.* (1973) found that the orebody at the San Manuel Copper Mine, Arizona, USA, contains three approximately orthogonal joint sets, two nearly vertical and one close to horizontal. However, rhyolite sills in the orebody contain two sets of closely spaced, tight subvertical fractures, but show high resistance to caving, requiring that they be drilled and blasted from extra undercut levels excavated within the sills. The low cavability of the rhyolite was attributed to the lack of a subhorizontal joint set and the tightness of the other joints, despite their low spacing and the low *RQD* of the rock mass.

Many of the earlier attempts to develop cavability indices did not account adequately, if at all, for all of the factors now known to influence cavability. Laubscher's modification for mining applications of Bieniawski's geomechanics classification (Table 3.8) does include all of these factors, and so might be expected to provide an improved prediction of cavability. Table 12.1 and the associated discussion in section 12.5, sets out a classification of caving performance for the various classes of rock mass rating given by Laubscher (1981). Because they do not take account of the mechanics of the problem, it is considered unlikely that further advances in cavability determination will be made using classification approaches alone.

It is clear that the state of stress in the block exerts a major influence on the initiation and development of caving. Mahtab and Dixon (1976) used experimental data obtained by Brown and Trollope (1970) (Figure 15.24) to demonstrate that even if the rock mass structure and geomechanical properties are well conditioned for slip, the existence of a high confining stress can inhibit rock mass failure. For this reason, high horizontal *in situ* stresses are an impediment to block caving initiation and development. Boundary slots relieve the caving block of these stresses and may weaken the abutments against which arching may develop in a broken rock mass. The development of arching in a caving rock mass contained within fixed boundaries is illustrated in Figure 15.25 which

Figure 15.24 Influence of discontinuity spacing and confining pressure on the peak strengths of block-jointed models in triaxial compression tests (after Brown and Trollope, 1970).

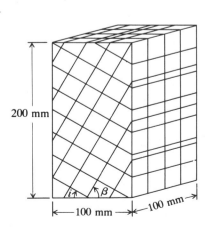

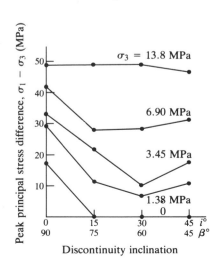

shows the results obtained by Voegele *et al.* (1978) in a digital simulation of block caving carried out using the distinct element method. Each pair of drawings in Figure 15.25 represents the geometric configuration and the interblock contact

Figure 15.25 Distinct element simulation of block caving (after Voegele *et al.*, 1978).

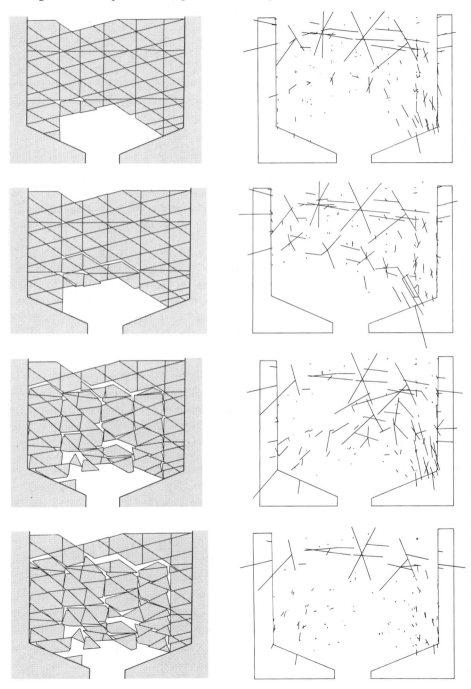

427

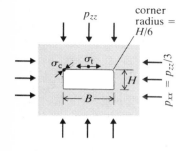

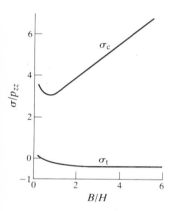

Figure 15.26 An example of the variation in boundary stresses at the centre (σ_t) and corners (σ_c) of an undercut roof with undercut width-to-height ratio.

force vectors at different stages in the progressive caving of the mass. Note that two apparently independent arches form where high levels of interblock force traverse the mass. The upper arch is stronger and is sustained longer than the lower arch, but eventually fails because of slip at the abutments.

If arching is able to develop to such an extent that a stable, self-supporting arch forms above a drawn-out area, several serious consequences can follow. Production will cease in the area concerned, the possibility of damage to installations and injury to personnel exists from the impact loads and air blast that can be produced when the arch eventually fails, and the expense of the measures that may be necessary to induce caving can render the operation uneconomic. One of the best accounts available of the difficulties caused by the arrest of cave propagation and the measures taken to re-initiate caving, is that given by Kendrick (1970) for the Urad Mine, Colorado, USA. Kendrick found that once a stable arch has formed across the minimum dimension of an undercut area, the maximum dimension can be extended considerably without causing the ore to cave.

The mechanisms by which caving initiates and propagates are not understood. However, it is apparent that caving may occur under the influence of both induced shear and tensile stresses. Elastic stress analyses (Mahtab and Dixon, 1976; Coates, 1981) show that the creation of the undercut can induce tensile stresses at the base of the block and shear stress concentrations around the ends of the undercut (Figure 15.26). The magnitudes of these induced tensile and shear stresses increase with increasing width of an undercut of a given height. The creation of boundary slots increases the volumes of rock under tension above the undercut and susceptible to shear failure around the ends of the undercut and the boundary slots.

Clearly, when conditions of induced lateral tension tend to develop above the undercut, blocks of rock isolated by near-vertical discontinuities will become free to displace under the influence of gravity. Slip on low-angle discontinuities and general shear failure of the rock mass in zones of high compressive stress are other mechanisms involved in **stress caving** (Heslop and Laubscher, 1981). A somewhat different mechanism is involved in **subsidence caving** in which the block can subside *en masse* rapidly and with little bulking, as a result of shear failure on the vertical or near-vertical boundaries of the block. This can occur when low or zero horizontal stress is transmitted to block boundaries which are isolated by boundary slots, are in areas of mass boundary weakening or are adjacent to a caved block. An example of such an occurrence at the Jenifer Mine, California, USA, is discussed in section 16.4.1 and illustrated in Figure 16.9.

15.5.2 Design of block caving layouts

The object of the layout design should be to achieve optimum extraction while minimising problems that may interrupt production and require expensive remedial measures. A mining area will be divided into blocks which are mined sequentially in a systematic manner. The area may be divided into a series of square or rectangular blocks separated by pillars, or into a series of long, parallel panels separated by pillars. Alternatively, the area may be mined continuously without pillars. Clearly, this approach gives higher extraction and promotes better control of caving and more even surface subsidence. The size of each block will be determined largely by the area of undercut required to promote satisfactory

428

caving. Table 12.1 may be used to give a preliminary indication of the undercut dimensions required for block caving of orebodies in Laubscher's five geomechanical classes. The block dimensions of representative cases are shown in Figures 15.27 and 15.28.

The selection of the draw system exerts a major influence on the design of the mining layout. The traditional gravity draw system (Figure 15.27) which results in a labour-intensive operation, requires that more development openings be made than in the alternative slusher (Figure 15.28) and load-haul-dump (LHD) (Figure 15.29) systems. Gravity draw is best suited to ore that fragments to a relatively fine size. Hang-up of larger particles often occurs in gravity draw systems. The sizes of these particles can be reduced by secondary blasting, but this can cause damage to the production openings.

The slusher system is more effective for medium to coarse ore than is the gravity system. However, it is difficult to maintain good draw control with this method. The method is inflexible in that it is difficult to re-establish a drawpoint if ground control problems exist. The more modern LHD system offers high productivity of coarse ore. This method requires wide drawpoints, which may be

Figure 15.27 Typical layout for continuous panel retreat block caving with pillars between the panels and gravity draw (after Pillar, 1981).

Grizzly level detail – plan view

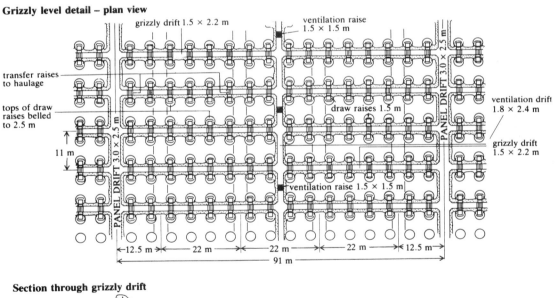

Section through grizzly drift

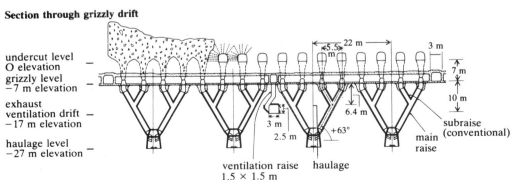

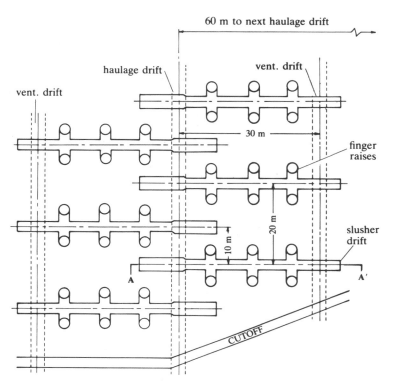

Section A–A': typical vertical section through panel

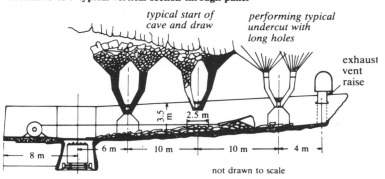

not drawn to scale

Figure 15.28 Typical layout for continuous panel retreat block caving without pillars and slusher draw (after Pillar, 1981).

a disadvantage with fine ores, wide haulageways, large-radius turnouts, and maintenance of trafficable floor conditions. The major advantage of mechanical systems over gravity draw is that the ore pass connections to each drawpoint can be eliminated. This advantage is offset by the need to maintain stability of larger excavations.

The numbers and sizes of the development openings should be as small as possible, consistent with extraction efficiency. Important aspects of the layout are the drawpoint size and spacing, the geometry of the extraction pillar, the orientations of the excavations and of the major apex of the pillar, the provision of reinforcement, and the sequencing and direction of mining. The excavations

430

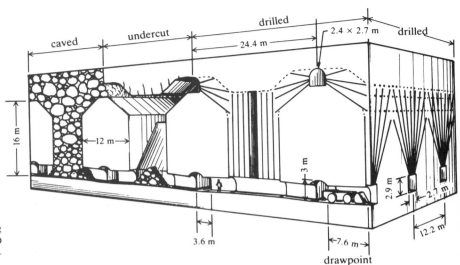

Figure 15.29 Layout and mining steps for block caving with LHD draw, Bell Mine, Canada (after Lacasse and Legast, 1981).

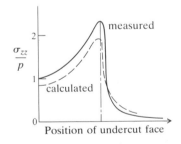

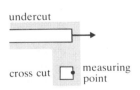

Figure 15.30 Variation in measured and calculated values of the vertical stress, σ_{zz}, at a measuring point in the wall of a cross cut expressed as functions of the *in situ* vertical stress, p, as the undercut approaches, and passes over, the measuring point (after Hult and Lindholm, 1967).

around the drawpoint horizon are subjected to a number of stress changes during development and operation. Stresses are redistributed as an opening is created, as the undercut approaches the opening (assuming that the drawpoint horizon was developed before the undercut drifts are mined), as the undercut passes over the opening, and during drawing of the ore. To preserve the long-term stability of production openings, the number and severity of stress changes to which they are subjected should be minimised.

By definition, block caving is generally carried out in poorer quality rock masses having low rock mass strengths. Wherever possible, the permanent mine installations and important production openings such as drawpoints, should be located in the better quality sections of the rock mass. Ferguson (1979) and Lacasse and Legast (1981) each give examples of improvements in ground control in drawpoint horizons being achieved by locating these installations in parts of the rock mass having the higher rock mass ratings as given by Laubscher's geomechanics classification. They also point to improvements in excavation stability being achieved by orienting excavations as closely perpendicular as possible to the strikes of the dominant, low shear strength structural features.

The development of the undercut level has a major influence on the stresses induced in, and the stability of, the extraction pillar and the drawpoint horizon openings. Figure 15.30 shows the increase, and subsequent decrease, in vertical stress measured in the wall of a cross cut at the Grängesberg Mine, Sweden, as an undercut approached, and passed over, the cross cut. A similar result may be predicted using elastic stress analysis. If the undercut is wide with respect to the vertical distance between the excavation and the undercut, the undercut will shield the excavation from the vertical field stress once it has passed over the excavation. Thus the excavation is subjected to only a horizontal field stress after the undercut has been formed. For this reason, it is desirable to position the undercut drift above the apex of the extraction pillar so as to unload it at an early stage.

431

Ferguson (1979) recommends that the drawpoint horizon should be developed after the undercut has been mined. Using this sequence will ensure that the pillar will be subjected to only the stress concentration arising from the completed undercut and will not be subjected to the high stresses induced by the advancing undercut face (Figure 15.30) superimposed on the stresses induced by the excavations on the drawpoint horizon. In some cases, it may not be practicable to adopt such a development schedule. Experience has shown that because the development of mining-induced stresses and displacements is likely to be time dependent to some extent, the effects of the stress concentration associated with an advancing undercut face can be minimised by ensuring that the undercut face advances at as high a rate as is practicable. Because the undercut will usually be in a poor to fair quality rock mass, instabilities in the undercut drifts may occur. The volumes of potentially unstable rock can be minimised by orienting the drifts as closely normal as possible to the strikes of the predominant low-strength structural features.

Ferguson (1979) suggests several geomechanics based guide-lines that should be observed when scheduling the mining of a given block. Where possible, it is preferable to mine from a weaker to a stronger, or from a failed to an intact, rock mass rather than vice versa. Undercut drifts should not be advanced towards an existing cave, i.e. the mining of a new block should retreat from an adjacent caved block rather than advance towards it. As shown in Figure 15.31, this avoids the creation of a potentially highly stressed pillar between the two blocks.

Even when the most soundly based layouts and schedules are adopted, it may still be necessary to reinforce development and production excavations. Rockbolts, dowels and mesh-reinforced shotcrete are widely used for this purpose. Figure 15.32 shows the details of the reinforcing system used on the extraction level at the Bell Mine, Quebec, Canada, with the LHD draw system illustrated in Figure 15.29.

Figure 15.31 The initiation of caving of a block adjacent to an existing caved block for (a) the preferred mining direction, and (b) an unfavourable mining direction (after Ferguson, 1979).

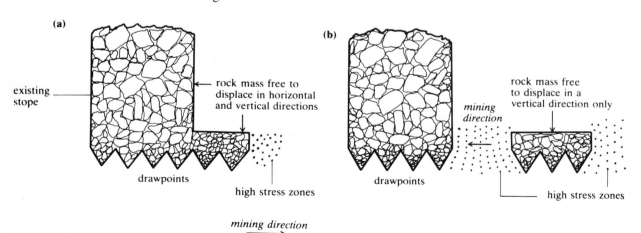

15.5.3 Fragmentation and draw control

The degree of fragmentation of the orebody achieved under the stresses induced during caving, has an important influence on the performance of a block caving operation. Fragmentation influences the selection of the draw system and the

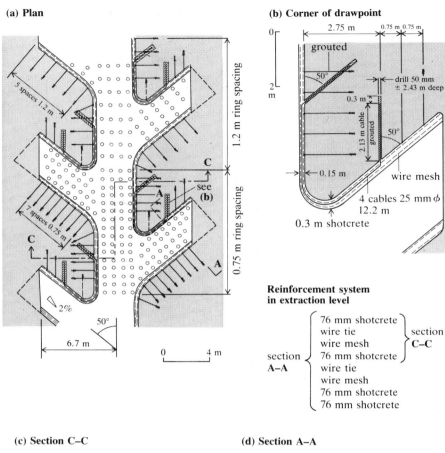

(a) Plan

(b) Corner of drawpoint

Reinforcement system in extraction level

section A–A {
76 mm shotcrete
wire tie
wire mesh
76 mm shotcrete
wire tie
wire mesh
76 mm shotcrete
76 mm shotcrete
} section C–C

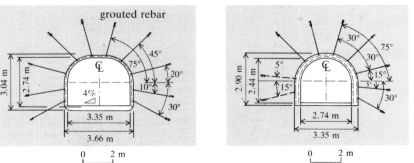

(c) Section C–C

(d) Section A–A

Figure 15.32 Reinforcement of the extraction level, Bell Mine, Canada, for the LHD draw system illustrated in Figure 15.29 (after Lacasse and Legast, 1981).

spacing of drawpoints, since it is one of the determinants of the shape of the flow ellipsoid of the broken ore. Poor fragmentation can lead to the hang-up of large blocks and the possible formation of stable arches within the caved mass. Such occurrences can impose high stresses on pillars and drawpoint installations. The secondary blasting required to re-initiate caving may have deleterious effects on production openings.

The degree of fragmentation likely to be achieved for a given orebody is difficult to predict. Fragmentation depends not only on the quality of the rock mass and on its variability, but also on the height and rate of draw. Table 15.2 sets out the block sizes expected by Heslop and Laubscher (1981) for subsidence and for stress caving of rock masses in the various geomechanical classes. Uniform block sizes of 0.5 m or less are generally preferred. Heslop and Laubscher suggest that if the geomechanics classification rating varies by less than 10 points throughout the block, then uniform fragmentation can be expected.

Table 15.2 Fragment sizes expected from the block caving of orebodies in each of the geomechanics classification classes (after Heslop and Laubscher, 1981).

| Class | Rating | Fragment size (m) | |
		Subsidence caving	Stress caving
1	80–100	8–12	4–9
2	60–79	3–9	3–6
3A	50–59	1.4–4	0.7–3
3B	40–49	0.9–1.5	0.3–1
4A	30–39	0.6–1.0	0.2–0.5
4B	20–29	0.6	0.4
5A	10–19	0.3	0.3
5B	0–9	0.1	0.1

The rate of removal of ore from a caving block can influence the fragmentation and the control of caving. At one extreme, if a caved stope is left undrawn, the ore can compact and re-establish locally stable structures. Conversely, if the caved ore is drawn too quickly, a large void may form beneath uncaved ground. In this case, fragmentation would be reduced because of reductions in the loads imposed by the overlying caved mass and in the number of stress reversals which serve to break up ore as it is drawn. To improve fragmentation, the height of the cave should be maximised and a slow initial rate of draw should be used. The rate should be such that the volume of ore removed during caving is equal to the volume increase or bulking of the caving mass.

Drawpoint spacing should be chosen so that the three-dimensional flow ellipsoids from adjacent drawpoints overlap slightly. This produces almost complete extraction and minimises dilution. The sequence and rate of drawing ore from the several drawpoints in a block should be selected so as to produce controlled development of the cave line and of the associated surface subsidence. It is the common practice not to attempt to draw down a complete block uniformly, but to retreat the cave line across the block at a controlled rate. The drawpoints removed from the developing cave line should be drawn uniformly to promote the even lowering of the caved ore. Funnelling can occur if ore is drawn from a given drawpoint at an excessive rate, and if the cap rock is weaker than the ore (section 16.4.1 and Figure 16.12).

Problems

1 Figure (a) below shows the rectangular cross section of a deep underground excavation. The height of the excavation is small compared with its width, and its length, perpendicular to the section, is large. The principal *in situ* stresses are

p_{xx} and p_{zz}. For these conditions, the elastic stresses in the rock surrounding the excavation are well represented by the expressions:

$$\sigma_{xx} = d - e \sin \beta + p_{xx} - p_{zz}$$

$$\sigma_{zz} = d + e \sin \beta$$

$$\sigma_{xz} = - e \cos \beta$$

where

$$d = \frac{rp_{zz}}{\sqrt{r_1 \, r_2}} \cos [\theta - \tfrac{1}{2} (\theta_1 + \theta_2)]$$

$$e = \frac{a^2 rp_{zz} \sin \theta}{(r_1 \, r_2)^{3/2}}$$

$$\beta = \frac{3}{2} (\theta_1 + \theta_2)$$

Figure (b) shows the dimensions and relative positions of two such excavations produced by the longwall mining of two parallel tabular orebodies.

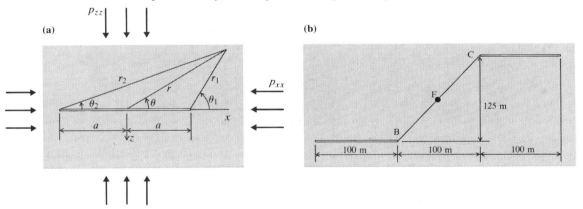

(a)   If $p_{zz} = 32$ MPa and $p_{xx} = 40$ MPa, show that each of the excavations is outside the other's 5% zone of influence.

(b)   Calculate the state of stress at F which is midway between B and C.

(c)   Using equation 10.84, calculate the mining-induced elastic displacements of the upper and lower surfaces at the midpoint of each of the excavations if $G = 16$ GPa and $\nu = 0.20$.

2 A 1.0 m thick, flat-lying, metalliferous orebody of large areal extent is to be extracted by a longwall method. The *in situ* vertical stress at the mining depth is 85 MPa, and Young's modulus and Poisson's ratio of the rock are 48 GPa and 0.20, respectively.

(a)   Calculate the critical span of a longwall face under these conditions. What total amount of energy is released per unit length of stope and per unit

435

volume of extraction, for this critical span? What values do these quantities approach as the span increases above the critical?

(b) Partial extraction is to be used to control the energy release rates in this case. Using equations 10.88 and 15.1, write down an expression for the total energy released per unit volume of extraction when the orebody is mined in a number of parallel panels of span, l, spaced on centres of S as in Figure 15.4a. Determine values of l and S that will restrict this energy release rate to 15 MJ m^{-3} and maintain an extraction ratio of at least 0.80. In order to ensure the long-term stability of the pillars it is necessary that they have width to height ratios of at least 20.

3. A horizontal coal seam, 2.0 m thick and at a depth of 600 m, is to be extracted in a series of parallel, advancing longwall panels each 250 m wide with rib pillars left between them.

Using Wilson's method of analysis

(a) plot the estimated distribution of vertical stress across the rib side and in the waste after mining, ignoring the influence of roadways;
(b) determine the pillar width required if a 3.5 m high roadway situated next to the rib side of one panel is to be outside the zone of influence of the adjacent panel.

Take $\gamma = 23$ kN m^{-3}, $C_0 = 4.0$ MPa, $b = 3.75$ and $p^* = 0.1$ MPa. Assume that the roof, seam and floor have similar strengths.

4 Pillar sizes in a longwall coal mining operation have been set using the empirical relation

$$\frac{h}{1 - R} = 2150$$

where h is the mining depth in metres and R is the area extraction ratio. This approach has given satisfactory results at mining depths of between 100 m and 600 m. However, it is now intended that mining will proceed to depths of up to 1500 m, and a more rational approach to pillar design is required.

Prepare a diagram showing the minimum pillar sizes predicted by the empirical formula and by Wilson's method for mining depths of from 100 m to 1500 m with an extraction width of 220 m and a roadway height of 2.5 m. Take $\gamma = 22.5$ kN m^{-3}, $C_0 = 3.4$ MPa, $b = 3.0$ and $p^* = 0.3$ MPa. Laboratory tests gave the average uniaxial compressive strengths of the roof, seam and floor rocks as 45, 17 and 60 MPa, respectively.

5. An estimate of the dimensions of the flow and limit ellipsoids is required for use in preparing a preliminary sublevel caving layout. The trial height of gravity flow is to be $h_N = 16$ m. Using Janelid and Kvapil's chart for estimating the eccentricity ε (Figure 15.19), and assuming the loosening factor for the material in the limit ellipsoid to be $\beta = 1.080$, calculate predicted values of the major and minor semi-axes and volumes of the flow and limit ellipsoids.

What values of ring burden and slice width do these results indicate are required to minimise dilution and ore loss in sublevel caving operations?

6 A 60 m high block of caved ore under a capping of waste is to be drawn using drawpoints spaced on 10 m centres. Field trials show that one drawpoint will pass the material originally contained within the ellipsoid

$$6x^2 + 0.5\,(y - 1.4\,V^{1/3})^2 = V^{2/3}$$

where x and y are measured horizontally and vertically from an origin on the centre-line at the top of the drawpoint, and V is the volume drawn. The caved ore has a unit weight of 25 kN m^{-3} and the loosening factor is $\beta = 1.066$.

(a) Will the waste-ore contact be displaced by drawing 1000 tonnes of ore from one drawpoint?

(b) How much ore can be drawn from one drawpoint before waste appears?

16 Mining-induced surface subsidence

16.1 Types and effects of mining-induced subsidence

In the present context, subsidence is the lowering of the ground surface following underground extraction of an orebody. Subsidence is produced, to a greater or a lesser degree, by almost all types of underground mining. Surface displacement may result from the redistribution of stresses associated with excavation or from mining-related activities such as de-watering.

Subsidence can be regarded as being of two types – continuous and discontinuous. **Continuous or trough subsidence** involves the formation of a smooth surface subsidence profile that is free of step changes (Figure 16.1). The resulting displacements of surface points may be of only elastic orders of magnitude when compared with the dimensions of the subsiding area or the mining depth. This type of subsidence is usually associated with the extraction of thin, horizontal or flat-dipping orebodies overlain by weak, non-brittle sedimentary strata. It results from the longwall mining of coal, but has also been associated with the extraction of a wide variety of other minerals such as sulphur and the evaporites deposited in sedimentary environments. Methods of predicting subsidence profiles in these cases are discussed in section 16.5.

Discontinuous subsidence is characterised by large surface displacements over limited surface areas and the formation of steps or discontinuities in the surface profile. It may be associated with a number of mining methods, may involve a range of mechanisms, may develop suddenly or progressively, and may occur on a range of scales. Figure 16.2 illustrates some of the forms of discontinuous subsidence.

Crown holes (Figure 16.2a) arise from the collapse of the roofs of generally abandoned, shallow open workings. Much of the surface damage in the anthracite mining region of northern Pennsylvania, USA, is due to this cause. Crown holes are also associated with old coal, ironstone and flint workings in England (Piggott and Eynon, 1978). Crown holes may be regarded as a special case of chimney caving which is discussed in detail in section 16.2.

Pillar collapse in old, shallow workings may lead to similar surface expressions of discontinuous subsidence as does crown hole formation. Such collapses may occur as a result of a deterioration in pillar strength with time or the imposition of additional load on the pillar by surface construction. Large-scale pillar collapse in a working mine can produce discontinuous subsidence over a larger area with more serious effects. A most catastrophic failure of this type occurred at the Coalbrook North Colliery, South Africa, on 21 January 1960, when a room-and-pillar mining area covering approximately 3 km^2 suddenly collapsed with the loss of 437 lives (Bryan *et al.*, 1964).

Chimney caving, piping or **funnelling** (Figure 16.2b) involves the progressive migration of an unsupported mining cavity through the overlying material to the surface. The surface subsidence area may be of a similar plan shape and area to

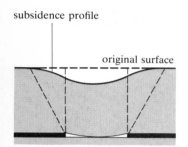

subsidence profile

original surface

Figure 16.1 Trough subsidence over a longwall extraction.

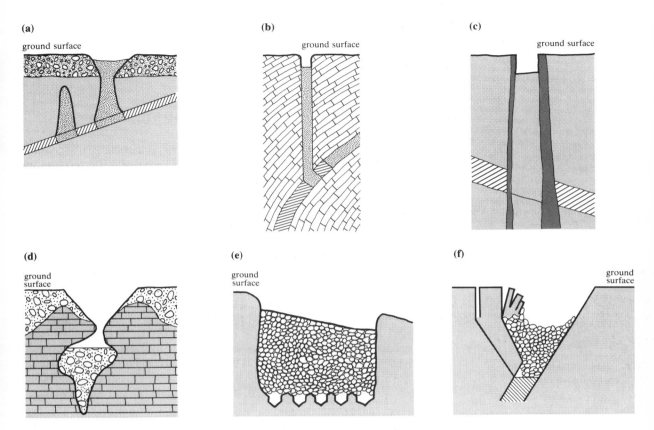

Figure 16.2 Types of discontinuous subsidence: (a) crown hole; (b) chimney caving; (c) plug subsidence; (d) solution cavities; (e) block caving; (f) progressive hangingwall caving.

the original excavation. Chimney caves may form in weak overburden materials as on the Zambian copper belt, in previously caved material or in regularly jointed rock which progressively unravels. Chimney caves have been known to propagate upwards to surface through several hundreds of metres.

If chimney formation is sudden rather than progressive, the phenomenon is sometimes known as **plug subsidence** (Figure 16.2c). Generally, plug subsidence is controlled by some structural feature such as a dyke or a fault which provides a plane of weakness whose shear strength is overcome at some critical stage of mining. Underground air blasts are generally produced by this form of subsidence.

Chimney caving was the cause of the Mufulira Mine disaster in which 89 men were killed when 450 000 m³ of mud (tailings) flooded part of the mine on 25 September 1970. A chimney cave propagated upwards by about 500 m to connect the sublevel caving mining area with the overlying tailings pond. The most disastrous consequence of this was the loss of so many lives, but a further important consequence was the sterilisation of a major part of the mine which was subsequently isolated between concrete bulkheads. The Final Report of the Commission of Inquiry (1971) makes salutory reading for aspiring mining engineers.

Chimney caves are sometimes known as **sinkholes**. However, this term is also used to describe the subsidence features associated with pre-existing solution cavities in dolomites and limestones (Figure 16.2d). The characteristics of these features and the mechanisms of their formation are discussed in section 16.3. Perhaps the best known examples of dolomite sinkholes associated with mining operations are those that have occurred in the Far West Rand gold mining district of South Africa. Jennings *et al.* (1965) describe the sudden collapse into a sinkhole of a three-storey crusher chamber at the West Driefontein Mine in December 1962; 29 lives were lost.

Discontinuous subsidence also occurs as a result of **caving** methods of mining. Figure 16.2e illustrates the large-scale mass subsidence associated with block caving operations, and Figure 16.2f shows the surface configuration that can be produced by progressive hangingwall failure in a sublevel caving operation. An analysis of the latter case is presented in section 16.4.2.

Some of the instances of mining-induced subsidence outlined above have had disastrous consequences. They have led to loss of life on a large scale, loss of parts of producing mines and loss of major surface installations. Such catastrophic consequences are usually associated with sudden discontinuous subsidence. An obvious objective of mine planning must be to limit discontinuous subsidence to areas over which the mine operator holds the surface rights. Having done this, it is also necessary to ensure that subsidence does not affect surface installations, transportation routes or underground access. As the Mufulira example shows, it is especially important to avoid discontinuous subsidence beneath surface accumulations of water or tailings.

The effects of continuous subsidence are generally not as dramatic as those of discontinuous subsidence. Because of the large surface areas affected, longwall coal mining often influences built-up areas and services. Differential vertical movements, horizontal compressive and tensile strains, and curvature of the ground surface, can cause distress to engineered structures, domestic buildings, roads, railways and pipelines. A wide variety of examples of damage due to mining subsidence is given in the National Coal Board's *Subsidence Engineer's Handbook* (1975) which also describes design features that have been used to limit damage to structures and pipelines.

It is clear from this introductory discussion of mining-induced subsidence and its effects, that the prediction of the subsidence profile in the case of continuous subsidence, and of the likely occurrence and areal extent of discontinuous subsidence, are vitally important to the planning of underground mining operations. Before rational predictions of these types can be made, it is necessary that the mechanisms involved be understood. Accordingly, the remainder of this chapter is devoted to considerations of the mechanisms involved in the major types of subsidence and to the few methods of analysis that are available for use as predictive tools.

16.2 Chimney caving

16.2.1 Chimney caving mechanisms

Three distinct chimney caving mechanisms may be identified, each associated with different geological environments.

The first mechanism occurs in weathered or weak rock, or in previously caved rock. It is a progressive mechanism that starts with failure of the stope roof or

440

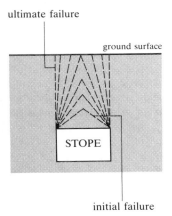

ultimate failure

ground surface

STOPE

initial failure

Figure 16.3 Progressive chimney cave formation.

hanging wall on inclined surfaces. If a stable, self-supporting arch cannot be formed, the failure may progressively propagate towards the surface as shown in Figure 16.3. As material falls from the roof or from the propagating cave, it will bulk and will tend to fill the stope void. Unless the stope is initially large and open, or unless sufficient material is progressively drawn from it, the stope will eventually become filled with caved material which will provide support for the upper surface and so arrest the development of the cave. It is for this reason that the development of chimney caving is so closely associated with draw control.

This progressive failure mechanism has been well established in model studies of the failure of shallow tunnels in sand and clay (Atkinson *et al.* 1975) and in model studies of the mining of steeply dipping, tabular orebodies. It is most likely to occur when the mechanical properties of the hangingwall material are similar to those of a soil. Once initiated, propagation of the failure to surface can be very rapid. This can give the impression that the cave reaches the surface instantaneously and that the mechanism is that of sudden plug subsidence rather than a progressive one.

This mechanism was first postulated by Crane (1931) who studied chimney caving development, particularly in the northern Michigan iron ore mining district. Much of the chimney caving that has occurred on the Zambian copper belt is of this type. Some of these caves appeared some time after the areas beneath them had been mined out. A possible explanation of this is the progressive deterioration with time of the narrow rib pillars that were left between the open stopes. Eventually the pillars collapse, leaving an unsupported span that is wider than the critical width at which progressive caving initiates. More typically, the chimney caves, or sinkholes as they are called locally, form almost simultaneously with the blasting of rib and crown pillars of two or three adjacent stopes. Here again, the maximum stable span may be suddenly exceeded.

The second mechanism is also progressive, but occurs as a result of the unravelling of a discontinuous rock mass. The rock material itself may be quite strong and may not fail except in flexure. The mechanism is controlled by the regular discontinuities in the rock mass. As in the previous case, a sufficient void must be maintained beneath the cave if it is to continue to propagate. Mechanisms of this type have been observed in physical model studies using the base friction apparatus, for example. They have also been studied using distinct element numerical models.

Rice (1934) describes an example of this type of chimney cave development over an opening having a roof area of $4.3 \times 8.6 \, \text{m}^2$ in a graphitic slate dipping at approximately 60°. The excavation was created to provide backfill material, and as the broken slate was progressively drawn, the cave propagated vertically through some 300 m of slates to surface in approximately one year, maintaining the cross-sectional area of the initial opening.

The third mechanism, that of plug subsidence, differs from the other two in that it is controlled by one or more major structural features which provide low shear strength surfaces on which the plug of undercut rock may slide under the influence of gravity. In this case, the mass of rock will undergo essentially rigid-body displacement without breaking up or dilating. Thus, a vertical displacement at the stope boundary will result in a vertical displacement of similar magnitude at the surface. Although an initial void must exist for this mechanism to occur, the development of this type of chimney cave is not as closely associated with draw control as are the other two types.

Allen (1934) describes an example of chimney caving that was controlled by geological planes of weakness, but which developed progressively as in the first mechanism. After three years of mining by a caving method at the Athens Mine in northern Michigan, a cave developed and rapidly progressed to surface through 600 m of cover. As shown in Figure 16.4, the subsidence zone was bounded on either side by subvertical dykes.

16.2.2 Limiting equilibrium analysis of chimney caving

Atkinson *et al.* (1975) found that limiting equilibrium analyses of the final vertically sided collapse configurations gave good approximations to the ultimate collapse conditions observed in their tests on model tunnels in sand. By analogy, limiting equilibrium calculations may be expected to be helpful in estimating

Figure 16.4 North–south section, Athens Mine, showing plug subsidence controlled by dykes. Coordinates and scale are shown in metres (after Boyum, 1961).

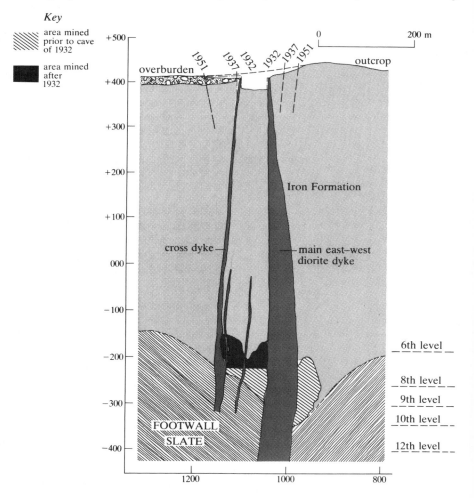

442

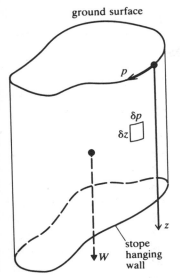

ground surface

p

δp

δz

z

W stope hanging wall

Figure 16.5 General block geometry for limiting equilibrium analysis of chimney caving.

ultimate collapse conditions when chimney caving develops by the first of the mechanisms described. They should also be useful in the case of plug subsidence, although in this case an assumption of zero dilation on the slip surface should be made. The approach cannot be expected to produce reliable results when the second chimney caving mechanism develops in strong discontinuous rock.

Consider the general case illustrated in Figure 16.5. The base and top of a vertically sided block coincide with the hanging wall of a stope and with the ground surface, respectively. It is assumed that rigid-body motion of the block occurs by sliding vertically downwards under the influence of gravity when the shear resistance that can be developed on the vertical block boundary is exceeded. This shear resistance will depend on the effective normal stress on the block boundary. This requires that assumptions be made about the distributions of normal stress and groundwater pressure with depth. It will be assumed that the vertical *in situ* stress is given by $\sigma_{xx} = \gamma z$ where z is the depth below surface and γ is the unit weight of the overburden, and that there is an equal all round horizontal normal stress of $\sigma_{zz} = \sigma_{yy} = k\gamma z$ where k is a constant.

If τ is the limiting vertical shear stress that can be developed on a boundary element $\delta p \times \delta z$ to which an effective normal stress of σ_n' is applied, then the available shear resistance for the entire surface will be

$$Q = \int_0^p \int_0^z \tau \, dz \, dp \tag{16.1}$$

where z and p are such that all surface elements are summed.

If W is the total weight of the block, then the factor of safety against shear failure on the vertical block boundaries is simply

$$F = \frac{Q}{W} \tag{16.2}$$

The complexity of the manipulations involved in obtaining solutions for particular cases will vary with the geometry of the block, the groundwater pressure distribution and the shear strength criterion used. If $u(z, p)$ is the groundwater pressure at an element and an effective stress Coulomb shear strength law (equation 4.11) is assumed, equation 16.1 becomes

$$Q = \int_0^p \int_0^z \{c' + [\gamma z - u(z, p)] \tan \phi' \} \, dz \, dp \tag{16.3}$$

If an effective stress form of the Hoek–Brown non-linear strength criterion (equation 4.37) is used, equation 16.1 becomes

$$Q = \int_0^p \int_0^z \{A \sigma_c^{1-B} (\sigma_n - u(z, p) - \sigma_t)^B\} \, dz \, dp \tag{16.4}$$

An example of wide applicability is that shown in Figure 16.6a. A block of width a and base length b has one pair of sides oriented in the direction of the

443

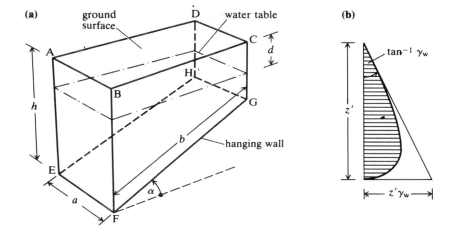

Figure 16.6 (a) Rectangular block geometry, and (b) assumed distribution of water pressure with depth, for limiting equilibrium analysis of chimney caving.

strike of the orebody and the other pair in the dip direction. The dip is α, the maximum height of the block is h, and the water table is a distance d below the horizontal ground surface. The groundwater pressure is assumed to be zero at the water table, to have an initially hydrostatic rate of increase with depth, to be zero at the stope hanging wall and to have an infinite rate of decrease with increasing depth at this point. The skewed parabolic water pressure distribution shown in Figure 16.6b satisfies these boundary conditions. For this distribution, the maximum water pressure is $z' \gamma_w/2$, and the total water pressure force generated over a depth $z' = z - d$, is $z'^2 \gamma_w/3$.

In this case, it is necessary to evaluate the shear resistance developed on each of the four vertical faces (Figure 16.7). The total shear resistance is then given by

$$Q = 2\,Q_{BCGF} + Q_{DCGH} + Q_{ABFE} \qquad (16.5)$$

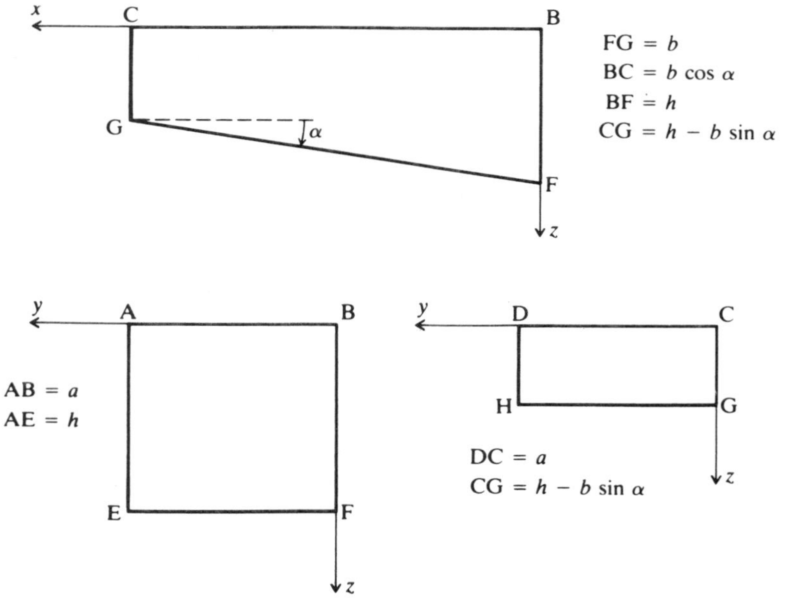

Figure 16.7 Geometry of vertical shear surfaces in rectangular block.,

$FG = b$
$BC = b \cos \alpha$
$BF = h$
$CG = h - b \sin \alpha$

$AB = a$
$AE = h$

$DC = a$
$CG = h - b \sin \alpha$

444

Consider the side face BCGF when $0 \leq d \leq h - b \sin \alpha$, i.e. groundwater level intersects the up-dip face DCGH. If Coulomb's shear strength criterion is used, equation 16.3 applies. This gives

$$Q_{BCGF} = \int_0^{b \cos \alpha} \left[\int_0^z (c' + k \gamma z \tan \phi') \, dz - \frac{(z-d)^2}{3} \gamma_w \tan \phi' \right] dx$$

From Figure 16.7, $z = h - x \tan \alpha$. Substitution for z and integrating gives

$$Q_{BCGF} = Q_1 - \frac{\gamma_w \tan \phi'}{3} b \cos \alpha \left\{ h^2 - hb \sin \alpha + \frac{b^2}{3} \sin^2 \alpha \right.$$

$$\left. - d[\, 2h - b \sin \alpha - d\,] \right\}$$

where

$$Q_1 = \frac{b \cos \alpha}{2} \left[c(2h - b \sin \alpha) + k\gamma \tan \phi' \left(h^2 - bh \sin \alpha + \frac{b^2 \sin^2 \alpha}{3} \right) \right]$$

For $h - b \sin \alpha \leq d \leq h$, i.e. groundwater level intersects the stope hanging wall, a similar calculation gives

$$Q_{BCGF} = Q_1 - \frac{\gamma_w \tan \phi' (h - d)^3}{9 \tan \alpha}$$

When $d \geq h$, no water pressures act on the block surface and

$$Q_{BCGF} = Q_1$$

Similar calculations for the up-dip face DCGH lead to the result

$$Q_{DCGH} = a \left[c'(h - b \sin \alpha) + \frac{k \gamma}{2} \tan \phi' (h - b \sin \alpha)^2 \right.$$

$$\left. - \frac{\gamma_w}{3} \tan \phi'(h - b \sin \alpha - d)^2 \right]$$

for $0 \leq d \leq h - b \sin \alpha$, and

$$Q_{DCGH} = a \left[c'(h - b \sin \alpha) + \frac{k\gamma}{2} \tan \phi'(h - b \sin \alpha)^2 \right]$$

for $d \geq h - b \sin \alpha$.

The results for the down-dip face ABFE are

$$Q_{ABFE} = a \left[c'h + \frac{k\gamma h^2}{2} \tan \phi' - \frac{(h - d)^2}{3} \gamma_w \tan \phi' \right]$$

445

for $0 \leq D \leq h$, and

$$Q_{ABFE} = a \left[c'h + \frac{k\gamma h^2}{2} \tan \phi' \right]$$

for $d \geq h$.

The total weight of the block is

$$W = \gamma ab \cos \alpha \left(h - \frac{b \sin \alpha}{2} \right)$$

Substitution into equation 16.2 of this value and of the value of Q given by equation 16.5, leads to the following results for the three cases of groundwater level considered.

(a) $d \geq = h$

$$F_1 = \frac{2c'(a + b \cos \alpha)}{\gamma ab \cos \alpha} + \frac{k \tan \phi'}{(2h - b \sin \alpha)} \left\{ \frac{h^2 + (h - b \sin \alpha)^2}{b \cos \alpha} \right.$$

$$\left. + \frac{2}{a} \left[h(h - b \sin \alpha) + \frac{b^2 \sin^2 \alpha}{3} \right] \right\} \qquad (16.6)$$

(b) $h \geq d \geq h - b \sin \alpha$

$$F = F_1 - \frac{2\gamma_w(h - d)^2 \tan \phi'}{3\gamma b(2h - b \sin \alpha)} \left[\sec \alpha + \frac{2(h - d)}{3a \sin \alpha} \right] \qquad (16.7)$$

(c) $0 \leq d \leq h - b \sin \alpha$

$$F = F_1 - \frac{2\gamma_w \tan \phi'}{3\gamma(2h - b \sin \alpha)} \left\{ \frac{h^2 + (h - b \sin \alpha)^2 - 2d(2h - b \sin \alpha - d)}{b \cos \alpha} \right.$$

$$\left. + \frac{2}{3a} [3h(h - b \sin \alpha) + b^2 \sin^2 \alpha - 3d(2h - b \sin \alpha - d)] \right\} \qquad (16.8)$$

Example. The applicability of the results given by the limiting equilibrium analysis can be illustrated by an example of chimney cave formation on the Zambian copper belt. Caving was largely through weathered rock for which shear strength properties were measured as $c' = 50$ kPa and $\phi' = 31°$. The unit weight of the material was $\gamma = 23.5$ kN m^{-3}, and because the material was largely a residual soil, a low value of k of 0.33 was assumed. The water level was below mining depth and stopes were separated by 2–5 m wide rib pillars. Typical values of the the other variables were $h = 140$ m, $a = 12$–20 m, $b = 80$–95 m (taken as 90 m), and $\alpha = 45°$.

Substitution of the values of all variables other than a into equation 16.6 and putting $F = 1.0$, gives the critical span as $a = 46.7$ m. This is close to the combined spans of two of the widest or three of the narrowest stopes after rib pillar blasting.

It was under such conditions that chimney caves were often found to develop. For the water table 50 m below the ground surface, equation 16.18 gives the critical span as 30.8 m.

The previous analysis used a linear Coulomb shear strength equation. If the non-linear Hoek-Brown equation is used, the evaluation of the integrals in equation 16.4 can be cumbersome. However, the simple example of a rectangular stope in a horizontal orebody with no groundwater pressures gives a ready solution that may be used for purposes of illustration.

From two equations of the form of equation 16.4 giving the vertical shear resistance developed on the faces of the block, the total shear resistance is found as

$$Q = 2(a + b)\, A\sigma_c^{1 - B} \int_0^h (k\gamma z - \sigma_t)^B \, dz$$

or

$$Q = \frac{2(a + b)\, A\sigma_c^{1 - B}}{\gamma k\,(1 + B)} \left[(k\gamma h - \sigma_t)^{1 + B} - |\sigma_t|^{1 + B} \right]$$

The weight of the block is $W = \gamma abh$, and so, at limiting equilibrium, the factor of safety against vertical sliding is

$$F = \frac{2(a + b)\, A\sigma_c^{1 - B}}{\gamma^2\, abhk\,(1 + B)} \left[(k\gamma h - \sigma_t)^{1 + B} - |\sigma_t|^{1 + B} \right] \tag{16.9}$$

Table 4.2 shows that for very poor quality rock masses of the type commonly involved in chimney caving, $\sigma_t = 0$. In this case, equation 16.9 reduces to

$$F = \frac{2(a + b)k^B h^B\, A\sigma_c^{1 - B}}{\gamma^{1 - B} ab(1 + B)} \tag{16.10}$$

This result shows that the factor of safety decreases as the stope dimensions increase, the mining depth decreases, k decreases, the rock mass quality decreases, the uniaxial compressive strength of the intact rock decreases and as the unit weight of the overburden increases.

As an example of the application of equation 16.10, consider the hypothetical case of a shallow abandoned room-and-pillar coal mine. The overburden is a very poor quality argillaceous rock mass with $\gamma = 0.023\ \mathrm{MN\ m^{-3}}$, $A = 0.050$, $B = 0.539$ and $\sigma_c = 15\ \mathrm{MPa}$. The room span is $a = b = 10\ \mathrm{m}$, and the depth of the roof below surface is $h = 30\ \mathrm{m}$. For $k = 1.0$, equation 16.10 gives $F = 1.61$. The factor of safety reduces to unity when h is reduced to 12.4 m.

It must be emphasised that the limiting equilibrium analyses presented here provide only approximate solutions for particular types of chimney caving. They are not applicable to the case in which caving occurs by progressive unravelling of discrete blocks of strong rock. They apply best to those cases in which geological planes of weakness form the vertical boundaries of the caving block, and those in which the caving takes place through a very weak rock mass, a residual soil, or rock previously disturbed by a caving method of mining. Even though a limiting equilibrium analysis may indicate that chimney caving to

447

surface is possible, the development of the cave can be influenced by draw control in the stope. Goel and Page (1982) use draw density and geometrical parameters in an empirical method for predicting the probability of chimney cave occurrence over a mining area.

16.3 Sinkholes in carbonate rocks

A form of discontinuous subsidence that has a similar surface expression but a different cause from the various types of chimney caving, can occur in carbonate rocks such as dolomites and limestones. These rocks are susceptible to solution by slightly acid waters percolating initially through discontinuities in the rock mass. With time, quite large volumes of rock can be dissolved leading to what are known as **karst** features. These carbonate rocks are also susceptible to deep and irregular surface weathering which produces a highly variable depth to rock head (Figure 16.8). The mantle of unconsolidated materials may include residual clay or residuum resulting from the weathering of the rock.

Cavities develop in carbonate rocks, generally above the water table, where surface water containing small amounts of dissolved gases which render it slightly acidic, flows downwards towards the water table. Experience in mining and civil engineering has been that sinkholes develop from these cavities when the rock mass is de-watered and the water table is lowered. Lowering the water table has several effects. It permits existing caverns to enlarge and may cause new ones to be developed, it removes buoyancy support forces, and it can increase the velocity of downward water movements which can then become sufficient to wash out unconsolidated surface materials.

Major examples of sinkhole formation have occurred in the dolomites and dolomitic limestones overlying the gold-bearing Witwatersrand system of rocks in the Far West Rand, South Africa. As discussed in section 3.2 and illustrated in Figure 3.5, the area is divided up into a number of essentially water-tight compartments by a series of vertical, syenite and diabase dykes. In 1960, a major

Figure 16.8 Section showing ground susceptible to solution-induced subsidence (after Jennings *et al.*, 1965).

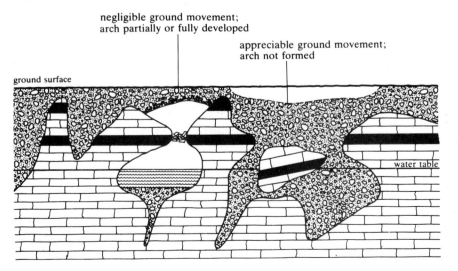

negligible ground movement; arch partially or fully developed

appreciable ground movement; arch not formed

ground surface

water table

de-watering program of the Oberholzer compartment was begun. A series of sinkholes developed co-incidentally with the lowering of the water table. As noted previously, the most disastrous case involved the loss of 29 lives when a crusher chamber suddenly collapsed in December 1962. It was found that sink-hole development was accentuated by the presence of excess surface water from heavy rainfall or a burst pipe.

The exact mechanism of sinkhole formation under these conditions is difficult to establish. A plausible set of conditions that must be necessarily and sequen-tially satisfied if sinkholes are to form was postulated by Jennings *et al.* (1965). The basic concept is that a sinkhole will form when the equilibrium of a stable arch of material above the void is disturbed (Figure 16.8).

(a) There must be adjacent stiff material to form the abutments for the roof of the void. These abutments may be provided by the dolomite pinnacles or the sides of steep-sided, subsurface canyons. The span must be appropriate to the properties of the bridging material. If the span is too large, the arch cannot form.

(b) Arching must develop in the residuum with part or all of the self-weight being transferred to the abutments as arching thrusts.

(c) A void must exist or develop below the arched residuum. The void may be quite small initially, but may be enlarged by percolating water following lowering of the water table.

(d) A reservoir must exist below the arch to accept the material which is removed as the void is enlarged. Some means of transporting this material, such as flowing water, is also essential.

(e) When a void of substantial size has been established in the residuum, some disturbing agency is required to cause the roof to collapse. Water can cause collapse of the arched material and the formation of a sinkhole by reducing the soil strength or by washing out critical keying or binding material.

16.4 Discontinuous subsidence associated with caving methods of mining

16.4.1 Block caving

The mechanics of block caving methods of mining were discussed in Chapter 15, where reference was made to the factors influencing the development of the cave. Obviously, block caving results in discontinuous subsidence which influences large areas at the surface (Figures 16.9 and 16.10). The extent of this influence, as defined by the **angle of break** (the angle made with the horizontal at a section by a straight line drawn from the undercut level to the extremity of surface disturbance), varies with

(a) the strength properties of the orebody;

(b) the strength properties of the overburden;

(c) the presence of major structural features such as faults;

(d) the depth of mining as defined by the undercut level;

(e) the use of backfill in the caved zone;

(f) the slope of the natural ground surface.

449

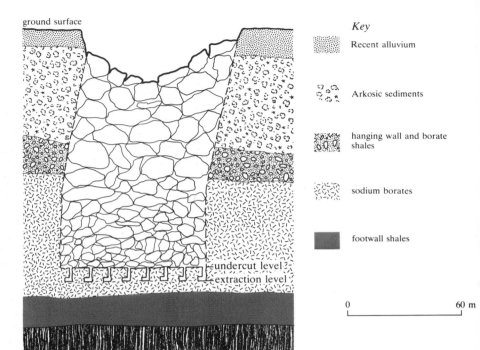

Figure 16.9 Idealised cross section through caved and subsided area, Jenifer Mine (after Obert and Long, 1962).

Vertical and near-vertical boundaries to the subsiding zone may be produced in relatively strong rock masses, where there are no dominant inclined discontinuities to influence cave development, where mining depths are low, and where there is little or no soil or very weak rock overburden in which unsupported near-vertical slopes cannot develop. Figure 16.9 shows an example of such a subsidence profile produced during the drawing of borate ore from a caving block 83 m long, 35–45 m wide and mined at a depth of 100 m at the Jenifer Mine, Kern County, California, USA. The ore was described as weak but as containing no joints.

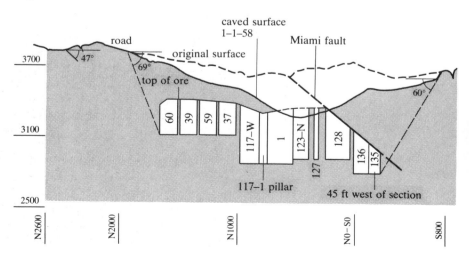

Figure 16.10 E800 section, Miami Mine, showing mined areas and subsidence, 14 February 1958. Co-ordinates are shown in feet (after Fletcher, 1960).

450

(a)

(b)

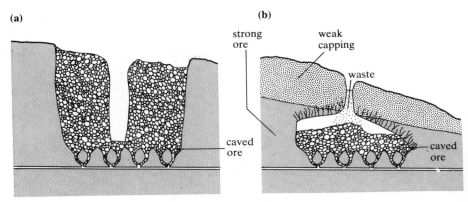

strong
ore

weak
capping

waste

caved
ore

caved
ore

Figure 16.11 Funnelling due to (a) uneven draw, and (b) weak capping.

A good example of the development of a block caving subsidence profile over a period of time is given by Fletcher (1960), who describes the development of subsidence at the Miami Mine, Miami, Arizona, USA, where the mixed copper ore had been mined by block caving since 1926. Initially, the cave had vertical sides, but the angle of break decreased as the mining depth increased. Tension cracks defined the extremities of the subsidence area. Toppling of blocks isolated by successive tension cracks occurred in some areas. A major fault, dipping at about 45° to the east, governed cave development on the west side of the mine. Figure 16.10 shows the surface profile and the location of tension cracks at the E800 section of the Miami Mine on 14 February 1958.

In this, and in other cases, chimney caving, funnelling or piping, as well as mass subsidence occurred. Funnelling may occur within the caved or caving rock, particularly if the rate of draw of the block is uneven, or the capping is weaker than the ore (Figure 16.11). An example of funnelling at the San Manuel Copper Mine, Pinal County, Arizona, USA, is given by Hatheway (1968) who describes the progressive development of block caving induced subsidence over a 10 year period. Caving was found to develop in the following sequence:

Figure 16.12 Progressive hanging-wall caving at Grangesberg (after Hoek, 1974).

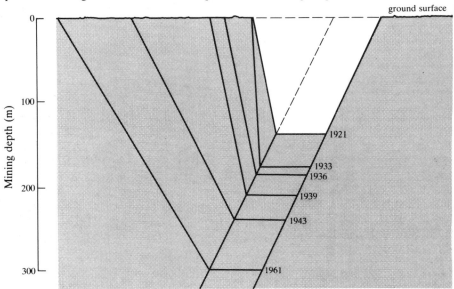

ground surface

Mining depth (m)

0

100

200

300

1921

1933
1936

1939

1943

1961

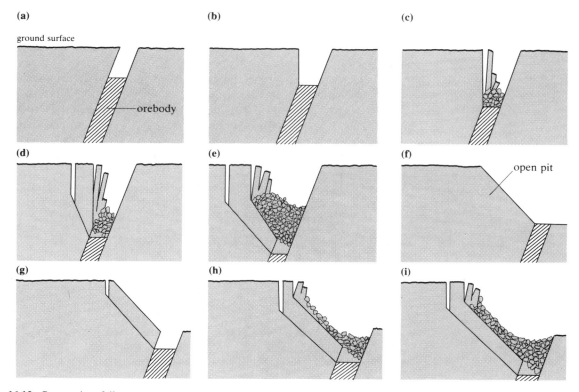

Figure 16.13 Progressive failure sequence with increase in depth of mining: (a) mining from outcrop; (b) failure of overhanging wedge; (c) formation of steep face; (d) development of tension crack and failure surface; (e) development of second tension crack and failure surface; (f) initial open pit; (g) development of tension crack and failure surface; (h) development of second tension crack and failure surface; (i) progressive failure with increase in mining depth. (After Hoek, 1974.)

1 Funnelling to surface vertically above the initial drawpoints.
2 Enlargement and coalescence of these pipes as draw proceeded.
3 Vertical tension cracks formed at the surface marking the boundary of the discontinuous subsidence zone.
4 Displacement of the rock on inclined shear surfaces joining the base of the block to the tension cracks.
5 Development of a new set of tension cracks and subsequent displacement on new shear surfaces with further stages of mining.

Although progressive mechanisms such as this have long been identified, no rational methods of predicting the development of block caving-induced subsidence have been developed. Generally, as in much mining practice, such predictions are made on an empirical basis using previous experience as a guide. With suitable modification, the limiting equilibrium method of analysis developed in the next section for the case of progressive hangingwall caving, could be applied to block caving as well.

16.4.2 Progressive hangingwall caving

When the orebody is not massive and is relatively steeply dipping, caving of only the hanging wall need be considered. In such cases, progressive caving of the hanging wall may result as mining progresses down-dip using sublevel caving methods of mining, for example. A classic example of this form of discontinuous subsidence is that at the Grängesberg iron ore mine in Sweden. Figure 16.12

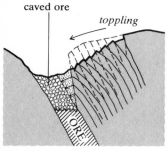

Figure 16.14 Toppling of steeply dipping hangingwall strata.

452

shows the progressive development of the subsidence zone at Grängesberg with increased mining depth over a 40-year period.

This form of subsidence is often associated with mines in which the near-surface portions of the orebody have been extracted previously using open-pit methods. Alternatively, underground mining operations may commence in previously unmined ground at shallow depths below the surface. Figure 16.13 shows the sequence of progressive hangingwall failure postulated by Hoek (1974) for the cases of mining from an outcrop (a–e) and from an open pit (f–i). It is assumed that at each new stage of mining, a tension crack and a shear failure surface form in the hangingwall rock mass at a critical location determined by the strength of the rock mass and the imposed stresses. In some cases, mechanisms other than this may occur. As noted above for the case of block caving, major discontinuities such as faults may provide preferential shear planes. Alternatively, if a major set of persistent discontinuities dips steeply in a similar direction to the orebody, toppling of the hangingwall rock mass may occur (Figure 16.14).

Hoek (1974) developed a limiting equilibrium analysis for predicting the progress, from a known initial position, of hangingwall failure with increasing mining depth. This analysis assumed a flat ground surface and full drainage throughout the caving mass. Brown and Ferguson (1979) extended Hoek's analysis to take account of a sloping ground surface and groundwater pressures in the tension crack and on the shear plane. The idealised model used for the extended analysis is shown in Figure 16.15. The variables involved in the analysis are:

A = base area of wedge of sliding rock mass
c' = effective cohesion of rock mass
H_1 = mining depth at which initial failure occurs

Figure 16.15 Idealised model used in limiting equilibrium analysis of progressive hangingwall caving (after Brown and Ferguson, 1979).

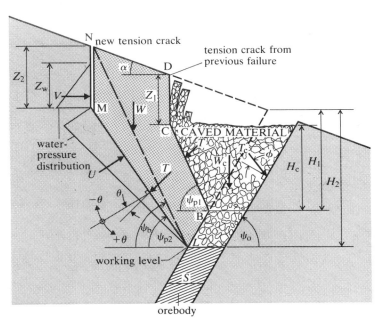

H_2 = mining depth at which subsequent failure occurs
H_c = depth of caved material
 S = width of orebody
 T = thrust on failure plane due to caved material
T_c = thrust on foot wall due to caved material
 U = water-pressure force on failure surface
 V = water-pressure force in tension crack
 W = weight of wedge of sliding rock
W_c = weight of caved material
Z_1 = depth of initial tension crack
Z_2 = depth of subsequent tension crack
Z_w = depth of water in tension crack
 α = dip of upper ground surface (shown positive in Figure 16.15 but may be negative)
 γ = unit weight of undisturbed rock mass
γ_c = unit weight of caved material
γ_w = unit weight of water
 θ = inclination of T to normal to failure surface
σ_n' = effective normal stress on failure plane
 τ = shear stress on failure plane
ϕ' = effective angle of friction of rock mass
ϕ_w = friction angle between caved and undisturbed rock
ψ_0 = dip of orebody
ψ_b = angle of break
ψ_{p1} = dip of initial failure plane
ψ_{p2} = dip of subsequent failure plane.

The analysis is based on the following assumptions:

(a) Mining and caving occur for a large distance along strike compared with the cross-sectional dimensions shown in Figure 16.15. As a consequence, the problem may be reduced to one of two dimensions. Calculations are carried out for unit thickness perpendicular to the plane of the cross section.

(b) The initial position of the hangingwall face is defined by known values of the geometrical parameters H_1, H_c, Z_1, ψ_{p1}.

(c) The extent of caving at the new mining depth, H_2, is defined by a tension crack which forms to a critical depth and strikes parallel to the orebody.

(d) Failure of the hangingwall rock mass occurs along a critical, planar, shear surface whose location is determined by the strength properties of the rock mass and the imposed effective stresses.

(e) The hangingwall rock mass has homogeneous and isotropic mechanical properties. Its shear strength can be defined by an effective stress form of Coulomb's criterion.

(f) Water may enter the tension crack and seep along the potential failure surface into the underground excavations, producing a triangular distribution of excess water pressure along the shear surface.

(g) In carrying out the limiting equilibrium calculations, simplified distributions of stress within the caved and caving masses are used. In particular, the

454

effective normal stresses and shear stresses acting on the failure plane are averaged using the methods of statics.

The development of the equations used in the solution is given in Appendix D which also sets out the stepwise sequence of calculations required to obtain numerical solutions. A simple iterative procedure is required to determine values of ψ_{p2} and ψ_b for a given stage of mining. The required calculations can be carried out most readily using a programmable calculator or microcomputer.

Example. Brown and Ferguson (1979) used the limiting equilibrium analysis to predict the progress of hangingwall caving at Gath's Mine, Zimbabwe. At several sections, the progress of caving had been monitored as mining had progressed down-dip from the 99 level to the 158 level (Figure 16.16). Using the dimensions and problem idealisation shown on Figure 16.16 with $\gamma = 28 \, \text{kN m}^{-3}$, $\gamma_c = 25 \, \text{kN m}^{-3}$, $\alpha = 0$, $U = V = 0$, $\phi_w = 35°$, and $c' = 200 \, \text{kPa}$ and $\phi' = 40°$ determined from Bieniawski's geomechanics classification scheme (Table 3.5), values of $\dot{Z}_l = 31.3 \, \text{m}$, $\psi_{p2} = 61.4°$ and $\psi_b = 66°$ were calculated for mining in increments from the 99 level to the 158 level. These values agreed remarkably well with field data, and provided some confidence in the applicability of the method in this case.

Using the 158 level as the starting point, successive calculations were then made of the locations of the tension crack and shear surface as mining advanced down-dip at this section. The results obtained are shown in Figure 16.16. The results were found to be sensitive to the value of H_c at the beginning of each mining lift. The results shown in Figure 16.16 were obtained using values of H_c that were determined from calculations of the volumes of caved materials and estimates of the volumes of material drawn historically and likely to be drawn in the future.

In practice, the geometry of the problem is rarely as simple as that used in the model. Care has to be taken in assigning values of geometrical parameters such as orebody dip and width and depth of caved material, and of rock mass properties. In the case of Gath's Mine, the real problem also involved the use of backfill in the cave (which served to steepen the angle of break), a three-dimensional effect at the end of the mined section of orebody, the influence of a hill, the

Figure 16.16 Idealisation of progressive hangingwall caving, Gath's Mine, Zimbabwe (after Brown and Ferguson, 1979).

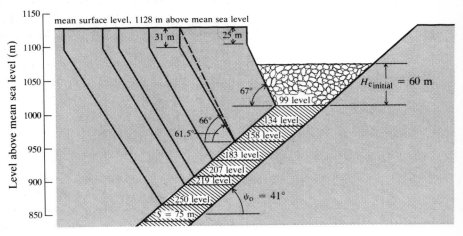

presence of a pillar designed to protect the mill area, and the influence of a major set of geological discontinuities which controlled cave development at one location. The effects of some of these factors could be assessed by sensitivity studies using the analysis, but assessing the influence of others had to remain a matter of judgement and experience.

16.5 Continuous subsidence due to the mining of tabular orebodies

16.5.1 Concepts and definitions

A continuous subsidence trough is produced at the surface when thin, flat-dipping, tabular orebodies are mined by longwall methods which give 100% extraction over relatively large panels. In order for the subsidence to be continuous rather than discontinuous, the relation of the depth of mining and the caving-induced stresses to the strength properties of the rock overlying the orebody must be such that fracture and discontinuous movement of the rock are restricted to the immediate vicinity of the orebody. This occurs in most longwall coal mining operations and in metalliferous longwall mining at depth as, for example, in the South African goldfields. It may also result from the mining of other minerals such as evaporites overlain by relatively weak sedimentary rocks.

Figure 16.17 shows a vertical section through the workings in a typical case. Superimposed on the diagram are the subsidence and surface slope profiles for three classes of panel width. The vertical displacement, s, of any point of the surface, is called the **subsidence**. The maximum subsidence in a given profile is denoted by S. For a given area and corresponding width of extraction known as the **critical area** and **critical width**, S will take the maximum value possible for the particular seam, S_{max}. Areas or widths for which $S < S_{max}$ are described as **subcritical**. For **supercritical** areas or widths, a subsidence of S_{max} is achieved over a finite width rather than at a single point as in the critical case.

The angle of draw, ζ, is defined as the angle made with the vertical by a line drawn from the base of the seam to the point of zero surface subsidence. For the UK coalfields, ζ is taken as 35°. Its values vary with the mechanical properties of the rocks, being lower for stronger rocks and higher for weaker rocks and soils. The value of ζ will also vary with the resolution of the instruments used to measure subsidence and with the cutoff value taken as being equivalent to 'zero' subsidence.

For a horizontal seam at depth h, the mining of a critical area having a critical width, W_c, as its diameter, will produce complete subsidence at a single point on the surface vertically above the centre of the area. In this case

$$W_c = 2h \tan \zeta$$

In addition to this vertical movement, each point in the subsidence trough will also undergo horizontal movement known as **displacement**. It is clear from the shapes of the subsidence profiles shown in Figure 16.17 that the slope of the ground surface and the induced strains must vary from point to point across

456

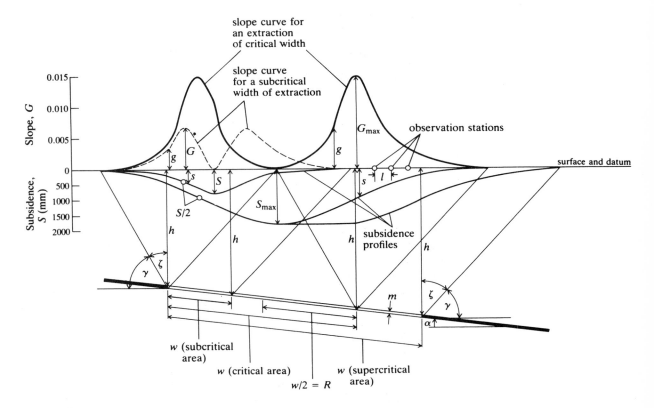

Figure 16.17 Typical section through workings, illustrating standard symbols for subsidence and slope (after National Coal Board, 1975).

the width of the panel. Since differential vertical movement, horizontal compressive or tensile strain, tilt and curvature can all adversely affect surface structures and utilities, in many cases more severely than the subsidence itself, it is essential that means be developed of predicting the values of these variables produced by trough subsidence.

16.5.2 Empirical prediction methods

The most comprehensive and widely used empirical method of predicting subsidence and surface strain profiles is that developed by the National Coal Board (NCB) of the United Kingdom, and reported in the NCB's *Subsidence Engineer's Handbook* (1975). This method is based on data collected in surveys of subsidence over 157 longwall panels at 50 collieries in many parts of Britain, but mainly in the Midlands and Yorkshire. The panels surveyed were at depths ranging from 24 m to 833 m, and had width/depth ratios of from 0.16 to more than 4.

From these data, several tables and graphs were compiled for use in predicting surface movements. The first graph (Figure 16.18) is used to give the maximum or central subsidence, S, as a fraction of extraction thickness, m, for a given depth, h, and panel width, W. If the panel is not sufficiently long for full subsidence to be developed, a correction given by Figure 16.19 must be applied. The complete subsidence profile is then determined from a graph (Figure 16.20) in which the width to depth ratio, W/h, is plotted against the ratio of distance from the centre

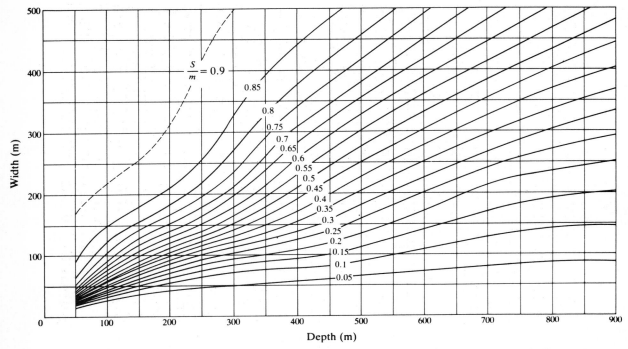

Figure 16.18 Relation of subsidence to width and depth (after National Coal Board, 1975).

of the panel to depth, d/h, in terms of the central subsidence, S. Corrections are given for the effects of seam inclination and surface slope. A method, similar to that described for subsidence, is given for the prediction of horizontal surface strain. The maximum slope occurs at the point of inflection of the subsidence trough where $s = S/2$. The maximum slope that any extraction will cause is said to be approximately $2.75 S_{max}/h$, and the maximum horizontal tensile and compressive strains $0.65 S_{max}/h$ and $0.51 S_{max}/h$, respectively. The tables and graphs apply to single, rectangular panels, but nomograms are given which permit subsidence to be estimated for cases of partial extraction in which rib pillars are used. The effects of a single pillar or of a centre gate are also allowed for.

The NCB method is widely used in the UK where the maximum subsidence is said to be given correctly to within 10% in the great majority of cases. The method makes no allowance for the influence of major geological features such as a fault intersecting the panel or the deforming strata. These effects must be allowed for separately. The fact that the method gives satisfactory predictions for subsidence profiles in collieries in a wide geographical area is probably due to the fact that the overall properties of the carboniferous strata involved are generally similar throughout the region. Attempts to apply the NCB method to longwall coal faces in other parts of the world have met with variable success (Hood et al., 1983).

Profile functions may be used to describe the shape of the subsidence profile. They take the general form

$$s = S_{max} f(B, x, c)$$

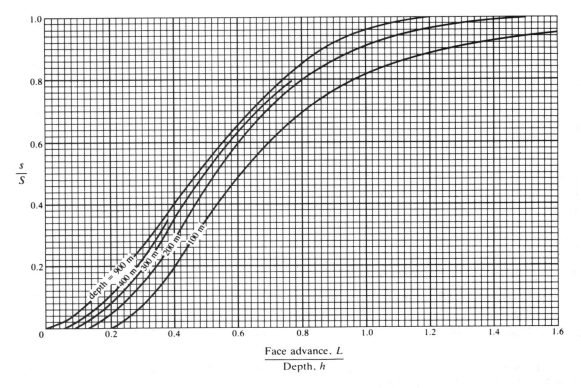

Figure 16.19 Correction graph for limited face advance based on Figure 16.18 (accuracy ±4%) (after National Coal Board, 1975).

where S_{max} is the subsidence at the centre of a panel of critical width, $B = h \tan \zeta$ is the critical radius of extraction, x is the horizontal distance of the point from the origin of co-ordinates, and c is some function or constant.

Exponential, trigonometric, hyperbolic and error functions have been used as profile functions. The function that appears to have given the best results is the hyperbolic tangent function

$$s(x) = \tfrac{1}{2} S_{max}\left[1 - \tanh\left(\frac{bx}{h}\right)\right] \tag{16.11}$$

where x is the horizontal distance measured from the point of inflection (where $s = S_{max}/2$) in the direction of decreasing subsidence, h is the depth of the seam, and b is a constant controlling the slope at the inflection point. For UK conditions, a value of $b = 5$ is used (King *et al.*, 1975, for example). Hood *et al.* (1983) found that a value of $b = 11.5$ applied for a number of transverse subsidence profiles at the Old Ben Number 24 Mine in Illinois, USA.

By differentiation of equation 16.11, the surface slope, or tilt, is given as

$$g = \frac{ds}{dx} = \frac{bS_{max}}{2h} \operatorname{sech}^2 \frac{bx}{h} \tag{16.12}$$

For $b = 5$, this gives the maximum slope at the point of inflection ($x = 0$) as

459

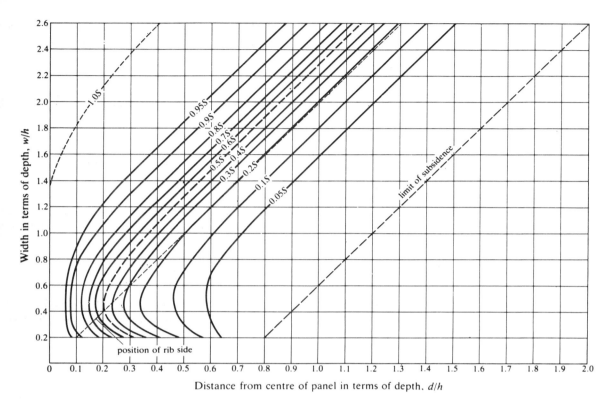

Figure 16.20 Graphs for predicting subsidence profiles (after National Coal Board, 1975).

Section

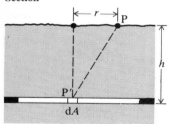

Plan

Figure 16.21 Extraction of an element dA (after Berry 1978).

$$G = \frac{2.5 S_{max}}{h}$$

a similar result to that given by the NCB (1975).

The surface curvature is given by differentiation of the expression for g given in equation 16.12 with respect to x. Methods using profile functions have been developed for estimating subsidence profiles for subcritical widths and for making allowance for the effects of seam inclination (Brauner, 1973).

Influence functions are used to describe the surface subsidence caused by the extraction of an element, dA. The principle of superposition is assumed to apply, so that the subsidence profile for the complete extraction can be found by integrating the influence function over the complete extraction area. The use of numerical integration permits subsidence predictions to be made for extraction areas of any shape.

The influence function $p(r)$ gives the contribution to subsidence at a point P on the surface due to an element of extraction dA at P' as a function of r, the horizontal projection of PP' (Figure 16.21). If P has co-ordinates x, y referred to a set of axes in the plane of the surface, and P' has co-ordinates ξ, η referred to similar axes vertically below in the seam, the influence function takes the form

$$p(r) = w(\xi, \eta) f(r) \tag{16.13}$$

460

where $w(\xi, \eta)$ is a weighting factor introduced to take account of variations in thickness of extraction or the effect of convergence control measures, and

$$r = [(x - \xi)^2 + (y - \eta)^2]^{1/2}$$

By writing $dA = d\xi \, d\eta$, the subsidence due to an area of extraction, A, can be found by integrating equation 16.13 to give

$$s(x, y) = \int \int_A w(\xi, \eta) f \{ [(x - \xi)^2 + (y - \eta)^2]^{1/2} \} \, d\xi \, d\eta$$

Brauner (1973) discusses a range of influence functions used mainly in Germany and eastern Europe. They are generally trigonometric or exponential functions of the form

$$p(r) = k_1 S f(B, r, k_2)$$

where S is the subsidence at the panel centre, $B = h \tan \zeta$ is the critical radius of extraction, and k_1, k_2 are constants. One of the most widely used functions is

$$p(r) = \frac{n S_{max}}{B^2} \exp\left[-n\pi \left(\frac{r}{B} \right)^2 \right]$$

where n is a parameter which characterises the strata properties.

It is apparent that by integration over a large area A, a profile function can be derived from an influence function. Thus the two types of function are not mathematically distinct. Profile functions appear to have the advantage of greater simplicity, but influence functions are more adaptable and can be more useful for irregularly shaped mining panels.

16.5.3 Trough subsidence analysed as elastic deformation

If in mining deep tabular deposits, fracture or plastic deformation of the rock mass is restricted to a relatively small zone surrounding the excavation, it may be assumed that most of the superincumbent strata deforms elastically, at least to a reasonable approximation. As a further idealisation, the problem of excavating a thin seam may be represented as one of a crack in an elastic medium. The problem is then one of determining the stresses and, through the stress–strain equations of elasticity, the strains and hence the displacements, induced by the creation of the crack or slit in a previously stressed semi-infinite elastic body.

If it is assumed that the thickness of the extracted seam is small compared with the other dimensions of the excavation and with the depth, a point on the lower boundary of the seam can be given the same co-ordinate as the nearest point on the upper boundary. The excavation is then located by a single plane, and the convergence of opposing points in the roof and floor can be treated as a discontinuity in displacement at a single point. Unless the excavated width is small,

461

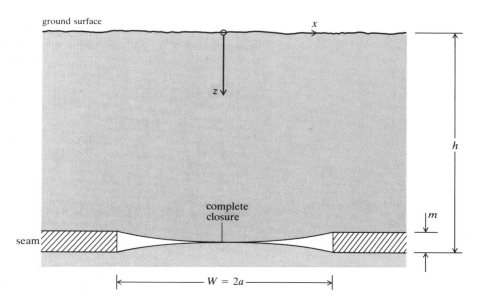

Figure 16.22 Problem definition for elastic analysis of trough subsidence.

the roof and floor will meet over some central area where the displacement discontinuity has its greatest magnitude, m, the thickness of the extraction (Figure 16.22). Where the roof and floor do not meet, the boundaries of the excavation are traction free. A further boundary condition is given by the fact that the upper plane surface remains traction free before, during and after mining.

Berry (1960) solved the simple two-dimensional case involving hydrostatic *in situ* stress and isotropic ground, exactly for complete closure, and approximately for less than complete closure. The calculated maximum settlements were found to be independent of the elastic constants and were less than the values recorded in UK coalfields. In order to give a better representation of the mechanical response of the sedimentary strata, Berry and Sales (1961) carried out a similar analysis using the stress–strain relations of a transversely isotropic medium with the planes of symmetry parallel to the ground surface (Figure 2.10).

As noted in section 2.10, a transversely isotropic material has five independent elastic constants. The stress–strain relations may be written in terms of the five elastic stiffnesses c_{11}, c_{12}, c_{13}, c_{33} and c_{44} as

$$\sigma_{xx} = c_{11}\,\varepsilon_{xx} + c_{12}\,\varepsilon_{yy} + c_{13}\,\varepsilon_{zz}$$

$$\sigma_{yy} = c_{12}\,\varepsilon_{xx} + c_{11}\,\varepsilon_{yy} + c_{13}\,\varepsilon_{zz}$$

$$\sigma_{zz} = c_{13}\,\varepsilon_{xx} + c_{13}\,\varepsilon_{yy} + c_{33}\,\varepsilon_{zz}$$

$$\sigma_{yz} = 2c_{44}\,\gamma_{yz}$$

$$\sigma_{xz} = 2c_{44}\,\gamma_{xz}$$

$$\sigma_{xy} = (c_{11} - c_{12})\gamma_{xy}$$

For the two-dimensional plane strain case, the number of equations reduces to three and the number of elastic stiffnesses to four. The solution of such problems involves two parameters α_1, α_2 which are those roots of the characteristic equation

$$c_{11}\, c_{44}\, \alpha^4 + [c_{13}\,(2c_{44} + c_{13}) - c_{11}\, c_{33}]\, \alpha^2 + c_{33}\, c_{44} = 0$$

having positive real parts. Sometimes it is more convenient to use two parameters k_1, k_2, which are always real and are defined by

$$k_1 = \alpha_1\, \alpha_2 = (1 - v_1^2)^{1/2}\, (E_1/E_2 - v_2^2)^{-1/2}$$

$$k_2 = \frac{1}{2}\,(\alpha_1^2 + \alpha_2^2) = \{\frac{1}{2}\, E_1/[G_2 - v_2\,(1 + v_1)]\}\,(E_1/E_2 - v_2^2)^{-1}$$

where the elastic constants are as defined in section 2.10.

Berry (1963) found that for plane strain and complete closure of an excavation parallel to the surface, the subsidence and horizontal strain at the surface for real α_1, α_2 are given by

$$s(x) = \frac{m}{\pi\,(\alpha_1 - \alpha_2)}\left[\alpha_1 \tan^{-1} \frac{2\,ah_1}{x^2 - a^2 + h_1^2} - \alpha_2 \tan^{-1} \frac{2\,ah_2}{x^2 - a^2 + h_2^2}\right]$$

$$\varepsilon(x) = \frac{2\,a\alpha_1\,\alpha_2\, m}{\pi(\alpha_1 - \alpha_2)}\left[\frac{x^2 - a^2 - h_2^2}{(x^2 - a^2 - h_2^2)^2 + 4h_2^2\, x^2} - \frac{x^2 - a^2 - h_1^2}{(x^2 - a^2 - h_1^2)^2 + 4h_1^2 x^2}\right]$$

where $h_{1,2} = h/\alpha_{1,2}$.

Berry found that reasonable fits to some subsidence and strain data from UK coalfields were obtained using $\alpha_1 = 4.45$, $\alpha_2 = 0.45$, or $k_1 = 2$, $k_2 = 10$. However, these values are probably unreasonable for the coal measures rocks concerned. This suggests that the results may be influenced by some of the assumptions made in the analysis, or alternatively, that the observed subsidence phenomena may not be regarded as being principally due to elastic deformation.

Salamon (1974) showed that the values $\alpha_1 = 9.0$, $\alpha_2 = 0.5$ give values of maximum tilt and maximum compressive and tensile strain that are in closer agreement with those given by the NCB observations. More recently, Avasthi and Harloff (1982) modified Berry and Sales' theory for the three-dimensional case, and introduced an empirical adjustment to account for incomplete closure for those cases in which $W/h < 0.6$. They found that the best fit to the NCB data base was given by $\alpha_1 = 2.74$, $\alpha_2 = 0.65$ which are much closer to values commonly measured for UK coal measures rocks (Berry, 1978) than the other sets of values postulated.

It appears likely, therefore, that the theory of elasticity for transversely isotropic materials can be used to give reasonable estimates of subsidence profiles in some circumstances. The use of suitable numerical methods such as the boundary element method outlined in section 6.5 could eliminate the influence of some of the simplifying assumptions made in Berry's original analysis. In particular,

463

the effects of the *in-situ* stress field and incomplete closure could be taken into account more fully. If this were done, elastic solutions could well prove to be more useful for subsidence prediction than they have been hitherto.

16.5.4 Relation of subsidence to face position and time

It is important to know at what stage subsidence begins and ends in relation to the time of the extraction and the position of the face. This is particularly important in the longwall mining of coal where the surface rights are held by others and sensitive issues of damage to surface structures and services may arise.

The essential fact is that any point on the surface will continue to subside for as long as extraction is taking place within a critical area below that point. In addition to this subsidence, often called **active subsidence**, there may be some time-dependent subsidence due to phenomena such as consolidation or visco-elastic behaviour of the strata, which continues to develop after the point is no longer in the zone of influence of the face. This additional subsidence is called **residual subsidence**. In coal mining, it is known to occur during stoppages in extraction.

Figure 16.23 shows the development of subsidence at a surface point P as a face passes below and beyond it. This curve represents the 'average' of a number of such curves plotted from data obtained from UK coalfields. Measurable subsidence starts when the face is within about $0.75h$ of P and reaches 15% of the eventual maximum value, S, when the face is below P. When the face is $0.4h$ in advance of P, about 75% of S has occurred, and when it is $0.7h$ in advance of P, the subsidence is $0.975S$.

The longitudinal subsidence profile associated with the advancing face produces a travelling horizontal strain wave in which an initial tensile phase is followed by a compressional phase. Transverse tensile or compressional horizontal strains will also result, depending on the location of the point with respect to the centre of the panel. Damage to buildings is likely to be most marked when they are subjected to the maximum tensile strain, which is produced when the surface point is vertically above the face for longitudinal strain or above the rib side for transverse strain.

Figure 16.23 Typical subsidence development curve (after National Coal Board, 1975).

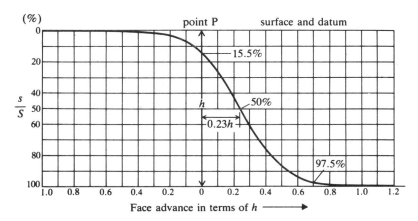

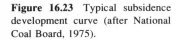

464

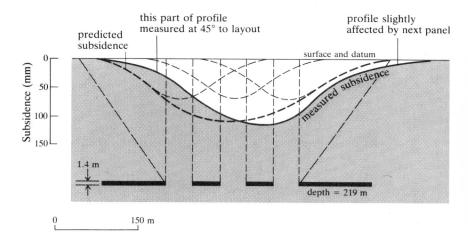

this part of profile
measured at 45° to layout

predicted
subsidence

profile slightly
affected by next panel

surface and datum

measured subsidence

1.4 m

depth = 219 m

0   150 m

Figure 16.24 Predicted and measured subsidence profiles for a case of partial extraction (after Orchard and Allen, 1970).

16.5.5 Design measures to limit subsidence effects

Partial extraction in which substantial rib pillars are left between panels, has been successfully used to limit maximum subsidence and to produce a net subsidence profile that is free from horizontal strain and tilt over most of its width. As shown in Figure 16.24, the estimated net subsidence profile is obtained by superposition of the profiles produced by the individual panels. In the UK coalfields, this method has been used to give up to 70% extraction with 30–100 m wide pillars left between panels extracted with $W/h < 1/3$. Depending on the layout and the extraction ratio, reductions in maximum subsidence in the order of 80% may be achieved.

Goaf treatment by strip packing or hydraulic or pneumatic solid stowing can reduce the subsidence in a single panel by up to 50% depending on the nature and timing of the treatment. The largest reductions are obtained for solid stowing carried out immediately after mining.

Harmonic extraction involves the phased removal of the mineral from a critical area such that the ground surface is lowered smoothly and horizontal strains are minimised. The technique may be used to protect structures that are especially important or susceptible to subsidence-induced damage. Harmonic extraction requires that the panel be advanced in at least two faces maintained at a carefully calculated distance apart. The orientation of the structure with respect to the direction of face advance determines whether protection against the transverse or the longitudinal surface wave is the more important.

17 Blasting mechanics

17.1 Blasting processes in underground mining

Reliable procedures for rock blasting are well established in mining engineering practice. Explosive charges are emplaced in blast holes suitably located relative to a free surface of an opening, and detonated. Rock surrounding the charges is fragmented and displaced by the impulsive loading in the medium, generated by the sudden release of the explosive's potential energy. Control and routine application of such an intrinsically violent process represent significant feats in both chemical and mining technology.

Underground mining in hard-rock settings is highly dependent on the successful execution of blasting procedures. Although shaft borers and raise borers are occasionally used in developing vertical access to and within an orebody, the great majority of mine development is still undertaken with well-established and efficient methods of drilling and blasting. Orebody rock is liberated from its natural surroundings, and subjected simultaneously to an episode of massive comminution, by primary blasting in the stopes. Other minor blasts, in stope drawpoints and ore passes, for example, may be required to maintain the free flow of ore in the ore-handling system. The significance of rock blasting in mine development and production indicates the value of a brief exposition of the fundamentals of blasting mechanics.

Preceding discussions of mine excavation design and excavation support design have been based, in the main, on consideration of static forces and stresses generated in the medium, and their effect on the *in situ* rock. Most mine openings are created incrementally, by segments excavated in a near-instantaneous process. The consequences of the rapidity of excavation have been discussed in Chapter 10, but the nature and effects of the excavation process per se have not been considered. A particular concern with blasting is its effect on the rock in the immediate vicinity of an excavation. Intense local fracturing, and disruption of the integrity of the interlocked, jointed assembly, can be produced in the near-field rock by poor blast design. More extensive adverse effects can be induced by the transmission to the far field of energy input to the rock by explosive action. In high-stress settings such as occur at depth, or in pillars supporting panels mined to a high extraction ratio, perturbations associated with blasting may trigger extensive instability in mine structures.

17.2 Explosives

An explosive is any material or device which can produce a sudden outburst of gas, applying a high impulsive loading to the surrounding medium. Chemical explosives are the most widely used in mining practice. Their composition and properties have been described in detail by Cook (1958). Industrial chemical explosives are of two main types: deflagrating explosives, such as black powder, which burn relatively slowly, and produce relatively low blasthole pressure; detonating (high) explosives, which are characterised by superacoustic reaction rate and high blasthole pressure. The detonating explosives are themselves considered in three categories, reflecting their respective sensitivities to ease of initiation of detonation. Primary explosives, such as lead azide, lead styphnate, or mercury fulminate, are initiated by spark or impact. They are highly unstable compounds, and used industrially only in initiating devices such as blasting caps, as the top charge. Secondary explosives require the use of a blasting cap for practical initiation, and in some cases may require an ancillary booster charge. Explosives in this category are formulated from chemicals such as nitroglycerin (NG), ethyleneglycoldinitrate (EGDN), or pentaerythrotetranitrate (PETN), mixed with other explosive materials and stabilizing agents. Tertiary explosives are insensitive to initiation by a standard (No. 6) strength blasting cap. Most explosives in this category are the dry blasting agents (DBAs) or slurry explosives and blasting agents.

Historically, most high explosives were manufactured from the organic nitrates, mixed with organic and inorganic chemicals to produce mechanically and chemically stable materials known as gelignites and dynamites. Now, the proportion of organic nitrates in these types of explosives has been reduced, by their partial replacement with such chemicals as ammonium nitrate (AN). Also, industrial consumption of these explosives has declined, due to the formulation of more convenient explosive mixtures.

A blasting agent such as ANFO (94% AN–6% fuel oil) is an oxygen-balanced mixture of an oxidiser and a fuel. This means that by achieving close admixture of these components, the material can be detonated, to yield H_2O, NO_2, and CO_2. Slurry blasting agents and explosives are also formulated as an oxygen-balanced mixture, but instead of a simple physical admixture, the oxidiser and fuel constituents of the explosive are suspended in a stable water–gel matrix. They are typically more stable chemically than dry blasting agents, and may be water resistant.

Detonation in an explosive column (contained, for example, in a blast hole) involves the passage through the column of a chemical reaction front. The front is driven through the column, by the products of the reaction, at superacoustic velocity, called the detonation velocity. Passage of the front causes a step rise in pressure in the explosive, achieving a detonation pressure which may exceed many GPa. The blasthole wall is subjected to a pressure approaching that in the detonation front. Pressure decay in the blast hole is relatively slow, due to the persistent action of the detonation product gases.

The response of rock to explosive attack is determined by both the explosive properties and the rock properties. The main explosive property is its strength, which is a standardised measure of the explosive's capacity to do useful work on its surroundings. It may be measured in standard tests, such as an underwater test, in which pressure pulses transmitted in the water after detonation of a charge are used to estimate the energy release from the charge. These tests indicate that the release of the explosive's chemical potential energy does not occur in a single episode. The initial energy release associated with detonation (the shock energy) is followed by oscillation of the product gas bubble. Successive phases of energy output are identified with the bubble energy of the explosive, which represents energy possessed by the detonation product gas due to its high pressure.

Another current method of assessing the strength of an explosive is by estimating its free-energy output, from the thermodynamics of its detonation reaction. The Absolute Strength Value (ASV) of an explosive is expressed in terms of the free-energy output (in joules) per 0.1 kg of explosive. The Relative Strength Value of an explosive is the ratio of its calculated useful work output, relative to that of ANFO, taken as a basis of 100 units. Details of explosive strength specification are discussed by Harries (1977).

The partitioning of explosive energy between shock energy and bubble energy suggests that the ASV of an explosive is an incomplete measure of its potential performance in rock. The term 'brisance' is used to indicate the potential 'shattering action' of an explosive. It is related directly to the detonation pressure of the explosive, which is itself related to detonation velocity. High-brisance explosives may be characterised by detonation velocities greater than about 5000 m s^{-1}. Explosives with detonation velocities less than 2500 m s^{-1} would be classed as low-brisance compounds.

Several rock properties determine the performance of a particular explosive in the medium. The capacity of the rock mass to transmit energy is related in part to the Young's modulus of the rock medium. The ease of generating new fractures in the medium is a function of the strength properties of the rock material, which may be represented for convenience by the uniaxial compressive strength C_0. The unit weight of the rock mass affects both the energy required to displace the fragmented rock, and the energy transmissive properties of the intact medium.

Efficient and successful performance of an explosive in a rock mass requires that its properties be compatible with those of the subject rock mass. An empirical correlation of the preferred explosive type, for a range of rock material and rock

Figure 17.1 An empirical matching of explosive and rock mass properties.

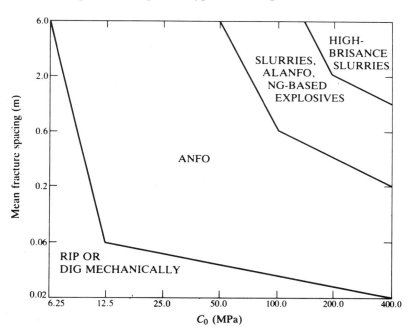

mass properties, is proposed in Figure 17.1. For very low average fracture spacing, or very low material compressive strength, the rock mass may be ripped or dug mechanically, rather than blasted. A significant observation from Figure 17.1 is that ANFO is a suitable explosive for use in a wide range of rock mass conditions. The application of high-brisance explosives is justified only in the strongest, more massive rock formations.

Other explosive properties in addition to strength and brisance need to be considered in explosive selection for a particular blasting operation. In underground applications, the fume properties of the explosive, i.e. the chemicals present in the detonation product gases, determine the air quality in the working area after blasting. The chemical properties of the explosive itself, and of the lithology in which it is to be used, can present safety problems. For example, some sulphide ores are subject to rapid, exothermic oxidation on contact with ammonium nitrate. The nature of the flame or flash produced by the explosive on detonation determines its potential for initiating detonation in the mine atmosphere. Such unfavourable atmospheric conditions can arise in gassy coal seams, or in highly pyritic metalliferous orebodies. Finally, explosive properties such as the air gap sensitivity and impact sensitivity have direct operational consequences related to their use in fractured ground and in development blasting.

17.3 Elastic models of explosive–rock interaction

Detonation of an explosive charge in rock results in dynamic loading of the walls of the borehole and generation of a stress wave which transmits energy through the surrounding medium. The generation of fractures can be assumed to be related in some way to the magnitudes of the transient stresses associated with the passage of the wave. The solutions discussed in Chapter 10 for elastic wave generation by spherical and cylindrical sources are the basis for considering explosive interaction with a rock mass.

The most important solution for the stress distribution around an explosive source is that due to Sharpe (1942) for a spherical charge. The charge is detonated in a spherical cavity of radius a, generating a spherically divergent P wave. The transient cavity pressure is taken to be represented by the expression

$$p = p_0 \, e^{-\alpha T}$$

where p_0 is the peak wall pressure, and α is a decay constant. Sharpe showed that, in the general solution of the spherical wave equation, the function f in equation 10.36 is given by

$$f = p_0 a / \{\rho \left[\omega_0^2 + (\alpha_0 - \alpha)^2 \right] \} \left[e^{-\alpha_0 T} \left(\cos \omega_0 T + \frac{\alpha_0 - \alpha}{\omega_0} \sin \omega_0 T \right) - e^{-\alpha T} \right] \qquad (17.1)$$

where

$$\alpha_0 = (C_p / a) \left[(1 - 2\nu)/(1 - \nu) \right]$$

$$\omega_0 = C_p \left[(1 - 2\nu)^{1/2}/(1 - \nu) \right]$$

$$T = t - (r - a)/C_p$$

An elaboration of the Sharpe solution by Duvall (1953) was based on the assumption that the cavity pressure could be represented by the expression

$$p = p_0 \left(e^{-\alpha t} - e^{-\beta t} \right)$$

Since this is merely the superposition of two separate temporal pressure variations, the form of the complete function f can be established simply by replacing α by β in equation 17.1, and subtracting the resulting expression from that given in equation 17.1.

When the form of the function f describing transient displacement is known, the procedures discussed in Chapter 2 can be used to establish expressions for the transient strain components. Application of the appropriate stress–strain relations produces expressions for the transient stress components.

Most explosive charges used in practice are cylindrical. Difficulties with obtaining a general solution for a cylindrical source have already been noted. Since explosive charges are neither infinitely long nor subject to instantaneous detonation, the several attempts to obtain a solution to this type of problem are of little practical interest. A useful solution for a cylindrical charge has been formulated by Starfield and Pugliese (1968). In their procedure, a cylindrical charge is discretised into a set of charge segments, each of which is represented by an equivalent spherical charge. With the unit solution given by equation 17.1, and a known detonation velocity for the charge, the temporal variation of displacements, strains and stresses at any point in the medium can be determined numerically. It is to be noted also that this numerical model correctly describes the divergence of the wave generated by the charge over a conical front, as illustrated in Figure 17.2. The results obtained by Starfield and Pugliese in field experiments with cylindrical charges were in general agreement with predictions from the simple finite difference model.

Figure 17.2 A finite difference model of detonation and wave generation by a long, cylindrical charge.

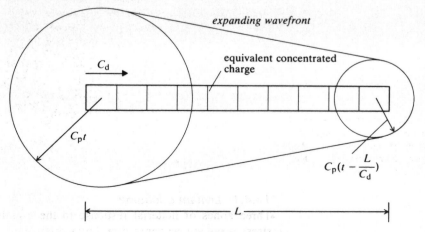

470

17.4 Phenomenology of rock breakage by explosives

Explosive attack on rock is an extremely violent process, and experimental attempts to define the mechanics of rock breakage by explosives have not been highly successful. The following qualitative account of explosive interaction with rock is based mainly on the accounts by Duvall and Atchison (1957), and Kutter and Fairhurst (1971).

In the period during and following the passage of a detonation wave along an explosive charge, the rock around the blast hole is subjected to the following phases of loading:

(a) dynamic loading, during detonation of the explosive charge, and generation and propagation of the body wave in the medium;

(b) quasi-static loading, under the residual blasthole pressure applied by the detonation product gases;

(c) release of loading, during the period of rock displacement and relaxation of the transient stress field.

The evolution of fracture patterns associated with these intervals of loading is illustrated in Figure 17.3.

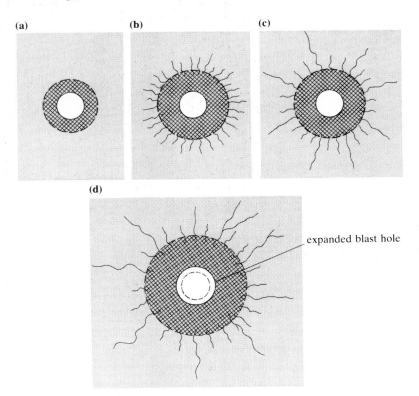

Figure 17.3 Successive stages of destruction of rock by an explosive charge under dynamic load (a,b,c) and quasi-static load (d) (after Kutter and Fairhurst, 1971).

17.4.1 Dynamic loading

Three zones of material response to the impulsive loading and high-intensity stress wave are recognised in the medium.

471

In the immediate vicinity of the blast hole, the high stress intensity results in the generation of a shock wave in the rock. In this so-called shock zone, the rock behaves mechanically as a viscous solid. Passage of the stress wave causes the rock to be crushed or extensively cracked, and the intensity of the wave is reduced by viscous losses. The attenuation process also results in reduction of the wave propagation velocity to the acoustic velocity. For a blast hole of radius r_h, the radius r_s of the shock zone may be about $2r_h$. In some cases, superficial observation may not, in fact, reveal a crushed zone around a blast hole.

The domain immediately outside the shock zone is called the transition zone. In this region, the rock behaves as a non-linear elastic solid, subject to large strain (i.e. the small strain elastic theory developed in this text is inadequate to describe rock behaviour in this zone). New fractures are initiated and propagated in the radially compressive stress field, by wave interaction with the crack population. Crack development is in the radial direction, resulting in a severely cracked annulus, called the Rose of Cracks. Generation of the radial cracks extracts energy from the radial P-wave, resulting in reduction in the stress intensity. The radius, r_t, of the transition zone is about 4–$6r_h$.

In the transition zone, the intensity of the state of stress associated with the radial wave is reduced to a level at which the rock behaves linearly elastically. The behaviour of the rock in this domain, called the seismic zone, can be explained adequately by elastic fracture mechanics theory. Although new cracks may be initiated in this region, crack propagation occurs exclusively by extension of the longest cracks of the transition zone, in accordance with the notions outlined in Chapter 4. Thus, a short distance outside the transition zone, a few cracks continue to propagate, at a velocity of about 0.20–$0.25C_p$. The P-wave therefore rapidly outruns the crack tips, and propagation ceases. It appears that at a radius of about $9r_h$, macroscopic crack generation by the primary radial wave ceases. However, during transmission of the wave towards the free face, fractures may be initiated at the Griffith cracks. The process may be considered as one of conditioning the rock mass for subsequent macroscopic rupture, or of an accumulation of damage in the rock fabric.

When the radial compressive wave is reflected at a free face, a tensile wave, whose apparent source is the mirror image of the blast hole reflected in the free

Figure 17.4 Reflection of a cylindrical wave front at a free face.

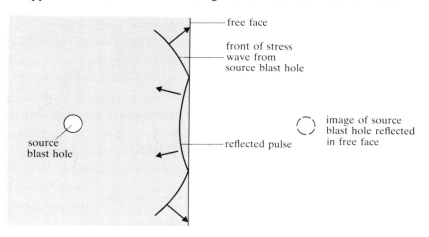

free face

front of stress wave from source blast hole

image of source blast hole reflected in free face

reflected pulse

source blast hole

face, is generated. The relevant geometry is shown in Figure 17.4. It is possible that, in massive rock, slabbing or spalling may occur at the free face, although there is no convincing evidence for this. In the interior of the medium, crack extension may be promoted by the more favourable mechanical environment of the tensile stress field, resulting in further accumulation of damage. In a jointed rock mass, the rôle of the reflected tensile pulse is limited, due to joint separation, trapping the wave near the free face.

17.4.2 Quasi-static loading

The dynamic phase of loading is complete when the radial wave propagates to the free face, is reflected, and propagates back past the plane of the blast hole. For an average rock mass ($C_p = 4000 \mathrm{\,m\,s}^{-1}$), this process is complete within 0.5 ms/m of burden. (The burden is the perpendicular distance from the blasthole axis to the free face.) Because mass motion of the burden does not occur for an elapsed time much greater than the dynamic load time, it appears that pressure exerted in the blast hole by the detonation product gases may exercise a significant role in rock fragmentation. Sustained gas pressure in the blast hole increases the borehole diameter, and generates a quasi-static stress field around the blast hole. Gas may also stream into the fractures formed by dynamic loading, to cause fracture extension by pneumatic wedging. An idea of the action of the gases may be obtained by considering the stress distribution around a pressurized hole, by applying the Kirsch equations (equations 6.18). In the following discussion, the effect of field stresses on the quasi-static stress distribution is neglected.

The simplest case of quasi-static loading involves a pressurized hole, of expanded radius a, subject to internal pressure p_0, as shown in Figure 17.5a. If the region around the hole boundary is uncracked, the state of stress at any interior point, of radius co-ordinate r, is given by

$$\sigma_{rr} = p_0 \, a^2/r^2, \quad \sigma_{\theta\theta} = -\, p_0 \, a^2/r^2, \quad \sigma_{r\theta} = 0 \qquad (17.2)$$

Figure 17.5 Conditions of quasi-static loading around a blast hole.

and the hole boundary stresses are given by

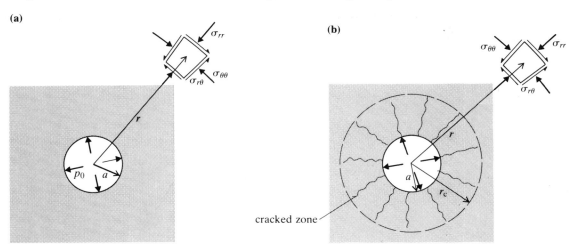

473

$$\sigma_{rr} = p_0, \quad \sigma_{\theta\theta} = -p_0 \qquad (17.3)$$

Thus, if the state of stress represented by Equation 17.3 is incapable of generating fractures at the hole boundary, that represented by equation 17.2 cannot generate fractures in the body of the medium. This suggests that the pattern of cracks produced during the dynamic phase may be important in providing centres from which crack propagation may continue under gas pressure.

Quasi-static loading may occur in the presence of radial cracks, as illustrated in Figure 17.5b, with no gas penetration of the cracks. The presence of radial cracks means that no circumferential tensile stresses can be sustained in the cracked zone. At any point within the cracked zone of radius r_c, the state of stress at any point r is defined by

$$\sigma_{rr} = p_0 \, a/r, \quad \sigma_{\theta\theta} = 0$$

and at the perimeter of the cracked zone by

$$\sigma_{rr} = p_0 \, a/r_c, \quad \sigma_{\theta\theta} = -p_0 \, a/r_c \qquad (17.4)$$

The implication of equations 17.4 is that existing radial cracks around a hole may extend so long as the state of stress at the boundary of the cracked zone satisfies the macroscopic failure criterion for the medium.

A third possible case of quasi-static loading involves radial cracks, but with full gas penetration. If the volume of the cracks is negligible, the state of stress at the boundary of the cracked zone is given by

$$\sigma_{rr} = p_0, \quad \sigma_{\theta\theta} = -p_0 \qquad (17.5)$$

In practice, the degree of diffusion of gas into the fractures is likely to lie somewhere between the second and third cases, described by equations 17.4 and 17.5. In any event, the existence of circumferential tensile stresses about the blast hole provides a satisfactory environment for radial fracture propagation.

Analysis of high-speed photographs of surface blasts suggests that radial fractures propagate to the free face, and that the elapsed time for generation of these fractures may be about 3 ms/m of burden (Harries, 1977).

17.4.3 Release of loading

Field observations of blasts suggest that the elapsed time between charge detonation and the beginning of mass motion of the burden may exceed ten times the dynamic phase of loading. At that stage, the burden is rapidly accelerated to a throw velocity of about 10–20 m s^{-1}. In the process of displacement, disintegration of the rock mass occurs. It has been suggested by Cook, M.A. *et al.* (1966) that impulsive release of the applied load may lead to over-relaxation of the displacing rock, generating tensile stresses in the medium. Some evidence to support this postulated mechanism of final disintegration is provided by Winzer and Ritter (1980). Their observations, made on linearly scaled test blasts, indicated that new fractures are generated in the burden during its airborne

474

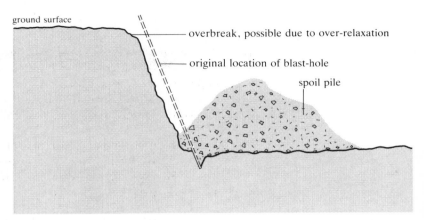

Figure 17.6 Overbreak in a surface bench blast, possibly due to an over-relaxation mechanism.

displacement. There is no similar evidence from full-scale blasts to support the mechanism, and attempts to generate fractures by impulsive unloading in laboratory tests have been unsuccessful. However, there is one aspect of blasting which can be explained most satisfactorily by the over-relaxation mechanism. Overbreak, illustrated in Figure 17.6, is readily comprehensible in terms of the rebound of the solid following rapid release of the blasthole pressure.

The preceding discussion was concerned with blasting in a medium at low states of stress. In underground mines, the state of stress at a blast site may be high. In such a situation, illustrated in Figure 17.7, where the local maximum principal stress is parallel to the free face, crack generation and propagation occur preferentially parallel to the free face. Crack propagation perpendicular to the free face is impeded.

17.5 Computational models of blasting

The qualitative description of explosive action in rock does not present a basis for either design of blasts or for control of the effect of blasting in the surrounding medium. The most promising approaches for the long-term resolution of these needs may lie in the further development of various computational models of the blasting process. The most advanced schemes have been described by Dienes and Margolin (1980), Grady and Kipp (1980) and Margolin (1981). Each is based on a finite difference representation of the explosive–rock interaction, within a continuum mechanics framework. The model described by Margolin is based on:

(a) a statistical description of Griffith crack distribution in the medium;
(b) a criterion for crack growth;
(c) a specification of modulus reduction with crack development;
(d) a statistical description of the crack distribution as it evolves in time, taking account of crack growth termination by intersection with other cracks.

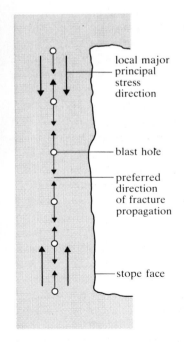

Figure 17.7 Local stress field near a stope face, and the imposed preferred direction of fracture propagation.

While these models are not now suitable for routine blast design, they have been used in special cases, such as the design of comminution blasts required for *in-situ* oil-shale retorting. A particular source of interest in these models is that they seek to predict the macroscopic behaviour of a rock mass from the microscopic theory of fracture and the global equations of continuum mechanics.

475

17.6 Perimeter blasting

The discussion in Chapters 7–9 was concerned with the design of excavations in various types of rock masses. The general assumption in the development of the design methods was that the final mechanical state of the excavation peripheral rock was not affected by the method of creating the opening. Since blasting is an intrinsically violent process, the potential exists in practice for significant local degradation of the mechanical properties of the rock. A well-conceived design philosophy not supported subsequently by appropriate excavation procedures is impossible to reconcile with sound engineering practice.

Perimeter blasting is the process in which closely controlled blasting practices are applied to produce a geometrically precise and relatively undisturbed ultimate surface. The objectives are to restrict the number and extent of new fractures in the rock, to prevent undue disturbance of the jointed mass, and therefore to preserve the inherent strength of the *in-situ* rock. The consequences of effective perimeter blasting are expressed in the operational performance of the opening. In permanent openings, lower support costs are achieved. In temporary openings, reduced maintenance costs are obtained by reduced scaling. In both cases, excavation development rates may be improved by reduced overbreak, leading to reduced effort in scaling the freshly generated surface, and reduced loading and haulage of development mullock. Finally, smooth walls result in reduced frictional resistance to airflow and improved mine ventilation capacity.

There are two techniques of perimeter blasting – pre-splitting and smooth blasting. Both methods are industrially important, and each complements the other in scope for practical application. They are based on the use of decoupled charges, in which the objective is to restrict the development of a rose of cracks around a hole. This is sought by isolating the explosive charge from the blasthole surface, using a charge diameter appreciably less than the blasthole diameter, and using spacers to locate the charge axis along the hole axis. Special explosives, based on NG, are formulated and packaged, to ensure stable detonation of the explosive at relatively low charge diameter.

17.6.1 Pre-split blasting

With this method, a continuous fracture which will form the final surface of an excavation is generated in the absence of a local free face. As an example, Figure 17.8a represents a cross section of a development heading. Parallel, close-spaced holes, which will define the excavation perimeter, are drilled in the direction of face advance. Prior to excavation of any material within the prospective final surface of the heading, charges in the perimeter holes are detonated nearly simultaneously. If hole spacings are sufficiently small, the explosive charges adequate, and the geomechanical conditions satisfactory, a fracture surface is developed over the smooth surface containing the axes of the blast holes.

An understanding of the mechanics of pre-split blasting is necessary to assure its successful implementation. A useful insight into pre-split mechanics is provided by Kutter and Fairhurst (1968). Fracture development along the centreline between adjacent blast holes is the result of interaction of detonation in one blast hole with the local stress field produced by explosive action in an adjacent hole. Kutter and Fairhurst demonstrate that fracture development does not occur when

476

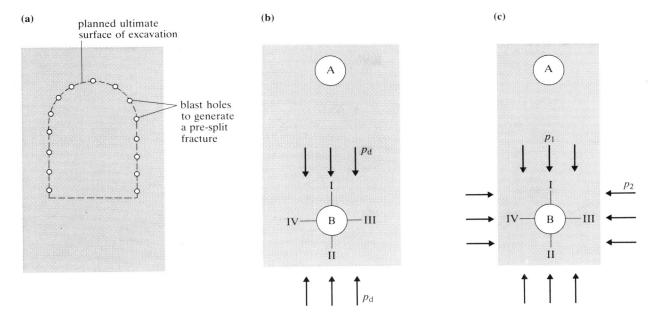

Figure 17.8 Layout of blast holes in a pre-split blast, and transient stress conditions around a blast hole for very short and short delay between initiation of adjacent blast holes.

either there is very long delay between the initiation of adjacent holes (i.e. the charges detonate independently), or when adjacent blast holes initiate simultaneously. For the sake of clarity, in the following discussion, two blast holes, A and B, are considered, with hole A initiated prior to hole B. The rock medium is stress free.

Consider the case where hole B initiates as the stress wave emitted from hole A passes over it, as illustrated in Figure 17.8b. The transient local stress field around hole B is effectively uniaxial, of magnitude p_d, and oriented parallel to the centreline of the holes. Since the wavelength of the pulse is relatively long with respect to hole diameter, the transient stress concentrations around hole B can be estimated from the Kirsch equations. At points I and II on the hole boundary, the boundary stress is

$$\sigma_{\theta\theta} = -p_d$$

and at positions III and IV, it is given by

$$\sigma_{\theta\theta} = 3p_d$$

Emission of the stress wave by detonation of hole B, and superposition on the transient boundary stresses, result in tensile stresses which are greatest at points I and II, and least at points III and IV. Radial cracks therefore initiate preferentially at points I and II (i.e. in the centreline direction). The effect of gas pressure in B is to promote the development of the initially longest cracks, i.e. those in the centreline direction.

A second feasible method for pre-splitting involves the initiation of hole B while quasi-static pressure operates in hole A. Suppose gas pressure in hole A produces a local biaxial stress field for hole B defined by components p_1 and p_2, oriented as shown in Figure 17.8c. A pressurised hole produces a local biaxial stress field approximated by

$$p_2 = -p_1$$

At points I and II on the hole boundary, the circumferential stress component is estimated from the Kirsch equations to be

$$\sigma_{\theta\theta} = -4p_1$$

and at points III and IV, to be

$$\sigma_{\theta\theta} = 4p_1$$

Thus, emission of the stress wave on detonation of hole B and superposition of these transient boundary stresses imposed by hole A, generates highest tensile stresses at points I, II. Radial cracks initiate preferentially at points I and II. Again, gas pressure in hole B promotes the development of the initially longest cracks, which are oriented in the centreline direction.

These two pre-split mechanisms, involving short delay and very short delay between the initiation of adjacent blastholes, have been confirmed experimentally. They suggest that, for the complete set of blast holes which will form the ultimate surface of an excavation, effective pre-splitting requires that all pairs of adjacent holes are initiated with a short delay with respect to one another.

The preceding discussion was concerned with pre-split blasting in isotropic rock with low field stresses. Real blasting settings underground may involve high *in situ* stress and stratified or jointed rock.

In Figure 17.9a two adjacent blast holes, members of a set of pre-split holes, are oriented so that their centreline is perpendicular to the major principal field stress. The static boundary stresses are a maximum at points c,d,e,f, and a minimum at points g,h,i,j. When either hole initiates, the longest initial cracks will form in the direction parallel to the major principal field stress, and they will propagate preferentially in that direction under gas pressure. For the case of holes oriented such that the centreline direction is parallel to the direction of the major principal field stress, as shown in Figure 17.9b, cracks will initiate preferentially at points g,h,i,j, and gas pressure will promote preferential crack development in the centreline direction. However, the effect of the minor principal stress is to impede crack development. If the absolute value of Kp is high relative to the gas pressure developed in the blast hole, crack propagation will be prevented. It is concluded that pre-split blasting will show variable results in a stressed medium, depending on the orientation of adjacent pairs of holes relative to the field stresses, and that the process becomes less effective as the field stresses increase. The practical consequence is that pre-splitting may be successful in near surface development work, but at even moderate depth, it may be completely ineffective.

In stratified rock, the rock fabric is populated by micro-cracks, oriented parallel to the visible texture. Considering the single blast hole shown in Figure 17.9c, drilled in the plane of stratification, the preferred direction of crack development is parallel to the stratification, exploiting the natural micro-structure as guide

(a) **(b)** **(c)**

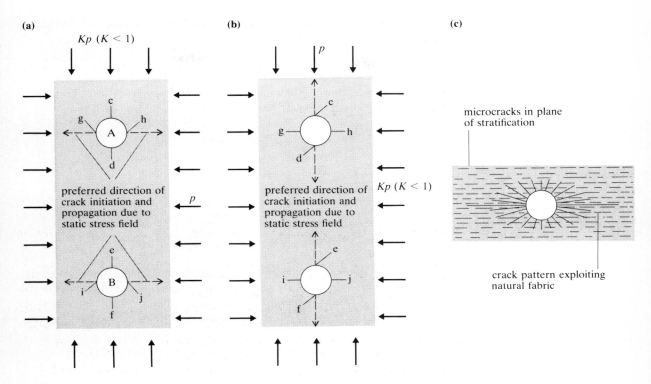

Figure 17.9 Influence of field static stresses and rock stratification on the development of pre-split fractures.

cracks. The general consequence is that pre-split fractures may develop in any anisotropic rock parallel to a dominant fabric element. Fracture development perpendicular to the fabric element may be difficult or practically impossible.

17.6.2 Smooth blasting

Smooth blasting practice involves the development of the ultimate surface of the excavation by controlled blasting in the vicinity of a penultimate free face. Holes are initiated with short delay between adjacent holes, and the burden on holes exceeds the spacing.

The mechanics of smooth blasting may be understood by examining the local state of stress around the penultimate boundary of the excavation. The situation is illustrated in Figure 17.10a. It has been noted, in Chapters 7–9, that the design of an opening should achieve a compressive state of stress in the excavation boundary and adjacent rock. Considering a typical perimeter blast hole near the free face, shown in Figure 17.10b, the local stress field is virtually uniaxial and directed parallel to the penultimate surface. This generates tensile boundary stresses around the blast hole at points a,b, and compressive stresses at points c,d. Thus, the stress wave emitted by detonation of the charge in the hole initiates radial fractures at points a,b, and these propagate preferentially parallel to the local major field stress. Both these factors favour generation of fractures parallel to the penultimate surface of the excavation.

It is to be noted that a high *in-situ* state of stress, or a high local state of stress around an excavation, promotes more effective smooth blasting. It is concluded that smooth blasting is the preferred method of perimeter blasting at underground

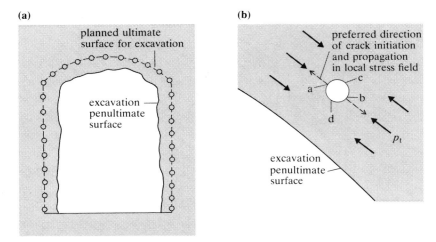

(a)

planned ultimate
surface for excavation

excavation
penultimate
surface

(b)

preferred direction
of crack initiation
and propagation
in local stress field

c

a

b

d

p_t

excavation
penultimate
surface

Figure 17.10 Penultimate stage in the execution of a smoothing blast, and the mechanism of control of fracture development by the local boundary stress.

sites, where high states of stress are common. In the design of a smoothing blast, however, particular stress environments and excavation geometries may require that the evolving boundary stress around an excavation be taken into account to assure success of the blast around the complete excavation periphery.

17.7 Transient ground motion

Blasts in an underground mine are conducted for two purposes: fragmentation and comminution of the rock mass and excavation of access and service openings. Blasting for fragmentation purposes is conducted on a large scale. For example, a major stope blast, or a large pillar-wrecking blast, may involve the sequential detonation of several hundred tonnes of explosive distributed through several hundred thousand cubic metres of rock. In such cases, two objectives are to control transient general motion in the far field, at the ground surface, and to prevent damage to mine access and service openings.

Figure 17.11a is a schematic illustration of a blast site at an underground mine. When the blast is executed, P waves are generated at the various blast sources. S waves are generated in the rock medium by internal reflections and refractions. Thus the blast site acts as an apparent source for both P and S body waves, which propagate in all directions. Some of the waves travel to the ground surface, where they are partly reflected as PP waves, PS waves etc. In addition, the waves are partly refracted in the ground surface, and a surface wave is generated in the upper layers of the rock medium. The point O directly above the blast site, on the ground surface, acts as the apparent source of the surface waves. The characteristic of these waves, of which there are two types, is that they are generated and sustained only near the ground surface.

The most generally occurring surface wave is the Rayleigh wave. Particle motion induced by passage of the wave is backward elliptical, in a vertical plane parallel to the direction of wave propagation. The motion is equivalent to coupled P and vertically polarised shear (SV) waves. The nature of the ground motion is illustrated in Figure 17.11b. The intensity of the motion dies out rapidly with

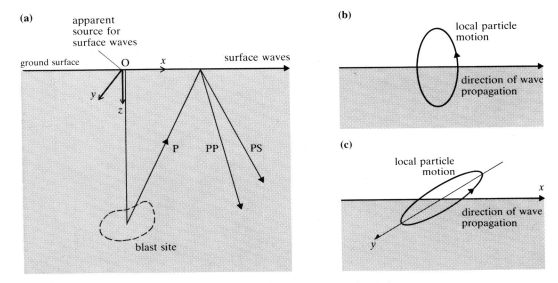

Figure 17.11 Generation of surface waves above a blast site, and the nature of particle motion associated with Rayleigh and Love waves.

depth. The propagation velocity of the R wave is given by the positive real root of the equation

$$4C_s^3 \left[(C_p^2 - C_R^2)(C_s^2 - C_R^2) \right]^{1/2} = C_P (2C_s^2 - C_R^2)^2 \tag{17.6}$$

For $\nu = 0.25$

$$C_R = 0.926\, C_s = 0.53\, C_p$$

A Love wave is generated at the ground surface when a layer of low modulus material overlies a higher modulus material. This is a relatively common condition, which can arise when a weathered layer overlies fresh rock, or when backfill has been placed to produce a level surface. In their refraction at the ground surface, waves are effectively trapped in the upper soft layer. A Love wave consists of coupled P and horizontally polarised shear (SH) waves, as illustrated in Figure 17.11c. A particle undergoes flat elliptical motion, so that the surface displacement corresponds to horizontal shaking. The propagation speed of the wave is a function of the wavelength of the motion, and is given by the solution of the equation

$$\frac{G_1 \rho_1}{G_2 \rho_2} \left(\frac{C_{s1}^2 - C_L^2}{C_L^2 - C_{s2}^2} \right) = \tan \left[\frac{2\pi H}{\lambda} \left(\frac{C_L^2}{C_{s2}^2} - 1 \right) \right] \tag{17.7}$$

where subscripts 1 and 2 refer to the hard and soft media respectively, and H is the depth of the soft layer.

Equation 17.7 indicates that the C_L must lie in the range

$$C_{s2} < C_L < C_{s1}$$

481

Also, for the case $\lambda \gg H$, C_L tends to C_{s1}, indicating that wave propagation is occurring mainly in the hard layer. For $\lambda \ll H$, C_L tends to C_{s2}, indicating wave transmission is occurring mainly in the soft layer.

The generation of surface waves above a blast site is important in that it affects the range of potential ground disturbance due to transient motion. Body waves are spherically divergent, so that the amplitude u_r (and particle velocity) at some range, r, is related to r by equation 10.43, i.e.

$$u_r \propto 1/r$$

Surface waves are cylindrically divergent in the near-surface rock, so that, from equation 10.46, u_r (and $\dot{u}_r$) are related to r by

$$u_r \propto 1/r^{1/2}$$

Thus, body waves are subject to attenuation due to geometric spreading at a greater rate than surface waves. At points remote from an underground blast site, the induced surface waves are therefore the main source of transient motion.

The components of ground motion induced by the various types of wave must be taken into account in construction of the system for its measurement. Consider the reference axes shown in Figure 17.11a, and a blast conducted at some point on the z axis. For waves propagating in any radial plane (e.g. the xz plane), a point on the ground surface will experience the following motion:

(a) radial motion measurable as x and z components, due to the P wave;
(b) transverse motion, measurable as x, y and z components, due to the S waves;
(c) longitudinal and lateral motion, measurable as x and y components, due to the Love wave;
(d) longitudinal and vertical motion, due to the Raleigh waves.

Thus, a satisfactory ground vibration measurement system must be based on a triaxial array of measurement transducers. Distinction between the various wave components of ground motion can be made, in the measurement process, on their relative arrival times and the associated motion in other co-ordinate directions.

The performance of surface and underground excavations subject to nearby dynamic events such as blasts, and also rockbursts and earthquakes, is related to the intensity of the associated ground motion. Most of the study of ground motion from these sources has been concerned with the effects of earthquakes and nuclear explosions. With the qualifications noted below, mine blasts and conventional explosions induce ground motion, outside the very near field, comparable with that from the the other seismic sources. Ground motion due to these various seismic events is quantified in several ways, including time histories of displacement, velocity or acceleration, the response spectrum, and seismic motion magnitude parameters. These descriptive methods have developed with different purposes in mind, and provide different degrees of information about the nature and damage potential of the dynamic loading imposed on an excavation or rock structure.

The most comprehensive description of ground motion is provided by time history records of the various motion parameters. The acceleration record is

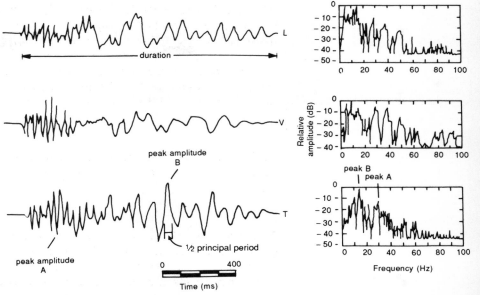

Figure 17.12 (a) Time histories of ground motion from a coal-mine blast; (b) frequency spectra for the components of the ground waves (after Stagg and Engler, 1980).

measured most conveniently, and from it the velocity and displacement can be obtained by successive integration. Three mutually orthogonal components of ground motion must be measured so that the magnitudes of the motion vectors can be defined as a time record. An example of a ground motion record is given in Figure 17.12a. As a general rule, similar patterns of ground motion are obtained from explosions, rockbursts and earthquakes when observed at similar distances from the source. However, most ground motion from mine blasts is observed close to the source, where the record appears more pulse-like.

A second method of characterising ground motion expresses the frequency content of the motion, by computation of the shock (or response) spectrum (Clough and Penzien, 1975). This is useful because the response of a structure to dynamic loading reflects the natural frequency of the structure and the dominant frequency in the ground wave. The frequency contents of the ground waves shown in Figure 17.12a are illustrated in Figure 17.12b. Consistent with the preceding description of time history records, ground motion near a blast site tends to have a higher frequency content than that experienced at a remote site. However, no information on duration of motion can be obtained from the response spectrum.

Several seismic magnitude parameters have been proposed to describe concisely the types of motions induced by dynamic events. The magnitude parameters defined in Table 17.1, due to Housner and Jennings (1982), illustrate the profusion of descriptions. The most commonly used parameters are local magnitude (M_l), surface wave magnitude (M_s), body wave magnitude (M_b), and moment magnitude (M_w). Moment magnitude is applicable only to earthquakes and rockbursts, where the source mechanism involves fault slip. It is observed that the local, surface wave and body wave magnitude parameters are derived from measurements of maximum amplitudes of the relevant ground motions, and therefore convey no descriptions of the duration of motion.

483

Table 17.1 Magnitude parameters for seismic events (after Housner and Jennings, 1982).

Magnitude	Definition	Application
Local, M_L	Logarithm of peak amplitude (in microns) measured on a Wood–Anderson seismograph at a distance of 100 km from source, and on firm ground. In practice, corrections made to account for different instrument types, distances, site conditions.	Used to represent size of moderate earthquake or rockburst. More closely related to damaging ground motion than other magnitude scales.
Surface wave, M_s	Logarithm of maximum amplitude of surface waves with 20 s period.	Used to represent size of large earthquakes.
Body wave, M_b	Logarithm of maximum amplitude of P waves with 1 s period.	Useful for assessing size of large, deep-focus earthquakes which do not generate strong surface waves.
Moment, M_w	Based on total elastic strain-energy released by fault rupture, which is related to seismic moment M_o ($M_o = G \cdot A \cdot D$, where G = modulus of rigidity of rock, A = area of fault rupture surface and D = average fault displacement).	Avoids difficulty associated with inability of surface wave magnitudes to distinguish between two very large events of different fault lengths (saturation).

17.8 Dynamic performance and design of underground excavations

The performance of underground excavations subject to ground motion from explosions, rockbursts and earthquakes is of interest in mine blast design, for the protection of access and service openings, and also in the design of other underground facilities for which impulsive loading may be important in their operating roles. The issues of concern are the modes of response and the types of damage of excavations under seismic loading, and design criteria for prevention or mitigation of damage.

The response of an excavation to an episode of seismic loading depends on the static condition of the excavation, as well as the transient effects associated with seismic loading. In assessments of seismic loading of excavations, Stevens (1977) and Owen and Scholl (1981) identified three modes of damage: fault slip, rock mass failure, and shaking. Excavation damage due to shaking appears to be most prevalent, and is expressed as slip on joints and fractures with displacement of joint-defined blocks, and local cracking and spalling of the rock surface. For lined excavations, cracking, spalling, and rupture of the liner may occur.

Although explosive loading of a rock mass results in transient loading of excavations, the resulting state of stress may be either dynamic or pseudo-static. As noted by Labreche (1983), the type of loading to be considered depends on the ratio (λ/D) of the wavelength (λ) of the stress or velocity waveform to the excavation diameter (D). When the duration of loading is short, corresponding to a small λ/D ratio, excavation response is dynamic. A large λ/D ratio corresponds to a relatively prolonged loading, and the response is effectively static.

Natural and induced rock structure and the duration of strong ground motion are critical determinants of rock response to dynamic loading. Field experience that damage by shaking is predominantly due to joint motion is consistent with the experimental observation that joints decrease in shear strength under cyclic shear loading (Brown and Hudson, 1974). Model studies of excavations in jointed rock under cyclic loading by Barton and Hansteen (1979) confirmed that excavation failure occurred by accumulation of shear displacements at joints. This is the basis of the conclusion by St John and Zahrah (1987) that it is the number of excursions of joint motion into the plastic range that determines dynamic damage to an excavation.

In spite of recognition of the importance of duration of ground motion on excavation dynamic response, current engineering practice correlates damage during an episode of dynamic loading to peak ground motion. For purposes of classifying excavation response due to earthquakes, Dowding and Rozen (1978) defined three levels of damage due to ground motion: no damage; minor damage, involving new cracks and minor rockfalls; substantial damage, involving severe cracking, major rockfalls and closure of the excavation. As proposed by McGarr *et al.* (1981), peak velocity is the most appropriate motion parameter with which to correlate damage, since it can be related directly to peak transient stress in the ground wave, and the second power of velocity is related to dynamic strain energy. Observations of excavation performance are correlated with peak velocity in Figure 17.13, and related to the threshold velocities for minor damage and substantial damage of 200 mm s^{-1} and 400 mm s^{-1} originally proposed by Dowding and Rozen (1978).

According to St John and Zahrah (1987), the damage thresholds proposed above are well below those observed in a major underground explosion test

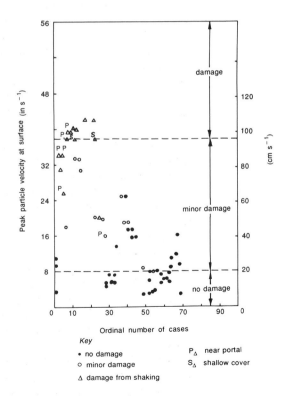

Figure 17.13 Calculated peak surface velocities and associated underground excavation damage for earthquakes (after Owen and Scholl, 1981).

program, involving the detonation of large charges adjacent to excavations with the purpose of establishing dynamic design criteria. Damage associated with intermittent spalling was observed at 900 mm s^{-1}, and continuous damage at 1800 mm s^{-1}. Because these observations were made for single explosions, they are probably of limited relevance to the performance of permanent mine excavations, which may be subject to many episodes of explosive loading during their duty lives.

It is clear that a criterion for dynamic design of permanent underground excavations based on the single parameter of peak tolerable velocity in the ground wave would be subject to a substantial margin of uncertainty. Nevertheless, the threshold value of 200 mm s^{-1} may be suitably conservative to allow routine application. Site specific empirical relations between peak velocity and explosive charge weight per delay, of the type described by Siskind *et al.* (1980), but determined for transmission of body waves, may be used to estimate probable peak velocity due to a particular blast.

Ultimately, more appropriate design for dynamic loading of an underground excavation must be based on an analysis of rock displacements induced by a synthesised history of likely ground motion, and take account of site conditions such as rock structure. Dynamic analysis of jointed and fractured rock also requires description of the dynamic properties of rough, dilatant joints under cyclic loading, which is a topic about which there is comparatively little information.

17.9 Evaluation of explosive and blast performance

17.9.1 General procedures

The current state-of-the-art of explosive engineering practice does not permit prediction of the performance of a given explosive or blast design in a particular application. The performances of different explosives in a given geomechanical setting are evaluated best in experimental blasts which may involve one or two blast holes for each explosive type. As noted in section 17.2, an underwater test is commonly used to estimate the energy released from a charge. In test blasts in rock, each trial explosive should be sampled as it is being loaded into the blast hole and its physical and chemical properties determined for comparison against specifications.

Measurement techniques used in the field to evaluate the performance of explosives and of experimental blasts include (McKenzie, 1987, 1988)

(a) velocity of detonation timing
(b) near-field vibration monitoring
(c) muckpile surveying
(d) fragmentation measurement
(e) high-speed photography
(f) gas penetration measurement.

Damage caused to the surrounding rock by a blast may be assessed by a variety of techniques including the direct measurement of fractures in blastholes and on rock surfaces and indirect methods including crosshole or tomographic seismic scans, in-hole acoustic or seismic profiling, ground probing radar, microseismic emission monitoring, blast vibration measurements, and *in situ* permeability testing (McKenzie, 1988). Because blast damage might be expected to reduce the static and dynamic Young's moduli of the rock surrounding the blast site, some methods rely on direct and indirect measurements of these properties (e.g. Holmberg *et al.*, 1983). Ground motions experienced in the far field, including those at the ground surface, may be measured using systems of the general type discussed in section 18.2.7 for monitoring micro-seismic activity.

Details of the methods used in making the many types of measurements involved in assessing explosive and blast performance are given in the papers and reports referenced in this section and in texts such as those by Johansson and Persson (1970) and Dowding (1985). Further consideration here will be restricted to near field ground motion monitoring, the application of which has produced widespread industrial benefit.

17.9.2 Near-field ground motion monitoring

The characteristics of the ground motions produced in the near field (say within 100 m) of a production blast will generally be represented by peak particle velocity = $10–1000 \, \text{mm s}^{-1}$, peak dynamic displacement = $0.01–0.5$ mm, frequency range = $10–1000$ Hz and dominant frequency = $50–500$ Hz. The instrumentation system required to measure these motions consists of a transducer array cemented into or onto the rock, a cable system to transmit the signals from the transducer array to the monitoring equipment, a signal recorder, and a waveform analyser or computer-based processor which will display the waveform and from which vibration levels and frequency spectra can be determined.

Figure 17.14 Analysis of ground motion waveforms using a triaxial geophone array (after McKenzie, 1988).

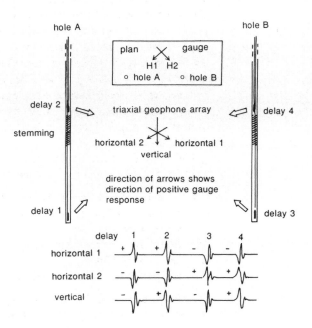

Obviously, the transducers selected must have responses in the ranges given above for blast-induced ground motions. Cost, reliability and signal-to-noise ratio are also important considerations in making the choice. Transducers may measure either particle velocity (geophones) or particle acceleration (accelerometers). Piezo-electric accelerometers easily meet the data specifications but are ·expensive, often require ancillary equipment such as power supplies and pre-amplifiers, and can introduce electrical noise problems.

Accelerometers are preferred where the transducers are to be surface mounted and are recoverable. Where transducers must be regarded as consumable items, as is usually the case in monitoring the near-field motions arising from underground blasts, velocity sensitive geophones are used. Geophones are unable to respond to higher frequencies but give acceptable responses in the frequency range 10–500 Hz.

Single transducers may be used if the information required is limited, say, to determining whether or not detonations have occurred at each delay. The numbers of transducers required increase with the complexity of the blast design and with the amount of information sought. Complete analysis of waveforms and the determination of the vector sum of the motions at a point require the use of triaxial arrays consisting of three mutually orthogonal geophones. As illustrated in Figure 17.14, the direction of movement of the geophone coil in response to the passage of each vibration determines whether the first peak of the record is

Figure 17.15 (a) Schematic cross section of a blast and geophone monitoring design, (b) the ground motion recorded by the upper radial geophone, and (c) the result expected from the blast (after McKenzie, 1988).

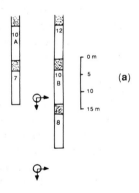

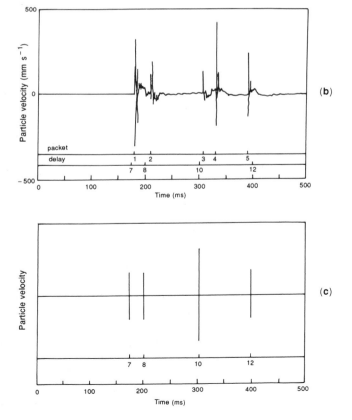

488

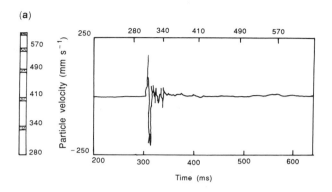

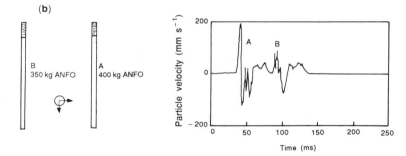

Figure 17.16 (a) Ground motion recorded during sympathetic detonation of charges; (b) reduction of vibration amplitude as a result of overbreak (after McKenzie, 1988).

positive or negative. This allows the direction of travel of the wave to be deduced and the location of the detonation relative to the detector to be determined.

The waveforms recorded near a production blast arise from a charge or group of charges. The first step in analysing a waveform is to determine which charge each vibration packet represents and hence the actual initiation sequence. This provides invaluable insight into the blast performance. The vibration amplitude provides a measure of the explosive energy transferred to the rock mass so that, for a given charge type and monitoring geometry, the relative amplitude can be used as a measure of charge efficiency.

Each charge is identified on a ground motion record by considering one or more of the nominal initiation sequence, the transducer polarity (the direction of the first break) and the relative vibration amplitude. Figure 17.15b shows the radial component of ground motion (the record obtained from the transducer whose axis is horizontal and parallel to the direction of incidence of the wave) recorded for the blast design shown in cross section in Figure 17.15a. The transducers were located about 10 m behind the plane of the blast holes. As shown in Figure 17.15c, the vibration waveform should have contained four packets appearing at 175, 200, 300 and 400 ms. Furthermore, because two charges were designed to initiate at 300 ms, this packet could be expected to be of greater magnitude than the others.

The actual record contains five vibration packets (Figure 17.15b). This is reasonable given the low probability that two nominally identical delay elements will actually detonate at the same time. In this case, departures from nominal

489

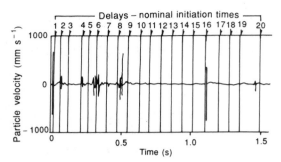

Figure 17.17 Malfunctions in an underground blast (after McKenzie, 1987).

initiation times are in the range 2–10%. In order to distinguish between the two charges on delay 10, it is necessary to consider transducer polarity and the records of both the upper and the lower transducer arrays. It transpires that packet 3 is associated with delay 10 in hole A and packet 4 with that in hole B.

One of the most significant outcomes of near-field vibration monitoring and analysis of the type outlined has been the identification of a high frequency of occurrence of charge malfunctions in production blasts (McKenzie, 1987, 1988). Traditionally, a misfire is defined as an explosive charge which does not detonate, typically due to an initiation system fault. Recent experience has shown that such misfires are relatively uncommon. More common causes of charge malfunction are delay inaccuracy, instantaneous detonation, sympathetic detonation (Figure 17.16a), explosive desensitisation or overbreak (Figure 17.16b). In the underground production blast recorded in Figure 17.17, only nine out of 20 charges initiated as designed. Detailed analysis showed that four charges (delays 13, 17, 18, 19) failed to detonate, while seven charges initiated sympathetically with delays 5 and 8 producing a tight bunching of ground motion pulses. Once identified, these malfunctions can be eliminated, usually by decreasing charge concentrations with resulting improvement in fragmentation and major savings in explosives costs (McKenzie, 1987, 1988).

490

18 Monitoring rock mass performance

18.1 The purposes and nature of monitoring rock mass performance

Monitoring is the surveillance of engineering structures, either visually or with the help of instruments. In a general geomechanics context, monitoring may be carried out for four main reasons:

(a) to record the natural values of, and variations in, geotechnical parameters such as water table level, ground levels and seismic events before the initiation of an engineering project;
(b) to ensure safety during construction and operation by giving warning of the development of excess ground deformations, groundwater pressures and loads in support elements, for example;
(c) to check the validity of the assumptions, conceptual models and values of soil or rock mass properties used in design calculations;
(d) to control the implementation of ground treatment and remedial works such as ground freezing during shaft sinking or tunnelling through water-bearing ground, grouting, drainage or the provision of support by tensioned cables.

In mining rock mechanics, most monitoring is carried out for the second and third of these reasons. Monitoring the safety of the mine structure is a clear responsibility of the mining engineer. Monitoring to check the rock mass response and, as a consequence, adjust the overall mine design or take remedial measures, is equally important. It will be appreciated from the discussions presented throughout this book, that rock masses are extremely complex media whose engineering properties are difficult, if not impossible, to predetermine accurately ahead of excavation. It will also be clear that the models used to predict the various aspects of rock mass response to different types of mining procedure, are based on idealisations, assumptions and simplifications. It is vitally necessary, therefore, to obtain checks on the accuracy of the predictions made in design calculations. As illustrated in the design flow diagram of Figure 1.7, monitoring the behaviour of the rock mass surrounding the mining excavation is an integral part of a mine rock mechanics programme, and provides the feedback necessary to close the design loop. In some cases, the design may be based largely on the results obtained by monitoring trial excavations or the initial behaviour of the prototype excavation, with little or no reliance being placed on pre-excavation design calculations. This use of field observations of the performance of structures is central to the general practice of geotechnical engineering in which it is known as the **observational method** (Peck, 1969).

Monitoring systems used in conjunction with modern large-scale underground mining operations can be very sophisticated and expensive. However, it should be remembered that valuable conclusions about rock mass response can often be reached from visual observations and from observations made using very simple

491

monitoring devices. Items that may be monitored in an underground mining operation include

(a) fracture or slip of the rock on the excavation boundary (observed visually);
(b) movement along or across a single joint or fracture (either monitored by a simple mechanical 'tell-tale' or measured more accurately);
(c) the relative displacement or convergence of two points on the boundary of an excavation;
(d) displacements occurring within the rock mass away from the excavation periphery;
(e) surface displacements or subsidence;
(f) changes in the inclination of a borehole along its length;
(g) groundwater levels, pressures and flows;
(h) changes in the normal stress at a point in the rock mass;
(i) changes in loads in support elements such as steel sets, props, rock bolts, cables and concrete;
(j) normal stresses and water pressures generated in fill;
(k) settlements in fill;
(l) seismic and microseismic emissions;
(m) wave propagation velocities.

Although it may appear from this list that a wide range of variables may be monitored, only two basic physical responses, displacement and pressure, can be measured relatively directly using current technology. Measurements can be made of the **absolute displacements** of a series of points on the boundaries of an excavation or, with more difficulty, within the rock mass. The relative displacement, or **convergence**, of two points on the boundary of an excavation is easier to measure than absolute displacement. Because the relative displacement of two points can usually be measured, a measurement of **normal strain** can be obtained by assuming that the strain is uniform over the base length of the measurement. **Pressures** in groundwater and normal stresses at rock-support contact or in fill can be measured by the pressures induced in fluid-filled pressure cells, often using a null method. An average pressure normal to the surface of the pressure cell sensor is obtained. **Time** is always recorded as a fundamental variable.

It is important to recognise that the 'measurement' of most other variables of interest, notably forces and stresses, requires the use of a mathematical model and material properties (e.g. elastic constants) to calculate the required values from measured displacements, strains or pressures. As a general rule, it is preferable to use directly measurable parameters for purposes of comparison and decision making rather than parameters calculated from a mathematical model using measured parameters as input.

18.2 Monitoring systems

18.2.1 General features of monitoring systems

The instrumentation system used to monitor a given variable will generally have three different components. **A sensor** or **detector** responds to changes in the variable being monitored. **A transmitting system** which may use rods, electrical cables, hydraulic lines or radiotelemetry devices, transmits the sensor output to the read-out location. **A read-out** and/or **recording** unit such as a dial gauge, pressure gauge, digital display or magnetic tape recorder, converts the data into a usable form and presents them to the engineer.

In order that the monitoring system should fulfil its intended function economically and reliably, it should satisfy a number of requirements:

(a) easy installation, if necessary under adverse conditions;
(b) adequate sensitivity, accuracy and reproducibility of measurements;
(c) robustness and suitable protection to ensure durability for the required period of operation;
(d) ease of reading and immediate availability of the data to the engineer;
(e) negligible mutual interference with mining operations.

The terms accuracy, error, precision and sensitivity as applied to measuring devices, require careful definition.

The stated **accuracy** of an instrument indicates the deviation of the output, or reading, from a known input. Accuracy is usually expressed as a percentage of the full-scale reading. For example, a 10 MPa pressure gauge having an accuracy of 1% would be accurate to within ± 100 kPa over the entire range of the gauge.

The **error** is the difference between an observed or calculated value and the true value; errors may be either systematic or random.

The **precision** is a measure of the ability of the instrument to reproduce a certain reading. It may be defined as the closeness of approach of each of a number of similar measurements to the arithmetic mean. Precision and accuracy are different concepts. Accuracy requires precision and an absence of bias, whereas precision implies a close grouping of results whether they are accurate or not.

The **sensitivity** of an instrument is variously defined as the ratio of the movement on the read-out unit to the change in the measured variable causing the output, the input to output ratio (the inverse of the previous ratio), or the smallest unit of the measurement detectable by the instrument. Sensitivity as defined in the third way may depend on the **readability** of the read-out unit (the closeness with which the scale may be read), and the **least count** (the smallest difference between two indications that can be detected on the read-out scale).

18.2.2 Modes of operation

The modes of operation of the sensing, transmission and read-out systems used in monitoring devices may be mechanical, optical, hydraulic or electrical.

Mechanical systems often provide the simplest, cheapest and most reliable methods of detection, transmission and read-out. Mechanical movement detectors use a steel rod or tape, fixed to the rock at one end, and in contact with a dial gauge or electrical measuring system at the other. The main disadvantage of mechanical systems is that they do not lend themselves to remote reading or to continuous recording. Mechanical convergence and displacement measuring systems are described in sections 18.2.3 and 18.2.4.

Optical systems are used in conventional, precise and photogrammetric surveying methods of establishing excavation profiles, measuring movements of

493

excavation boundaries, and recording natural and mining-induced fractures. These methods are also widely used in monitoring surface subsidence associated with underground extraction. In this application, electro-optical distance measuring equipment is particularly useful. Examples of these various surveying techniques are given by St John and Thomas (1970), Fellows (1976), Franklin (1977) and Hagan (1980).

In the past, considerable use was made of photoelastic plugs and discs to monitor stress changes in the rock surrounding excavations and in support elements. Details of several of these photoelastic instruments are given by Roberts (1977). Perhaps the most common remaining instrument of this type is the photoelastic rockbolt load cell. In this cell, the rockbolt load is transmitted across the diameters of two glass discs mounted between steel loading platens. To take readings, the glass discs are illuminated with polarised light and are viewed through a polaroid analyser. The load applied to the disc can be determined from the isochromatic fringe count and an elastic solution for the distribution of stresses in a line loaded disc.

Hydraulic and pneumatic diaphragm transducers are used for measuring water pressures, support loads, cable anchor loads, normal components of stress and settlements. In all cases the method of operation is the same. The quantity measured is a fluid pressure which acts on one side of a flexible diaphragm made of a metal, rubber or plastic. Twin tubes connect the read-out instrument to the other side of the diaphragm. To take readings, air, nitrogen or hydraulic oil pressure is supplied from the read-out unit through one of the tubes to the diaphragm. When the supply pressure is sufficient to balance the pressure to be measured, the diaphragm acts as a valve and allows flow along the return line to a detector in the read-out unit. The balance pressure is recorded, usually on a standard Bourdon pressure gauge or a digital display.

Electrical devices probably provide the most common basis of the instruments presently used to monitor the performance of rock masses surrounding mining structures, although mechanical systems still find widespread use in displacement monitoring. Electrical systems generally operate on one of three basic principles.

Electric resistance strain gauges operate on a principle discovered by Lord Kelvin, namely that the resistance of a wire changes proportionately with strain. In strain gauge systems, strain gauges made from thin wire or foil, are bonded to a rock, concrete or steel surface. Changes in the strains in the host material are accompanied by changes in the strains, and resistances, of the strain gauges which are read using a Wheatstone bridge circuit.

The use of electrical resistance strain gauges in the measurement of *in-situ* stresses was discussed in section 5.3. Most load cells, many pressure transducers and some types of inclinometer also use electrical resistance strain gauges. The major disadvantages associated with the use of electrical resistance strain gauges for rock mechanics applications are

(a) it is difficult to obtain and maintain a good bond between the strain gauge and the rock,
(b) the strains are measured over relatively short gauge lengths, and
(c) temperature effects cannot always be eliminated.

494

Vibrating wire sensors are based on the fact that the natural frequency of vibration, f, of a tensioned wire of length l_w and density ρ is related to the tensile stress in the wire, σ_w, by the equation

$$f = \left(\frac{1}{2l_w}\right)\left(\frac{\sigma_w}{\rho}\right)^{1/2}$$

If this frequency is measured by electromagnetic plucking, the tension σ_w can be determined. This value may be used to determine the pressure acting on a diaphragm to which one end of the wire is attached, or the axial load on a load cell in which the vibrating wire is mounted. Thus vibrating wire systems are used in piezometers, soil pressure cells, stressmeters and load cells. A vibrating wire instrument used for monitoring unidirectional stress changes in rock is described in section 18.2.6.

The outputs of vibrating wire sensors are frequencies rather than voltages or currents. This is an advantage in that frequency is easier to transmit over long distances without distortion than are analogue signals. Frequency counts can usually be transmitted and detected reliably even in the presence of heavy background noise. Vibrating wire sensors also have excellent long-term stability.

Self-inductance instruments are based on the mutual inductance of a pair of coaxial solenoids forming a circuit of resonant frequency

$$f = \frac{(L\,C)^{-1/2}}{2\pi}$$

where L is the self-inductance and C the capacitance of the solenoids. A relative displacement, d, of the inner and outer solenoids produces a change in the measured resonant frequency, f, such that

$$d = K\left(1 - \frac{f_0^2}{f^2}\right)^{1/2}$$

where f_0 is the calibrated frequency for the null position and K is a constant. Londe (1982) describes a series of borehole extensometers and inclinometers that are based on this principle. These instruments may be read remotely using radio-telemetry. The self-inductance multiple-point extensometer is described in section 18.2.4.

Linear variable differential transformers (LVDTs) are sometimes used as sensors in displacement monitoring instruments. The LVDT also operates on a mutual inductance principle. However, it produces as the output signal not a frequency, but a voltage which is directly proportional to the linear displacement of an iron core along the axis of the instrument.

In the following sections, some examples are given of individual instruments used to monitor a range of variables. The selection of instruments described is not exhaustive, but examples of the major instrument types are included.

(a)

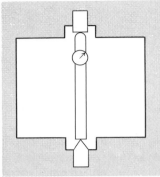

(b)

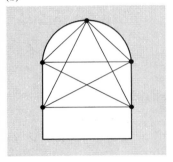

Figure 18.1 (a) Convergence between roof and floor measured with a rod convergence gauge; (b) typical five-point convergence array.

Figure 18.2 The Distometer ISETH, a high-precision mechanical convergence measuring system.

18.2.3 Convergence measurement

Convergence, or the relative displacement of two points on the boundary of an excavation, is probably the most frequently made underground measurement. The measurement is variously made with a telescopic rod, invar bar or tape under constant tension, placed between two measuring points firmly fixed to the rock surface (Figure 18.1). A dial gauge, micrometer, or an electrical device such as an LVDT, is used to obtain the measurement of relative displacement.

Figure 18.2 shows a high precision convergence measuring system developed by Kovari *et al.* (1974). The instrument is used to provide a constant tension of 78.5 kN to a 1.0 mm diameter invar wire stretched between an anchoring point and the instrument which is itself attached to the second anchoring point at the rock surface. The tensioning device moves the end of the wire towards the displacement gauge with a precision thread. The applied tension is read on the dynamometer. The displacement gauge has a readability of 0.01 mm and a range of 100 mm. The overall accuracy of the convergence measurements is ± 0.02 mm.

18.2.4 Multiple-point borehole extensometers

Among the most useful measurements of rock mass performance are those made using multiple-point borehole extensometers (MPBXs) (Figure 18.3). A single-point borehole extensometer gives the relative displacement between an anchor point inside the rock mass and a measuring point, generally located at the excavation boundary. A multiple-point extensometer can also give the relative displacements between several points at different depths in the borehole. In this way, the distribution of displacements in relatively large volumes of rock can be recorded. These data are generally more useful than the results of convergence measurements which only give relative surface displacements and may be influenced by surface conditions. Sometimes convergence measurements are made between the heads of MPBX installations as illustrated in Figure 18.3.

Suggested methods for monitoring rock movements using borehole extensometers are given by the International Society for Rock Mechanics Commission on Standardization of Laboratory and Field Tests (1978b). Instruments using both electrical and mechanical sensors are described, and guidelines for determining the precisions, sensitivities and measuring ranges required for a number of applications are given. It is suggested that for large underground excavations, the minimum measuring range should be 50 mm (300 mm with reset), the precision should be in the range 0.25–2.5 mm, and the instrument sensitivity should be typically 0.25–1.00 mm.

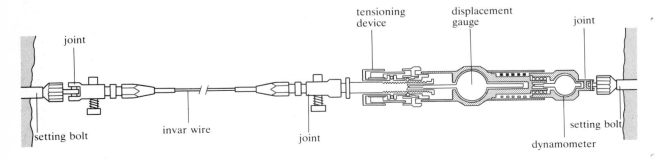

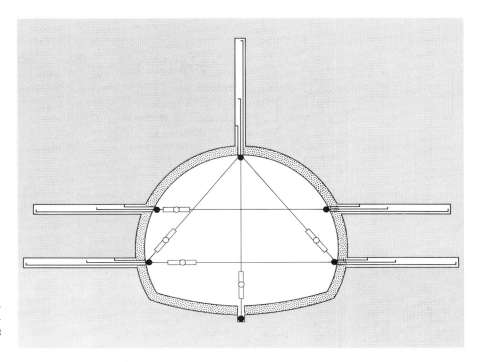

Figure 18.3 Multiple-point borehole extensometer (MPBX) installations with convergence measurement between MPBX heads.

Figure 18.4 shows a modern mechanical multiple-point borehole extensometer. Up to six measuring points may be used in a 86 mm diameter borehole. Read-out may be by a dial gauge or by a permanently fitted inductance transducer. With the measuring head near the rock surface (the usual case), the dial gauge system

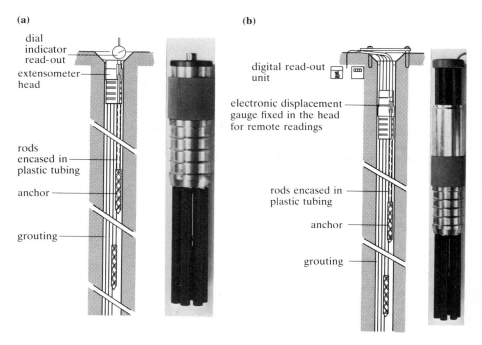

(a)

dial indicator read-out

extensometer head

rods encased in plastic tubing

anchor

grouting

(b)

digital read-out unit

electronic displacement gauge fixed in the head for remote readings

rods encased in plastic tubing

anchor

grouting

Figure 18.4 Multiple-point borehole extensometer: (a) near-surface measuring head with dial gauge measurements; (b) measurements read with a permanently fitted inductance transducer (after Amstad and Koppel, 1977).

497

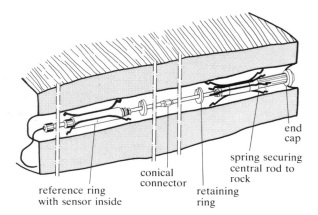

end
cap

spring securing
central rod to
rock

conical
connector

retaining
ring

reference ring
with sensor inside

Figure 18.5 Self-inductance multiple-point borehole extensometer
(after Londe, 1982).

gives a measuring range of 150 mm and a sensitivity of 0.01 mm. The electrical
transducer has a measuring range of 40 mm and a sensitivity of 0.01 mm. A
special feature of this extensometer is that the measuring head may be placed up
to 70 m from the collar of the borehole, with readings being transmitted to the
read-out point by electrical cable. Several measuring heads may be located in the
same borehole.

Figure 18.5' shows a multiple-point extensometer that operates on the self-inductance principle. The extensometer consists of a central rod fixed at one end
of the borehole, and carrying a set of inductance displacement sensors passing
through coaxial rings fixed by springs to the rock at selected points. Relative
movement between the sensor and the ring at any point along the axis modifies
the resonant frequency of the circuit. The read-out unit is calibrated to give a
direct reading of displacement. The sensitivity is 0.01 mm, the precision is
± 0.02 mm and the travel range of each sensor is 125 mm. Londe (1982) reports
that this instrument has been used to make measurements in boreholes up to 80
m deep. Sensor spacings can be as close as 1 m and the borehole may be at any
inclination. The fundamental advantage of this instrument over conventional rod
and wire extensometers is that it is not necessary to provide a permanent mechanical connection between the sensors and the rock.

18.2.5 Hydraulic pressure cells

The hydraulic pressure cell consists of a flatjack connected to a hydraulic or
pneumatic diaphragm transducer which in turn is connected by flexible tubing to
a read-out unit. Normal stress transferred from the surrounding soil, rock or
concrete, is measured by balancing the fluid pressure in the cell by a pressure
applied to the reverse side of the diaphragm. Hydraulic pressure cells are used
to measure changes in total normal stress in materials such as fill, or at interfaces
between materials, e.g. at a rock–shotcrete interface. If effective stresses are
required, a piezometer should be installed alongside the pressure cell.

Procedures for monitoring normal stresses with hydraulic pressure cells are
given by the International Society for Rock Mechanics Commission on Standardization of Laboratory and Field Tests (1980). The most widely used hydraulic
pressure cell is probably the Glötzl cell which is described by Franklin (1977)
and by Sauer and Sharma (1977). Examples of Glötzl pressure cells are shown

498

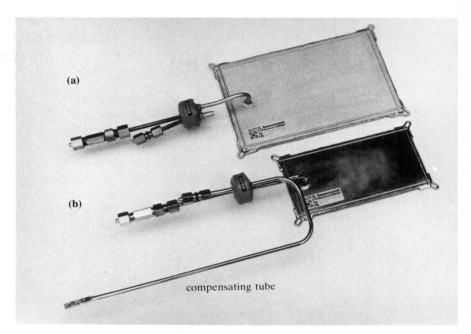

Figure 18.6 Glötzl hydraulic pressure cells for use in (a) soil, and (b) concrete. (Photograph by Glötzl Gesellschaft für Baumesstechnik mbH.)

in Figure 18.6. The installation of Glötzl pressure cells to measure circumferential stress in a shotcrete lining and normal stress at the shotcrete–rock interface is illustrated in Figure 18.7.

For installation in concrete, the Glötzl cell is provided with a compensating tube (Figure 18.6b) connected to the flatjack. As the concrete sets, it contracts

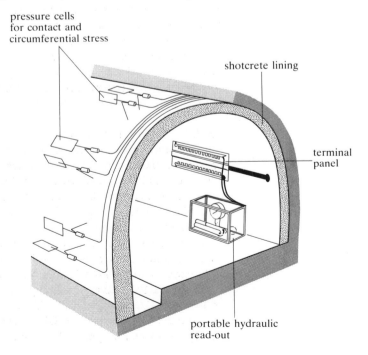

Figure 18.7 Glötzl pressure cell installation in a shotcrete lining.

499

and so an air gap may develop along the faces of the flatjack. A small volume of fluid injected through the compensating tube shortly after installation expands the faces of the flatjack against the concrete. A small positive pressure is restored within the flatjack, and all subsequent pressure readings are referred to this initial value.

The fluid used to fill the cell depends on the material in which the cell is installed. The compressibility of the cell should be similar to that of the surrounding material if the cell is to give a correct measure of the undisturbed normal stress in the material. A cell that is too stiff for its surroundings will register an excessive pressure, and one that is insufficiently stiff will register a pressure that is too low. Cells installed in soils are usually filled with hydraulic oil; those installed in rock or concrete are filled with mercury.

The fluid pressure is measured by applying an air or oil pressure to one of the twin tubes that connect the hydraulic transducer to the read-out. When this applied pressure is sufficient to balance the pressure in the cell, a return flow of air or oil will be registered at the read-out unit. The normal pressure is then given as

$$P = (P_r - P_i - P_h - P_f)E$$

where P_r = indicated pressure, P_i = initial cell pressure (adjusted for concrete shrinkage compensation, if necessary), P_h = static head correction for the difference in elevation between the cell and read-out, P_f = correction for friction losses in the delivery line, and E = multiplication factor (less than 1.0) to compensate for cell edge effects.

If only pressure differences are to be monitored, and P_i is set after installation and compensation, P can be calculated as

$$P = (P_r - P_i)E$$

18.2.6 Stress change measurements

A wide range of instruments has been developed for monitoring the stress changes induced in rock by mining activity. These have included photoelastic plugs and discs described by Roberts (1977), instruments based on the hydraulic pressure cell described by Sauer and Sharma (1977), the vibrating wire stress meter (Hawkes and Hooker, 1974; Sellers, 1977), and rigid inclusion instruments which use electric resistance strain gauges to measure the strains induced in the plug (Wilson, 1961; Worotnicki *et al.*, 1980). These are usually borehole instruments and suffer from the disadvantage that they monitor the stress change in one direction only. Three measurements are sometimes taken in mutually perpendicular directions, but even these measurements cannot give the complete change in the stress tensor unless the three directions chosen are the principal stress directions. The difficulties caused by stiffness incompatibility, discussed for the Glötzl pressure cell, also arise with these instruments. In order to obtain the desired quantity, a stress change, a mathematical model and material properties or, alternatively, an experimentally determined calibration curve, must be used. Accordingly, the accuracy of the measurements is often considerably less than that obtainable for displacement measurements.

500

A newer generation of instruments permit changes in the components of stress in more than one direction to be monitored in one borehole. Walton and Worotnicki (1986) describe three such instruments – the yoke gauge which is a purpose designed, non-reusable three component borehole deformation gauge, and the conventional and thin-walled versions of the CSIRO hollow inclusion triaxial strain cell discussed in section 5.3.2. The conventional hollow inclusion cell has been provided with three additional strain gauges to a total of twelve to render the cell more suitable for making measurements in anisotropic rock or to provide additional strain measurement redundancy. Walton and Worotnicki (1986) present results of laboratory tests and field trials carried out to investigate the effects of moisture absorption, temperature variation, polymer shrinkage and time on instrument stability and sensitivity.

The vibrating wire stress meter is used widely for stress monitoring, especially in the USA. This instrument was developed originally as a low-cost device for monitoring stress changes in coal, but it has since found use in a variety of hard- and soft-rock applications. It will be discussed in some detail here because it illustrates a number of the principles and difficulties associated with the monitoring of stress changes in rock.

The main components of the vibrating wire stressmeter are shown in Figure 18.8. The instrument consists of a hollow, hardened steel body which, in use, is pre-loaded diametrically between the walls of a 38 mm diameter borehole by means of a sliding wedge and platen assembly. The gauge is set using a manual or hydraulic setting tool which pulls the wedge forward relative to the platen and gauge body and thus pre-loads the platen and gauge against the borehole walls. Gauges have been installed in boreholes up to 100 m deep.

Figure 18.8 Exploded view of a vibrating wire stressmeter (after Hawkes and Hooker, 1974).

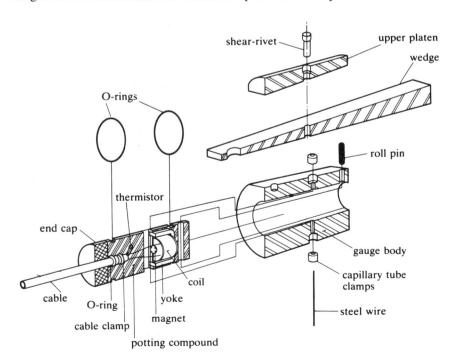

501

Stress changes in the rock in the pre-load direction cause small changes in the diameter of the gauge cylinder. These changes are measured in terms of the change in the frequency of vibration of a high tensile steel wire stretched across the cylinder in the pre-load direction. To make the measurements, the wire is 'plucked' by an electromagnetic coil which also acts as an electronic pick up to count the number of vibrations of the wire in a known time. A read-out box displays the period of vibration of the stressmeter wire which can be used to determine the stress change in the rock from calibration tables prepared for a range of rock types.

The sensitivity of the gauge varies with rock type but more especially with the period of the vibration. For the same basic gauge design, by setting the initial wire tension within different predetermined ranges during manufacture, it is possible to have either a highly sensitive gauge with a limited range or a less sensitive gauge with a wide range. The sensitivity and calibration of the gauge also vary with the contact angle made by the platen with the borehole wall. Different platens have been used in 'hard-rock' and 'soft-rock' versions of the instrument. The difficulties involved in treating this problem analytically and in obtaining universally applicable calibration data have been set out by Pariseau and Eitani (1977). Although these important questions will not be pursued here, it should be noted that this case provides an excellent illustration of the difficulties involved in converting a measurable variable (in this case vibration frequency) into a different variable (rock stress).

18.2.7 Monitoring microseismic activity

Rock noise is often heard by miners working underground and taken as a warning of imminent danger from rock failure. Laboratory and field studies have shown that these audible noises are preceded by subaudible energy emissions from the failing rock. The monitoring of such microseismic or acoustic emissions induced by mining activity forms an essential part of the monitoring programmes in a number of mines, particularly those susceptible to rockburst activity.

Audible or acoustic wave frequencies are in the range 20 Hz to 20 kHz; seismic waves are of lower frequency. The frequencies of waves radiated by events associated with mining activity range from less than 1 Hz to more than 10 kHz. That part of this frequency range in which most of the energy is concentrated depends on the size of the event. The frequency decreases with increasing magnitude of the energy release. The higher frequencies, which are those recorded in laboratory tests, tend to be attenuated rapidly with increase in distance from the focus of the event. This wide frequency range has led to some confusion in the terminology used to describe these events. Some authors refer to the phenomenon as acoustic emission, while others describe it as seismic or microseismic activity. Because the latter terminology is the more common, and because the events of lower frequency are probably the more significant in underground mining, the phenomenon will be termed seismic or micro-seismic activity in the present discussion.

Research which was initiated at the United States Bureau of Mines in the early 1940s, has shown that rocks under stress emit detectable seismic energy at a rate which increases with increasing stress level. An increase in emission rate generally precedes major fracturing of the rock. The energy source for a given event can be located using a triangulation method based on differences in wave travel times to several receiving stations.

Disparate examples of microseismic monitoring systems based on these principles are given by Blake *et al.* (1974), Hardy and Mowrey (1976), Langstaff (1977), Godson *et al.* (1980), Pattrick (1984) and Labuc *et al.* (1987). In these systems, the wave propagating from the source of a microseismic or seismic event is detected by a geophone (a velocity gauge suitable for detecting frequencies in the range 1–100 Hz) or accelerometer which converts the mechanical vibration into its electrical analogue. This electrical signal is then amplified and transmitted to the monitoring station. After signal conditioning, the signal goes to the computer interface which determines if it is strong enough to trigger the timing and control component of the system. If so, the first arrival times of the seismic wave at each of the geophones are determined. These data are used to locate the source of the event using geometric and seismic velocity relations. The amplitude and duration of the signal are used to determine the relative magnitude of the event. Data such as the time of day, source location, event magnitude and geophone first-arrival times may be printed out and the computer reset to receive further signals.

In most microseismic monitoring systems developed for use in hard-rock metalliferous mines, the geophones or accelerometers are placed on the faces of the underground excavations, and the data collection is carried out underground. For some longwall coal mining and underground storage cavern applications, the sensors may be located in boreholes drilled from the ground surface with all data being collected on the surface (Hardy and Mowrey, 1976).

Figure 18.9 illustrates the monitoring systems installed in the Lucky Friday and Star Mines in the Coeur d'Alene mining district of northern Idaho, USA, in the 1970s. The Lucky Friday system used 24 geophones placed on the lower six levels of the mine. The array covered 400 m along strike, 250 m across strike and 300 m vertically. The minimum distance between any two geophones was 45 m. The geophone used had a natural, undamped frequency of 10 Hz and an intrinsic sensitivity of 0.32 V per mm s^{-1}. It performed very well in the mid-range of emission frequencies, and was resistant to the warm, damp mine environment. The geophone could also detect audible noises associated with mining activity. Geophones were encapsulated in purpose-made plastic holders and attached to the rock surface with a fast-setting gypsum cement.

More recently developed systems may make use of optical fibre for signal transmission from the sensor array to the central receiver (Labuc *et al.*, 1987). Modern digital technology has also been introduced into microseismic monitoring systems in place of the analogue methods used in earlier systems such as that illustrated in Figure 18.9 (Green, 1984; Pattrick, 1984).

18.3 Examples of monitoring rock mass performance

18.3.1 Crown pillar behaviour at New Broken Hill Consolidated Mine, Australia

Worotnicki *et al.* (1980) report the monitoring of the behaviour of a crown pillar in an open stoping operation at the New Broken Hill Consolidated Mine, Austra-

503

lia. The design height of the stope was 120 m and the maximum length at midheight was 44 m. The crown pillar above the stope was planned to be 12 m thick. However, overbreak reduced the thickness of the pillar to 6 m over part of its length. During extraction of the stope, the crown pillar began to 'bump', and drives contained within the pillar suffered spalling and timber damage. The pillar was then reinforced by fully grouted cables and monitoring of the pillar was commenced. The primary purpose of monitoring in this case was to ensure the

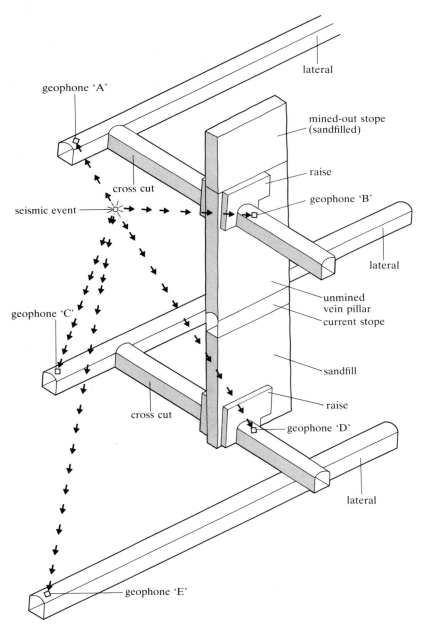

continued safety of the operation and the stability of the mine structure, by giving warning of the development of excess stresses or deformations in the crown pillar. The results obtained also permitted a rational appraisal to be made of the manner in which stresses were redistributed in the crown pillar as. mining proceeded.

Table 18.1 Results of absolute stress measurements (MPa), panel 10 crown pillar, NBHC Mine (after Worotnicki *et al.*, 1980).

Date	σ_1 (approx. E–W)	σ_2 (approx. vertical)	σ_3 (approx. N–S)
30 Aug. 1976	33	5	3
21 Oct. 1976	58	6	−5
18 Feb. 1977	56	26	25

Figure 18.9 Seismic detection system used at Lucky Friday and Star Mines (after Langstaff, 1977).

The geometry of the overbroken section of the crown pillar and the arrangement of the instrumentation are shown on east–west and north–south sections in Figure 18.10. Absolute stress measurements were made by the overcoring method using the CSIRO hollow inclusion stress cell described by Worotnicki and

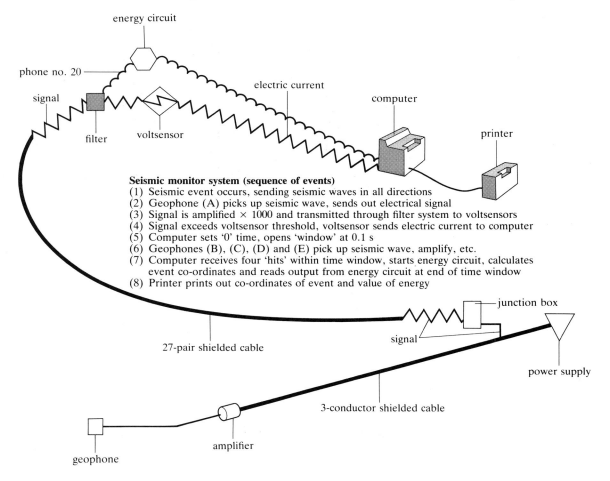

Seismic monitor system (sequence of events)
(1) Seismic event occurs, sending seismic waves in all directions
(2) Geophone (A) picks up seismic wave, sends out electrical signal
(3) Signal is amplified × 1000 and transmitted through filter system to voltsensors
(4) Signal exceeds voltsensor threshold, voltsensor sends electric current to computer
(5) Computer sets '0' time, opens 'window' at 0.1 s
(6) Geophones (B), (C), (D) and (E) pick up seismic wave, amplify, etc.
(7) Computer receives four 'hits' within time window, starts energy circuit, calculates event co-ordinates and reads output from energy circuit at end of time window
(8) Printer prints out co-ordinates of event and value of energy

505

(a) E–W section

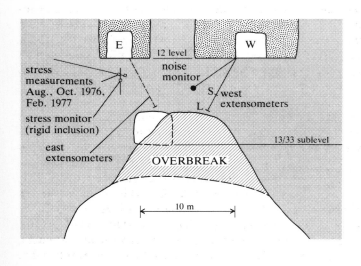

(b) N–S section

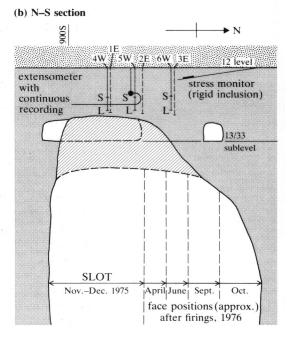

Figure 18.10 (a) East–west section, and (b) north–south section of panel 10 crown pillar, NBHC Mine, showing overbreak and monitoring installations (after Worotnicki *et al.*, 1980).

Walton (1976). Table 18.1 shows the results of stress measurements made at the commencement of monitoring (August 1976), at the completion of excavation of the stope (October 1976) and after a further period of approximately four months (February 1977). In early November 1976, a single National Coal Board rigid inclusion stressmeter of the type described by Wilson (1961) was installed to monitor changes in the vertical component of stress in the pillar.

A number of rod extensometers were installed to monitor dilation of the pillar. These extensometers were read manually at periodic intervals, and while they

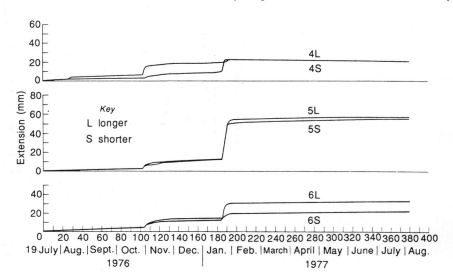

Figure 18.11 Pillar dilation measured on the west side of panel 10 crown pillar, NBHC Mine. Two extensometers were used in each hole – S = shorter, L = longer (after Worotnicki *et al.*, 1980).

506

provided a measure of the development of pillar deformations, it was not possible to establish precisely the times at which sudden movements occurred. For this purpose, one extensometer was fitted with an LVDT to sense movements continuously which were then recorded on an analogue chart recorder. The instrumentation system was completed by a microseismic detector installed near the anchor of the continuously recording extensometer (Figure 18.10). Microseismic noises detected by this device were transmitted to a recording station on 12 level where background noise below a frequency of 10 kHz was filtered out and each remaining analogue signal was converted to a digital count representative of the intensity of the signal.

Figure 18.11 shows the results obtained from the extensometers on the west side of the stope over a one year period, and Figure 18.12 shows a combined plot of extensometer movement, vertical stress and microseismic noise count recorded over a shorter period. Figure 18.11 shows that for the period of active mining up to the end of October 1976, all extensometers recorded displacements of modest magnitude, increasing approximately linearly with time. This was attributed to the quasi-elastic response of the pillar to the increase in east–west compressive stress in the crown pillar generated by mining (Table 18.1). At the end of October 1976, and in mid-January 1977, there were two sudden increases in displacement, interpreted as episodes of violent pillar failure. Between these two events, pillar displacement increased at a higher rate with time than it had done previously.

Figure 18.12 Extensometer movement, vertical stress and microseismic noise count measured in panel 10 crown pillar, NBHC Mine (after Worotnicki *et al.*, 1980).

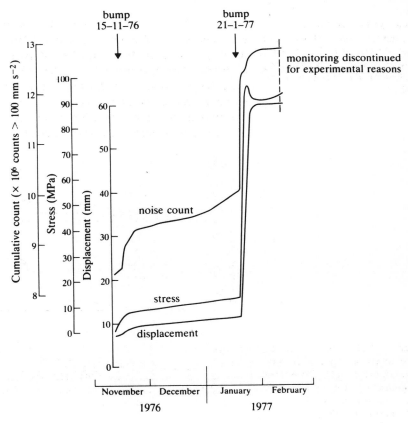

After the event of mid-January 1977 neglegible further displacement was recorded.

Figure 18.12 shows that the event of mid-January 1977, was reflected by sudden increases in the levels of vertical stress and microseismic activity in the pillar. This event was preceded by a period of about two weeks in which the background noise count noticeably increased without corresponding increases in the vertical stress or the displacement. For approximately half an hour immediately before the event, the noise count dropped to a negligible level. Such behaviour has also been observed in other cases.

The absolute stress measurement data shown in Table 18.1 are consistent with the other observations. The first two measurements reflect the increase in predominantly east–west stress associated with mining. The second measurement was taken before, and the third measurement after, the two failure events. Worotnicki *et al.* (1980) postulated that late in October 1976, the east–west stress locally reached the level of the compressive strength of the rock mass and that fracture or yield of the pillar occurred. Yielding of the pillar in the east–west direction under approximately constant stress could have led to consequent increases in the north–south and vertical stresses if free deformation in these directions was partially restrained. As this process continued with a further major fracturing event in mid-January 1977, the pillar became stable again due to the strength-enhancing effect of the new confining stresses.

18.3.2 *Performance of stabilising pillars at ERPM, South Africa*

It was noted in section 10.8 that a marked deterioration in ground conditions around the work places in longwall stopes in the deep-level gold mines of South Africa occurs as the volume rate of energy release with mining increases. In these cases, one of the objectives of mine design and excavation sequencing should be to minimise the spatial rate of energy release. This rate can be controlled most effectively by limiting the displacement of the excavation peripheral rock in the mined-out area. As was shown in section 15.2, elastic analysis suggests that partial extraction using regularly spaced stabilising pillars, should provide a highly effective means of limiting convergence volume and energy release rates.

Salamon and Wagner (1979) describe an example of monitoring carried out to verify these theoretical predictions and so provide a sound basis for future mine design. Based on theoretical reasoning, the management of one of South Africa's deepest gold mines, East Rand Proprietary Mines Limited (ERPM), decided to leave regularly spaced strike pillars in the deeper sections of the mine to reduce the rockburst hazard. Several 70 m wide strike pillars were left on a centre spacing of about 270 m to reduce the convergence volume. In view of the importance of this decision to future mining, it was decided to install an underground seismic network to monitor seismic activity in the area of the stabilising pillars and in other areas of the mine not affected by the pillars.

A total of 10 geophones was installed over an area of about 6 km². After being amplified by a factor of 200, the geophone signals were transmitted via telephone cables to a central underground recording station and recorded on a 16-channel tape recorder. The output of a digital clock was also recorded on one of the recorder channels. The sources of all seismic events which were recorded on five or more geophones, were located using a digital technique described by Salamon

508

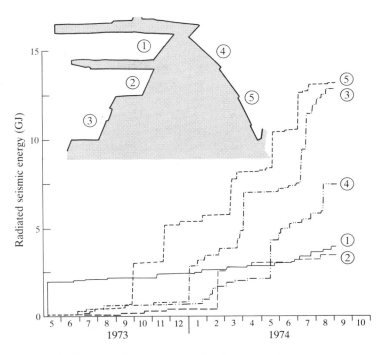

Figure 18.13 Cumulative radiated seismic energies for five locations at ERPM (after Salamon and Wagner, 1979).

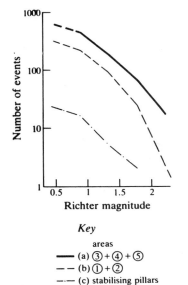

Key

areas
— (a) ③ + ④ + ⑤
— — (b) ① + ②
—·— (c) stabilising pillars

Figure 18.14 Frequency distribution of seismic event magnitudes in areas (a) not protected by stabilising pillars; (b) protected by stabilising pillars; (c) in and around stabilising pillars (after Salamon and Wagner, 1979).

and Weibols (1974). Location accuracy in the area of concern was typically ± 30 m in the plane of the reef and ± 50 m in the normal direction.

Figure 18.13 shows a cumulative plot of the seismic energies released in five areas of interest over an 18 month period of continuous recording. Most seismic events at ERPM occur within a short distance of the working face. For the purposes of this study, all events located up to 100 m ahead of or behind a stope face were taken to be associated with the mining of the face.

Areas 1 and 2 within the sphere of influence of the stabilising pillars display markedly different behaviour from that of the other three areas. The overall gradients of the cumulative seismicity curves for areas 1 and 2 are significantly lower than those for areas 3, 4 and 5. The frequency of occurrence of the larger seismic events, identified by the large steps in the radiated seismic energy, is also lower for areas 1 and 2. According to Figure 18.13, the level of seismicity in the areas protected by the stabilising pillars is only about one quarter to one half of that in the other areas. Figure 18.14 clearly shows the differences between the numbers and magnitudes of seismic events recorded in the unprotected and protected areas and in and around the pillars themselves.

It must be remembered that the volumes of rock mined in the various areas differed considerably. To account for these differences, Salamon and Wagner (1979) prepared Table 18.2 which shows the seismic energies released at the protected and unprotected faces per unit of reef area mined, and the mining-induced energy changes per unit area calculated from elastic theory. Close agreement was obtained between the measured specific seismic energy and the calculated specific mining-induced energy change ratios.

The results obtained not only demonstrate the effectiveness of stabilising pillars in reducing energy release rates with mining, but also show that, for the

509

conditions at ERPM, elastic theory can be used to provide useful predictions of some aspects of rock mass performance.

Table 18.2 Summary of radiated seismic energy and calculated mining-induced energy change at ERPM (after Salamon and Wagner, 1979).

	Stope faces protected by stabilising pillars (areas 1 and 2)	Other stope faces (areas 4, 5 and 6)	Relative seismic energy and mining-induced energy change
total mined area (m^2)	63 200	130 400	
total radiated seismic energy (GJ)	10.05	47.83	
specific seismic energy (MJ m^{-2})	0.159	0.360	0.44
calculated specific mining-induced energy change (MJ m^{-2})	21.9	51.8	0.42

18.3.3 The Näsliden Project, Sweden

The Näsliden Mine in northern Sweden commenced production in 1970. The orebody which consists of a complex sulphide ore, dips at approximately 70° to the west, is 110–220 m long, has an average width of 18 m and extends vertically to a depth of 770 m. Mining is by a modern mechanised cut-and-fill method. Present mining is being carried out in four stopes above the 460 m level (Figure 18.15); each stope will be about 100 m high when completed.

After the Näsliden Mine had been in operation for a few years, the owners, Boliden Mineral AB, concluded that better knowledge of the rock mechanics of

Figure 18.15 Vertical section, Näsliden Mine (after Krauland, 1981).

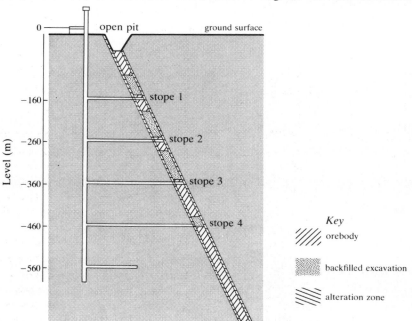

cut-and-fill mining and improved predictive techniques for use in mine planning were required. Accordingly, it was decided that a major project, based on the Näsliden Mine, should be undertaken with the following objectives:

(a) to examine the degree to which present methods of calculation can be used as predictive tools in mine planning;

(b) to develop a method of providing, principally for cut-and-fill mining,

(i) a successive and continuous forecast of mining conditions,

(ii) a quantification of the effects of different stabilisation measures and a means of choosing and optimising these measures.

It was decided that the tools to be developed for making the calculations should use the finite element method. In order to check the predictions of rock mass and fill performance made using the finite element programs developed, and to provide data that could be used to give values of rock mass properties, an extensive programme of monitoring was undertaken at the Näsliden Mine. Figure 18.16 shows the locations and types of measurements made to 1980. In addition to these measurements, *in situ* stresses were measured on several levels, the development of stresses in the roof of stope 3 was monitored, and visual observations were made of rock mass failure mechanisms and water conditions in various parts of the mine. The results of these measurements and observations are presented in detail in the volume edited by Stephansson and Jones (1981). Only the results of

Figure 18.16 Vertical section, Näsliden Mine, showing locations of monitoring points and mining sequence (after Nilsson and Krauland, 1981).

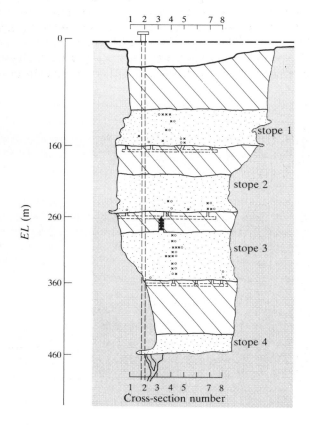

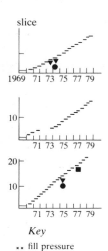

Key

×× fill pressure

○○ convergence across backfilled stope

▲ convergence across alteration zone and/or orebody

▼ roof deflection

● photogrammetric displacement measurement

■ open stope convergence

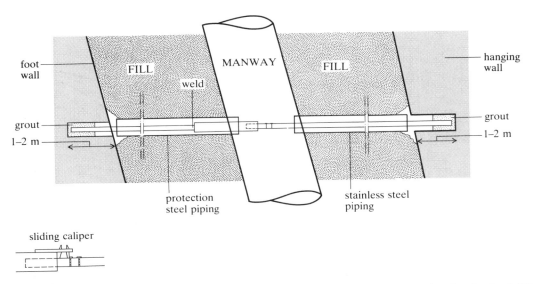

Figure 18.17 Convergence measurement across a backfilled stope from a manway in the fill, Näsliden Mine (after Nilsson and Krauland, 1981).

the convergence measurements and pressure monitoring in the backfill will be discussed here.

Convergence of the filled stopes was monitored using a rod extensometer system. Readings were taken manually from a manway in the fill (Figure 18.17) or from a ramp in the foot wall. Glötzl pressure cells were used to measure fill pressures normal to the plane of the orebody at the sites identified in Figure 18.16. Figure 18.18 shows the development of horizontal convergence, ΔL, and transverse fill pressure, P, at various locations as filling proceeded in stopes 1, 2 and 3. At three locations, vertical, longitudinal and transverse fill pressures were monitored. Figure 18.19 shows the development of fill pressure in three directions at a point in the centre of slice 15, stope 3.

The measured convergence varied with depth below ground surface and with the height of the stope. The rate of increase of convergence with further mining increased with the height of the stope. Total convergence was less in the lower third or half of the stope than it was higher in the stope. The greatest cumulative convergence measured at any point was 100 mm.

The maximum value of transverse fill pressure measured was less than 0.4 MPa. The greatest fill pressures in each stope were recorded at the lowest stations when the fill was at its maximum height. The measurements made in three directions showed that vertical fill pressures in the centres of stopes were less than 50% of those calculated from the weight of the overlying fill. This suggests that an arching mechanism was developed in the fill with part of the self-weight stresses being transferred to the stope walls. On the basis of these and other measurements, Knutsson (1981) concluded that, on average, 70–80% of the backfill pressure arises from self-weight and 20–30% from stope convergence. This suggests that, in this case, the presence of the fill will have little influence on the stress distribution in the mining area. Its role will be to control excessive convergence, particularly of detached or overstressed rock.

As a result of the work done on the Näsliden project up to mid-1980, Krauland *et al.* (1981) concluded that an elastic model was adequate for use in those mine

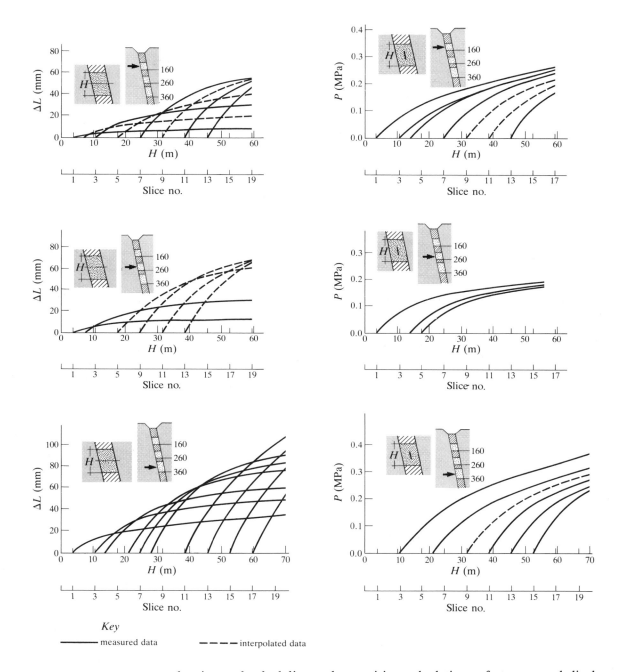

Figure 18.18 Development of horizontal convergence, ΔL, and transverse fill pressure, P, as mining proceeded in stopes 1, 2, and 3, Näsliden Mine (after Nilsson and Krauland, 1981).

planning and scheduling tasks requiring calculations of stresses and displacements over large areas of the mine. For the detailed prediction of the performance of the rock mass near the periphery of a stope, better results were obtained using a finite element model incorporating joint elements at the boundaries between the alteration zones and the hanging wall and foot wall, and between the alteration zones and the orebody.

513

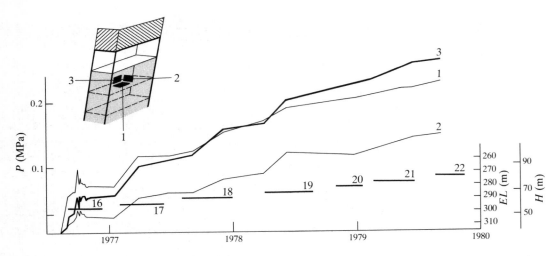

Figure 18.19 Development of normal stresses in three orthogonal directions in fill in slice 15, stope 3, section 4, Näsliden Mine (after Nilsson and Krauland, 1981).

18.3.4 Behaviour of crown pillars between cut and fill stopes, the Mount Isa Mine, Australia

At the Mount Isa Mine, Australia, silver–lead–zinc mineralisation occurs as distinct concordant beds within the Urquhart Shale – a sequence of well-bedded dolomitic and pyritic shales and siltstones. The bedding strikes north–south and dips at 65° to the west. The most important geological structures in the Urquhart Shale are bedding-plane breaks, fractures striking north–south and dipping at about 60° to the east, two sets of orthogonal extension fractures orientated normal to the bedding, and conjugate shear fractures. Bedding-plane breaks are frequently planar, smooth, graphite-coated and slickensided, and can have continuities of up to several hundred metres. The major principal *in situ* stress is perpendicular to the bedding. The other two principal *in situ* stresses have similar magnitudes and act in the plane of the bedding. The measured vertical component of *in situ* stress is equal to the weight of the overlying rock.

The silver–lead–zinc orebodies are disposed in an en echelon pattern down dip and along strike. The Racecourse orebodies to the east or footwall side are narrow and, at the time of concern here, were mined mainly by a mechanised cut-and-fill method known locally as MICAF. The widths of the MICAF stopes varied between 3 and 11 m. The minimum separation between stopes was 3–4 m. Mining advance was up-dip by a breast stoping or 'flat-backing' method. Both the hangingwall and crowns (or backs) of stopes were rock bolted routinely, and pre-placed, untensioned cable dowels were installed as required (Lee and Bridges, 1981).

The stability of the crown pillars is of major concern in this type of mining. A justifiable mining objective is to recover as much ore as possible from the pillars. However, men and machines work in the advancing stopes and their safety must be ensured. Potential hazards in the stopes are rock falls from the crown and buckling failures in the hanging- and footwalls. For the 11–9 level and 13–11 level lifts the stopes were advanced up-dip in a sequence that ensured that all stope backs were kept in a line perpendicular to bedding (Figure 18.20). This method was adopted to ensure that the major principal stress would always act normal to the bedding and so eliminate the possibility of crown instability being induced by slip on the bedding planes.

514

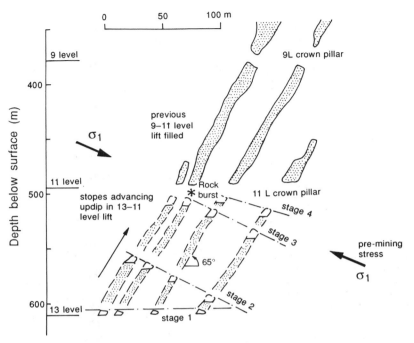

Figure 18.20 Cross section through Racecourse orebodies at 6570N, Mount Isa mine, Australia, showing mining sequence in 13–11 level lift leading to formation of 11 level crown pillar (after Lawrence and Bock, 1982).

At stage 3 of the 13–11 level lift (Figure 18.20) very high stresses developed in the stope crowns. This produced spalling of intact rock, rock falls, audible rock noise and rockbursts in the 11 level crown pillar above 7 and 8 orebodies. A series of stress measurements made at various locations on 11 level showed that the induced stresses were very high. At 6650N (Figure 18.21) a major principal stress of 95 MPa was measured perpendicular to the bedding.

Because of the bad ground conditions in the crowns of the leading hangingwall stopes mining in these stopes ceased, and further mining was undertaken in the footwall orebodies, which were 'lagging' behind under the mining strategy that was being used. It was noted that, where this was done, shear displacement occurred on a few bedding planes in the crown pillars of the hangingwall orebodies and ground conditions improved. Figure 18.21 shows the shear displacements measured in such a case at 6650 N between December 1975, when the high stresses previously referred to were measured on 11 level, and December 1977, when much lower stresses were measured.

The mechanical explanation of this destressing phenomenon is that, by advancing the footwall stopes, the principal stress directions in the crown pillar became inclined to the bedding planes. Because of the very low shear strengths of the bedding planes slip could occur readily, the relative displacement being footwall side down. The angle of friction on the bedding planes was 10° and their dilation angles were less than 2°. As the stopes were progressively mined to and through 11 level, non-dilatant shear displacement on bedding planes continued, but tensile cracking sub-perpendicular to the bedding began to occur. Slip on the bedding planes decreased the stiffness of the whole crown pillar. This produced a regional redistribution of stress away from the crown pillar to the north and south abutments of the stopes where high stresses could still be sustained.

515

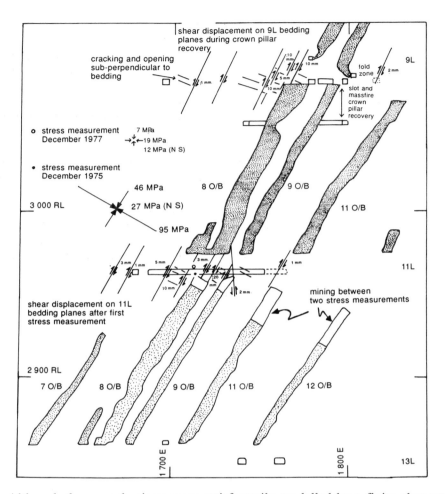

Figure 18.21 Cross section at 6650N, looking north, showing results of two stress measurements and shear displacements on bedding planes (after Lee and Bridges, 1981).

Although these mechanisms were satisfactorily modelled by a finite-element analysis that incorporated suitable yield mechanisms (Lawrence and Bock, 1982), such sophisticated analyses are not required to develop an appreciation of the fundamental mechanics of the problem. The keys to an understanding of why slip occurred on the bedding planes and ground conditions improved with a modified extraction schedule are

(a) a general appreciation of the directions that would be taken by the principal stress trajectories in the crown pillar at the different stages of mining and the implications for slip on bedding planes;

(b) a recognition of the low frictional resistance likely to be generated on smooth, continuous, graphite-coated and slickensided bedding planes; and

(c) an understanding of the rôle played by rock mass stiffness in governing induced stress magnitudes.

18.3.5 Concluding remarks

The results of the monitoring programmes undertaken in the three case histories discussed were used to

516

(a) provide warning of the development of excessive stresses and displacements in an important part of the mine structure (NBHC);

(b) aid the development of an understanding of the re-distribution of stresses in mine structures as mining proceeded (NBHC, Näsliden and Mount Isa);

(c) verify the effectiveness of a major design innovation in reducing energy release rates and the associated rockburst hazard (ERPM);

(d) elucidate the stabilising rôle of fill (Näsliden);

(e) confirm the applicability of design models based on the theory of elasticity (ERPM and Näsliden).

In these and many other comparable cases, monitoring was an essential component of a rock mechanics programme used successfully to develop, verify or improve mine design procedures. Such rock mechanics programmes are central to modern underground mining practice. They provide the tools and understanding needed to develop safe and economic mining methods for new mining areas, and to improve the efficiency and competitiveness of existing operations by modifying mining practice, often by the introduction of large-scale mechanisation.

Appendix A Basic constructions using the hemispherical projection

A.1 Projection of a line

The construction described here is used to represent the orientations in space of lines such as the normals to planes or force vectors. The orientation of a line in space is represented by its trend and plunge.

The **plunge** of the line, $\beta(-90° \leq \beta \leq 90°)$, is the acute angle, measured in a vertical plane, between the line and the horizontal. A line directed in a downward sense has a positive plunge; a line directed upwards has a negative plunge.

The **trend** of the line, $\alpha (0° \leq \alpha \leq 360°)$, is the azimuth, measured by clockwise rotation from north, of the vertical plane containing the line. The trend is measured in the direction of the plunge. Thus the downwards directed and upwards directed ends of a line have trends that differ by $180°$.

The trend and plunge of a line are analogous to the dip direction and dip of a plane. They are also represented by a three-digit number (trend) and a two-digit number (plunge) separated by a slash, e.g. 045/73.

The following steps are used to plot the equatorial or meridional projection of a line:

(1) Locate a piece of tracing paper over the meridional net (Figure 3.24) by means of a centre pin. Mark the north point and the centre and circumference of the net on the tracing paper. When a large number of stereographic analyses are to be carried out, it is useful to prepare several sheets of tracing paper on which the north point and the centre and circumference of the circle have been marked.

(2) Using the gradations around the perimeter of the net, mark the azimuth corresponding to the trend of the line. (This procedure is illustrated in Figure A.1a for the comparable case of the dip direction of a plane.)

(3) Rotate the tracing paper about the centre pin until this mark lies on the east–west diameter of the net.

(4) Count from the perimeter of the net, along the diameter, the required plunge angle. Mark and label the point using some convenient notation.

(5) Rotate the tracing paper so that the north point is returned to its home position.

A.2 Projection of the great circle and pole to a plane

The construction used to plot the equatorial or meridional projections of the great circle and pole of the plane 130/50 is illustrated in Figure A.1. The following steps are used:

(1) Locate a piece of tracing paper over the meridional net (Figure 3.24) by means of a centre pin. Mark the north point and the centre and circumference of the net.

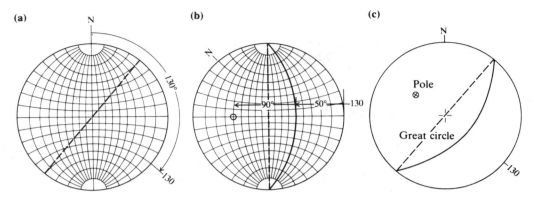

Figure A.1 Plotting the great circle and pole to a plane on the meridional projection.

(2) Measure off the dip direction (130°) clockwise from the north point around the perimeter of the net and mark this direction on the tracing paper (Figure A.1a). Alternatively, measure $130° - 90° = 40°$ from the north point and mark in the strike line, shown dashed in Figure A.1a.

(3) Rotate the tracing about the centre pin until the dip direction lies on the east–west diameter of the net.

(4) Count from the perimeter of the net, along the east–west diameter, the dip (50°) and trace in the great circle passing through this point (Figure A.1b).

(5) Plot the pole to the plane by counting a further 90° along the east–west diameter with the dip direction still aligned with the east–west axis (Figure A.1b). Alternatively, the pole may be plotted as the projection of the line 310/40 using the construction described in section A.1.

(6) Rotate the tracing paper so that the north point is returned to its home position. The final appearance of the stereographic projection of the great circle and pole to the plane 130/50 is shown in Figure A.1c.

A.3 Determination of the line of intersection of two planes

Figure A.2 Determining the trend and plunge of the line of intersection of two planes.

Figure A.2 illustrates the construction used to determine the trend and plunge of the line of intersection of the two planes 130/50 and 250/30.

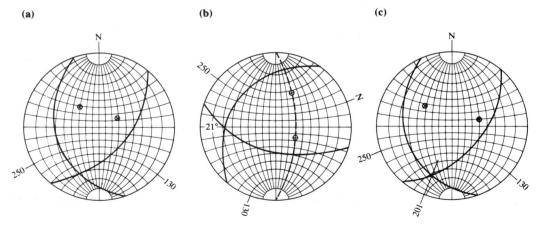

(1) Plot the great circle and pole of the plane 130/50 on tracing paper, using the procedure described in section A.2.

(2) With the north point on the tracing paper in its home position, mark the dip direction of the second plane (250°).

(3) Rotate the tracing paper about the centre pin until the dip direction lies on the east–west diameter. This is most conveniently done in this case by aligning the dip direction (250°) with the west (270°) point.

(4) Plot the great circle and pole to the plane 250/30 using the procedure described in section A.2, this time counting the dip (30°) along the west–east diameter from its western end. Figure A.2a shows the appearance of the plot with the north point returned to its home position.

(5) Rotate the tracing about the centre pin until the intersection of the two great circles which defines the line of intersection of the two planes, lies on the west–east diameter of the net (Figure A.2b).

(6) The plunge of the line of intersection is measured as 21° by counting along the west–east axis from the 270° point to the great circle intersection (Figure A.2b). In other problems, it may be more convenient to align the intersection point with the east–west diameter on the eastern side of the net and count the plunge from the east or 90° point.

(7) With the tracing in this position, the poles of the two planes lie on the same great circle (Figure A.2b). This provides an alternative means of locating the line of intersection as the pole to the great circle passing through the poles to the two planes.

(8) Rotate the tracing to return the north point to its home position.

(9) Draw a straight line through the centre of the net and the point of intersection of the two great circles, to the perimeter of the net. This line defines the trend of the line of intersection, measured as 201° clockwise from the north point (Figure A.2c).

A.4   Determination of the angle between two lines in a plane

Figure A.3 illustrates the construction used to determine the angle between lines with orientations 240/54 and 140/40 in the plane containing the two lines. If these lines are the normals to two planes, then the construction may be used to determine the angle between the two planes.

(a)　　　　　　　**(b)**

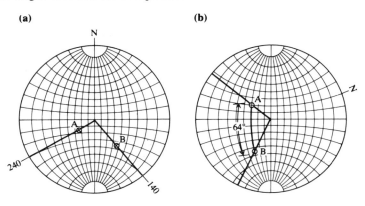

Figure A.3 Determining the angle between two lines.

(1) Plot the projections of the two lines using the procedure described in section A.1. These points are marked A and B in Figure A.3a.

(2) Rotate the tracing about the centre point until points A and B lie on the same great circle of the stereonet (Figure A.3b). The dip and dip direction of the plane which contains the two lines are measured from the stereonet as 60° and 200°, respectively.

(3) The angle between the lines is found to be 64° by counting the small circle divisions between A and B along the great circle (Figure A.3b).

A.5 Determination of dip direction and true dip

A common problem encountered in mapping geological features underground is the determination of the orientation of a feature from the orientations of the traces made by the intersection of the feature with the boundaries of an excavation. Figure A.4 illustrates the steps involved in the determination of the true dip and dip direction of a discontinuity plane in a simple case. A square tunnel has vertical side walls which trend in the direction 140° to 320°. The apparent dip of a discontinuity where it intersects the side wall is 40° SE. The same discontinuity intersects the horizontal roof of the tunnel in a line trending 020–200. The construction uses the following steps:

Figure A.4 Determining the dip direction and true dip of a plane.

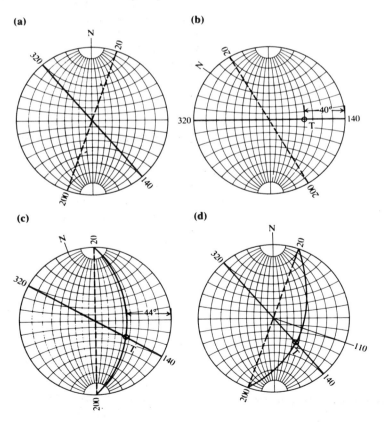

521

(1) On a piece of tracing paper located over the meridional net, mark the trends of the tunnel side wall (140–320) and the intersection of the discontinuity with the horizontal roof (020–200) (Figure A.4a). By definition, the latter line represents the strike of the discontinuity.

(2) Rotate the tracing so that the 140–320 line lies on the east–west diameter of the net.

(3) Count off the apparent dip (40°) from the perimeter of the net along the east–west diameter, and mark the point T (Figure A.4b).

(4) Rotate the tracing so that the 020–200 line, defining the strike of the discontinuity, lies on the north–south diameter of the net.

(5) With the tracing in this position, draw in the great circle on which the line of apparent dip, T, lies (Figure A.4c).

(6) The true dip of the discontinuity which, by definition, is at right angles to the strike of the plane, is measured as 44° from the periphery of the net along the east–west diameter (Figure A.4c). Mark in the line of true dip.

(7) Rotate the tracing so that the north point is in the home position.

(8) Read the dip direction, clockwise from the north point, as 110° (Figure A.4d).

The problem solved in Figure A.4 was simplified by the fact that an observation could be made of the intersection of the discontinuity with a horizontal plane. In the more general case, the dip direction and true dip of a plane can be found if the trend and plunge of its line of intersection with two planes are known. The projections of these two lines are plotted on the meridional net as described in section A.1. Since both of these lines lie in the plane of the discontinuity, the plane must be given by the great circle that passes through the two points.

A.6 Rotation about an inclined axis

In many problems encountered in structural geology and in geological data collection for rock mechanics purposes, it becomes necessary to rotate poles and lines about an axis that is inclined to the plane of the projection. Such problems arise particularly in the interpretation of data obtained from non-vertical and non-parallel boreholes. It is usual to measure the apparent dips and dip directions of discontinuities intersecting the core with respect to a reference line marked on the core and assumed to be vertical and, say, aligned with the north point. If the borehole is inclined and true orientation of the reference line is known, it is necessary, in the general case, to carry out rotations about both vertical and horizontal axes to orient correctly the reference line and the poles to the discontinuities intersected by the drill core.

Rotations of points may be conveniently carried out on the stereographic projection using the small circles centred on the north and south poles of the net. However, if the small circles are to be used to achieve rotation about an inclined axes, it is necessary to first apply an **auxiliary rotation** to bring the axis of rotation into a horizontal position. After the required rotation has been effected, the auxiliary rotation must be reversed to bring the axis back to its original orientation. Full details of the concepts and procedures used are given by Priest (1985).

522

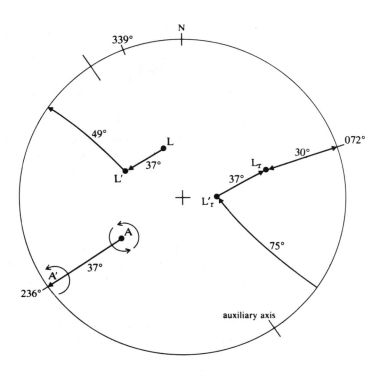

Figure A.5 Rotation about an inclined axis (after Priest, 1985).

Figure A.5 shows the construction used to determine the orientation of a line of initial orientation 339/51, after rotation about an axis of orientation 236/37, by 124° in an anticlockwise direction looking down the axis. The grid markings on the net have been omitted from Figure A.5 in the interest of clarity. The following steps are used in this construction:

(1) Locate a piece of tracing paper over the meridional net by means of a centre pin and mark the north point and the centre and circumference of the net.

(2) Plot the projections of the axis of rotation, A, and the line to be rotated, L, drawing a small arrow around A indicating the direction of the required rotation.

(3) Rotate the tracing on the centre pin so that A lies on the west-east diameter of the net.

(4) Rotate the axis to the horizontal by moving A through 37° to A′ on the perimeter of the net. At the same time, L must be rotated through 37° along a small circle to L′.

(5) Rotate the tracing on the centre pin to bring A′ to the north point of the net.

(6) Rotate L′ through the prescribed angle (124°) by moving it along a small circle in the direction indicated by the small direction arrow near A′. After a rotation of 49°, the point leaves the perimeter of the net and re-enters at a position diametrically opposite, to complete the remaining 75° of rotation to point L'_r.

(7) Rotate the tracing on the centre pin to bring A′ back to the west point.

(8) Reverse the auxiliary rotation by moving L'_r along a small circle by 37° in the opposite direction to the initial auxiliary rotation, to the point L_r.

523

(9)   Measure the trend and plunge of the rotated point, L_r, as $072°$ and $30°$, respectively.

If the axis of rotation, A, has a high plunge, it is more convenient to rotate this axis to the centre of the projection in the auxiliary rotation, rather than to the perimeter. As before, during this auxiliary rotation, all other points are rotated along small circles by the same amount and in the same direction as A. When the auxiliary rotation has been completed, the specified rotation is achieved by moving the point through a circular arc centred on the centre of the projection. The auxiliary rotation is then reversed and the orientation of the rotated line read off.

Appendix B   Stresses and displacements induced by point and infinite line loads in an infinite, isotropic, elastic continuum

B.1   A point load (the Kelvin equations)

For a point load, P, applied at the x, y, z co-ordinate origin, in the positive x direction, and tensile normal stresses reckoned positive, etc:

$$\sigma_{xx} = -\frac{P}{8\pi\,(1-v)}\frac{x}{R^3}\left[(1-2v)+\frac{3\,x^2}{R^2}\right]$$

$$\sigma_{yy} = -\frac{P}{8\pi\,(1-v)}\frac{x}{R^3}\left[-(1-2v)+\frac{3\,y^2}{R^2}\right]$$

$$\sigma_{zz} = -\frac{P}{8\pi\,(1-v)}\frac{x}{R^3}\left[-(1-2v)+\frac{3\,z^2}{R^2}\right]$$

$$\sigma_{xy} = -\frac{P}{8\pi\,(1-v)}\frac{y}{R^3}\left[(1-2v)+\frac{3\,x^2}{R^2}\right]$$

$$\sigma_{yz} = -\frac{P}{8\pi\,(1-v)}\frac{3xyz}{R^5}$$

$$\sigma_{zx} = -\frac{P}{8\pi\,(1-v)}\frac{z}{R^3}\left[(1-2v)+\frac{3\,x^2}{R^2}\right]$$

524

$$u_x = \frac{P}{16\pi G(1-v)}\left[\frac{x^2}{R^3} + (3-4v)\frac{1}{R}\right]$$

$$u_y = \frac{P}{16\pi G(1-v)}\frac{xy}{R^3}$$

$$u_z = \frac{P}{16\pi G(1-v)}\frac{zx}{R^3}$$

where $R^2 = x^2 + y^2 + z^2$.

B.2 An infinite line load

For a line load, of intensity p_x per unit length, applied at the xy co-ordinate origin, in the x direction, and tensile normal stresses reckoned positive, etc.

$$\sigma_{xx} = -\frac{p_x}{4\pi(1-v)}\frac{x}{r^2}\left[(1-2v) + \frac{2x^2}{r^2}\right]$$

$$\sigma_{yy} = -\frac{p_x}{4\pi(1-v)}\frac{x}{r^2}\left[-(1-2v) + \frac{2y^2}{r^2}\right]$$

$$\sigma_{xy} = -\frac{p_x}{4\pi(1-v)}\frac{y}{r^2}\left[(1-2v) + \frac{2x^2}{r^2}\right]$$

$$u_x = \frac{p_x}{8\pi G(1-v)}\left[\frac{x^2}{r^2} - (3-4v)\ln r\right]$$

$$u_y = \frac{p_x}{8\pi G(1-v)}\frac{xy}{R^2}$$

where $r^2 = x^2 + y^2$.

Appendix C Calculation sequences for rock–support interaction analysis

C.1 Scope

The stepwise calculation sequences presented in this appendix permit rock–support interaction analyses to be carried out for the axisymmetric problem defined

525

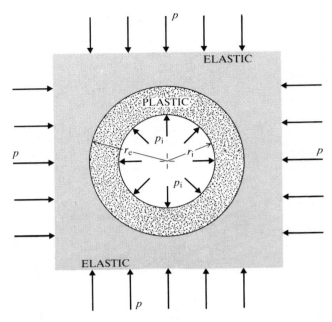

Figure A.6 Axisymmetric tunnel problem showing the development of a plastic zone around the tunnel.

in Figure A.6. Further details of these analyses are given by Daemen (1975), Hoek and Brown (1980), and Brown *et al.* (1983).

C.2 Required support line calculations

C.2.1 Solution using the elastic-brittle rock mass behaviour model of Figure 11.5

Input data

σ_c = uniaxial compressive strength of intact rock pieces;
m, s = material constants for the original rock mass;
E, ν = Young's modulus and Poisson's ratio of the original rock mass;
m_r, s_r = material constants for the broken rock mass;
f = gradient of $-\varepsilon_3^p$ vs, ε_1^p line (Figure 11.5);
p = hydrostatic field stress;
r_i = internal tunnel radius.

Calculation sequence

$$(1) \quad M = \frac{1}{2}\left[\left(\frac{m}{4}\right)^2 + \frac{mp}{\sigma_c} + s\right]^{1/2} - \frac{m}{8}$$

$$(2) \quad G = \frac{E}{2(1+\nu)}$$

526

(3) For $p_i \geqslant p - M\sigma_c$, deformation around the tunnel is elastic,

$$\frac{\delta_i}{r_i} = \frac{(p - p_i)}{2G}$$

(4) For $p_i < p - M\sigma_c$, plastic deformation occurs around the tunnel

$$\frac{u_e}{r_e} = \frac{M\sigma_c}{2G}$$

(5) $N = 2\left(\dfrac{p - M\sigma_c}{m_r\sigma_c} + \dfrac{s_r}{m_r^2}\right)^{1/2}$

(6) $\dfrac{r_e}{r_i} = \exp\left[N - 2\left(\dfrac{p_i}{m_r\,\sigma_c} + \dfrac{s_r}{m_r^2}\right)^{1/2}\right]$

(7) $\dfrac{\delta_i}{r_i} = \dfrac{M\sigma_c}{G(f+1)}\left[\dfrac{(f-1)}{2} + \left(\dfrac{r_e}{r_i}\right)^{f+1}\right]$

C.2.2 Solution using the strain-softening rock mass behaviour model of Figure 11.7

In this case, a closed-form solution cannot be obtained and an iterative finite difference solution procedure is used. The plastic zone surrounding the tunnel is divided into a number of thin annuli as shown in Figure A.7. The solution takes account of the possible existence around the tunnel of three different zones:

Figure A.7 Plastic zone divided into thin annuli for finite difference solution.

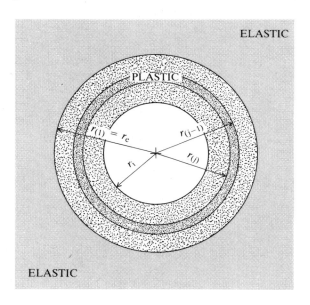

527

(a)   an elastic zone remote from the tunnel;
(b)   an intermediate plastic zone in which the stresses and strains fall on the strain-softening portion of Figure 11.7; and
(c)   an inner plastic zone in which stresses are limited by the residual strength of the rock mass.

Input data

σ_c = uniaxial compressive strength of intact rock pieces;
m, s = material constants for the original rock mass;
E, ν = Young's modulus and Poisson's ratio of the original rock mass;
m_r, s_r = material consitants for the broken rock mass;
f, h = gradients of $-\varepsilon_3^p$ vs. ε_1^p lines (Figure 11.7);
α = constant defining the strain at which residual strength is reached (Figure 11.7);
p = *in-situ* hydrostatic field stress;
r_i = internal tunnel radius.

Preliminary calculations

$$(1) \; M = \frac{1}{2}\left[\left(\frac{m}{4}\right)^2 + \frac{mp}{\sigma_c} + s\right]^{1/2} - m/8$$

$$(2) \; G = \frac{E}{2(1+\nu)}$$

$$(3) \; \varepsilon_{\theta(1)} = \varepsilon_{\theta_e} = \frac{M\sigma_c}{2G}$$

$$(4) \; \varepsilon_{r(1)} = \varepsilon_{r_e} = -\frac{M\sigma_c}{2G}$$

$$(5) \; \sigma_{r(1)} = p_{cr} = p - M\sigma_e$$

$$(6) \; \sigma_{\theta(1)} = 2p - \sigma_{r(1)}$$

$$(7) \; m_{(1)} = m$$

$$(8) \; s_{(1)} = s$$

$$(9) \; \lambda_{(1)} = \frac{r_{(1)}}{r_e} = 1$$

Sequence of calculations for each ring

$$(1) \; d\varepsilon_{\theta(j)} = 0.01\varepsilon_{\theta(j-1)}$$

528

(2) $\varepsilon_{\theta(j)} = \varepsilon_{\theta(j-1)} + d\varepsilon_{\theta(j)}$

(3) $\varepsilon_{r(j)} = \varepsilon_{r(j-1)} - h\, d\varepsilon_{\theta(j-1)}$

(4) If $\varepsilon_{\theta(j)} > \alpha\, \varepsilon_{\theta(1)}$ then $\varepsilon_{r(j)} = \varepsilon_{r(j-1)} - f\, d\varepsilon_{\theta(j-1)}$

(5) $\lambda_{(j)} = \left[\dfrac{2\varepsilon_{\theta(j-1)} - \varepsilon_{r(j-1)} - \varepsilon_{r(j)}}{2\varepsilon_{\theta(j)} - \varepsilon_{r(j-1)} - \varepsilon_{r(j)}} \right] \lambda_{(j-1)}$

where $\lambda_{(j)} = \dfrac{r_{(j)}}{r_e}$, $\lambda_{(j-1)} = \dfrac{r_{(j-1)}}{r_e}$

(6) $\dfrac{u_{(j)}}{r_e} = -\varepsilon_{\theta(j)} \times \dfrac{r_{(j)}}{r_e}$

(7) $\overline{m}_{(j)} = m + (m_r - m) \dfrac{(\varepsilon_{\theta(j)} - \varepsilon_{\theta_e})}{(\alpha - 1)\varepsilon_{\theta_e}}$

(8) $\overline{s}_{(j)} = s + (s_r - s) \dfrac{(\varepsilon_{\theta(j)} - \varepsilon_{\theta_e})}{(\alpha - 1)\varepsilon_{\theta_e}}$

(9) If $\varepsilon_{\theta(j)} > \alpha\varepsilon_{\theta(1)}$ then $\overline{m}_{(j)} = m_r$ and $\overline{s}_{(j)} = s_r$

(10) $\overline{m}_a = \frac{1}{2}(\overline{m}_{(j-1)} + \overline{m}_{(j)})$

(11) $\overline{s}_a = \frac{1}{2}(\overline{s}_{(j-1)} + \overline{s}_{(j)})$

(12) $k = \left[\dfrac{\lambda_{(j-1)} - \lambda_{(j)}}{\lambda_{(j-1)} + \lambda_{(j)}} \right]^2$

(13) $a = \sigma_{r(j-1)}^2 - 4k\left(\frac{1}{2}\overline{m}_a\sigma_c\sigma_{r(j-1)} + \overline{s}_s\,\sigma_c^2 \right)$

(14) $b = \sigma_{r(j-1)} + k\overline{m}_a\,\sigma_c$

(15) $\sigma_{r(j)} = b - \sqrt{b^2 - a}$

(16) $\sigma_{\theta(j)} = \sigma_{r(j)} + (\overline{m}_{(j)}\,\sigma_{r(j)}\,\sigma_c + \overline{s}_{(j)}\,\sigma_c^2)^{1/2}$

(17) If $\sigma_{r(j)} > p_i$ then increment j by 1 and repeat the calculation sequence for next ring.

(18) If $\sigma_{r(j)} \approx p_i$ then $r_{(j)} = r_i$, and $r_e = \dfrac{r_{(j)}}{\lambda_{(j)}}$

529

(19) The radii of all the rings may now be calculated, using $r_{(j)} = \lambda_{(j)} r_e$

(20) The values of the displacement $u_{(j)}$ may be determined from the previously computed values of $u_{(j)}/r_e$.

C.3 Available support line calculations

C.3.1 Support stiffness and maximum support pressure for a concrete or shotcrete lining

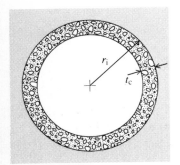

Figure A.8 Shotcrete or concrete lining.

Input data

E_c = Young's modulus of concrete or shotcrete;
ν_c = Poisson's ratio of concrete or shotcrete;
t_c = lining thickness (Figure A.8);
r_i = internal tunnel radius;
σ_{cc} = uniaxial compressive strength of concrete or shotcrete.

Support stiffness

$$k_c = \frac{E_c \left[r_i^2 - (r_i - t_c)^2 \right]}{(1 + \nu_c) \left[(1 - 2\nu_c) r_i^2 + (r_i - t_c)^2 \right]}$$

Maximum support pressure

$$p_{sc\,max} = \frac{\sigma_{cc}}{2} \left[1 - \frac{(r_i - t_c)^2}{r_i^2} \right]$$

C.3.2 Support stiffness and maximum support pressure for blocked steel sets

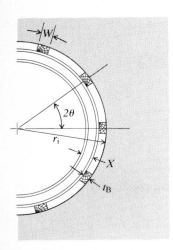

Figure A.9 Blocked steel set.

Input data (Figure A.9)

W = flange width of steel set and side length of square block;
X = depth of section of steel set;
A_s = cross-sectional area of steel set;
I_s = second moment of area of steel set;
E_s = Young's modulus of steel;
σ_{ys} = yield strength of steel;
r_i = internal tunnel radius;
S = steel set spacing along tunnel axis;
θ = half angle between blocking points (radians);
t_B = thickness of block;
E_B = Young's modulus of block material.

530

Support stiffness

$$\frac{1}{k_s} = \frac{Sr_i}{E_s A_s} + \frac{Sr_i^3}{E_s I_s}\left[\frac{\theta(\theta + \sin\theta\cos\theta)}{2\sin^2\theta} - 1\right] + \frac{2S\,\theta t_B}{E_B\,W^2}$$

Maximum support pressure

$$p_{ss\,max} = \frac{3\,A_s I_s \sigma_{ys}}{2Sr_i\theta\,\{3\,I_s + XA_s[r_i - (t_B + 0.5\,X)]\,(1 - \cos\theta)\}}$$

C.3.3 Support stiffness and maximum support pressure for ungrouted mechanically or chemically anchored rockbolts or cables

Input data

Important items of data required in this case are obtained from the results of pull-out tests. A typical load-extension curve is shown in Figure A.10. The variables used to determine the constant, Q, are defined on this figure.

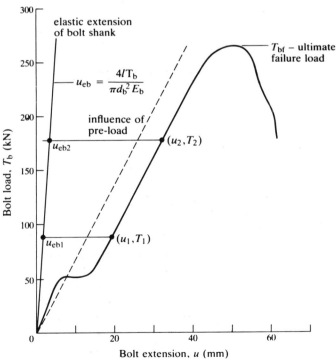

Figure A.10 Bolt load–extension curve obtained in a pull-out test on a 25 mm diameter, 1.83 m long rock-bolt anchored by a four-leaf expansion shell (after Franklin and Woodfield, 1971).

l = free bolt or cable length;
d_b = bolt diameter or equivalent cable diameter;
E_b = Young's modulus of bolt or cable;

T_{bf} = ultimate failure load in pull-out test;
r_i = internal tunnel radius;
s_c = circumferential bolt spacing;
s_ℓ = longitudinal bolt spacing;
Q = load-deformation constant for anchor and head, given (from Figure A.10) by

$$Q = \frac{(u_2 - u_{eb2}) - (u_1 - u_{eb1})}{T_2 - T_1}$$

where u_1, T_1 and u_2, T_2 are two points on the linear portion of the load-extension plot.

Support stiffness

$$\frac{1}{k_b} = \frac{s_c s_\ell}{r_i}\left(\frac{4l}{\pi d_b^2 E_b} + Q\right)$$

Maximum support pressure

$$p_{sb\,max} = \frac{T_{bf}}{s_c\,s_\ell}$$

C.3.4 Available support line for a single support system

If k is the stiffness of the support system, $p_{s\,max}$ is the maximum support pressure that the system can provide, and δ_{i0} is the radial tunnel deformation before installation of the support, then for $p_i < p_{s\,max}$, the available support line is given by

$$\frac{\delta_i}{r_i} = \frac{\delta_{i0}}{r_i} + \frac{p_i}{k}$$

C.3.5 Available support line for combined support systems

Input data

k_1 = support stiffness of system 1;
$p_{s\,max\,1}$ = maximum support pressure for system 1;
k_2 = support stiffness of system 2;
$p_{s\,max\,2}$ = maximum support pressure for system 2;
δ_{i0} = initial tunnel deformation before installation of support. (The two systems are assumed to be installed simultaneously.)

Available support line

(1) $\quad \delta_{max\,1} = \dfrac{r_i\,p_{smax1}}{k_1}$

(2) $\quad \delta_{\max 2} = \dfrac{r_i \, p_{\text{smax} 2}}{k_2}$

(3) $\quad \delta_{12} = \dfrac{r_i \, p_i}{(k_1 + k_2)}$

(4) For $\delta_{12} < \delta_{\max 1} < \delta_{\max 2}$

$$\frac{\delta_i}{r_i} = \frac{\delta_{i0}}{r_i} + \frac{p_i}{(k_1 + k_2)}$$

(5) For $\delta_{12} > \delta_{\max 1} < \delta_{\max 2}$

$$p_{\max 12} = \frac{\delta_{\max 1}(k_1 + k_2)}{r_i}$$

(6) For $\delta_{12} > \delta_{\max 2} < \delta_{\max 1}$

$$p_{\max 12} = \frac{\delta_{\max 2}(k_1 + k_2)}{r_i}$$

Appendix D Limiting equilibrium analysis of progressive hangingwall caving

D.1 Derivation of equations

The assumptions made, and the variables involved in the limiting equilibrium analysis of the problem illustrated in Figure 16.16, are set out in section 16.4.2.

Weight of wedge. The weight of the wedge of rock BCDNML in Figure 16.15 is

$$W = \frac{\gamma}{2} \left[\frac{H_2^2 \sin(\alpha + \psi_0) \sin(\psi_{p2} + \psi_0)}{\sin^2 \psi_0 \sin(\psi_{p2} - \alpha)} - \frac{H_1^2 \sin(\alpha + \psi_0) \sin(\psi_{p1} + \psi_0)}{\sin^2 \psi_0 \sin(\psi_{p1} - \alpha)} \right.$$

$$\left. + Z_1^2 \frac{\cos \alpha \cos \psi_{p1}}{\sin(\psi_{p1} - \alpha)} - Z_2^2 \frac{\cos \alpha \cos \psi_{p2}}{\sin(\psi_{p2} - \alpha)} \right] \tag{D.1}$$

Base area of wedge. The area of unit thickness of the surface LM (Figure 16.15) on which failure takes place is

$$A = \frac{H_2 \left(\sin \alpha \cot \psi_0 + \cos \alpha \right) - Z_2 \cos \alpha}{\sin (\psi_{p2} - \alpha)} \tag{D.2}$$

Thrust due to caved material.   The thrust acting on the wedge BCDNML due to the caved material left in the crater is one of the most difficult parameters to estimate with confidence in this analysis. A simplified system of forces used to calculate T is shown in Figure 16.15. Resolving the forces W_c, T_c and T in the horizontal and vertical directions and applying the equations of equilibrium of forces gives the solution

$$T = \tfrac{1}{2} \gamma_c H_c^2 K \tag{D.3}$$

where

$$K = \frac{\left\{ \left(\cot \psi_{p1} + \cot \psi_0 \right) + 2 \dfrac{S}{H_c} \right\}}{\cos (\psi_{p1} - \phi_w) + \sin (\psi_{p1} - \phi_w) \cot (\psi_0 - \phi_w)} \tag{D.4}$$

The weight of caved material below the level of point B has been ignored in this calculation.

Inclination of thrust to failure surface.   It is assumed that the thrust T is transmitted through the wedge BCDNML to the failure surface without loss or deviation. Hence, the inclination of T to the normal to the failure surface LM is

$$\theta = \psi_{p2} + \phi_w - \psi_{p1} \tag{D.5}$$

As shown in Figure 16.15, the angle θ may be either positive or negative. If θ is negative, the thrust T has a shear component that acts up the failure plane, and tends to stabilise rather than activate slip of the wedge.

Water-pressure forces.   The water-pressure force due to water in the tension crack is

$$V = \tfrac{1}{2} \gamma_w Z_w^2 \tag{D.6}$$

The water-pressure force U that acts normal to the failure surface is

$$U = \tfrac{1}{2} \gamma_w Z_w A$$

$$= \tfrac{1}{2} \gamma_w Z_w \left[\frac{H_2 \left(\sin \alpha \cot \psi_0 + \cos \alpha \right) - Z_2 \cos \alpha}{\sin (\psi_{p2} - \alpha)} \right] \tag{D.7}$$

Conditions of limiting equilibrium.   It is assumed that the shear strength of the rock mass in the direction of failure is given by the linear Coulomb criterion

$$\tau = c' + \sigma_n' \tan \phi' \tag{D.8}$$

534

The effective normal and shear stresses acting on the failure surface are

$$\sigma_n' = \frac{W \cos \psi_{p2} + T \cos \theta - U - V \sin \psi_{p2}}{A} \tag{D.9}$$

and

$$\tau = \frac{W \sin \psi_{p2} + T \sin \theta + V \cos \psi_{p2}}{A} \tag{D.10}$$

The conditions for limiting equilibrium are found by substituting for σ_n' and τ into equation D.8, which, on rearrangement, gives

$$W \sin (\psi_{p2} - \phi') + T \sin (\theta - \phi') + V \cos (\psi_{p2} - \phi') + U \sin \phi'$$
$$- c' A \cos \phi' = 0 \tag{D.11}$$

Mining depth for new failure. Substitution for W, A, T, V and U from equations D.1, 2, 3, 6 and 7 into equation D.9 and rearrangement gives a quadratic equation for H_2, the new mining depth at which failure will occur

$$\left(\frac{\gamma H_2}{c'}\right)^2 \frac{\sin (\alpha + \psi_0) \sin (\psi_{p2} + \psi_0) \sin (\psi_{p2} - \phi')}{\sin^2 \psi_0}$$

$$- 2 \left(\frac{\gamma H_2}{c'}\right)^2 \left[\frac{\sin (\alpha + \psi_0) \cos \phi'}{\sin \psi_0} - \frac{\gamma_w Z_w \sin (\alpha + \psi_0) \sin \phi}{2c' \sin \psi_0} \right]$$

$$- \left(\frac{\gamma H_1}{c'}\right)^2 \frac{\sin (\alpha + \psi_0) \sin (\psi_{p1} + \psi_0) \sin (\psi_{p2} - \phi') \sin (\psi_{p2} - \alpha)}{\sin^2 \psi_0 \sin (\psi_{p1} - \alpha)}$$

$$+ \left(\frac{\gamma Z_1}{c'}\right)^2 \frac{\cos \alpha \cos \psi_{p1} \sin (\psi_{p2} - \phi) \sin (\psi_{p2} - \alpha)}{\sin (\psi_{p1} - \alpha)}$$

$$- \left(\frac{\gamma Z_2}{c'}\right)^2 \cos \alpha \cos \psi_{p2} \sin (\psi_{p2} - \phi')$$

$$+ \frac{\gamma_c}{\gamma} \left(\frac{\gamma H_c}{c'}\right)^2 K \sin (\theta - \phi') \sin (\psi_{p2} - \alpha)$$

$$+ \frac{2\gamma Z_2 \cos \alpha}{c'} \left(\cos \phi' - \frac{\gamma_w Z_w}{2c'} \sin \phi' \right)$$

$$+ \frac{\gamma_w}{\gamma} \left(\frac{\gamma Z_w}{c'}\right)^2 \cos (\psi_{p2} - \phi') \sin (\psi_{p2} - \alpha) = 0 \tag{D.12}$$

535

Equation D.12 gives a solution for the dimensionless group $\gamma H_2/c'$ in the form

$$\frac{\gamma H_2}{c'} = a + (a^2 + b)^{1/2} \tag{D.13}$$

where

$$a = \frac{\sin \psi_0 \left(\cos \phi' - \dfrac{\gamma_w Z_w \sin \phi'}{2c'} \right)}{\sin (\psi_{p2} + \psi_0) \sin (\psi_{p2} - \phi')} \tag{D.14}$$

and

$$b = \left(\frac{\gamma H_1}{c'}\right)^2 \frac{\sin (\psi_{p1} + \psi_0) \sin (\psi_{p2} - \alpha)}{\sin (\psi_{p1} - \alpha) \sin (\psi_{p2} + \psi_0)}$$

$$- \left(\frac{\gamma Z_1}{c'}\right)^2 \frac{\cos \alpha \cos \psi_{p1} \sin (\psi_{p2} - \alpha) \sin^2 \psi_0}{\sin (\alpha + \psi_0) \sin (\psi_{p2} - \psi_0)}$$

$$+ \left(\frac{\gamma Z_2}{c'}\right)^2 \frac{\cos \alpha \cos \psi_{p2} \sin^2 \psi_0}{\sin (\alpha + \psi_0) \sin (\psi_{p2} - \psi_0)}$$

$$- \frac{\gamma_c}{\gamma} \left(\frac{\gamma H_c}{c'}\right)^2 \frac{K \sin (\theta - \phi') \sin (\psi_{p2} - \alpha) \sin^2 \psi_0}{\sin (\alpha + \psi_0) \sin (\psi_{p2} + \psi_0) \sin (\psi_{p2} - \phi')}$$

$$- 2 \left(\frac{\gamma Z_2}{c'}\right) \frac{\cos \alpha \left(\cos \phi' - \dfrac{\gamma_w Z_w}{2c'} \sin \phi' \right) \sin^2 \psi_0}{\sin (\alpha + \psi_0) \sin (\psi_{p2} + \psi_0) \sin (\psi_{p2} - \phi')}$$

$$- \frac{\gamma_w}{\gamma} \left(\frac{\gamma Z_w}{c'}\right)^2 \frac{\cos (\psi_{p2} - \phi') \sin (\psi_{p2} - \alpha) \sin^2 \psi_0}{\sin (\alpha + \psi_0) \sin (\psi_{p2} + \psi_0) \sin (\psi_{p2} - \phi')} \tag{D.15}$$

Critical tension crack depth. The left-hand side of equation D.12 can be differentiated with respect to Z_2 holding ψ_{p2} constant; the result is equated to zero to obtain the critical value of Z_2 as

$$\frac{\gamma Z_2}{c'} = \frac{\cos \phi'}{\cos \psi_{p2} \sin (\psi_{p2} - \phi')} \tag{D.16}$$

Critical failure plane inclination. By holding Z_2 constant, differentiating equation D.12 with respect to ψ_{p2}, putting $\partial H_2 / \partial \psi_{p2} = 0$ and rearranging, an expression for the critical failure plane angle may be obtained as

$$\psi_{p2} = \frac{1}{2} \left(\phi' + \cos^{-1} \frac{X}{(X^2 + Y^2)^{1/2}} \right) \tag{D.17}$$

where

$$X = \left(\frac{\gamma H_1}{c'}\right)^2 \frac{\sin(\alpha + \psi_0)\sin(\psi_{p1} + \psi_0)\cos\alpha}{\sin^2\psi_0 \sin(\psi_{p1} - \alpha)}$$

$$-\left(\frac{\gamma H_2}{c'}\right)^2 \frac{\sin(\alpha + \psi_0)\cos\psi_0}{\sin^2\psi_0}$$

$$-\left(\frac{\gamma Z_1}{c'}\right)^2 \frac{\cos\psi_{p1}\cos^2\alpha}{\sin(\psi_{p1} - \alpha)} - \frac{\gamma_c}{\gamma}\left(\frac{\gamma H_c}{c'}\right)^2 K\cos(\psi_{p1} + \alpha - \phi_w)$$

$$+\frac{\gamma_w}{\gamma}\left(\frac{\gamma Z_w}{c'}\right)^2 \cos\alpha \tag{D.18}$$

and

$$Y = \left(\frac{\gamma H_2}{c'}\right)^2 \frac{\sin(\alpha + \psi_0)}{\sin\psi_0} + \left(\frac{\gamma H_1}{c'}\right)^2 \frac{\sin(\alpha + \psi_0)\sin(\psi_{p1} + \psi_0)\sin\alpha}{\sin^2\psi_0 \sin(\psi_{p1} - \alpha)}$$

$$-\left(\frac{\gamma Z_1}{c'}\right)^2 \frac{\cos\alpha \sin\alpha \cos\psi_{p1}}{\sin(\psi_{p1} - \alpha)} - \left(\frac{\gamma Z_2}{c'}\right)^2 \cos\alpha$$

$$-\frac{\gamma_c}{\gamma}\left(\frac{\gamma H_c}{c'}\right)^2 K\sin(\psi_{p1} + \alpha - \phi_w) + \frac{\gamma_w}{\gamma}\left(\frac{\gamma Z_w}{c'}\right)^2 \sin\alpha \tag{D.19}$$

Angle of break. By simple trigonometric manipulation it is found that

$$\tan\psi_b = \tan\psi_{p2} + \frac{Z_2 \sin(\psi_{p2} - \alpha)}{\cos\psi_{p2}\left[\dfrac{H_2 \sin(\alpha + \psi_0)}{\sin\psi_0} - Z_2 \cos\alpha\right]} \tag{D.20}$$

D.2 Calculation sequence

The following sequence of calculations allows angles of failure and angles of break to be estimated using the equations derived above. For dry conditions, $U = V = 0$. If the upper surface is horizontal, $\alpha = 0$. Positive values of α are as shown in Figure 16.15. The equations listed also apply for negative values of α. The sequence of calculations is:

(1) Calculate K from equation D.4.

(2)   Let $\psi_{p2e} = \frac{1}{2} (\psi_{p1} + \phi')$

(3)   Calculate $\dfrac{\gamma Z_2}{c'}$ from equation D.16

(4)   Calculate a from equation D.14.

(5)   Calculate b from equation D.15.

(6)   Calculate $\dfrac{\gamma H_2}{c'}$ from equation D.13.

(7)   Calculate X from equation D.18.

(8)   Calculate Y from equation D.19.

(9)   Calculate the estimated value of ψ_{p2} from equation D.17.

(10)   Compare ψ_{p2} from step 9 with ψ_{p2e} from step 2. If $\psi_{p2} \neq \psi_{p2e}$, substitute ψ_{p2} for ψ_{p2e} in steps 3, 4 and 5 and repeat the calculation cycle until the difference between successive values of ψ_{p2} is $< 0.1\%$.

(11) Calculate ψ_b from equation D.20.

Answers to problems

Chapter 2

1. (a) 200 kN, with clockwise moment.
   (b) $\sigma_{xx} = -15$ MPa, $\sigma_{yy} = 16$ Mpa, $\sigma_{xy} = 10$ MPa.
   (c) $\sigma_{ll} = 1.41$ MPa, $\sigma_{mm} = -0.41$ Mpa, $\sigma_{lm} = 18.42$ MPa.
   (d) $\sigma_1 = 8.95$ MPa, $\sigma_2 = -17.95$ Mpa, $\theta = 73.6°$.
   (e) $\theta = 0°$; $t_x = 15$ MN m^{-2}, $t_y = -10$ MN m^{-2}.
   $\theta = 60°$; $t_x = -1.16$ MN m^{-2}, $t_y = -18.86$ MN m^{-2}, $R = 18.90$ MPa
   $\theta = 90°$; $t_x = -10$ MN m^{-2}, $t_y = -16$ MN m^{-2}

2. (a) $\sigma_{xx} = 7.825$, $\sigma_{yy} = 6.308$, $\sigma_{zz} = 7.866$;
   $\sigma_{xy} = 1.422$, $\sigma_{yz} = 0.012$, $\sigma_{zx} = -1.857$.
   (b) $\sigma_{mm} = 9.824$; $\sigma_{nl} = 0.079$.
   (c) $I_1 = 22.0$, $I_2 = 155.0$, $I_3 = 350.0$;.
   $\sigma_1 = 10.0$ ($\alpha_1 = 43.4°$, $\beta_1 = 73.9°$, $\gamma_1 = 129.05°$) ;
   $\sigma_2 = 7.0$ ($\alpha_2 = 108.5°$, $\beta_2 = 131.8°$, $\gamma_2 = 132.4°$) ;
   $\sigma_3 = 5.0$ ($\alpha_3 = 51.8°$, $\beta_3 = 132.4°$, $\gamma_3 = 66.2°$) .

4. (a)
$$\varepsilon_{xx} = \frac{1}{2G}\left(-\frac{y}{r^2} + \frac{2x^2 y}{r^4}\right)$$

$$\varepsilon_{yy} = \frac{1}{2G}\left[(1-4v)\frac{y}{r^2} + \frac{2y^3}{r^4}\right],$$

$$\gamma_{xy} = \frac{1}{2G}\left[2(1-2v)\frac{x}{r^2} + \frac{4xy^2}{r^4}\right].$$

   (b)
$$\sigma_{xx} = -(1-2v)\frac{y}{r^2} + \frac{2x^2 y}{r^4}.$$

$$\sigma_{yy} = (1-2v)\frac{y}{r^2} + \frac{2y^3}{r^4},$$

$$\sigma_{xy} = (1-2v)\frac{x}{r^2} + \frac{2xy^2}{r^4}.$$

5. (a) $\sigma_1 = 25.0$, $\sigma_2 = 7.1$, $\alpha_1 = 58°$, $\alpha_2 = 148°$.
   $\sigma_n = 21.0$, $\tau = 7.2$.
   (b) $\sigma_{ll} = 20.9$, $\sigma_{lm} = 7.46$.
   (c) $\sigma_1 = 24.94$, $\sigma_2 = 7.10$, $\alpha_1 = 58.3°$, $\alpha_2 = 148.3°$.

Chapter 3

1. (a) $\bar{x}_{0A} = 0.369$ m; $\bar{x}_{0B} = 0.274$ m.
   (b) $\bar{x} = 0.288$ m.
   (c) $RQD = 95.2$.

2. (a) 0.087–0.100 m; 0.083–0.104 m.
   (b) 26 m.

539

3 48°.
4 038/20.
5 71° measured in the plane normal to the line of intersection.
6 275/50.
7 312/17.
8 RMR = 69.

Chapter 4

1 $E_t = 14.3$ GPa, $v = 0.23$.
2 $\sigma_1 = 133.3$ MPa; $\sigma_1 = 85$ MPa.
3 (a) $\sigma_1 = 133 + 3.0\sigma_3$ MPa.

4 $\dfrac{\sigma_1}{100} = 1.0 + 4.75 \left(\dfrac{\sigma_3}{100} \right)^{0.75}$ is one of several possible solutions.

5 $s = 15.0 + \sigma_n \tan 27.5°$ MPa.

7 (a) $\phi_b = 30°$, $i = 15°$.
   (b) Peak strength envelope bilinear with $c = 0$, $\phi = 45°$ for $0 < \sigma_n < 1$ MPa, and $c = 0.5$ MPa, $\phi = 26.5°$ for $1 < \sigma_n < 5$ MPa. Residual strength envelope bilinear with $c_r = 0$, $\phi_r = 30°$ for $0 < \sigma_n < 1$ MPa, and $c_r = 0.1$ MPa, $\phi_r = 26°$ for $1 < \sigma_n < 5$ MPa.

8 $u = 1.247$ MPa; $u = 0.642$ MPa.
9 $\sigma_1' = 0.346 + 3.0\sigma_3'$ MPa.

Chapter 5

1 (b) $\sigma_{xx}^i = 1.35$ MPa, $\sigma_{zz}^i = 1.35$ MPa, $\sigma_{zx}^i = 0.86$ MPa.
   (c) $\sigma_{xx} = 3.51$ MPa, $\sigma_{zz} = 4.59$ MPa, $\sigma_{zx} = 0.86$ MPa.
       $\sigma_1 = 5.07$ MPa, $\sigma_2 = 3.03$ Mpa, $\alpha_1 = 61.1°$

2 $p_{zz} = 20.80$ MPa, fissure water pressure $u = 6.86$ MPa.
       $13.69 < p_{xx} < 43.29$ MPa.

3 (a) Plane principal stresses are vertical and horizontal.
   (b) $\sigma_1 = 20.0$ MPa (vertical), $\sigma_2 = 15.0$ MPa (horizontal),
       $\sigma_3 = 12.0$ MPa (parallel to tunnel axis).

4 All stresses are in MPa.
   (b) $p_{ll} = 10.22$, $p_{mm} = 17.09$, $p_{nn} = 10.68$, $p_{lm} = -1.30$, $p_{mn} = -4.86$, $p_{nl} = 1.12$.
   (c) $p_{xx} = 15.59$, $p_{yy} = 11.06$, $p_{zz} = 11.35$, $p_{xy} = 3.05$, $p_{yz} = 3.12$, $p_{zx} = 3.99$.
   (d) $p_1 = 20.0$ MPa, oriented $030°/30°$ (dip direction/dip);
       $p_2 = 10.0$ MPa, oriented $135°/24°$;
       $p_3 = 8.0$ MPa, oriented $257°/50°$.

5 $\phi_{mob} = 31.3°$ (cf. $\phi_{act} = 20°$).

Chapter 7

1 (a) Compressive failure in side wall, tensile failure in crown.
   (b) Increase W/H ratio of excavation.

2 (a) Maximum boundary stress of 55 MPa located 30° below x axis in right-hand wall, and 30° above horizontal in left-hand wall. Minimum boundary stress of 11 MPa, located across diameter perpendicular to that defining maximum stresses.

540

(b) Angular range of boundary failure is $\pm 35.5°$ about the location of the maximum stress.

3 Boundary stresses are approximately those for isolated openings. At A, $\sigma_1 = 21.31$ MPa $\sigma_2 = 20$ MPa, $\sigma_3 = 18.69$ MPa, and $\alpha_1 = -54.1°$ in problem plane.

4 $D_{1,2} = 10$ m.

5 Maximum and minimum boundary stresses are 112.9 MPa and -2.4 MPa, both located near the ends of the ellipse.

6 On fault, $\tau_{max} = 1.41$ MPa.

Chapter 13

1 Initial F of S = 1.1. For F of S = 1.6, $w_p = 9.2$ m.

2 For existing layout, F of S = 1.98. If roof coal is stripped, F of S = 1.69.

3 F of S = 5.47.

4 Pillar is stable.

Chapter 15

1 (b) $\sigma_{xx} = 32.224$ MPa, $\sigma_{zz} = 40.224$ MPa, $\tau_{zx} = 2.409$ MPa,
   (c) $u_z = \pm 0.080$ m.

2 (a) $L_0 = 294.1$ m.
       For $L = L_0$, $W_r = 9817.5$ MJ per unit length and 33.38 MK m^{-3}.
       For $L \gg L_0$, $W_r = 85L$ MJ per unit length and 85 MJ m^{-3}.
   (b) $W_{rp} = \dfrac{-(1-v)\,p^2 S}{\pi G} \ln\left(\cos\dfrac{\pi l}{2S}\right)$

       $l = 100$ m, $S = 125$ m.

3 (a) $p = 13.8$ MPa, $\sigma_y = 55.75$ MPa, $x_b = 5.0$ m, $A_b = 82.7$ MN m^{-1}, $A_w = 1126.0$ MN m^{-1}, $C = 26.52$ m.
   (b) $w_p = 70$ m.

4 Wilson's method gives the following results:

h(m)	w_p(m)	
100	12.2	} $w > 0.6h$
250	34.6	
400	57.2	
500	68.2	
750	84.4	} $w > 0.6h$
1000	93.2	
1250	98.9	
1500	102.8	

5 $a_N = 8.0$ m, $b_N = 1.85$ m, $E_N = 114.2$ m^3.
   $a_G = 19.05$ m, $b_G = 4.40$ m, $E_G = 1542.1$m^3;
   $A \le 8.68$ m, $V \ge 1.85$ m

6 (a) No. Assuming the same eccentricity for the flow and limit ellipsoids, the height of the limit ellipsoid above the top of the drawpoint is calculated as 52.4 m.
   (b) 24 288 tonne.

References

Adler, L. and Sun, M. (1968) Ground control in bedded formations, *Bulletin 28, Research Div.,* Virginia Polytechnic Institute.

Airy, G. B. (1862) On the strains in the interior of beams, *Rep. 32nd Mtg Br. Assoc. Adv. Sci.,* Cambridge, 82–6.

Aki, K. and Richards, P. G. (1980) *Quantitative Seismology,* W. H. Freeman, San Francisco.

Allen, C. W. (1934) Subsidence arising from the Athens system of mining at Nauganee, Mich. *Trans. Am. Inst. Min. Metall. Petrolm Engrs,* **109**, 195–202.

Almgren, G. (1981) Rock mechanics and the economics of cut and fill mining, in *Applications of Rock Mechanics to Cut and Fill Mining* (eds O. Stephansson and M.J. Jones), Instn Min. Metall., London, pp. 28–35.

Amadei, B. (1988) Strength of a regularly jointed rock mass under biaxial and axisymmetric loading conditions. *Int. J. Rock Mech. Min. Sci. & Geomech. Abstr.,* **25**, 3–13.

Amadei, B. and Goodman, R. E. (1981) Formulation of complete plane strain problems for regularly jointed rocks, in *Rock Mechanics from Theory to Application, Proc. 22nd US Symp. Rock Mech.,* Mass. Inst. Technol., Cambridge, Mass. pp. 245–51.

Amstad, C. and Koppel, J. (1977) A multihead borehole rod extensometer design, in *Field Measurements in Rock Mechanics* (ed. K. Kovari), **1**, A. A. Balkema, Rotterdam, pp. 429–36.

Anon (1973) Mine filling, in *Jubilee Symposium on Mine Backfilling,* Mount Isa, Aust. Inst. Min. Metall., Melbourne.

Anon (1974) Use of shotcrete for underground structural support, in *Proc. Engng Foundn Conf.,* ASCE, New York.

Anon (1978) *Mining with Backfill, Proc. 12th Can. Rock Mech. Symp.,* Sudbury, Can. Inst. Min. Metall., Montreal.

Anon (1982) Equipment for shotcreting. *Min. Mag.,* **147**, 547–57.

Ashwin, D. P., Campbell, S. G., Kibble, J. D., Haskayne, J. D., Moore, J. F. A. and Shepherd, R. (1970) Some fundamental aspects of face powered support design. *Min. Engr,* **129**, 659–71.

Atkinson, J. H., Brown, E. T. and Potts, D. M. (1975) Collapse of shallow unlined tunnels in dense sand. *Tunnels and Tunnelling,* **7**, 81–7.

Avasthi, J. M. and Harloff, G. J. (1982) Subsidence associated with single and multi-cavities for underground coal gasification. *J. Energy Resources Technol., Trans. Am. Soc. Mech. Engrs,* **104**, 99–104.

Bandis, S. C., Lumsden, A. C. and Barton, N. R. (1981) Experimental studies of scale effects on the shear behaviour of rock joints. *Int. J. Rock Mech. Min. Sci. & Geomech. Abstr.,* **18**, 1–21.

Bandis, S. C., Lumsden, A .C. and Barton, N. R. (1983) Fundamentals of rock joint deformation. *Int. J. Rock Mech. Min. Sci. & Geomech. Abstr.,* **20**, 249–268.

Bandis, S. C., Barton, N. R. and Christianson, M. (1985) Application of a new numerical model of joint behaviour to rock mechanics problems, in *Proc. Int. Symp. Fundamentals of Rock Joints* (ed. O. Stephansson), Centek, Lulea, pp. 345–55.

Barker, R. M. and Hatt, F. (1972) Joint effects in bedded formation roof control, in *New Horizons in Rock Mechanics, Proc. 14th US Symp. Rock Mech.* (eds H. R. Hardy and R. Stefanko), ASCE, New York, pp 247–61.

Barr, M. V. and Brown, E. T. (1983) A site exploration trial using instrumented diamond drilling, in *Proc. 5th Congr., Int. Soc. Rock Mech.,* Melbourne, **1**, A51–7, A. A. Balkema, Rotterdam.

Barrett, J. R. (1973) Structural aspects of cemented fill behaviour, in *Mine Filling, Proc. Jubilee Symp. Mine Filling,* Mount Isa, Aust. Inst. Min. Metall, Melbourne, pp. 97–104.

Barrett, J. R., Coulthard, M. A. and Dight, P. M. (1978) Determination of fill stability, in *Mining with Backfill, Proc. 12th Can. Rock Mech. Symp.,* Sudbury, Can. Inst. Min. Metall, Montreal, pp. 85–91.

Barton, N. R. (1973) Review of a new shear strength criterion for rock joints. *Engng Geol.,* **8**, 287–332.

Barton, N. R. (1983) Application of Q-system and index tests to estimate shear strength and deformability of rock masses, in *Proc. Int. Symp. on Engineering Geology and Underground Construction,* **2**, Lisbon, pp. II.51–70.

Barton, N. and Hansteen, H. (1979) Very large span openings at shallow depth: deformation magnitudes from jointed models and finite element analysis, in *Proc. 4th Rapid Excavation and Tunnelling Conf.,* Atlanta, 1979, **2**, pp. 1331–53.

542

Barton, N. R., Bandis, S. and Bakhtar, K. (1985) Strength, deformation and conductivity coupling of rock joints. *Int. J. Rock Mech. Min. Sci. & Geomech. Abstr.*, **22**, 121–40.

Barton, N. R., Lien, R. and Lunde, J. (1974) Engineering classification of rock masses for the design of tunnel support. *Rock Mech.*, **6**, 189–239.

Beer, G. (1984) BEFE – A combined boundary–finite element computer program. *Advances in Engineering Software*, **6** (2), 103–9.

Beer, G. and Meek, J. L (1982) Design curves for roofs and hangingwalls in bedded rock based on voussoir beam and plate solutions. *Trans Instn Min. Metall.*, **91**, A18–22.

Berry, D. S. (1960) An elastic treatment of ground movement due to mining – 1. Isotropic ground. *J. Mech. Phys. Solids*, **8**, 280–92.

Berry, D. S. (1963) Ground movement considered as an elastic phenomenon. *Min. Engr*, **123**, 28–41.

Berry, D. S. (1978) Progress in the analysis of ground movements due to mining, in *Large Ground Movements and Structures* (ed. J.D. Geddes), Pentech, London, pp. 781–811.

Berry, D. S. and Sales, T. W. (1961) An elastic treatment of ground movement due to mining – II. Transversely isotropic ground. *J. Mech. Phys. Solids*, **9**, 52–62.

Bieniawski, Z. T. (1973) Engineering classification of jointed rock masses. *Trans S. Afr. Inst. Civ. Engrs*, **15**, 335–44.

Bieniawski, Z. T. (1974) Estimating the strength of rock materials. *J. S. Afr. Inst. Min. Metall.*, **74**, 312–20.

Bieniawski, Z .T. (1975) The point load test in engineering practice. *Engng Geol.*, **9**, 1–11.

Bieniawski, Z. T. (1976) Rock mass classifications in rock engineering, in *Exploration for Rock Engineering*, **1**, A. A. Balkema, Cape Town, pp. 97–106.

Bieniawski, Z. T. (1978) Determining rock mass deformability: experience from case histories, *Int. J. Rock Mech. Min. Sci. & Geomech. Abstr.*, **15**, 237–47.

Bieniawski, Z. T. (1984) *Rock Mechanics Design in Mining and Tunneling*, A. A. Balkema, Rotterdam.

Bjurstrom, S. (1979) Shear strength on hard rock joints reinforced by grouted untensioned bolts, in *Proc. 4th Congr., Int. Soc. Rock Mech.*, **2**, A .A. Balkema, Rotterdam, pp. 1194–99.

Blake, W., Leighton, F. and Duvall, W. I. (1974) *Microseismic Techniques for Monitoring the Behaviour of Rock Structures*, US Bur. Mines Bull. 665.

Blight, G. E. (1984) Soil mechanics principles in underground mining. *J. Geotech. Engng Div., ASCE*, **110**, 567–81.

Blyth, F. G. H. and de Freitas, M. H. (1974) *A Geology for Engineers*, 6th edn, Edward Arnold, London.

Bock, H. (1986) In-situ validation of the borehole slotting stressmeter, in *Proc. Int. Symp. on Rock Stress and Rock Stress Measurement*, Stockholm, Centek, Lulea, pp. 261–70.

Borquez, G. V. (1981) Sequence of the analysis of the block caving method, in *Design and Operation of Caving and Sublevel Stoping Mines* (ed. D. R. Stewart), Soc. Min. Engrs, AIME, New York, pp. 283–97.

Boussinesq, J. (1883) *Applications des potentials a l'étude de l'équilibre et du mouvement des solides élastiques*, Gauthier-Villars, Paris.

Boyd, J. M. (1975) *The Interpretation of Geological Structure for Engineering Design in Rock*, PhD thesis, University of London.

Boyum, B. H. (1961) Subsidence case histories in Michigan mines, in *Proc. 4th US Symp. Rock Mech.*, *Bull. Mineral Ind. Expt Stn*, Penn. State Univ., No. 76, pp. 19–57.

Brace, W. F. and Byerlee, J. D. (1966) Stick-slip as a mechanism for earthquakes. *Science*, **153**, 990–92.

Brace, W. F. and Martin, R. J. (1968) A test of the law of effective stress for crystalline rocks of low porosity. *Int. 'J. Rock Mech. Min. Sci. & Geomech. Abstr.*, **5**, 415–26.

Brady, B. H. G. (1975) Rock mechanics aspects of mining the 1100 orebody, *Mount Isa Mines Limited Tech. Rep.*

Brady, B. H. G. (1977) An analysis of rock behaviour in an experimental stoping block at the Mount Isa Mine, Queensland, Australia. *Int. J. Rock Mech. Min. Sci. & Geomech Abstr.*, **14**, 59–66.

Brady, B. H. G. (1981) Determination of stability of underground mine structures, in *Design and Operation of Caving and Sublevel Stoping Mines* (ed. D. R. Stewart), Soc. Min. Engrs, AIME, New York, pp. 427–435.

543

Brady, B. H. G. and Brown, E. T. (1981) Energy changes and stability in mine structures: design applications of boundary element methods. *Trans Instn Min. Metall.*, **90**, A61–8.

Brady, B. H. G. and Wassyng, A. (1981) A coupled finite element–boundary element method of stress analysis. *Int. J. Rock Mech. Min Sci. & Geomech. Abstr.*, **18**, 475–85.

Brady, B. H. G., Cramer, M.L. and Hart, R. D. (1985) Preliminary analysis of a loading test on a large basalt block. *Int. J. Rock Mech. Min. Sci. & Geomech. Abstr.*, **22**, 345–48.

Brady, B. H. G., Friday, R. G. and Alexander, L. G. (1976) Stress measurement in a bored raise at the Mount Isa Mine, Queensland, Australia, in *Proc. Int. Soc. Rock Mech. Symp. on Investigation of Stress in Rock*, Instn Engrs Aust., Sydney, pp. 12–16.

Brady, B. and Lorig, L. (1988) Analysis of rock reinforcement using finite difference methods. *Computers and Geotechnics*, **5**(2), 123–49.

Brauner, G. (1973) *Subsidence due to Underground Mining – 1. Theory and Practices in Predicting Surface Deformation*, US Bur. Mines Inf. Circ. 8571.

Bray, J. W. (1977) Unpublished note, Imperial College, London.

Bray, J. W. (1979) Unpublished note, Imperial College, London.

Bray, J. W. (1986) Some applications of elastic theory, in *Analytical and Computational Methods in Engineering Rock Mechanics*, (ed. E. T. Brown), Allen & Unwin, London, pp. 32–94.

Bray, J. W. and Goodman, R. E. (1981) The theory of base friction models. *Int. J. Rock Mech. Min. Sci. & Geomech. Abstr.*, **18**, 453–68.

Brinch Hansen, J. (1970) Bearing capacity. *Danish Geotech. Inst. Bull.*, No. 28, 5–11.

Broch, E. and J. A. Franklin (1972) The point-load strength test. *Int. J. Rock Mech. Min. Sci. & Geomech. Abstr.*, **9**, 669–97.

Brown, E. T. (1970) Strength of models of rock with intermittent joints. *J. Soil Mech. Foundns Div., ASCE*, **96**, 1935–49.

Brown, E. T. (1974) Fracture of rock under uniform biaxial compression, in *Advances in Rock Mechanics, Proc. 3rd Congr., Int Soc. Rock Mech.*, Denver, **2A**, Nat. Acad. Sciences, Washington, DC, pp. 111–17.

Brown, E. T. and Ferguson, G.A. (1979) Progressive hangingwall caving at Gath's mine, Rhodesia. *Trans Instn Min. Metall.*, **88**, A92–105.

Brown, E. T. and Gonano L.P. (1974) Improved compression test technique for soft rock. *J. Geotech. Engng Div., ASCE*, **100**, 196–9.

Brown, E. T. and Gonano, L. P. (1975) An analysis of size effect behaviour in brittle rock, in *Proc. 2nd Aust.–NZ Conf. Geomech.*, Instn Engrs Aust., Sydney, 139–43.

Brown, E. T. and Hoek, E. (1978) Trends in relationships between measured in-situ stresses and depth. *Int. J. Rock Mech. Min. Sci. & Geomech. Abstr.*, **15**, 211–15.

Brown, E. T. and Hudson, J. A. (1972) Progressive collapse of simple block-jointed systems. *Aust. Geomech. J.*, **G2**, 49–54.

Brown, E. T. and Hudson, J. A. (1974) Fatigue failure characteristics of some models of jointed rock. *Earthquake Eng. and Struct. Dyn.*, **2**, 379–386.

Brown, E. T. and Trollope, D.H. (1970) Strength of a model of jointed rock. *J. Soil Mech. Foundns Div., ASCE*, **96**, 685–704.

Brown, E. T., Bray, J.W., Ladanyi, B. and Hoek, E. (1983) Characteristic line calculations for rock tunnels. *J. Geotech. Engng, ASCE*, **109**, 15–39.

Brown, E. T., Richards, L. R. and Barr, M. V. (1977) Shear strength characteristics of Delabole slates, in *Proc. Conf. Rock Engng*, Univ. Newcastle-upon-Tyne, Newcastle-upon-Tyne, 33–51.

Brown, S. M., Leijon, B. A. and Hustrulid, W. A. (1986) Stress distribution within an artificially loaded block, in *Proc. Int. Symp. on Rock Stress and Rock Stress Measurement*, Stockholm (ed. O. Stephansson), Centek, Lulea, pp. 429–440.

Brune, J.N. (1970) Tectonic stress and the spectra of seismic shear waves from earthquakes. *J. Geophys. Res.*, **75**, 4997–5009; correction (1971) *J. Geophys. Res.*, **76**, 5002.

Bryan, A., Bryan, J. G. and Fouche, J. (1964) Some problems of strata control and support in pillar workings. *Min. Engr*, **123**, 238–66.

Bucky, P. B. (1956) Fundamental considerations in block caving. *Q. Colo. School Mines*, **51**(3), 129–46.

Bucky, P. B. and Taborelli, R. V. (1938) Effects of immediate roof thickness in longwall mining as determined by barodynamic experiments. *Trans Am. Inst. Min. Metall.*, **130**, 314–32.

Bunting, D. (1911) Chamber pillars in deep anthracite mines. *Trans Am. Inst. Min. Engrs*, **42**, 236–45.

REFERENCES

Carlsson, A. and Christiansson R. (1986) Rock stresses and geological structures in the Forsmark area, in *Proc. Int. Symp. on Rock Stress and Rock Stress Measurement*, Stockholm (ed. O. Stepphansson), Centek, Lulea, pp. 457–66.

Cartwright, A. P. (1969) *West Dreifontein – Ordeal by Water*, Goldfields of S. Afr., Johannesburg.

Charlton, T. M. (1959) *Energy Methods in Applied Statics*, Blackie, London.

Chatterjee, P. K., Just, G. D. and Ham, G. I. (1974) Design and evaluation of sublevel caving by dynamic digital simulation, in *Proc. 12th Symp. on the Applications and Mathematics in the Minerals Industry* (eds T. B. Johnson and D. W. Gentry), **2**, Colo. School Mines, Golden, G153–77.

Chatterjee, P. K., Just, G. D. and Ham, G. I. (1979) Sub-level caving simulation of 3000 pillar recovery operation at Mount Isa Mine, Australia. *Trans Instn Min. Metall.*, **88**, A147–55.

Clough, R. W. and Penzien, J. (1975) *Dynamics of Structures*, McGraw Hill, New York.

Coates, D. F. (1981) *Rock Mechanics Principles* (3rd edn), Mines Branch Monograph 874, Information Canada, Ottawa.

Commission of Inquiry (1971) *The Mufulira Mine Disaster*, Republic of Zambia, Lusaka.

Cook, J. F. and D. Bruce (1983) Rockbursts at Macassa mine and the Kirkland Lake mining area, in *Rockbursts: Prediction and Control*, Instn Min. Metall., London, pp. 81–9.

Cook, M. A. (1958) *The Science of High Explosives*, Reinhold, New York.

Cook, M. A., Cook, V. D., Clay, R. B., Keyes, R. T. and Udy, L. L. (1966) Behaviour of rock during blasting. *Trans Soc. Min. Engrs, AIME*, **235**, 383–92.

Cook, N. G. W. (1965) A note on rockbursts considered as a problem of stability. *J. S. Afr. Inst. Min. Metall.*, **65**, 437–46.

Cook, N. G. W. (1967a) The design of underground excavations, in *Failure and Breakage of Rock, Proc. 8th US Symp. Rock Mech.*, (ed. C. Fairhurst), AIME, New York, 167–93.

Cook, N. G. W. (1967b) Contribution to discussion on pillar stability. *J. S. Afr. Inst. Min. Metall.*, **68**, 192–5.

Cook, N. G. W. (1978) *Rockbursts and Rockfalls*, Chamber of Mines of S. Afr. Publn No. 216.

Cook, N. G. W., Hodgson, K. and Hojem, J. P. M. (1971) A 100 MN jacking system for testing coal pillars underground. *J. S. Afr. Inst. Min. Metall.*, **71**, 215–24.

Cook, N. G. W., Hoek, E., Pretorius, J. P. G., Ortlepp, W.D. and Salamon, M.D.G. (1966) Rock mechanics applied to the study of rockbursts. *J.S. Afr. Inst. Min. Metall.*, **66**, 436–528.

Cording, E. J., Hendron, A. J. and Deere, D. U. (1971) Rock engineering for underground caverns, in *Proc. Symp. Underground Rock Chambers*, ASCE, New York, 567–600.

Coulomb, C. A. (1776) Essai sur une application des règles de maximis et minimis à quelque problèmes de statique, relatifs à l'architecture. *Mémoires de Mathématique et de Physique, L'Académie Royale des Sciences*, **7**, 343–82.

Crane, W. E. (1931) *Essential Factors Influencing Subsidence and Ground Movement*, US Bur. Mines Inf. Circ. 6501.

Croll, J. G. and Walker, A. G. (1973) *Elements of Structural Stability*, Halsted Press, New York.

Crotty, J. M. (1983) *User's Manual for BITEMJ – Two-dimensional Stress Analysis for Piecewise Homogeneous Solids with Structural Discontinuities*, CSIRO (Australia) Division of Geomechanics, Geomechanics Computer Program No. 5, 104 pp.

Crouch, S. L. (1970) Experimental determination of volumetric strains in failed rock. *Int. J. Rock Mech. Min. Sci. & Geomech. Abstr.*, **7**, 589–603.

Crouch, S. L. and Fairhurst (1973) *The Mechanics of Coal Mine Bumps*, US Bur. Mines Open File Report 53/73.

Cundall, P. A. (1971) A computer model for simulating progressive large scale movements in blocky rock systems, in *Rock Fracture, Proc. Int. Symp. Rock Fracture*, Nancy, Paper 2–8.

Cundall, P. and Board, M. (1988) A microcomputer program for modelling large-strain plasticity problems, in *Proc. 6th Int. Conf. Num. Meth. Geomech.*, Innsbruck, A. A. Balkema, Rotterdam, pp. 2101–08.

Cundall, P. A. and Lemos, J. V. (1990) Numerical simulation of fault instabilities with a continuously-yielding joint model, in *Rockbursts and Seismicity in Mines, Proc. 2nd Int. Symp. on Rockbursts and Seismicity in Mines*, (ed. C. Fairhurst), A. A. Balkema, Rotterdam, pp. 147–52.

Daemen J. J. K. (1975) Tunnel support loading cause by rock failure, *Tech. Rep. MRD-3-75*, Missouri River Div., US Corps Engrs, Omaha.

Daemen J. J. K. (1977) Problems in tunnel support mechanics. *Underground Space*, **1**, 163–72.

545

Deere, D. U. (1968) Geological considerations, in *Rock Mechanics in Engineering Practice* (eds K. G. Stagg and O. C. Zienkiewicz), Wiley, London, pp. 1–20.

Deere, D. U. (1979) Applied rock mechanics – the importance of weak geological features, in *Proc. 4th Congr., Int. Soc. Rock Mech.*, Montreux, **3**, pp. 22–5.

Desai, C. S. and Salami, M. R. (1987) Constitutive models for rocks. *J. Geotech. Engng Div., ASCE*, **113**, 407–23.

Dickhout, M. H. (1973) The role and behaviour of fill in mining, in *Mine Filling, Proc. Jubilee Symp. Mine Filling*, Mount Isa, Aust. Inst. Min. Metall., Melbourne, pp. 1–11.

Dienes, J. K. and Margolin, L. G. (1980) A computational approach to rock fragmentation, in *The State of the Art in Rock Mechanics, Proc. 21st US Symp. Rock Mech.* (ed. D. A. Summers), University of Missouri, Rolla, pp. 390–8.

Dieterich, J. H. (1978) Time-dependent friction and the mechanics of stick-slip. *PAGEOPH*, **116**, 790–806.

Dieterich, J. H. (1979) Modeling of rock friction: 1. experimental results and constitutive equations. *J. Geophys. Res.*, **84**(B5), 2161–68.

Donath, F. A. (1972) Effects of cohesion and granularity on deformational behaviour of anisotropic rock, in *Studies in Mineralogy and Precambrian Geology* (eds B. R. Doe and D. K. Smith), Geol. Soc. Am. Memoir, **135**, pp. 95–128.

Donovan, K., Pariseau, W. G. and Cepak, M. (1984) Finite element approach to cable bolting in steeply dipping VCR stopes, in *Geomechanics Applications in Underground Hardrock Mining* (ed. W. G. Pariseau), AIME, New York, pp. 65–90.

Dowding, C. H. (1985) *Blast Vibration Monitoring and Control*, Prentice-Hall, Englewood Cliffs.

Dowding, C. H. and Rozen, A. (1978) Damage to rock tunnels from earthquake loading, *J. Geotech. Engng Div., ASCE*, **104**(GT2), 175–91.

Dravo Corporation (1974) *Analysis of Large-scale Non-coal Underground Mining Methods*, US Bur. Mines Open File Rep., pp. 36–74.

Drescher, A. and Vardoulakis, I. (1982) Geometric softening in triaxial tests on granular material. *Géotechnique*, **32**, 291–303.

Duvall, W. I. (1953) Strain wave shapes in rock near explosions. *Geophys.*, **18**, 310–23.

Duvall, W. I. and Atchison, T. C. (1957) *Rock Breakage by Explosives*, US Bur. Mines Rep. Invest. 5356.

Eissa, E. A. (1980) *Stress Analysis of Underground Excavations in Isotropic and Stratified Rock using the Boundary Element Method*, PhD thesis, University of London.

Elliott, G. M. (1982) *An Investigation of a Yield Criterion for Rock*, PhD thesis, University of London.

Elliott, G. M. and Brown, E. T. (1985) Yield of a soft, high porosity rock. *Géotechnique*, **35**, 413–23.

Endersbee, L. A. and Hofto, E.O. (1963) Civil engineering design and studies in rock mechanics for Poatina underground power station, Tasmania. *J. Instn Engrs Aust.*, **35**, 187–209.

Enever, J. R. and Chopra, P. N. (1986) Experience with hydraulic fracture stress measurements in granite, in *Proc. Int. Symp. on Rock Stress and Rock Stress Measurement*, Stockholm, (ed. O. Stephansson), Centek, Lulea, pp. 411–20.

Evans, W. H. (1941) The strength of undermined strata. *Trans Instn Min. Metall.*, **50**, 475–532.

Everling, G. (1973) Die Vorausberechnung des Gebirgsdrucks fur einen Abbauplan. *Gluckauf*, **109** (23), 1–3.

Ewy, R. T., Kemeny, J. M., Zheng, Z. and Cook, N. G. W. (1987) Generation and analysis of stable excavation shapes under high rock stresses, in *Proc. 6th Congr., Int. Soc. Rock Mech.*, (Montreal), A. A. Balkema, Rotterdam, 875–881.

Farmer, I. W. and Shelton, P. D. (1980) Review of underground reinforcement systems. *Trans Inst. Min. Metall.*, **89**, A68–83.

Fayol, M. (1885) Sur les mouvements de terrain provoques par l'exploitation des mines. *Bull. Soc. l'Industrie Minérale, 2nd series*, **14**, 818.

Fellows, S. (1976) Tunnel profiling by photography. *Tunnels and Tunnelling*, **8**(4), 70–73.

Fenner, R. T. (1974) *Computing for Engineers*, MacMillan, London.

Ferguson, G. A. (1979) Optimisation of block caving within a complex environment. *Min. Mag.*, **140**, 126–39.

Fletcher, J. B. (1960) Ground movement and subsidence from block caving at Miami mine. *Trans Soc. Min. Engrs, AIME*, **217**, 413–22.

Folinsbee, J. C. and Clarke, R. W. D. (1981) Selecting a mining method, in *Design and Operation of Caving and Sublevel Stoping Mines* (ed. D. R. Stewart), AIME, New York, pp. 55–65.

Franklin, J. A. (1977) The monitoring of structures in rock. *Int. J. Rock Mech. Min. Sci. & Geomech. Abstr.*, **14**, 163–92.

Franklin, J. A. and Woodfield, P. F. (1971) Comparison of a polyester resin and a mechanical rockbolt anchor. *Trans. Instn Min. Metall.*, **80**, A91–100.

Fuller, P. G. (1981) Pre-reinforcement of cut and fill stopes, in *Applications of Rock Mechanics to Cut and Fill Mining* (eds O. Stephansson and M. J. Jones), Instn Min. Metall., London, pp. 55–63.

•Gay, N. C. (1975) In-situ stress measurements in Southern Africa. *Tectonophysics*, **29**, 447–59.

Gay, N. C. and Ortlepp, W. D. (1979) Anatomy of a mining-induced fault zone. *Bull. Geol. Soc. Am.*, **90**, 47–58.

Gaziev, E. G. and Erlikmann, S. A. (1971) Stresses and strains in anisotropic rock foundations (model studies), in *Rock Fracture, Proc. Int. Symp. Rock Fracture*, Nancy, Paper 2–1.

Gen Hua Shi and Goodman, R. E. (1981) A new concept for support of underground and surface excavations in discontinuous rocks based on a keystone principle, in *Rock Mechanics from Research to Application, Proc. 22nd US Symp. Rock Mech.*, Mass. Inst. Technol., Cambridge, Mass., pp. 290–96.

Gerdeen, J. C., Snyder, V. W., Viegelahn G. L. and Parker, J. (1977) Design criteria for rockbolting plans using fully resin-grouted non-tensioned bolts to reinforce bedded mine roof, *USBM OFR 46(4)–80*.

Gerrard, C. M. (1977) Background to mathematical modelling in geomechanics, in *Finite Elements in Geomechanics* (ed. G. Gudehus), Wiley, London, pp. 33–120.

Gerrard, C. M. (1982) Elastic models of rock masses having one, two, and three sets of joints. *Int. J. Rock Mech. Min. Sci. & Geomech. Abstr.*, **19**, 15–23.

Gibowicz, S. J. (1988) The mechanism of seismic events induced by mining – a review, in *Rockbursts and Seismicity in Mines, Proc. 2nd Int. Symp. on Rockbursts and Seismicity in Mines* (ed. C. Fairhurst), A. A. Balkema, Rotterdam, pp. 3–27.

Goddard, I. A. (1981) The development of open stoping in lead orebodies at Mount Isa Mines Ltd, in *Design and Operation of Caving and Sublevel Stoping Mines* (ed. D. R. Stewart), Soc. Min. Engrs, AIME, New York, pp. 509–28.

Godson, R. A., Bridges, M. C. and McKavanagh, B. M. (1980) A 32-channel rock noise source location system, in *Proc. 2nd Conf. on Acoustic Emission/Microseismic Activity in Geological Structures and Materials* (eds H. R. Hardy anf F. W. Leighton), Trans. Tech. Publications, Clausthal, pp. 117–52.

Goel, S.C. and Page, C. H. (1982) An empirical method for predicting the probability of chimney cave occurrence over a mining area. *Int. J. Rock Mech. Min. Sci. & Geomech. Abstr.*, **19**, 325–37.

Gonano, L. P. (1975) In-situ testing and size effect behaviour of cemented mine fill, in *Proc. Symp. on In-situ Testing for Design Parameters*, Aust. Geomech. Soc., Melbourne.

Gonano, L.P. (1977) Mechanical properties of cemented hydraulic fill pillars, *Tech. Rept. No. 36, Div. App. Geomech., Aust Comm. Sci. Ind. Res. Org.*, Melbourne.

Goodman, R. E. (1976) *Methods in Geological Engineering in Discontinuous Rock*, West, St Paul.

Goodman, R. E. (1980) *Introduction to Rock Mechanics*, Wiley, New York.

Grady, D. E. and Kipp, M. E. (1980) Continuum modelling of explosive fracture in oil shale. *Int. J. Rock Mech. Min. Sci. & Geomech. Abstr.*, **17**, 147–57.

Green, R. W. E. (1984) Design considerations for an underground seismic network, in *Rockbursts and Seismicity in Mines* (eds N.C. Gay and E.H. Wainwright), Sth Afr. Inst. Min. Metall., Johannesburg, pp. 67–73.

Greenelsh, R. W. (1985) The N663 stope experiment at the Mount Isa Mine, *Int. J. Min. Eng.*, **3**, 183–94.

Greenwald, H. P., Howarth, H. C. and Hartmann, I. (1939) *Experiments on Strength of Small Pillars of Coal of the Pittsburgh Bed*, US Bur. Mines Tech. Paper No. 605.

Greenwald, H. P., Howarth, H. C. and Hartmann, I. (1941) *Progress Report: Experiments on the Strength of Small Pillars of Coal of the Pittsburgh Bed*, US Bur. Mines Rep. Invest. 3575.

Griffith, A. A. (1921) The phenomena of rupture and flow in solids. *Phil. Trans Roy. Soc.*, **A221**, 163–97.

Griffith, A. A. (1924) Theory of rupture, in *Proc. 1st Congr. Appl. Mech.*, Delft, pp. 55–63.

547

Haas, C. J. (1981) Analysis of rock bolting to prevent shear movement in fractured ground. *Min. Engng.* (June), 698–704.

Hagan, T. O. (1980) A case for terrestrial photogrammetry in deep-mine rock structure studies, *Int. J. Rock Mech. Min. Sci. & Geomech. Abstr.*, **17**, 191–8.

Hagan, T. O. (1988) Mine design strategies to combat rockbursting at a deep South African gold mine, in *Key Questions in Rock Mechanics, Proc. 29th U.S. Rock Mech. Symp.*, (eds P.A. Cundall, R. L. Sterling and A. M. Starfield), A. A. Balkema, Rotterdam, pp. 249–60.

Haimson, B. C. (1978) The hydrofracturing stress measuring method and recent field trials. *Int. J. Rock Mech. Min. Sci. & Geomech. Abstr.*, **15**, 167–78.

Hamrin, H. O. (1982) Choosing an underground mining method, in *Underground Mining Methods Handbook* (ed. W.A. Hustrulid), AIME, New York, pp. 88–112.

Hanks, T. C. and Kanamori, H. (1979) A moment amplitude scale. *J. Geophys. Res.*, **84**, 2348–50.

Hanks, T. C. and Wyss, M. (1972) The use of body-wave spectra in the determination of seismic source parameters. *Bull. Seism. Soc. Am.*, **62**, 561–89.

Hardy, H. R. and Mowrey, G. L. (1976) Study of microseismic activity associated with a longwall mining operation using a near-surface array. *Engng Geol.*, **10**, 263–81.

Hardy, M. P. (1973) *Fracture Mechanics applied to Rock*, PhD thesis, University of Minnesota.

Hardy, M. P. and Agapito, J. F. T. (1977) Pillar design in underground oil shale mines, in *Design Methods in Rock Mechanics, Proc. 16th US Symp. Rock Mech.* (eds C. Fairhurst and S. L. Crouch), ASCE, New York, pp. 257–66.

Harries, G. (1977) Theory of blasting, in *Drilling and Blasting Technology*, Aust. Min. Foundn, Adelaide.

Hatheway, A. W. (1968) Subsidence at San Manuel Copper Mine, Pinal Country, Arizona, in *Engineering Geology Case Histories, Geol. Soc. Am.*, No. 6, pp. 65–81.

Hawkes, I. and Hooker, V. E. (1974) The vibrating wire stressmeter, in *Advances in Rock Mechanics, Proc. 3rd Congr., Int. Soc. Rock Mech.*, Denver, **2A**, Nat. Acad. Sci., Washington, DC, pp. 439–44.

Hawkes, I. and M. Mellor (1970) Uniaxial testing in rock mechanics laboratories. *Engng Geol.*, **4**, 177–285.

Haycocks, C. (1973) Minor stoping systems – top slicing, breast stoping, underhand and overhand stoping, in *SME Mining Engineering Handbook* (eds A.B. Cummings and I.A. Given), **1**, AIME, New York, pp. 12–150 – 12–159.

Hedley, D. G. F. and Grant, F. (1972) Stope and pillar design for the Elliot Lake uranium mines. *Bull. Can. Inst. Min. Metall.*, **65**, 37–44.

Hedley, D. G. F., Roxburgh, J. W. and Muppalaneni, S. N. (1984) A case history of rockbursts at Elliot Lake, in *Stability in Underground Mining II*, University of Kentucky, AIME, New York, pp. 210–34.

Heslop, T. G. and Laubscher, D. H. (1981) Draw control in caving operations on southern African chrysotile asbestos mines, in *Design and Operation of Caving and Sublevel Stoping Mines* (ed. D. R. Stewart), Soc. Min. Engrs, AIME, New York, pp. 755–74.

Heunis, R. (1980) The development of rockburst control strategies for South African gold mines. *J.S. Afr. Inst. Min. Metall.*, **80**, 139–50.

Hill, R. (1950) *The Mathematical Theory of Plasticity*, Oxford University Press, Oxford.

Hills, E. S. (1972) *Elements of Structural Geology*, 2nd edn, Chapman & Hall, London.

Hino, K. (1956) Fragmentation of rock through blasting. *Q. Colo. School Mines*, **51**, 89–209.

Hobbs, B. E. (1976) *An Outline of Structural Geology*, Wiley, New York.

Hodgson, K. and Joughin, N. C. (1967) The relationship between energy release rate, damage and seismicity in mines, in *Failure and Breakage of Rock, Proc. 8th US Symp. Rock Mech.* (ed. C. Fairhurst), AIME, New York, pp. 194–203.

Hoek, E. (1974) Progressive caving induced by mining an inclined orebody. *Trans. Instn Min. Metall.*, **83**, A133–9.

Hoek, E. (1982) Geotechnical considerations in tunnel design and contract preparation. *Trans. Instn Min. Metall.*, **91**, A101–9.

Hoek, E. and Bray, J. W. (1981) *Rock Slope Engineering*, 3rd edn, Instn Min. Metall., London.

Hoek, E. and Brown, E. T. (1980) *Underground Excavations in Rock*, Instn Min. and Metall., London.

Hoek, E. and Brown, E. T. (1988) The Hoek–Brown criterion – a 1988 update, in *Proc. 15th Can. Rock Mech. Symp.*, Toronto University Press, Toronto, pp. 31–38.

Hoek, E. and Franklin, J. A. (1968) A simple triaxial cell for field and laboratory testing of rock. *Trans. Instn Min. Metall.*, **77**, A22–6.

548

Holland, C. J. and Gaddy, F. L. (1957) Some aspects of permanent support of overburden on coal beds, in *Proc. West Virginia Coal Min. Inst.*, 43–66.

Holmberg, R., Maki, K., Hustrulid, W. and Sellden, H. (1983) Blast damage and stress measurement at LKAB – Malmberget Fabian orebody, in *Proc. 5th Congr., Int. Soc. Rock Mech.*, Melbourne, **2**, A. A. Balkema, Rotterdam, E231–38.

Hood, M., Ewy, R. T. and Riddle, L. R. (1983) Empirical methods of subsidence prediction – a case study from Illinois. *Int. J. Rock Mech. Min. Sci. & Geomech. Abstr.*, **20**, 153–70.

Hooker, V. E., Bickel, D. L. and Aggson, J. R. (1972) *In-situ Determination of Stresses in Mountainous Topography*, US Bur. Mines Rep. Invest. 7654.

Hopkins, H. G. (1960) Dynamic expansion of spherical cavities in metals. *Prog. Solid Mech.*, **1**, 84–164.

Hoskins, E. R. (1969) The failure of thick-walled cylinders of isotropic rock. *Int. J. Rock Mech. Min. Sci. & Geomech. Abstr.*, **6**, 99–125.

Housner, G. W. and Jennings, P. C. (1982) *Earthquake Design Criteria*, EERI Monograph Series, Berkeley, California.

Hudson, J. A., Brown, E. T. and Fairhurst, C. (1972a) Shape of the complete stress–strain curve for rock, in *Stability of Rock Slopes, Proc. 13th Symp. Rock Mech.* (ed. E. J. Cording), ASCE, New York, pp. 773–95.

Hudson, J. A., Brown, E. T. and Fairhurst, C. (1972b) Soft, stiff, and servo-controlled testing machines: a review with reference to rock failure, *Engng Geol.*, **6**, 155–89.

Hudson, J. A. and Priest, S. D. (1983) Discontinuity frequency in rock masses. *Int. J. Rock Mech. Min. Sci. & Geomech. Abstr.*, **25**, 3–13.

Hult, J. and Lindholm, H. W. (1967) Stress change during undercutting for block caving at the Grangesberg mine, in *Proc. 4th Can. Rock Mech. Symp.*, Mines Branch, Dept Energy, Mines and Resources, Ottawa, pp. 155–67.

Hunt, R. E. B. and Askew, J.E. (1977) Installation and design guidelines for cable dowel ground support at ZC/NBHC, in *Proc. Underground Operators Conf.*, Broken Hill, pp. 113–22.

Hustrulid, W. and Moreno, O. (1981) Support capabilities of fill – a non-linear analysis, in *Application of Rock Mechanics in Cut-and-Fill Mining*, Instn Min. Metall., London, pp. 107–118.

International Society for Rock Mechanics Commission on Standardization of Laboratory and Field Tests (1974) Suggested methods for determining shear strength. Doc. No. 1. Reprinted in *Rock Characterization, Testing and Monitoring – ISRM Suggested Methods*, 1981 (ed. E. T. Brown), Pergamon, Oxford, pp. 129–40.

International Society for Rock Mechanics Commission on Standardization of Laboratory and Field Tests (1978a) Suggested methods for the quantitative description of discontinuities in rock masses. *Int. J. Rock Mech. Min. Sci. & Geomech. Abstr.*, **15**, 319–68. Reprinted in *Rock Characterization, Testing and Monitoring – ISRM Suggested Methods*, 1981 (ed. E.T. Brown), Pergamon, Oxford, pp. 3–52.

International Society for Rock Mechanics Commission on Standardization of Laboratory and Field Tests (1978b) Suggested methods for monitoring rock movements using borehole extensometers. *Int. J. Rock Mech. Min. Sci. & Geomech. Abstr.*, **15**, 319–68. Reprinted in *Rock Characterization, Testing and Monitoring – ISRM Suggested Methods*, 1981 (ed. E. T. Brown), Pergamon, Oxford, pp. 171–83.

International Society for Rock Mechanics Commission on Standardization of Laboratory and Field Tests (1979) Suggested methods for determining the uniaxial compressive strength and deformability of rock materials. *Int. J. Rock Mech. Min. Sci. & Geomech. Abstr.*, **16**, 135–40. Reprinted in *Rock Characterization, Testing and Monitoring – ISRM Suggested Methods*, 1981 (ed. E. T. Brown), Pergamon, Oxford, pp. 111–16.

International Society for Rock Mechanics Commission on Standardization of Laboratory and Field Tests (1980) Suggested methods for pressure monitoring using hydraulic cells. *Int. J. Rock Mech. Min. Sci. & Geomech. Abstr.*, **17**, 117–27. Reprinted in *Rock Characterization, Testing and Monitoring – ISRM Suggested Methods*, 1981 (ed. E.T. Brown), Pergamon, Oxford, pp. 201–11.

Jaeger, J. C. (1960) Shear fracture of anisotropic rocks. *Geol. Mag.*, **97**, 65–72.

Jaeger, J. C. (1971) Friction of rocks and stability of rock slopes. *Géotechnique*, **21**, 97–134.

Jaeger, J. C. (1978) *Elasticity, Fracture and Flow*, Chapman & Hall, London.

Jaeger, J. C. and Cook, N.G.W. (1979) *Fundamentals of Rock Mechanics*, 3rd edn, Chapman & Hall, London.

Jaeger, J. C. and Rosengren, K. J. (1969) Friction and sliding of joints. *Proc. Aust. Inst. Min. Metall.*, No. 229, 93–104.

Jager, A. J., Piper, P. S. and Gay, N. C. (1987) Rock mechanics aspects of backfill in deep South African gold mines, in *Proc. 6th Congr., Int. Soc. Rock Mech.*, Montreal, **2**, A. A. Balkema, Rotterdam, pp. 991–8.

Janelid, I. and Kvapil, R. (1966) Sublevel caving. *Int. J. Rock Mech. Min. Sci. & Geomech. Abstr.*, **3**, 129–53.

Jenike, A. W. (1966) Storage and flow of solids. *Trans Soc. Min. Engrs, AIME*, **235**, 267–75.

Jennings, A. (1977) *Matrix Computation for Engineers and Scientists*, Wiley, London.

Jennings, J. E., Brink, A. B. A., Louve, A. and Gowan, G. D. (1965) Sinkholes and subsidences in the Transvaal dolomites of South Africa, in *Proc. 6th Int. Conf. Soil Mech. Foundn Engng*, Montreal, **1**, 51–54.

Johansson, C. H. and Persson, P. A. (1970) *Detonics of High Explosives*, Academic Press, London.

Jolley, D. (1968) Computer simulation of the movement of ore and waste in an underground mine. *Can. Min. Metall. Bull.*, **61**, 854–9.

Jones, O. T. and Llewellyn-Davies, E. (1929) Pillar and stall working under a sandstone roof. *Trans Instn Min. Engrs*, **76**, 313–29.

Just G. D. (1981) The significance of material flow in mine design and production, in *Design and Operation of Caving and Sublevel Stoping Mines* (ed. D.R. Stewart), Soc. Min. Engrs, AIME, New York, pp. 755–74.

Just, G. D., Free, G. D. and Bishop, G. A. (1973) Optimization of ring burden in sublevel caving. *Int. J. Rock Mech. Min. Sci. & Geomech. Abstr.*, **10**, 119–31.

Kendorski, F. S. (1978) The cavability of ore deposits. *Min. Engng*, **30**, 628–31.

Kendrick, R. (1970) Induction caving of the Urad mine. *Min. Congr. J.*, **56**(10), 39–44.

King, H. J., Whittaker, B.N. and Shadbolt, C.H. (1975) Effects of mining subsidence on surface structures, in *Mining and the Environment* (ed. M.J. Jones), Instn Min. Metall., London, pp. 617–42.

Kirsch, G. (1898) Die theorie der elastizitat und die bedürfnisse der festigkeitslehre. *Veit. Ver. Deut. Ing.*, **42**, 797–807.

Knutsson, S. (1981) Stresses in hydraulic backfill from analytical calculations and in-situ measurement, in *Applications of Rock Mechanics to Cut and Fill mining*, (eds O. Stephansson and M.J. Jones), Instn Min. Metall., London, pp. 261–8.

Kolsky, H. (1963) *Stress Waves in Solids*, Dover, New York.

Korf, C. W. (1978–9) Stick and pillar support on Union Section, Rustenburg Platinum Mines Ltd. *Assn Mine Managers S. Afr., Papers and Discns*, 71–6.

Koskela, V. A. (1983) Consolidated backfilling at Outokumpu Oy's Vihanti, Keretti and Vammala mines, in *Proc. Int. Symp. on Mining with Backfill*, A. A. Balkema, Rotterdam, pp. 151–9.

Kovari, K., Amsted, C. and Grob, H. (1974) Displacement measurements of high accuracy in underground openings, in *Advances in Rock Mechanics, Proc. 3rd Congr., Int. Soc. Rock Mech.*, Denver, **2A**, Nat. Acad. Sci., Washington, DC, pp. 445–50

Krauland, N. (1981) FEM model of Nasliden mine – requirements and limitations at start of project, in *Applications of Rock Mechanics to Cut and Fill Mining* (eds O. Stephansson and M.J. Jones), Instn Min. Metall., London, pp. 141–4.

Krauland, N., Nilsson, G. and Jonasson, P. (1981) Comparison of rock mechanics observations and measurements with FEM calculations, in *Applications of Rock Mechanics to Cut and Fill Mining* (eds O. Stephansson and M.J. Jones), Instn Min. Metall., London, pp. 250–60.

Kutter, H. K. and Fairhurst, C. (1968) The roles of stress waves and gas pressure in presplitting, in *Status of Practical Rock Mechanics, Proc. 9th US Symp. Rock Mech.* (eds N.E. Grosvenor and B.W. Paulding), AIME, New York, pp. 265–84.

Kutter, H. K. and Fairhurst, C. (1971) On the fracture process in blasting. *Int. J. Rock Mech. Min. Sci. & Geomech. Abstr.*, **8**, 181–202.

Kvapil, R. (1965) Gravity flow of granular material in hoppers and bins. *Int. J. Rock Mech. Min. Sci. & Geomech. Abstr.*, **2**, 35–41 and 277–304.

Labreche, D. A. (1983) Damage mechanisms in tunnels subjected to explosive loads, in *Seismic Design of Embankments and Caverns*, ASCE, New York, pp. 128–41.

550

Labuc, V., Bawden, W. and Kitzinger, F. (1987) Seismic monitoring system using fiber-optic signal transmission, in *Proc. 6th Congr., Int. Soc. Rock Mech.*, Montreal, **2**, A.A. Balkema, Rotterdam, pp. 1051–5.

Lacasse, M. and Legast, P. (1981) Change from grizzly to LHD extraction system, in *Design and Operation of Caving and Sublevel Stoping Mines* (ed. D.R. Stewart), Soc. Min. Engrs, AIME, New York, pp. 107–18.

Ladanyi, B. (1974) Use of the long-term strength concept in the determination of ground pressure on tunnel linings, in *Advances in Rock Mechanics, Proc. 3rd Congr., Int. Soc. Rock Mech.*, Denver, **2A**, Nat. Acad. Sci., Washington, DC, pp. 1150–6.

Ladanyi, B. and Archambault, G. (1977) Shear strength and deformability of filled indented joints, in *Proc. Int. Symp. Geotechnics of Structurally Complex Formations*, Associazione Geotechnica Italiana, **1**, pp. 317–26.

Lamb, H. (1956) *Infinitesimal Calculus*, Cambridge University Press, Cambridge.

Lang, T. A. (1961) Theory and practice of rock bolting. *Trans Soc. Min. Engrs*, AIME, **220**, 333–48.

Lang, T. A. and Bischoff, J. A. (1982) Stabilization of rock excavations using rock reinforcement, in *Issues in Rock Mechanics, Proc. 23rd U.S. Symp. Rock Mech.*, (eds R.E. Goodman and F.E. Heuze), AIME, New York, pp. 935–43.

Langstaff, J. T. (1977) Hecla's rock burst monitoring system. *Min. Congr. J.*, **63**(1), 46–52.

Lappalainen, P. and Antikainen, J. (1987) Mechanized cable bolting in stoping and tunnelling at the Pyhasalmi Mine, in *Improvement of Mine Productivity and Overall Economy by Modern Technology*, A. A. Balkema, Rotterdam, pp. 793–96.

Laubscher, D. H. (1977) Geomechanics classification of jointed rock masses – mining applications. *Trans. Instn Min. Metall.*, **86**, A1–8.

Laubscher, D. H. (1981) Selection of mass underground mining methods, in *Design and Operation of Caving and Sublevel Stoping Mines* (ed. D.R. Stewart), Soc. Min. Engrs, AIME, New York, pp. 23–8.

Laubscher, D. H. (1984) Design aspects and effectiveness of support systems in different mining conditions. *Trans. Instn Min. Metall.*, **93**, A70–81.

Laubscher, D. H. and Taylor, H. W. (1976) The importance of geomechanics classification of jointed rock masses in mining operations, in *Exploration for Rock Engineering*, **1**, A. A. Balkema, Cape Town, pp. 119–28.

Lawrence, W. J. C. and Bock, H. F. (1982) Numerical modelling of the yielding 11-level crown pillar at Mount Isa. *Aust. Geomechs. News*, No.5, 9–16.

Lee, M. F. and Bridges, M.C. (1981) Rock mechanics of crown pillars between cut-and-fill stopes at Mount Isa Mine, in *Applications of Rock Mechanics to Cut and Fill Mining* (eds O. Stephansson and M. J. Jones), Instn Min. Metall., London, pp. 316–29.

Lee, M. F., Beer, G. and Windsor, C. R. (1990) Interaction of stopes, stresses and geologic structure at the Mount Charlotte mine, Western Australia, in *Rockbursts and Seismicity in Mines, Proc. 2nd Int. Symp. on Rockbursts and Seismicity in Mines* (ed. C. Fairhurst), A. A. Balkema, Rotterdam, pp. 337–43.

Leeman, E. R. and Hayes, D. J. (1966) A technique for determining the complete state of stress in rock using a single borehole, in *Proc. 1st Congr., Int. Soc. Rock Mech.*, Lisbon, **2**, pp. 17–24.

Londe, P. (1982) Concepts and instruments for improved monitoring. *J. Geotech. Engng Div., ASCE*, **108**, 820–34.

Lorig, L. J. (1985) A simple numerical representation of fully bonded passive reinforcement for hard rocks. *Computers and Geotechnics*, **1**, 79–97.

Lorig, L. J. and Brady, B. H. G. (1982) A hybrid discrete element–boundary element method of stress analysis, in *Issues in Rock Mechanics, Proc. 23rd US Symp. Rock Mech.*, (eds R. E. Goodman and F.E. Heuze), AIME, New York, pp. 628–36.

Lorig L.J. and Brady, B. H. G. (1983) An improved procedure for excavation design in stratified rock, in *Rock Mechanics – Theory – Experiment – Practice, Proc. 24th US Symp. Rock Mech.*, College Station, Texas, AEG, New York, pp. 577–85.

Love, A. E. H. (1944) *A Treatise on the Mathematical Theory of Elasticity*, Dover, New York.

McGarr, A. (1976) Seismic moments and volume changes. *J. Geophys. Res.*, **81**, 1487–94.

McGarr, A. (1984) Some applications of seismic source mechanism studies to assessing underground hazard, in *Rockbursts and Seismicity in Mines* (eds N. C. Gay and E. H. Wainwright), Sth Afr. Inst. Min. Metall., Johannesburg, pp. 199–208.

McGarr, A., Green, R. W. E. and Spottiswoode, S. M. (1981) Strong ground motion of mine tremors: Some implications for near-source ground motion parameters. *Bull. Seis. Soc. Am.*, **71**(1), 295–319.

McKenzie, C. K. (1987) Blasting in hard rock – techniques for diagnosing and modelling for damage and fragmentation, in *Proc. 6th Congr., Int. Soc. Rock Mech.*, Montreal, **3**, A. A. Balkema, Rotterdam, pp. 1425–31.

McKenzie, C. K. (1988) *Blasting Research for Rock Engineering*, Report to Australian Mineral Industries Research Association Ltd by Julius Kruttschnitt Mineral Research Centre, University of Queensland.

McLamore, R. and Gray, K. E. (1967) The mechanical behaviour of anisotropic sedimentary rocks. *J. Engng for Industry, Trans. Am. Soc. Mech. Engrs Ser. B*, **89**, 62–73.

McNay, L. M. and Corson, D. R. (1975) *Hydraulic Sandfill in Deep Metal Mines*, US Bur. Mines Inf. Circ. 8663.

Mahtab, M. A. and Dixon, J. D. (1976) Influence of rock fractures and block boundary weakening on cavability. *Trans Soc. Min. Engrs, AIME*, **260**, 6–12.

Mahtab, M. A., Bolstad, D. D., Alldredge, J. R. and Shanley, R. J. (1972) *Analysis of Fracture Orientations for Input to Structural Models of Discontinuous Rock*, US Bur. Mines Rep. Invest. 7669.

Mahtab, M. A., Bolstad, D. D. and Kendorski, F. S. (1973) *Analysis of the Geometry of Fractures in San Manuel Copper Mine, Arizona*, US Bur. Mines Rep. Invest. 7715.

Margolin, L. (1981) Calculations of cratering experiments with the bedded rock model, in *Proc. Am. Phys. Soc. Symp. on Shock Waves in Condensed Matter*, Menlo Park, Calif., pp. 465–9.

Mathews, K. E. (1978) Design of underground mining layouts, *Underground Space*, **2**, 195–209.

Mathews, K. E. and Edwards, D. B. (1969) Rock mechanics practice at Mount Isa Mines Limited, Australia, in *Proc. 9th Commonw. Min. Metall. Congr.*, London, **1**, pp. 321–88.

Mathews, K. E. and Meek, J. L. (1975) Modelling rock reinforcement systems in cut-and-fill mining, in *Proc. 2nd Aust.–NZ Conf. Geomech.*, Brisbane, pp. 42–7.

Matthews, S. M., Worotnicki, G. and Tillmann, V. H. (1983) A modified cable bolt system for the support of underground openings, in *Proc. Aust. Inst. Min. Metall. Conf.*, Broken Hill, pp. 243–55.

Maxwell, J. C. (1864) On the calculation of the equilibrium and stiffness of frames. *Phil Mag.*, Series 4, **27**, 294–9.

Mitchell, S. T. (1981) Vertical crater retreat stoping as applied at the Homestake mine, in *Design and Operation of Caving and Sublevel Stoping Mines* (ed. D. R. Stewart), Soc. Min. Engrs, AIME, New York, pp. 609–26.

Moy, D. (1975) *The design and monitoring of rock bolt and dowel systems in underground excavations*, PhD thesis, University of London.

Morrison, D. M. (1987) Rockburst research at Falconbridge Limited, presented at the Can. Inst. Min. Metall., Montreal, Annual Meeting, May.

Muskhelishvili, N. I. (1963) *Some Basic Problems of the Mathematical Theory of Elasticity*, 4th edn, trans. J. R. M. Radok, Noordhof, Gronigen.

National Coal Board (1975) *Subsidence Engineers Handbook*, 2nd (rev.) edn, National Coal Board Mining Dept, London.

National Coal Board (1979) *Underground Support Systems*, National Coal Board Industrial Training Branch, London.

Nilsson, G. and Krauland (1981) Rock mechanics observations and measurements in Nasliden mine, in *Applications of Rock Mechanics to Cut and Fill Mining* (eds O. Stephansson and M.J. Jones), Instn Min. Metall., London, pp. 233–49.

Nixon, D. W. and Mills, P. S. (1981) Pump packing developments at Hem Heath Colliery. *Min. Engr*, **140**, 645–51.

Obert, L. and Long, A. E. (1962) *Underground Borate Mining, Kern County, Calif.*, US Bur. Mines Rep. Invest. 6110.

Orchard, R. J. and Allen, W. S. (1970) Longwall partial extraction systems. *Min. Engr*, **129**, 523–32.

Ortlepp, W. D. (1978) The mechanism of a rockburst, in *Proc. 19th US Symp. Rock Mech.*, University of Nevada, Reno, pp. 476–83.

Ortlepp, W. D. (1983) Considerations in the design of support for deep hard-rock tunnels, in *Proc. 5th Congr., Int. Soc. Rock Mech.*, Melbourne, **2**, A. A. Balkema, Rotterdam, D179–87.

Ortlepp, W. D. and Gay, N. C. (1984) Performance of an experimental tunnel subjected to stresses ranging from 50 MPa to 230 MPa, in *Design and Performance of Underground Excavations*, Cambridge, British Geotech. Soc., London, pp. 337–46.

Ortlepp, W. D., More O'Ferrall, R. C. and Wilson, J. W. (1972–3) Support methods in tunnels, *Assn Mine Managers S. Afr., Papers and Discns*, pp. 167–94.

Otter, J. R. H., Cassell, A. C. and Hobbs, R. E. (1966) Dynamic relaxation. *Proc. Inst. Civ. Engrs*, **35**, 633–65.

Owen, G. N. and Scholl, R. E. (1981) *Earthquake Engineering of Large Underground Structures*, FHWA/RD-80-195, US Department of Transportation, Washington, DC, pp. 171–4.

Pariseau, W.G. and Eitani, I.M. (1977) Post-elastic vibrating wire stress measurements in coal, in *Field Measurements in Rock Mechanics* (ed. K. Kovari), **1**, A. A. Balkema, Rotterdam, pp. 255–73.

Pariseau, W. G. and Pfleider, E. P. (1968) Soil plasticity and the movement of material in ore passes. *Trans. Soc. Min. Engrs, AIME*, **241**, 42–56.

Pariseau, W. G., McDonald, M. M. and Hill, J. R. M. (1973) Support performance prediction for hydraulic fills, in *Mine Filling, Proc. Jubilee Symp. Mine Filling*, Mount Isa, Aust. Inst. Min. Metall., Melbourne, pp. 213–19.

Park, D. W. and Gall, V. (1989) Supercomputer assisted three- dimensional finite element analysis of a longwall panel, in *Rock Mechanics as a Guide for Efficient Utilization of Natural Resources, Proc. 30th US Rock Mech. Symp.*, A. A. Balkema, Rotterdam, pp. 133–40.

Paterson, M. S. (1978) *Experimental rock deformation – the brittle field*, Springer, Berlin.

Patton, F. D. (1966) Multiple modes of shear failure in rock, in *Proc. 1st Congr., Int. Soc. Rock Mech.*, Lisbon, **1**, 509–13.

Pattrick, K. W. (1984) The instrumentation of seismic networks at Doornfontein Gold Mine, in *Rockbursts and Seismicity in Mines* (eds N.C. Gay and E.H. Wainwright), Sth Afr. Inst. Min. Metall., Johannesburg, pp. 337–40.

Peck, R. B. (1969) Advantages and limitations of the observational method in applied soil mechanics. *Géotechnique*, **19**, 171–87.

Pells, P. J. N. (1974) The behaviour of fully bonded rockbolts, in *Advances in Rock Mechanics, Proc. 3rd Congr., Int. Soc. Rock Mech.*, Denver, **2A**, Nat. Acad. Sci., Washington, DC, pp. 1212–17.

Peng, S. S. (1978) *Coal Mine Ground Control*, Wiley, New York.

Perrett, W. R. (1972) Seismic source energies of underground nuclear explosions. *Bull. Seism. Soc. Am.*, **62**, 763–74.

Phillips, F. C. (1971) *The Use of the Stereographic Projection in Structural Geology*, 3rd edn, Edward Arnold, London.

Piggott, R. J. and Eynon, P. (1978) Ground movements arising from the presence of shallow abandoned mine workings, in *Large Ground Movements and Structures*, (ed. J. D. Geddess), Pentech, London, pp. 749–80.

Pillar, C. L. (1981) A comparison of block caving methods, in *Design and Operation of Caving and Sublevel Stoping Mines* (ed. D. R. Stewart), Soc. Min. Engrs, AIME, New York, pp. 87–97.

Poulos, H. G. and Davis, E. H. (1974) *Elastic Solutions for Soil and Rock Mechanics*, Wiley, New York.

Prager, W. (1959) *An Introduction to Plasticity*, Addison-Wesley, Reading, Mass.

Price, N. J. (1966) *Fault and Joint Development in Brittle and Semi-Brittle Rock*, Pergamon, Oxford.

Priest, S. D. (1980) The use of inclined hemisphere projection methods for the determination of kinematic feasibility, slide direction and volume of rock blocks. *Int. J. Rock Mech. Min. Sci. & Geomech. Abstr.*, **17**, 1–23.

Priest, S. D. (1985) *Hemispherical Projection Methods in Rock Mechanics*, George Allen & Unwin, London.

Priest, S. D. (1992) *Discontinuity Analysis for Rock Engineering*, Chapman & Hall, London.

Priest, S. D. and Hudson, J. A. (1976) Discontinuity spacings in rock. *Int. J. Rock Mech. Min. Sci. & Geomech. Abstr.*, **13**, 135–48.

Priest, S. D. and Hudson, J. A. (1981) Estimation of discontinuity spacing and trace length using scanline surveys. *Int. J. Rock Mech. Min. Sci. & Geomech. Abstr.*, **18**, 183–97.

Proctor, R. J. and White, T. L. (1977) *Rock Tunnelling with Steel Supports*, rev. edn, Commercial Shearing and Stamping Co., Youngstown, Ohio.

Rice, G. S. (1934) Ground movement from mining in Brier Hill Mine, Norway. *Mich. Trans Am. Inst. Min. Metall. Engrs*, **109**, 118–44.

Rice, J. R. (1983) Constitutive relations for fault slip and earthquake instabilities. *PAGEOPH*, **121**, 443–75.

Richardson, A. M., Brown, S. M., Hustrulid, W. A. and Richardson, D. L. (1986) In *Proc. Int. Symp. on Rock Stress and Rock Stress Measurement*, Stockholm, (ed. O. Stephansson), Centek, Lulea, pp. 441–8.

Roberts, A. (1977) *Geotechnology*, Pergamon, Oxford.

Robinson, L. H. (1959) The effect of pore and confining pressure on the failure process in sedimentary rock. *Q. Colo. School Mines*, **54**(3), 177–99.

Rorke, A. J. and Roering, C. (1984) Source mechanism studies of mining-induced seismic events in a deep-level gold mine, in *Rockbursts and Seismicity in Mines* (eds N. C. Gay and E. H. Wainwright), Sth Afr. Inst. Min. Metall., Johannesburg, pp. 51–6.

Rosengren, K. J. (1968) *Rock mechanics of the Black Star Open Cut, Mount Isa*, PhD thesis, Aust. Nat. Univ.

Rosengren, K. J. (1970) Diamond drilling for structural purposes at Mount Isa. *Ind. Diamond Rev.*, **30**, 388–95.

Roux, A. J. A., Leeman, E. R. and Denkhaus, H. G. (1957) De-stressing: a means of ameliorating rock-burst conditions. Part 1 – the conception of de-stressing and the results obtained from its application. *J. S. Afr. Inst. Min. Metall.*, **58**, 101–19.

Rudnicki, J. W. and Rice, J.R. (1975) Conditions for the localization of deformation in pressure-sensitive dilatant materials. *J. Mech. Phys. Solids*, **23**, 371–94.

Ruina, A. (1983) Slip instability and state variable friction laws. *J. Geophys. Res.*, **88**, 10359–70.

Ryder, J. A. (1987) Excess shear stress (ESS): an engineering criterion for assessing unstable slip and associated rockburst hazards, in *Proc. 6th Congr., Int. Soc. Rock Mech.*, Montreal, **2**, A.A. Balkema, Rotterdam, pp. 1211–15.

St John, C. M. and Thomas, T. L. (1970) The NPL Mekometer and its application in mine surveying and rock mechanics. *Trans. Instn Min. Metall.*, **79**, A31–6.

St John, C M. and Van Dillen, D.E. (1983) Rockbolts: a new numerical representation and its application in tunnel design, in *Rock Mechanics – Theory – Experiment – Practice, Proc. 24th US Sym. Rock Mech.*, AEG, New York, pp. 13–26.

St John, C. M. and Zahrah, T. F. (1987) Aseismic design of underground structures. *Tunnelling and Underground Space Technology*, **2**(2), 165–97.

Salamon, M. D. G. (1964) Elastic analysis of displacements and stresses induced by mining of seam or reef deposits. Part II. *J. S. Afr. Inst. Min. Metall.*, **64**, 197–218.

Salamon, M. D. G. (1967) A method of designing bord and pillar workings. *J. S. Afr. Inst. Min. Metall.*, **68**, 68–78.

Salamon, M. D. G. (1970) Stability, instability and the design of pillar workings. *Int. J. Rock Mech. Min. Sci. & Geomech. Abstr.*, **7**, 613–31.

Salamon, M. D. G. (1974) Rock mechanics of underground excavations, in *Advances in Rock Mechanics, Proc. 3rd Congr., Int. Soc. Rock Mech.*, Denver, **1B**, Nat. Acad. Sci., Washington, DC, pp. 951–1099.

Salamon, M. D. G. (1983) Rockburst hazard and the fight for its alleviation in South African gold mines, in *Rockbursts: Prediction and Control*, Instn Min. Metall., London, pp. 11–36.

Salamon, M. D. G. and Munro, A. H. (1967) A study of the strength of coal pillars. *J. S. Afr. Inst. Min. Metall.*, **68**, 55–67.

Salamon, M. D. G. and Wagner, H. (1979) Role of stabilizing pillars in the alleviation of rockburst hazard in deep mines, in *Proc. 4th Congr. Int. Soc. Rock Mech.*, Montreux, **3**, pp. 561–6.

Salamon, M. D. G. and Wiebols, G. A. (1974) Digital location of seismic events by an underground network of seismometers using arrival times of compressional waves. *Rock Mech.*, **6**, 141–66.

Salehy, M. R., Money, M. S. and Dearman, W. R. (1977) The occurrence and engineering properties of intra-formational shears in Carboniferous rocks, in *Proc. Conf. on Rock Engng*, Newcastle-upon-Tyne, pp. 311–28.

Sarin, D. K. (1981) A review of sub-level caving practices in Canada, in *Design and Operation of Caving and Sublevel Stoping Mines* (ed. D. R. Stewart), Soc. Min. Engrs. AIME, New York, pp. 373–85.

Sauer, G. and B. Sharma (1977) A system for stress measurement in construction in rock, in *Field Measurements in Rock Mechanics* (ed. K. Kovari), **1**, A. A. Balkema, Rotterdam, pp. 317–29.

554

Savin, G. N. (1961) *Stress Concentrations around Holes*, Pergamon, London.

Schach, R., Garshol, K. and Heltzen, A. M (1979) *Rock Bolting – a Practical Handbook*, Pergamon, Oxford.

Schofield, A. N. and Wroth, C. P. (1968) *Critical State Soil Mechanics*, McGraw-Hill, London.

Scott, J. J. (1977) Friction rock stabilizers – a new rock reinforcement method, in *Monograph on Rock Mechanics Applications in Mining, Papers presented at 17th US Symp. Rock Mech.* (eds W.S. Brown, S. J. Green and W. A. Hustrulid), AIME, New York, pp. 242–9.

Sellers, J. B. (1977) The measurement of stress changes in rock using the vibrating wire stressmeter, in *Field Measurements in Rock Mechanics* (ed. K. Kovari), **1**, A. A. Balkema, Rotterdam, pp. 275–88.

Serafim, J. L. and Periera, J. P. (1983) Considerations of the geomechanical classification of Bieniawski, in *Proc. Int. Symp. on Engineering Geology and Underground Construction*, Lisbon, **1**, II.33–42.

Sharpe, (J. A.) 1942. The production of elastic waves by explosion pressures. *Geophys.*, **7**, 144–54.

Siskind, D. E., Staff, M. S., Kopp, J. W. and Dowding, C. H. (1980) *Structure Response and Damage Produced by Ground Vibration from Surface Mine Blasting*, US Bur. Mines Rep. Invest. 8507.

Skempton, A. W. (1960) Terzaghi's discovery of effective stress, in *From Theory to Practice in Soil Mechanics* (eds L. Bjerrum, A. Casagrande, R. B. Peck and W. A. Skempton), Wiley, New York, pp. 42–53.

Skinner, W. J. (1959) Experiments on the compressive strength of anhydrite. *The Engineer*, London, **207**, 255–9 and 288–92.

Sneddon, I. N. (1946) The distribution of stress in the neighbourhood of a crack in an elastic solid, *Proc. Roy. Soc.*, **A187**, 229–60.

Southwell, R. V. (1940) *Relaxation Methods in Engineering Science*, Oxford University Press, Oxford.

Spottiswoode, S. M. (1984) Source mechanisms of mine tremors at Blyvooruitzicht Gold Mine, in *Rockbursts and Seismicity in Mines* (eds N. C. Gay and E. H. Wainwright), Sth Afr. Inst. Min. Metall., Johannesburg, pp. 29–38.

Spottiswoode, S. M. and McGarr, A. (1975) Source parameters of tremors in a deep-level gold mine. *Bull. Seis. Soc. Am.*, **65**, 93–112.

Stagg, M. S. and A. J. Engler (1980) *Measurement of Blast-Induced Ground Vibrations and Seismograph Calibrations*, US Bur. Mines Rep Invest. 8506.

Starfield, A. M. and Fairhurst, C. (1968) How high-speed computers advance design of practical mine pillar systems. *Engng Min. J.*, **169**, May, 78–84.

Starfield, A. M. and Pugliese, J. M. (1968) Compression waves generated in rock by cylindrical explosive charges: a comparison between a computer model and field measurements. *Int. J. Rock Mech. Min. Sci. & Geomech. Abstr.*, **5**, 65–77.

Starfield, A. M. and Wawersik, W. (1972) Pillars as structural components in room-and-pillar design, in *Basic and Applied Rock Mechanics, Proc. 10th US Symp. Rock Mech.* (ed. K. E. Gray), AIME, New York, pp. 793–809.

Steart, F. A. (1954) Strength and stability of pillars in coal mines. *J. Chem. Metall. Min. Soc. S. Afr.*, **54**, 309–25.

Steinhobel, M. W. L. and Klokow, J. (1978–9) Pipe stick support at West Driefontein Gold Mining Company Limited. *Assn Mine Managers S. Afr., Papers and Discns*, pp. 31–43.

Stephansson, O. and M. J. Jones (eds) (1981) *Applications of Rock Mechanics to Cut and Fill Mining*, Instn Min. Metall., London.

Sterling, R. L. (1980) The ultimate load behaviour of laterally constrained rock beams, in *The State of the Art in Rock Mechanics, Proc. 21st US Symp. Rock Mech.* (ed. D. A. Summers), University of Missouri, Rolla, pp. 533–42.

Stevens, P. R. (1977) *Review of Effects of Earthquakes on Underground Mines*, US Geological Survey, Open File Report 77–313.

Stiller, H., Hustig, E., Grosser, H. and Knoll, P. (1983) On the nature of mining tremors. *J. Earthquake Predict. Res.* (Tokyo), **2**, 61.

Stone, S. (1978) Introduction of concrete sausage packs at Rustenburg Platinum Mines Limited. *J. S. Afr. Inst. Min. Metall.*, **78**, 243–8.

Swan, G. (1985) Notes on Denison Mine backfill confinement tests, *CANMET Div. Report*, MRP/MRL 85.

555

Swan, G. and Board, M. (1989) Fill-induced post-peak pillar stability, in *Innovations in Mining Backfill Technology*, A. A. Balkema, Rotterdam, pp. 81–8.

Swanson, S. R. and Brown, W. S. (1971) An observation of loading path independence of fracture in rock. *Int. J. Rock Mech. Min. Sci. & Geomech. Abstr.*, **8**, 277–81.

Talobre, J. (1957) *La Mécanique des Roches*, Dunod, Paris.

Terzaghi, R. D. (1965) Sources of error in joint surveys. *Géotechnique*, **15**, 287–304.

Thomas, E. G. and Cowling, R. (1978) Pozzolanic behaviour of ground Isa Mine slag in cemented hydraulic mine fill at high slag/cement ratios, in *Mining with Backfill, Proc. 12th Can. Rock Mech. Symp.*, Sudbury, Can. Inst. Min. Metall, Montreal, pp. 129–32.

Thomas, E. G., Nantel, L. H. and Notley, K. R. (1979) *Fill Technology in Underground Metalliferous Mines*, Int. Academic Services, Kingston.

Thomas, L. J. (1978) *An Introduction to Mining*, rev. edn, Methuen, Sydney.

Thompson, A. G., Matthews, S. M., Windsor, C. R., Bywater, S. and Tillmann, V. H. (1987) Innovations in rock reinforcement in the Australian mining industry, in *Proc. 6th Congr., Int. Soc. Rock Mech.*, Montreal, **2**, A. A. Balkema, Rotterdam, pp. 1275–8.

Thompson, J. M. and Hunt, G. W. (1973) *A General Theory of Elastic Stability*, Wiley, London.

Timoshenko, S. P. and Goodier, J. N. (1970) *Theory of Elasticity*, 3rd edn, Wiley, New York.

Townsend, P. (1988) Personal communication.

Tucker, L. J. (1981) Sublevel caving at Pea Ridge, in *Design and Operation of Caving and Sublevel Stoping Mines* (ed. D. R. Stewart), Soc. Min. Engrs, AIME, New York, pp. 387–91.

van Heerden, W. L. (1975) In-situ complete stress–strain characteristics of large coal specimens. *J. S. Afr. Inst. Min. Metall.*, **75**, 207–17.

Vardoulakis, I. (1979) Bifurcation analysis of the triaxial test on sand samples. *Acta Mechanica*, **32**, 35–54.

Vardoulakis, I., Sulem, J. and Guenot, A. (1988) Borehole instabilities as deformation phenomena. *Int. J. Rock Mech. Min. Sci. & Geomech. Abstr.*, **25**, 159–70.

Voegele, M., Fairhurst, C. and Cundall, P. (1978) Analysis of tunnel support loads using a large displacement, distinct block model, in *Storage in Excavated Rock Caverns* (ed. M. Bergman), **2**, Pergamon, Oxford pp. 247–52.

Vutukuri, V. S., Lama, R. D. and Saluja, S. S. (1974) *Handbook on Mechanical Properties of Rocks*, **1**, Trans. Tech. Publications, Clausthal.

Wagner, H. (1974) Determination of the complete load deformation characteristics of coal pillars, in *Advances in Rock Mechanics, Proc. 3rd Congr., Int. Soc. Rock Mech.*, Denver, **2B**, Nat. Acad. Sci., Washington, DC, pp. 1076–81.

Wagner, H. (1980) Pillar design in coal mines. *J. S. Afr. Inst. Min. Metall.*, **81**, 37–45.

Wagner, H. and Madden, B. J. (1984) Fifteen years' experience with the design of coal pillars in shallow South African collieries: an evaluation of the performance of the design procedures and recent improvements, in *Design and Performance of Underground Excavations*, Cambridge, British Geotech. Soc., London, pp. 391–399.

Wahlstrom, E. E. (1973) *Tunneling in Rock*, Elsévier, Amsterdam.

Walton, R. J. and Worotnicki, G. (1986) A comparison of three borehole instruments for monitoring the change of stress with time, in *Proc. Int. Symp. on Rock Stress and Rock Stress Measurement*, *Stockholm* (ed. O. Stephansson), Centek, Lulea, pp. 479–88.

Warburton, P. M. (1981) Vector stability analysis of an arbitrary polyhedral rock block with any number of free faces. *Int. J. Rock Mech. Min. Sci. & Geomech. Abstr.*, **18**, 415–28.

Wells, J. H. (1929) Kinetic boundary friction. *The Engineer*, (London), **147**, 454–62.

Wawersik, W. R. and Fairhurst, C. (1970) A study of brittle rock fracture in laboratory compression experiments. *Int. J. Rock Mech. Min. Sci & Geomech. Abstr.*, **7**, 561–75.

Whittaker, B. N. (1974) An appraisal of strata control practice. *Min. Engr*, **134**, 9–24.

Whittaker, B. N. and Pye, J. H. (1977) Design and layout aspects of longwall methods of coal mining, in *Design Methods in Rock Mechanics, Proc. 16th US Symp. Rock Mech.* (eds C. Fairhurst and S. L. Crouch), ASCE, New York, pp. 303–14.

Wilson, A. H. (1961) A laboratory investigation of a high modulus borehole plug gage for the measurement of rock stress, in *Proc. 4th US Symp. Rock Mech.*, *Bull. Mineral Ind. Expt. Stn*, Pennsylvania State University, No. 76, pp. 185–95.

REFERENCES

Wilson, A. H. (1977) The effect of yield zones on the control of ground, in *Proc. 6th Int. Strata Control Conf.*, Banff, Paper 3.

Wilson, A. H. (1981) Stress and stability in coal ribsides and pillars, in *Proc. 1st Ann. Conf. on Ground Control in Mining*, West Virginia University (ed. S.S. Peng), pp. 1–12.

Winzer, S. R. and Ritter, A. P. (1980) The role of stress waves and discontinuities in rock fragmentation: a study of fragmentation in large limestone blocks, in *The State of the Art in Rock Mechanics, Proc. 21st US Symp. Rock Mech.* (ed. D. A. Summers), University of Missouri, Rolla, pp. 362–70.

Worotnicki, G. and Walton, R. J. (1976) Triaxial 'hollow inclusion' gauges for determination of rock stresses in situ, in *Proc. ISRM Symp. on Investigation of Stress in Rock*, Instn Engrs Aust., Sydney, Supplement, 1–8.

Worotnicki, G., Enever, J. R., McKavanagh, B., Spathis, A. and Walton, R. (1980) Experience with the monitoring of crown pillar performance in two Australian mines, in *Proc. 3rd Aust.–NZ Conf. Geomech.*, Wellington, **2**, NZ Instn Engrs, Wellington, pp. 161–8.

Wright, F. D. (1974) *Design of Roof Bolt Patterns for Jointed Rock*, US Bur. Mines Open File Rep. 61–75.

Yenge, L. I. (1980) Analysis of bulk flow of materials under gravity caving process. Part 1 – Sublevel caving in relation to flow in bins and bunkers, *Q. Colo. School Mines*, **75**(4), 1–45.

Zienkiewicz, O. C. (1977) *The Finite Element Method*, 3rd edn, McGraw-Hill, London.

557

Index

Page numbers appearing in **bold** refer to figures and page numbers appearing in *italic* refer to tables.